Microbial Biodiversity of Natural Ecosystem

— *Modern Trends* —

Microbial Biodiversity of Natural Ecosystem

— *Modern Trends* —

Editors
Asha Sinha
B.K. Sarma
Manisha Srivastava

BIOTECH
2012
BIOTECH BOOKS®

ISBN 978-81-7622-259-4

Published by: **BIOTECH BOOKS®**
4762-63/23, Ansari Road, Darya Ganj,
New Delhi - 110 002
Phone: +91-011-23262132
E-mail: biotechbooks@yahoo.co.in

Printed at: **Chawla Offset Printers**
Delhi - 110 052

PRINTED IN INDIA

Preface

The microbe means an organism that is so small that normally cannot be seen without aid of a microscope. Microbial biodiversity includes Prokaryotes and Eukaryotes *viz.*, bacteria, fungi and viruses. They live in a wide variety of habitats. Microbes are essential to life they use to produce food as well as industrial compounds and they cause many diseases of plants and animals and are responsible for food spoilage. Modern trends are being appropriately used resulted into our knowledge of understanding diverse interactions of host with pathogens, mechanisms of disease development, host defend responses against the invading pathogens and identification of pathogens, specifically bacteria and viruses. Utilization of resistance genes in crops are also based on diversity analysis of the pathogen population in specific localities. "Modern Trends in Microbial Biodiversity of Natural Ecosystem" covering over 35 chapters is a comprehensive compilation of all those issues related to ecological, morphological, pathogenic and molecular diversity in phytopathogens, impact of environmental factors on fungal diversity in crop fields, utilization of microbial diversity for selection of efficient strains of entomopathogenic fungi, prospects of microbial diversity for seeds protection at storage, diversity analysis of mushrooms for nutritional and medicinal properties, diversity and potentiality of Actinomycetes for biological control, exploitation of soil microbial diversity for disease suppression, crop production and sustainable agriculture, utilization of molecular techniques for exploring microbial diversity, which are useful for undergraduate and post graduate students, research scholars, scientists working in the areas of Agriculture, Botany, Microbiology and Plant Pathology. We acknowledge our sincere thanks to the authors contributed the chapters.

Asha Sinha

B.K. Sarma

Manisha Srivastava

Contents

Chapter 1

Ecology and Genetic Diversity of *Fusarium* spp. Associated with Bakanae Disease of Rice

B.M. Bashyal, Rashmi Aggarwal, Sangeeta Gupt and Sagar Banerjee

Division of Plant Pathology, Indian Agricultural Research Institute, New Delhi – 110 012

Bakanae disease is one of the emerging diseases of rice (*Oryza sativa* L.). The causal agent of bakanae disease of rice is *Gibberella fujikuroi*, Sawada, Wollenworth (teleomorph) (anamorph: *Fusarium moniliforme* Sheld.). Bakanae disease causes up to 40 per cent losses in rice (Ou, 1987). This disease has been reported from the rice tracts of South Asia, European countries and America (Desjardins *et al.*, 2000). It is emerging as a potential threat in Japan, Taiwan, Thailand and India (Webster and Gunnell, 1992; Kini *et al.*, 2002; Saremi, 2005; Anonymous, 2007). The typical symptoms of bakanae are slender, chlorotic and abnormally elongated primary leaves, however, not all infected seedlings show these symptoms, as crown rot is also seen, resulting in stunted rice plants (Ou, 1987; Amoah *et al.*, 1995).

The name bakanae means "bad" or "naughty" seedlings in Japanese, referring to the elongation symptoms specific for the disease, and caused by gibberellins production by the pathogen upon infection of the host. The fungus produces gibberellins and other secondary metabolites such as carotenoids, bikaverin and fusarin, which directly affect the rice growth. Bakanae is a monocyclic disease and the pathogen often spreads predominantly with infected seeds, although infected crop residue from the previous season may also serve as a source of inoculum.

Bakanae is traditionally associated with rice, but water grass plants such as *Echinochloa* spp., with classic symptoms of bakanae were also observed in California in 2002 (Carter *et al.*, 2008). Although the bakanae disease usually causes dieback or sterility of rice, mycotoxin contamination also poses a concern since the pathogen is seed-borne.

Pathogens Associated with Bakanae Disease and their Distribution

Although bakanae disease was first described over 100 years ago in Japan, it is still not clear which *Fusarium* species are associated with different symptoms. Early work in Japan identified the pathogen as *Fusarium moniliforme* in a broad sense (Ou, 1985); however, this taxon comprises a number of distinct species, now collectively termed the *Gibberella fujikuroi* species complex. Sun and Snyder (1981) produced perithecia by crossing strains of *F. fujikuroi* in the laboratory; then four reproductively isolated groups of *G. fujikuroi* were designated as groups A, B, C, and D, with genetically interfertile strains from rice designated as mating group C (Hseieh *et al.*, 1977; Kuhlman, 1982). Additional genetic studies have identified nine biological species or mating populations, designated A-I, within the *G. fujikuroi* species complex (Leslie, 1995; Viljoen *et al.*, 1997; Leslie and Summerell 2006). *Gibberella fujikuroi* species complex is generally designated as section Liseola, comprising of nine biological species (Table 1.1) designated as mating populations A to I (Leslie and Summerell 2006).

Table 1.1: Mating Populations of *Gibberella fujikuroi* Species Complex

Sl.No.	Anamorph	Telomorph	Mating population
1.	*F. verticilloides*	*G. fujikuroi*	MP-A
2.	*F. sacchari*	*G. fujikuroi*	MP-B
3.	*F. fujikuroi*	*G. fujikuroi*	MP-C
4.	*F. proliferatum*	*G. fujikuroi*	MP-D
5.	*F. subglutinans*	*G. fujikuroi*	MP-E
6.	*F. thapsinum*	*G. fujikuroi*	MP-F
7.	*F. nygnamai*	*G. fujikuroi*	MP-G
8.	*F. circinatum*	*G. fujikuroi*	MP-H
9.	*F. konzum*	*G. fujikuroi*	MP-I

Three mating populations of section Liseola (A, C and D) of the *G. fujikuroi* complex have been associated with bakanae disease of rice. Mating population C (MP-C) (anamorph, *Fusarium fujikuroi*) (Nirenberg, 1976) was first identified in 1977 among strains from rice from Taiwan (Hseieh *et al.*, 1977). It has been found responsible for bakanae disease in Italy (Amatulli *et al.*, 2010). Mating population A (MP-A) (anamorph, *Fusarium verticillioides* [synonym, *F. moniliforme*]) and mating population D (MP-D) (anamorph, *Fusarium proliferatum*) have been isolated from rice from Asia, and MP-D has been isolated from rice from Africa, Australia, and the

United States (Desjardins *et al.*, 1997; Amoh *et al.*, 1996; Voigt *et al.*, 1995). Thus, more than one species of *Fusarium* may be able to infect rice and cause symptoms of bakanae disease.

Wulff *et al.* (2010) isolated and characterized African and Asian populations of *Fusarium* spp. (*Gibberella fujikuroi* species complex) associated with bakanae of rice (*Oryzae sativa* L.) with respect to ecology, phylogenetics, pathogenicity and mycotoxin production. Independent of the origin, *Fusarium* spp. were detected in the different rice seed samples with varied infection rate ranging from 0.25 per cent to 9 per cent (Table 1.2). Four *Fusaria* (*F. andiyazi, F. fujikuroi, F. proliferatum* and *F. verticillioides*) were found associated with bakanae of rice. While three of the *Fusaria* were found in both African and Asian seed samples, *F. fujikuroi* was only detected in seed samples from Asia. Phylogenetic studies showed a broad genetic variation among the strains that were distributed into four different genetic clades. Pathogenicity tests showed that all strains reduced seed germination and possessed varying ability to cause symptoms of bakanae on rice, some species (*i.e. F. fujikuroi*) being more pathogenic than others. These findings provide new information on the variation within the *G. fujikuroi* species complex associated with rice seed and bakanae disease.

Table 1.2: Infection Rate (Range) of Seed Infected with *Gibberella fujikuroi* Species Complex

Origin	*Infection Rate Range* (per cent)*
Bangladesh	0.50-7.50
Burkina Faso	0.25-2.00
China	0.50-3.50
Ghana	0.50-4.50
India	0.50-1.00
Ivory Coast	0.50-6.50
Nepal	1.50-3.00
Tanzania	1.00-2.25
Vietnam	0.50-9.00

* Lowest and highest percentages of sample (400 seeds) found infected with species of the *Gibberella fujikuroi* species complex.

Gibberella fujikuroi species complex were detected in popularly grown rice varieties with infection percentage ranges from 1 per cent to 24 per cent in India (Bashyal *et al.*, unpublished) (Table 1.3). Three *Fusarium* spp. *viz. F. moniliforme, F. fujikuroi* and *F. proliferatum* were found associated with bakanae disease of rice in India. Maximum incidence of slender and chlorotic leaves were produced by *F. fujikuroi* (90 per cent) while maximum incidence of crown rot and stem rot was produced by *F. moniliforme* (50 per cent).

Table 1.3: Rice Varieties Infected with *Gibberella fujikuroi* Species Complex from India

Sl.No.	*Rice Variety*	*Infection (per cent)*
1.	Pusa Basmati 1121	24
2.	Pusa Basmati-1	4
3.	Tarori Basmati	20
4.	Samba Mahsuri	2
5.	Jaya	10
6.	Ajaya	4
7.	Vikramaraya	6
9.	Swarnadhan	1
10.	IR 50	4
11.	IR 64	4
12.	MTU 1010	4

Secondary Metabolite Production

Members of mating populations associated with bakanae disease can be found preferentially on different host plants (Leslie and Plattner, 1991; Zeller *et al.*, 2003) and differ in their ability to produce secondary metabolites, such as fumonisins (Desjardins *et al.*, 1995, 2000; Kedera *et al.*, 1999; Leslie *et al.*, 1992), fusaric acid (Bacon *et al.*, 1996), beauvericin (Logrieco *et al.*, 1998; Torres *et al.*, 2001; Reynoso *et al.*, 2004), fusaproliferin (Reynoso *et al.*, 2004), moniliformin (Leslie *et al.*, 1996; Desjardins *et al.*, 2000), fusarins (Wiebe and Bjeldanes, 1981), and gibberellins (GAs). Gibberellin and fumonisin are most important secondary metabolites for disease production.

GA Production

GA production was described so far only for the species, *F. fujikuroi* (MP-C), the causative agent of bakanae disease of rice. The genetic and biochemical background of GA biosynthesis by *F. fujikuroi* has been well characterized (Tudzynski and Holter, 1998; Tudzynski *et al.*, 2001, 2002, 2003; Rojas *et al.*, 2001; Bhalla *et al.*, 2010). As is the case for many fungal secondary metabolites, GA biosynthetic genes are organized in a gene cluster (Tudzynski and Holter, 1998; Linnemanstons *et al.*, 1999). In addition to genes encoding a pathway-specific geranylgeranyl diphosphate synthase (ggs2) and the bifunctional ent-copalyl diphosphate/ent-kaurene synthase (cps/ks), the GA gene cluster consists of four cytochrome P450 monooxygenase genes (P450-1 to P450-4) and a GA4 desaturase gene (des) (Rojas *et al.*, 2001; Tudzynski *et al.*, 2001, 2002, 2003).

At least 16 enzymatic steps are involved in the biosynthesis of gibberellic acid (GA3), and most of the genes encode multifunctional enzymes. Apart from *F. fujikuroi* (MP-C), some strains of MP-A (*F. verticillioides*) and MP-D (*F. proliferatum*) have also been isolated from infected rice seedlings in different geographic regions (Desjardins *et al.*, 1997). However Malonick *et al.* (2005) reported that none of these species except for *F. fujikuroi* (MP-C) and one strain of *F. konzum* (MP-I) express the GA biosynthetic genes at detectable levels (Table 1.4), and none of them except for *F. fujikuroi* and so far one *F. konzum* isolate are able to produce GAs. Contrary to that, Fusarium strains not related to the *G. fujikuroi* species complex contain none of the genes.

Table 1.4: The Sexually Fertile Species (MPs) of the *Gibberella fujikuroi* Species Complex: Host Plants, First Description, Presence of the GA Biosynthetic Genes and GA Production

Anamorph	*Telomorph*	*Host Plant*	*References*	*GA Cluster Genes*	*GA Production*
F. verticilloides (F. moniliforme)	*G. fujikuroi* MP-A	Rice	Nelson (1992)	Only des and P450-4	No
F. fujikuroi	*G. fujikuroi* MP-C	Rice	Hsieh *et al.* (1977)	Entire GA gene cluster	Yes
F. proliferatum	*G. fujikuroi* MP-D	Rice, maize, sorghum	Kuhlman (1982)	Entire GA gene cluster	No

Fumonisin Production

Fumonisins are a family of polyketide-derived mycotoxins that can accumulate in rice. These mycotoxins consist of several subfamilies, the A-, B-, and C-series, that differ in structure from one another by modifications of the nitrogen-containing group and by the presence or absence of a terminal methyl group two carbons from the nitrogen group (Powell and Plattner 1995). For example, in the B- and C-series fumonisins, the nitrogen group is a free amine, but in the B-series, the terminal methyl group is present and in the C series it is absent. Fumonisins within a series differ in structure by the presence or absence of hydroxyl groups at various positions along their linear backbone. The B-series fumonisins have been implicated as the cause of several fatal animal diseases (Gelderblom *et al.*, 1988). According to Amoah and colleagues (1995), pathogenicity of *G. fujikuroi* species complex on rice (and maize) may depend on the balance of toxins and growth regulators that are, on the other hand, affected by the strain of the pathogen, the environment and the nutritional status of the plant.

Table 1.5: Fumonisin (FB1 and FB2) Production in Different *Fusarium* spp. Associated with Bakanae Disease

Fungi	*Fuminosin Production*	
	FB1	*FB2*
F. verticilloids (F. moniliforme)	1-10 μg ml^{-1}	1-10 μg ml^{-1}
F. fujikuroi	Trace	Trace
F. proliferatum	10-25 μg ml^{-1}	10-25 μg ml^{-1}

Molecular Detection

Molecular methods are among the most precise tools for differentiation between species, and identification of new strains/isolates collected from infected samples. They differ regarding discriminatory power, reproducibility, ease of use, and interpretation. DNA fingerprinting of Fusarium has been successfully used for characterization of individual isolates and grouping them into standard racial classes and groups. This is particularly useful when any unknown fungal sample is to be identified. A comparison at the DNA sequence level provides accurate clasșification of fungal species and is beginning to elucidate the evolutionary and ecological relationships among diverse species (Mule *et al.*, 2005).

The sequences most commonly used to distinguish *Fusarium* spp. are portions of the genomic sequences encoding the translocation elongation factor 1-α (TEF) (Wulff *et al.*, 2010), β-tubulin (tub2) (O'Donnell *et al.*, 1998), calmodulin (O'Donnell *et al.*, 2000), internally transcribed spacer regions in the ribosomal repeat region (ITS1 and ITS2) (Waalwijk *et al.*, 1996; O'Donnell and Cigelnik 1997), and the intergenic spacer region (IGS) (Yli- Mattila and Gagkaeva 2010). Other molecular techniques, such as RAPDs (Du- Teau and Leslie, 1991; Mitter *et al.*, 2002; Voigt *et al.*, 1995), mitochondrial RFLPs (Correll *et al.*, 1992), AFLPs (Chulze *et al.*, 2000; Zeller *et al.*, 2003), and CHEF-gel karyotypes (Xu *et al.*, 1995) have been also used to differentiate members of the *G.*

fujikuroi species complex. Based on the results of these analyses, the *G. fujikuroi* complex has been delineated into three lineages, designated as the African, Asian, and American clades (O'Donnell *et al.*, 1998). Not all sequences work equally well for all species, with *TEF1* gene [primer ef1 (5'-ATGGGTAAGGA (A?G) GACAAGA C-3') and primer ef2 (5'- GGA (G?A) GTACCAGT (G? C) ATCATGTT-3')] being the most widely accepted across the genus. The ITS regions do not work well within the Liseola section.

Conclusion

Bakanae disease is one of the emerging diseases of rice (*Oryza sativa* L.). More than one pathogen of *Gibberella fujikuroi* spp. complex (mostly *Fusarium fujikuroi, Fusarium verticilloides, Fusarium proliferatum*) have been associated with bakanae disease of rice. Only *Fusarium fujikuroi* was found associated with typical symptoms of bakanae disease *i.e.* slender and elongated leaves and abnormal growth of the plant and it is only able to produce gibberellic acid. In the last decade, the organization of *Fusarium* spp. into well defined lineages and their mapping to specific geographic locations have been achieved by analysis of genes involved in mycotoxin biosynthesis or other metabolic processes to study the pathogen populations prevalent in those regions. Knowledge of the distribution and aggressiveness pattern of *Fusarium* spp. is very useful for understanding and developing strategies to control the bakanae disease of rice.

References

Amatulli, M.T., Spadaro, D., Gullino, M.L. and Garibaldi, A. (2010). Molecular identification of *Fusarium* spp. associated with bakanae disease of rice in Italy and assessment of their pathogenicity. *Plant Pathology* **59(5):** 839-44.

Amoah, B.K., Macdonald, M.V., Rezanoor, N. and Nicholson, P. (1996). The use of the random amplified polymorphic DNA technique to identify mating groups in the *Fusarium* section *Liseola. Plant Pathology* **45:**115–25.

Amoah, B.K., Rezanoor, H.N., Nicholson, P. and Mac-Donald, M.V. (1995). Variation in the *Fusarium* section *Liseola*: pathogenicity and genetic studies of *Fusarium moniliforme* Sheldon from different hosts in Ghana. *Plant Pathology* **44**: 563–72.

Anonymous (2007). Bakanae disease an emerging problem of Basmati rice. NCIPM Newsletter **13**:4.

Bacon, C.W. and Hinton, D.M. (1996). Symptomless endophytic colonization of maize by *Fusarium moniliforme. Can J Bot* **74:**1195–1202.

Bhalla, K., Singh, S.B. and Aggarwal, R. (2010). Quantitative determination of gibberellins by high performance liquid chromatography from various gibberellins producing Fusarium strains. *Environ Monit Assess* **167:** 515-520.

Carter, L.L.A., Leslie, F.J. and Webster, R.K. (2008). Population Structure of *Fusarium fujikuroi* from California Rice and Water Grass, *Phytopathology* 992–998.

Chulze, S.N., Ramirez, M.L., Torres, A. and Leslie, J.F. (2000). Genetic variation in Fusarium section Liseola from no-till maize in Argentina. *Appl. Environ. Microbiol.* **66:** 5312–5315.

Correll, J.C., Gordon, T.R. and McCain, A.H. (1992). Genetic diversity in California and Florida populations of the pitsch canker fungus *Fusarium subglutinans f. sp. pini. Phytopathology* **82:** 415–420.

Desjardins, A.E., Manandhar, H.K., Plattner, R.D., Manandhar, G.G., Poling, S.M. and Maragos, C.M. (2000). *Fusarium* species from Nepalese rice and production of mycotoxins and gibberellic acid by selected species. *Applied Environtal Microbiology* **66:** 1020–25.

Desjardins, A.E., Plattner, R.D. and Nelson, P.E. (1997). Production of fumonisin B1 and moniliformin by *Gibberella fujikuroi* from rice from various geographic areas. *Applied Enviromental Microbiology* **63**:1838–42.

Desjardins, A.E., Plattner, R.D., Nelsen, T.C. and Leslie, J.F. (1995). Genetic analysis of fumonisin production and virulence of Gibberella fujikuroi mating population A (*Fusarium moniliforme*) on maize (*Zea mays*) seedlings. *Appl. Environ. Microbiol.* **61:** 79–86.

Du Teau, N.M. and Leslie, J.F. (1991). RAPD markers for *Gibberella fujikuroi* (Fusariu section Liseola). *Fung. Genet. Newsl.* **38:** 37.

Gelderblom, W.C.A., Jaskiewicz, K., Marasas, W.F.O., Thiel, P.G., Horak, R.M., Vleggaar, R., and Kriek, N.P.J. (1988). Fumonisins-novel mycotoxins with cancer-promoting activity produced by *Fusarium moniliforme. Appl Environ Microbiol* **54:** 1806–1811.

Hseieh, W.H., Smith, S. N. and Snyder, W.C. (1977). Mating groups in *Fusarium moniliforme. Phytopathology* **67**:1041–43.

Kedera, D.J., Plattner, R.D. and Desjardings, A.D. (1999). Incidence of *Fusarium* spp. and levels of fumonisin B1 in maize in Western Kenya. *Appl. Environ. Microbiol.* **65:** 41–44.

Kini, K.R., Let, V. and Mathur, S.B. (2002). Genetic variation in *Fusarium moniliforme* isolated from seeds of different host species from Burkina Faso based on random amplified polymorphic DNA analysis. *Journal of Phytopathology* **150**: 209-12.

Kuhlman, E.G. (1982). Varieties of *Gibberella fujikuroi* with anamorphs in *Fusarium* section *Liseola. Mycologia* **74:**759–768.

Leslie, J.F. (1995). *Gibberella fujikuroi*: available populations and variable traits. *Canadian Journal of Botany* **73**:S282–91.

Leslie, J.F. and Plattner, R.D. (1991). Fertility and fumonisin B1 production by strains of *Fusarium moniliforme (Gibberella fujikuroi)*. In: Proceedings of the 17th Biennial Grain Sorghum Research and Utilization Conference, Lubbock, TX, USA, pp. 80–84

Leslie, J.F., Marasas, W.F.O., Shephard, G.S., Sydenham, E.W., Stockenstrom, S. **and Thiel, P.G.** (1996). Duckling toxicity and the production of fumonisin and moniliformin by isolates in the A and F mating populations of *Gibberella fujikuroi (Fusarium moniliforme). Appl. Environ. Microbiol.* **62:** 1182–1187.

Leslie, J.F., Plattner, R.D., Desjardins, A.E. and Klittich, C.J.R. (1992). Fumonisin B1 production by strains from different mating populations of *Gibberella fujikuroi* (Fusarium section liseola). *Mycotoxicology* **82:** 341–345.

Leslie, J.F. and Summerell, B.A. (2006). *The Fusarium Laboratory Manual*. Oxford, UK: Blackwell Publishing.

Linnemanstons, P., Voß, T., Hedden, P., Gaskin, P. and Tudzynski, B. (1999). Deletions in the gibberellin biosynthesis gene cluster of *Gibberella fujikuroi* by restriction enzyme-mediated integration and conventional transformation-mediated mutagenesis. *Appl. Environ. Microbiol.* **65:** 2558–2564.

Logrieco, A., Moretti, A., Castella, G., Kostecki, M., Golinski, P., Ritieni, A. and Chelkowski, J. (1998). Beauvericin production by Fusarium species. *Appl. Environ. Microbiol.* **64:** 3084–3088.

Malonek, S., Bömke, C., Bornberg-Bauer, E., Rojas, M.C., Hedden, P., Hopkins, P. and Tudzynski, B. (2005). Distribution of gibberellin biosynthetic genes and gibberellins production in the *Gibberella fujikuroi* species complex. *Phytochemistry* **66:** 1296–1311.

Mitter, N., Srivastava, A.C., Renu, A.S., Sarbhoy, A.K. and Agarwal, D.K. (2002). Characterization of gibberellin producing strains of *Fusarium moniliforme* based on DNA polymorphism. *Mycopathologia* **153:** 187–193.

Mule, G., Gonzalez-Jaen, M.T., Hornok, L., Nicholson, P and Waalwijk, C. (2005). Advances in molecular diagnosis of toxigenic Fusarium species: a review. *Food Addit Contam* 22:316–323.

Nelson, P.E. (1992). Taxonomy and biology of *Fusarium moniliforme*. *Mycopathologia* **117:** 29 36.

Nirenberg, H. (1976). Untersuchungen uber die morphologische und biologische Differenzierung in der Fusarium-Sektion *Liseola*. Mitt. Biol. Bundesanst. Land-Forstwirtsch. Berlin-Dahlem. **169**:1–117.

O'Donnell, K., Cigelnik, E. and Nirenberg, H.I. (1998). Molecular systematics and phylogeography of the *Gibberella fujikuroi* species complex. *Mycologia* **90**:465–93.

O'Donnell, K. and Cigelnik, E. (1997). Two divergent intragenomic rDNA ITS2 types within a monophyletic lineage of the fungus *Fusarium* are nonorthologous. *Mol. Phylo. Evol.* **7**:103-116.

O'Donnell, K., Nirenberg, H.I., Aoki, T. and Cigelnik, E. (2000). A multigene phylogeny of the *Gibberella fujikuroi* species complex: detection of additional phylogenetically distinct species. *Mycoscience* **41**:61–78

Ou, S.H. (1985). Rice diseases. Commonwealth Mycological Institue, Kew, England.

Ou, S.H. (1987). Rice Diseases. Second edition. Slough, UK: CAB International.

Powell, R.G. and Plattner, R.D. (1995) Fumonisins. In Alkaloids: chemical and biological perspectives (S. W. Pelletier, ed.): 247–278. Pergamon Press, Oxford

Reynoso, M.M., Torres, A.M. and Chulze, S.N. (2004). Fusaproliferin, beauvericin and fumonisin production by different mating populations among the *Gibberella fujikuroi* complex isolated from maize. *Mycol. Res.* **108:** 154–160.

Rojas, M.C., Hedden, P., Gaskin, P. and Tudzynski, B. (2001). The P450-1 gene of *Gibberella fujikuroi* encodes a multifunctional enzyme in gibberellin biosynthesis. *Proc. Natl. Acad. Sci USA* **98:** 5838–5843.

Saremi, H. (2005). Fusarium, Biology, Ecology and Taxonomy. 1st Edn. Jehad Daneshgahi Press University of Mashhad, Iran, pp: 153.

Sun, S.K. and Snyder, W.C. (1981). The bakanae disease of the rice plant. p. 104–113. In: *Fusarium*: Disease, Biology and Taxonomy (P.E. Nelson, T.A. Toussoun, R.J. Cook, eds.). The Pennsylvania State University Press, University Park.

Torres, A.M., Reynoso, M.M., Rojo, F.G., Ramirez, M.L. and Chulze, S.N. (2001). Fusarium species (section Liseola) and its mycotoxins in maize harvested in northern Argentina. *Food Addit. Contam.* **18:** 836–843.

Tudzynski, B. and Holter, K. (1998). The gibberellin biosynthetic pathway in *Gibberella fujikuroi*: evidence or a gene cluster. *Fungal Genet. Biol.* **25:** 157–170.

Tudzynski, B., Hedden, P., Carrera, E. and Gaskin, P. (2001). The P450-4 gene of *Gibberella fujikuroi* encodes ent-kaurene oxidase in the gibberellin biosynthetic pathway. *Appl. Environ. Microbiol.* **67:** 3514–3522.

Tudzynski, B., Mihlan, M. and Malonek, S. (2002). Secondary metabolism in *Gibberella fujikuroi*: gene clusters, regulatory circuits and evolutionary aspects. *Proc. Int. Congr. Mycol.*, 10th, Paris, France. Abstr. M-131

Tudzynski, B., Mihlan, M., Rojas, M.C., Linnemannstons, P., Gaskin, P. and Hedden, P. (2003). Characterization of the final two genes of the gibberellin biosynthesis gene cluster of *Gibberella fujikuroi*–des and P450-3 encode GA4 desaturase and the 13-hydroxylase, respectively. *J. Biol. Chem.* **278:** 28635–28643.

Viljoen, A., Marasas, W.F.O., Wingfield, M.J. and Viljoen, C.D. (1997). Characterization of *Fusarium subglutinans* f. sp. *pini* causing root disease of *Pinus patula* seedlings in South Africa. *Mycological Research* **101**:437–45.

Voigt, K., Schleier, S. and Bruckner, B. (1995). Genetic variability in *Gibberella fujikuroi* and some related species of the genus *Fusarium* based on random amplification of polymorphic DNA (RAPD). *Current Genetics* **27:** 528–35.

Waalwijk, C., de Koning, J.R.A., Baayen, R.P. and Gams, W. (1996). Discordant groupings of *Fusarium* spp. from sections Elegans, Liseola, and Dlaminia based on ribosomal ITS1 and ITS2 sequences. *Mycologia* **88**:361–368.

Webster, R.K. and Gunnell, P.S. (1992). Compendium of rice diseases. The American Phytopathological Society Press, St. Paul, Minn.

Wiebe, L.A. and Bjeldanes, L.F. (1981). Fusarin C, a mutagen from *Fusarium moniliforme*. *J. Food. Sci.* **46:** 1424–1426.

Wulff, E.D., Sorensen, J.L., Lubeck, M., Nlelson, K.F., Thrane, U. and Torp, J. (2010). *Fusarium* spp. associated with rice bakanae: ecology, genetic diversity, pathogenicity abd toxigenicity. *Environmental microbiology* **12(3)**: 649-57.

Xu, J.R., Yan, K., Dickman, M.B. and Leslie, J.F. (1995). Electrophoretic karyotypes distinguish the biological species of *Gibberella fujikuroi* (Fusarium section Liseola). *Mol. Plant-Microbe Interact.* **8:** 74–84.

Yli-Mattila, T. and Gagkaeva, T. (2010).Molecular chemotyping of *Fusarium gramineaum, F. culmorum* and *F. cerealis* isolates from Finland and Russia. In: Gherbawy Y, Voigt K (eds) Molecular identification of fungi. Springer, Berlin.

Zeller, K.A., Summerell, B.A., Bullock, S. and Leslie, J.F. (2003). *Gibberella konza* (*Fusarium konzum*) sp nov from prairie grasses, a new species in the *Gibberella fujikuroi* species complex. *Mycologia* **95:** 943–954.

Chapter 2

DNA Barcoding: A Modern Tool to Explore Microbial Diversity

Deeba Kamil, T. Prameela Devi, N. Prabhakaran and Jyoti Lekha Borah*

Indian Agricultural Research Institute, New Delhi-110 012, *

The description and identification of species are fundamental to biology. Without taxonomy, biologists in various disciplines would be unable to report their empirical findings or to access available information on their target organisms because they would not be sure of their identities. Taxonomy lays the foundations for the construction of the tree of life, makes baseline data available for conservation and ecology studies, and affords humans the possibility to take advantage of the underutilized resources offered by the earth's biodiversity (Wilson, 2004). Despite its importance as a foundation for other disciplines, taxonomy is one of the most neglected fields of research, suffering from low financial investment from fomenting agencies and, probably as a reflex of this, low interest from biology students (Godfray, 2002; Wilson, 2003). Two and a half centuries after Linnaeus, there are between 1.5 to 1.8 millions of described species, with an estimate that between five and 100 million species await discovery and description (Wilson, 2003). For this reason, the advent of new approaches to stimulate and advance taxonomy. both in terms of investment and popularity, were inevitable (Godfray, 2002; Hebert *et al.*, 2003a; Tautz *et al.*, 2003; Wheeler, 2007; La Salle *et al.*, 2009).

DNA methods aiming to modernize taxonomy were then proposed. Hebert *et al.* (2003a) incited the study of molecular diversity as a means to recognize and identify organisms by bringing up the inherent limitations of morphology, and the steady decrease in the number of specialists available for the task of uncovering our yet unknown diversity. The following limitations of morphology-based taxonomy were mentioned by them:

- ✰ Phenotypic plasticity in the characters employed for species recognition lead to incorrect identifications;
- ✰ Morphologically cryptic species are often overlooked;
- ✰ There is a lack of taxonomic keys to identify immature specimens of many species; and
- ✰ Traditional taxonomy requires high levels of expertise in any given group and is therefore restricted to specialists.

Because DNA sequences are unique for each species, they can be viewed as genetic 'barcodes' and have the potential to solve the problems inherent to the kind of taxonomy practiced so far. With a possible nucleotide variation of four nitrogenous bases (A, T, C, G) at each site, there are 4n (where "n" corresponds to the number of nucleotides surveyed) possible codes for any given sequence, making it possible to identify every taxon. The survey of just 15 nucleotide positions can identify 1 billion (4^{15}) species. The identifications can be performed quickly and at low cost, without the need of a taxonomist in the group. Additional advantages of the method would be the possibility of identifying individuals at any stage of development, and the prospect of discriminating between morphologically identical species.

DNA barcoding is a taxonomic system structured on sequence information from a short stretch of a core DNA sequence. A region of approximately 648-bp of the mitochondrial gene cytochrome c oxidase I (COI) was initially proposed as the barcode source to identify and delimit all animal species. The methodology involves the sequencing of that portion of DNA, followed by a comparison with other sequences previously deposited in a database. Species are identified by matching the obtained sequence with sequences of known identity already in the database (Hebert *et al.*, 2003a).

Currently, the molecular identification of species in Fungi is based primarily on nuclear DNA marker (nuclear ribosomal internal transcribed spacer; ITS). But the potential use of mitochondrial markers has also been considered due to their favorable features, among which, above all, their high copy number, the possibility of an easier and cheaper recovering of their sequences. Moreover, the results presented by Seifert *et al.* (2007), would strongly suggest that a mitochondrial gene could really be a good species molecular marker for Fungi, thanks to its appropriate intra and inter-species variability features. Unfortunately, a serious difficulty in the PCR and bioinformatics surveys is due to the presence of mobile introns in almost all the fungal mitochondrial genes and like other mitochondrial genes in these groups, it evolves too slowly for species-level discriminations (Chase and Fay, 2010). For this reason, alternative stretches of DNA have been proposed as target sequences for the barcoding of these organisms.

DNA Barcoding Initiatives

The Barcode of Life project was proposed to promote DNA barcoding as a global standard for sequence-based identification of eukaryotes. In 2004, this project was formally initiated by the establishment of the Consortium for the Barcode of Life (CBOL), which aims to develop a standard protocol for DNA barcoding and to construct a comprehensive DNA barcode library. Recently, the Barcode of Life project entered a new phase with the launch of the International Barcode of Life Project (iBOL; International Barcode of Life, 2010). The iBOL is a huge international collaboration of 26 countries that aims to establish an automated identification system based on a DNA barcode library of all eukaryotes. In the first five years, the iBOL will focus mainly on developing a barcode library, including five million specimens of 500000 species. The iBOL will also address the development of technologies, including new or improved protocols, informatics, equipment, DNA extraction methods and faster information systems.

The CBOL and iBOL have launched campaigns to build DNA barcode libraries of each animal group. The major targets are fish (Fish-BOL; Ward *et al.*, 2009), birds (ABBI; Hebert *et al.*, 2004a), mammals (Mammalia Barcode of Life), marine life (MarBOL) and insects. The Canadian Barcode of Life Network (BOLNET.ca) was the first national network for DNA barcoding. Subsequently, the following regions or countries have also initiated projects as a part of the iBOL: Europe (ECBOL; http://www.ecbol.org/), Norway (NorBOL; http://dnabarcoding.no/en/), Mexico (MexBOL; http://www.mexbol.org/) and Japan (JBOLI; http://www.jboli.org/). JBOLI provides information and promotes collaborative projects on DNA barcoding in Japan (see http://www.jboli.org/en/projects for relevant projects). There are also thematic programs, such as polar life (PolarBOL), quarantine and plant pathogens (QBOL, as a part of the ECBOL; Bonants *et al.*, 2010) and human health (HealthBOL).

Large barcoding projects for both trees (TreeBOL) and fungi (All Fungi Barcoding) have been launched. To construct an automated identification support system for DNA barcoding initiatives, it is necessary to accumulate comprehensive DNA barcode records for all organisms. Recent advances in information technology have made it possible to manage huge datasets. The Barcode of Life Data Systems (BOLD) is the official informatics work bench for the Barcode of Life project (Ratnasingham and Hebert, 2007), developed by the Canadian Center for DNA Barcoding (CCDB). BOLD provides a data repository for DNA barcodes, an identification support system based on them, and web services for other system developers. BOLD is freely available to any researcher via the Internet, although registration is required to create private databases and/or access restricted data. To identify unknown samples, researchers simply search for their sequenced barcode regions on the BOLD website. Importantly, the BOLD system is open to the public.

DNA Barcoding and Taxonomy

There is considerable controversy regarding the taxonomic perspective of molecular data, including DNA barcoding (Meier, 2008). There are two principal issues: (*i*) species identification; and (*ii*) species discovery. These are sometimes confused. Species identification using barcodes depends on the number of representatives of each species included in the database. The most reliable way to obtain a DNA barcode that accurately represents a species is to base it on the type specimen of that species. The first description of a new species using a DNA barcode from the holotype was by Brown coworkers (2003), who used this method to describe a new species of *Xenothictis* (Lepidoptera: Tortricidae). Since then, many new species have been described with DNA barcodes from the holotype or paratypes, not only in arthropods, but also in other animals (Burns *et al.*, 2007; Badek *et al.*, 2008; Dabert *et al.*, 2008a,b; Vaglia *et al.*, 2008; Yassin *et al.*, 2008; Yoshitake *et al.*, 2008; Adamski *et al.*, 2009).

On the contrary, species discovery is defined as the taxonomic process of recognizing a cluster of individuals and/or populations as a single species. The DNA barcode can accelerate species discovery. First, DNA barcoding can be used to identify cryptic, previously overlooked species (Hebert *et al.*, 2004b; Janzen *et al.*, 2005). Second, DNA barcode information helps to sort all specimens of related taxa, especially when taxonomic studies of these taxa are inadequate (Smith *et al.*, 2006, 2007, 2008). However, as discussed below, it should be noted that DNA barcoding cannot detect all candidates of undescribed species, especially for recently divergent groups. Some researchers have envisioned "DNA taxonomy", a concept of adopting DNA sequencing as a central criterion for taxonomic decisions and descriptions, and have proposed using DNA barcodes as the standard method of analysis (Blaxter, 2003; Tautz *et al.*, 2003; Vogler and Monaghan, 2007). However, there is concern over adopting one specific sequence region as the only criterion for taxonomic studies (Lipscomb *et al.*, 2003; DeSalle, 2006; Rubinoff, 2006). In addition, it is quite apparent that the DNA barcode itself is not a new species concept (*i.e.* a species cannot be defined based on the barcode only); neither does it provide enough information to describe unknown specimens as a new species. The results of barcoding can only suggest new species candidates (Witt *et al.*, 2006; Hajibabaei *et al.*, 2007; Miller, 2007; Waugh, 2007) as well as other valuable supporting information (*e.g.* distribution, life history, host plants) for taxonomic studies (*e.g.* integrative taxonomy: Dayrat, 2005, Yoshitake *et al.*, 2008 and Schlick-Steiner *et al.*, 2010). Species descriptions using barcodes based on type specimens will become more common and important in the near future.

Towards Integrative Taxonomy

Everyone knows that traditional taxonomy has serious problems that hinder it progress (May, 2004). Numerous species remain unknown because of the lack of specialists in their groups. Others are only known from their original descriptions, from just the holotype, or have had their type material lost or destroyed. The amount of type material deposited in museums is waiting for a specialist to take interest in it (Padial *et al.*, 2010). Within this context, DNA barcoding cannot be viewed as a threat to taxonomy because it is able to attract interest toward biodiversity studies (Smith, 2005). It is however necessary to understand that this initiative is not a panacea that will overcome all problems faced by traditional taxonomy. Even though, it can be successively applied when morphology is insufficient.

In fact, one of the main objectives of this initiative, the discovery and description of new taxa, cannot be accomplished with sequence data alone (Ebach and Holdrege, 2005b). As previously mentioned, the superposition of intra- and interspecific variation is a serious problem (Meyer and Paulay, 2005; Cognato, 2006; Meier *et al.*, 2006; Whitworth *et al.*, 2007). This difficulty, however, is not unique to molecular data, and is encountered with other sets of data such as morphology, ecology and other sources (Will *et al.*, 2005). The problems with the sole use of morphology in taxonomy work are also well-known (Packer *et al.*, 2009). Phenotypic plasticity, cryptic species and identification of immature stages are good examples (Hebert *et al.*, 2003b). From this perception that any character system used in taxonomy to the exclusion of others will fall short of the task, the practice of an integrative taxonomy that draws data from different sources is promising. This practice would certainly be superior to the current chaos generated by single data sources and illuminate taxonomic results with complementary sources of data (Will *et al.*, 2005).

A formal proposal for an integrative taxonomy already exists (Dayrat, 2005). This science aims to delimit the units of life's diversity from multiple and complementary perspectives, such as phylogeography, comparative morphology,

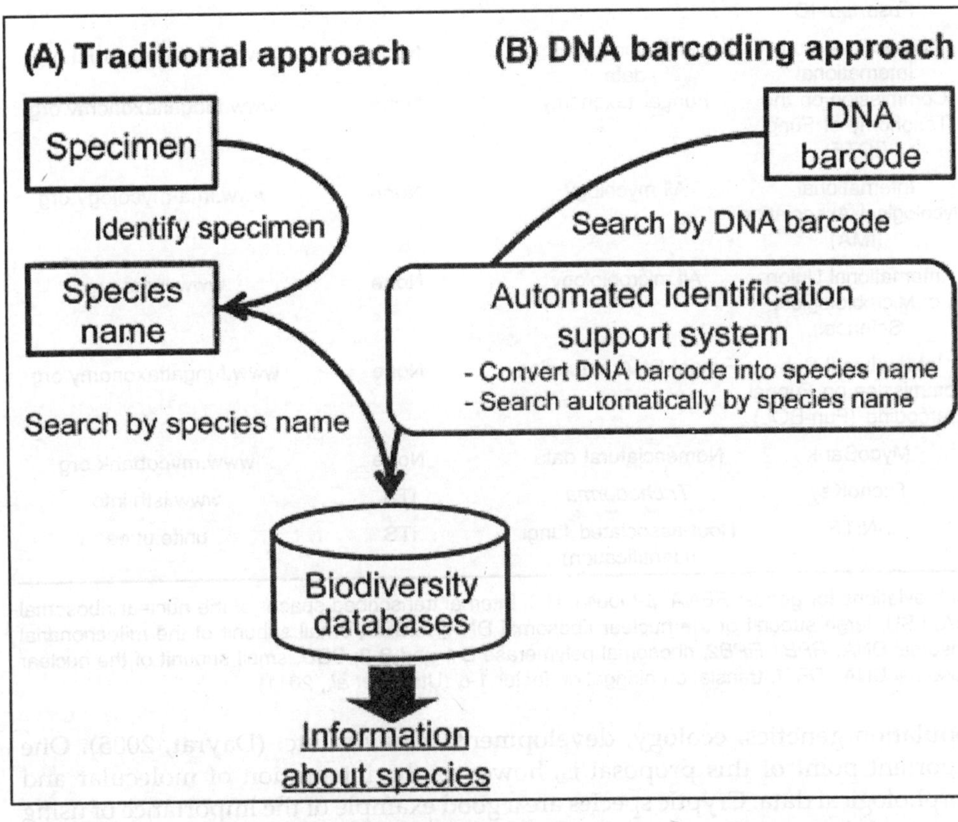

Figure 2.1: Workflow for Retrieving Biodiversity Information from Databases by (A) Traditional Approach and (B) DNA Barcoding (Seifert, 2009)

Table 2.1: Selected DNA Barcode-like Internet Resources for Fungi, and Organizations and Research Coordinating Networks Active in Fungal DNA Narcoding

Name	*Focus*	*Sequence Data Online**	*URL*
Assembling the Fungal Tree of Life (AFTOL)	Fungal phylogeny	LSU, SSU, RPB1, RPB2, TEF1, mtSSU ITS	aftol.org
CBS Fungal Biodiversity Centre (CBS) Identification databases	Dermatophytes *Penicillium* *Phaeoacremonium* *Yeasts* Medical fungi	ITS *BENA* *BENA* SSU, LSU, ITS SSU, LSU, ITS *BENA, TEF1*	www.cbs.knaw.nl → Databases
Deep Hypha Fungal Environmental Sampling and Informatics Network (FESIN) Fungi in Boreal Forests network *Fusarium*-ID	Fungal phylogeny Fungal ecology Mycorrhizal fungi *Fusarium*	(see AFTOL) (see UNITE) *TEF1 RPB2* (planned)	ocid.nacse.org/research/deephyphae www.bio.utk.edu/fesin www.bio.uio.no/meb/fbfs/index.html isolate.fusariumdb.org
Index Fungorum	Nomenclatural data	None	www.indexfungorum.org
International Commission on the Taxonomy of Fungi (ICTF)	Fungal taxonomy	None	www.fungaltaxonomy.org
International Mycological Association (IMA)	All mycology	None	www.ima-mycology.org
International Union of Microbiological Sciences	All microbiology	None	www.iums.org
International Sub-commission on Fungal Barcoding (Fun-BOL)	Fungal DNA barcoding	None	www.fungaltaxonomy.org
MycoBank	Nomenclatural data	None	www.mycobank.org
TrichoKey	*Trichoderma*	ITS	www.isth.info
UNITE	Root-associated fungi (identification)	ITS	unite.ut.ee

* Abbreviations for genes: *BENA*, β-tubulin; ITS, internal transcribed spacer of the nuclear ribosomal DNA; LSU, large subunit of the nuclear ribosomal DNA; mtSSU, small subunit of the mitochondrial ribosomal DNA; *RPB1 RPB2*, ribosomal polymerase B-1 and B-2; SSU, small subunit of the nuclear ribosomal DNA; *TEF1*, translation elongation factor 1-α (Utsugi *et al.*, 2011)

population genetics, ecology, development, behavior etc. (Dayrat, 2005). One important point of this proposal is, however, the integration of molecular and morphological data. Cryptic species are a good example of the importance of using integrated datasets whenever possible. This procedure can reveal species groups that had not been detected when a given species was initially described based on morphology alone. The use of DNA in addition to morphology helps the recognition of cryptic species that consequently become distinguished based on both sources of characters (Fisher and Smith, 2008, Wiedenbrug *et al.*, 2009, Hamada *et al.*, 2010).

The combination of different sources of data in taxonomy is not new. One good example that dates back to as early as 2003, is Wilson's "encyclopedia of life", a database created with the goal to include information on the natural history, morphology and DNA (EOL, available at http://www.eol.org/) of every species. Lately, the value of an integrative taxonomy has been recognized by the very proponents of the barcoding of life (Smith *et al.*, 2008; Fisher and Smith, 2008; Ward *et al.*, 2009).

By looking at taxonomy from a less radical point of view, it is possible to see that the integration of information is an appropriate way to promote this scientific endeavor. The adoption of new technologies becomes common-place and morphology can be appreciated. In this context, it is important to support alpha taxonomy, in a manner that maintains descriptions based on morphology in equilibrium with the application of new technologies that have been advanced to overcome the taxonomic impediment (De Salle *et al.*, 2005; Ebach and Holdrege, 2005a). The increasing number of contributions in this area is a good indication that integrative taxonomy has already been embraced.

Conclusion

DNA barcoding has become increasingly common since it was proposed in 2003. Currently, more than one million records are available in the BOLD system, which is the official depository of DNA barcode data. The new large-scale project, iBOL, will accelerate the creation of reference barcode libraries and will facilitate the application of this simple identification method. In the near future, DNA barcoding will become a standard identification protocol for various organisms. As reviewed above, fungal DNA barcoding approach has become less controversial and more supportive in field of taxanomy. However, clear limitations arise from the incomplete coverage of the existing diversity, the inherent characteristics of the mitochondrial DNA (evolutionary rate, inheritance, introns and neutrality) and the single-locus initial strategy. With its enlargement to all eukaryote taxa, the Barcode of life project has also evolved to a more flexible framework. The multi-locus barcoding approach is now commonly accepted, particularly to discriminate between low level taxa and to increase the power of the sequence assignments.

DNA barcoding projects are strongly related to other biodiversity and genetic database projects. Together with the identification support system, DNA barcode will be a new keyword to explore biodiversity and will serve as a bridge between research in the fields of biodiversity and genomics. Some taxonomists are concerned that DNA barcoding will compete with traditional taxonomic studies (Ebach and Holdrege, 2005a,b). However DNA barcoding is inseparably linked to taxonomy as a powerful tool for complements taxonomic studies (Schindel and Miller, 2005; Hajibabaei *et al.*, 2007). The integration of various types of data, such as morphological, ecological, physiological and molecular data, including DNA barcodes, will improve species discovery and description processes (Waugh, 2007; Padial *et al.*, 2010). This integrative approach will be strengthened by various biodiversity databases.

References

Adamski, D., Boege, K., Landry, J.F. and Sohn, J.C. (2009). Two new species of *Wockia heinemann* (Lepidoptera: Urodidae) from coastal dry-forests in Western Mexico. *Proceedings of the Entomological Society of Washington* **111**: 166–182.

Badek, A., Dabert, M., Mironov, S.V. and Dabert, J. (2008). A new species of the genus Proctophyllodes (Analgoidea: Proctophyllodidae) from cetti's warbler Cettia cetti (Passeriformes: Sylviidae) with DNA barcode data. *Annales Zoologici* **58**: 397–402.

Blaxter, M. (2003). Counting angels with DNA. *Nature* **421:** 122–124.

Bonants, P., Groenewald, E., Rasplus, J.Y. *et al.* (2010). QBOL: a new EU project focusing on DNA barcoding of quarantine organisms. *EPPO Bulletin* **40:** 30–33.

Brown, J.W., Miller, S.E. and Horak, M. (2003). Studies on New Guinea moths. 2. Description of a new species of *Xenothictis meyrick* (Lepidoptera: Tortricidae: Archipini). *Proceedings of the Entomological Society of Washington* **105**: 1043–1050.

Burns, J.M., Janzen, D.H., Hajibabaei, M., Hallwachs, W. and Hebert, P.D.N. (2007). DNA barcodes of closely related (but morphologically and ecologically distinct) species of skipper butterflies (Hesperiidae) can differ by only one to three nucleotides. *Journal of the Lepidopterists Society* **61:** 138–153.

Chase, M.W. and Fay, M.F. (2010). Barcoding of plants and fungi. *Science* **325:** 682-683.

Cognato, A.I. (2006). Standard percent DNA sequence difference for insects does not predict species boundaries. *J. Econ. Entomol.* **99:** 1037-1045.

Dabert, J., Ehrnsberger, R. and Dabert, M. (2008a). *Glaucalges tytonis* sp. n. (Analgoidea, Xolalgidae) from the barn owl *Tyto alba* (Strigiformes, Tytonidae): compiling morphology with DNA barcode data for taxon descriptions in mites (Acari). *Zootaxa* **1719:** 41–52.

Dabert, J., Natress, B. and Dabert, M. (2008b). *Xoloptes blaszaki* sp. nov. (Pterolichoidea: Pterolichidae)–A new species of feather mite from *Alectoris rufa* (Galliformes: Phasianidae) from Europe with DNA barcode data. *Annales Zoologici* **58**: 391–396.

Dayrat, B. (2005). Towards integrative taxonomy. *Biological Journal of the Linnean Society* **85:** 407–415.

DeSalle, R. (2006). Species discovery versus species identification in DNA barcoding efforts: response to Rubinoff. *Conservation Biology* **20:** 1545–1547.

Desalle, R., Egan, M.G. and Siddal, M. (2005). The unholy trinity: taxonomy, species delimitation and DNA barcoding. *Philos. T. R. Soc.* B **360:** 1905-1916.

Ebach, M.C., Holdrege, C. (2005a). DNA barcoding is no substitute for taxonomy. *Nature* **434:** 697.

Ebach, M.C. and Holdrege, C. (2005b). More taxonomy, not DNA barcoding. *BioScience* **55:** 822–823.

Fisher, B.L. and Smith, M.A. (2008). A revision of Malagasy Species of *Anochetus mayr* and *Odontomachus latreille* (Hymenoptera: Formicidae). *Plos One* **3:** 1787

Godfray, H.C.J. (2002). Challenges for taxonomy. *Nature* **417:** 17-18.

Hajibabaei, M., Singer, G.A.C., Hebert, P.D.N. and Hickey, D.A. (2007). DNA barcoding: how it complements taxonomy, molecular phylogenetics and population genetics. *Trends in Genetics* **23**: 167–172.

Hamada, N., Pepinelli, M., Mattos-glória, A. and Luz, S.L.B. (2010). A new black fly species from Brazil, closely related to *Simulium guianense* Wise (Diptera, Simuliidae), revealed by morphology and DNA barcoding. *Zootaxa* **2428:** 22-36.

Hebert, P.D.N., Cywinska, A., Ball, S.L. and deWaard, J.R. (2003a). Biological identifications through DNAbarcodes. *Proc. R. Soc. Lond.* B **270**: 313–321.

Hebert, P.D.N., Penton, E.H., Burns, J.M., Janzen, D.H. and Hallwachs, W. (2004b). Ten species in one: DNA barcoding reveals cryptic species in the neotropical skipper butterfly *Astraptes fulgerator*. *Proceedings of the National Academy of Sciences of the United States of America* **101:** 14 812–14 817.

Hebert, P.D.N., Ratnasingham, S. and deWaard, J.R. (2003b). Barcoding animal life: cytochrome c oxidase subunit 1 divergences among closely related species. *Proceedings of the Royal Society of London, Series B* **270:** S96–S99.

Hebert, P.D.N., Stoeckle, M.Y., Zemlak, T.S. and Francis, C.M. (2004a) Identification of birds through DNA Barcodes. *PLoS Biology* **2:** e312.

International Barcode of Life (2010). International Barcode of Life. [Cited 15 October 2010.] Available from URL: http://ibol.org/

Janzen, D.H., Hajibabaei, M., Burns, J.M., Hallwachs, W., Remigio, E. and Hebert, P.D.N. (2005). Wedding biodiversity inventory of a large and complex Lepidoptera fauna with DNA barcoding. *Philosophical Transactions of the Royal Society of London, Series* B **360**: 1835–1845.

La sale, J., Wheeler, Q., Jackway, P., Winterton, S., Hobern, D. and Lovell, D. (2009). Accelerating taxonomic discovery through automated character extraction. *Zootaxa* **2217:** 43-55.

Lipscomb, D., Platnick, N. and Wheeler, Q. (2003). The intellectual content of taxonomy: a comment on DNA taxonomy. *Trends in Ecology and Evolution* **18:** 65–66.

May, R.M. (2004). Tomorrow's taxonomy: collecting new species in the field will remain the rate-limiting step. *Philos. T. Roy. Soc.* B **359**: 733-734.

Meier, R. (2008). DNA sequences in taxonomy, opportunities and challenges. In: Wheeler, Q.D. (ed.) The New Taxonomy, pp 65–127. CRC Press, Boca Raton, FL.

Meier, R., Shiyang, K., Vaidya, G. and NG, P.K.L. (2006). DNA Barcoding and taxonomy in Diptera: a tale of high intraspecific variability and low identification success. *Syst. Biol.* **55:** 715-728.

Meyer, C.P. and Paulay, G. (2005). DNA barcoding: error rates based on comprehensive sampling. *Plos Biol.* **3:** 422.

Miller, S.E. (2007). DNA barcoding and the renaissance of taxonomy. *Proceedings of the National Academy of Sciences of the United States of America* **104:** 4775–4776.

Packer, L., Gibbs, J., Sheffield, C. and Hanner, R. (2009). DNA barcoding and the mediocrity of morphology. *Mol. Ecol. Resour.* **9:** 42-50.

Padial, J.M., Miralles, A., De la Riva, I. and Vences, M. (2010). The integrative future of taxonomy. *Frontiers in Zoology* **7:** 16.

Ratnasingham, S. and Hebert, P.D.N. (2007). BOLD: the Barcode of Life Data System (www.barcodinglife.org). *Molecular Ecology Notes* **7:** 355–364.

Rubinoff, D. (2006). Utility of mitochondrial DNA barcodes in species conservation. *Conservation Biology* **20:** 1026–1033.

Schindel, D.E. and Miller, S.E. (2005). DNA barcoding a useful tool for taxonomists. *Nature* **435:** 17.

Schlick-Steiner, B.C., Steiner, F.M., Seifert, B., Stauffer, C. *et al.* (2010). Integrative taxonomy: a multisource approach to exploring biodiversity. *Annual Review of Entomology* **55:** 431–438.

Seifert, K.A. (2009). Barcoding fungi: Progress towards DNA barcoding of fungi. *Molecular Ecological Resources* **9:** 83-89.

Seifert, K.A., Samson, R.A., Dewaard, J.R., Houbraken, J., Levesque, C.A., Moncalvo, J.M., Louis-Seize, G., Hebert, P.D.N. (2007). Prospects for fungus identification using C01 DNA barcodes, with Penicillium as a test case. *Proc. Natl. Acad. Sci.* U.S.A. **104:** 3901–3906.

Smith, M.A., Rodriguez, J.J., Whitfield, J.B. *et al.* (2008) Extreme diversity of tropical parasitoid wasps exposed by iterative integration of natural history, DNA barcoding, morphology, and collections. *Proceedings of the National Academy of Sciences of the United States of America* **105:** 12359–12364.

Smith, M.A., Wood, D.M., Janzen, D.H., Hallwachs, W. and Hebert, P.D.N. (2007). DNA barcodes affirm that 16 species of apparently generalist tropical parasitoid flies (Diptera, Tachinidae) are not all generalists. *Proceedings of the National Academy of Sciences of the United States of America* **104:** 4967–4972.

Smith, M.A., Woodley, N.E., Janzen, D.H., Hallwachs, W. and Hebert, P.D.N. (2006). DNA barcodes reveal cryptic host-specificity within the presumed polyphagous members of a genus of parasitoid flies (Diptera: Tachinidae). *Proceedings of the National Academy of Sciences of the United States of America* **103:** 3657–3662.

Smith, V.S. (2005). DNA barcoding: perspectives from a "Partnerships for Enhancing expertise in taxonomy" (PEET) debate. *Syst. Biol.* **54:** 841-844.

Tautz, D., Arctander, P., Minelli, A., Thomas, R.H. and Vogler, A.P. (2003). A plea for DNA taxonomy. *Trends in Ecology and Evolution* **18:** 70–74.

Utsugi, J., Toshihide, K. and Motoni, I. (2011). current progress in DNA barcoding and future implications for entomology. *Entomological Science* 14: 107-124.

Vaglia, T., Haxaire, J., Kitching, I.J., Meusnier, I. and Rougerie, R. (2008). Morphology and DNA barcoding reveal three cryptic species within the *Xylophanes neoptolemus* and loelia species-groups (Lepidoptera: Sphingidae). *Zootaxa* **1923:** 18–36.

Vogler, A.P. and Monaghan, M.T. (2007). Recent advances in DNA taxonomy. *Journal of Zoological Systematics and Evolutionary Research* **45:** 1–10.

Ward, R.D., Hanner, R. and Hebert, P.D.N. (2009). The campaign to DNA barcode all fishes, FISH-BCL. *Journal of Fish Biology* **74:** 329–356.

Waugh, J. (2007). DNA barcoding in animal species: progress, potential and pitfalls. *Bioessays* **29:** 188–197.

Wheeler, Q.D. (2007). Invertebrate systematics or spineless taxonomy? *Zootaxa* **1668:** 11-18.

Whitworth, T.L., Dawson, R.D., Magalon, H. and Baudry, E. (2007). DNA barcoding cannot reliably identify species of the blowfly genus *Protocalliphora* (Diptera: Calliphoridae). *P. R. Soc.* B **274:** 1731-1739.

Wiedenbrug, S., Mendes, H.F., Pepinelli, M. and Trivinhostrixino, S. (2009). Review of the genus *Onconeura* Andersen et Saether (Diptera: Chironomidae), with the description of four new species from Brazil. *Zootaxa* **2265:** 1-26.

Will, K.W., Mishler, B.D. and Wheeler, Q.D. (2005). The perils of DNA barcoding and the need for integrative taxonomy. *Syst. Biol.* **54:** 844–851.

Wilson, E.O. (2003). The encyclopedia of life. *Trends Ecol. Evol.* **18:** 77-80.

Wilson, E.O. (2004). Taxonomy as a fundamental discipline. *Philos. T. Roy.Soc.* B **359:** 739.

Witt, J.D.S., Threloff, D.L. and Hebert, P.D.N. (2006). DNA barcoding reveals extraordinary cryptic diversity in an amphipod genus: implications for desert spring conservation. *Molecular Ecology* **15:** 3073–3082.

Yassin, A., Capy, P., Madi-Ravazzi, L., Ogereau, D. and David, J.R. (2008). DNA barcode discovers two cryptic species and two geographical radiations in the invasive drosophilid *Zaprionus indianus*. *Molecular Ecology Resources* **8:** 491–501.

Yoshitake, H., Kato, T., Jinbo, U. and Ito, M. (2008). A new *Wagnerinus* (Coleoptera: Curculionidae) from northern Japan: description including a DNA barcode. *Zootaxa* **1740:** 15–27.

Chapter 3

Exploitation of Soil Microbial Diversity in Suppressing Soil Borne Plant Pathogens and Enhancing Crop Productivity

G. Sangeetha, S. Sundaramoorthy and V. Kurucheve*

Department of Plant Pathology, Faculty of Agriculture, Annamalai University, Annamalainagar – 608002. Tamil Nadu

Microbial diversity is just a subset of biodiversity involving fungi, bacteria, viruses, algae, protozoans, actinomycetes and nematodes etc. This complex soil microbial diversity reflects a great variability among the microbes. Microbes are integral component of soil. A handful of soil contains different kinds of microbes, even a single soil particle represents ecological niches for different types of microbes to develop (Nee, 2002; Morin, 2000). Soils without microbes are merely a dead material. The top 6 inches of fertile soil contains more than 2 tonnes of fungal and bacterial biomass (Torsvik *et al.*, 1990). The bacterial populations in soil top layers can go up to more than 10^9 cells per g of soil, but most of these are unculturable (Torsvik and Ovreas, 2002). During the last 50 years, many beneficial effects of microbes in soil have been discovered (Subba Rao and Gaur, 2000). It is now widely being recognized that the microbial wealth provide soil richness in terms of crop growth enhancement, making available slow-release nutrients, decomposition of organic matter, toxin removal and protecting crop plants by suppressing the invading plant pathogens (Haas and Defago, 2005).

Among the microorganisms, some are harmful plant pathogens while others are neutral or beneficial in their effects on plant growth. For the past two decades plant pathologists have been investigating the exploitation of microbial biodiversity to manage the harmful plant pathogenic organisms with beneficial antagonistic microbes. Sustainable agriculture can be directed towards maximizing the quality of soil microbial diversity in terms of disease suppression, by culturing and reinoculating the beneficial microbes to soil. Many microbes extracted from soil are explored in industrial production such as food processing, development of biocides, bio-control agents, medicines and other natural products, hence pharmaceutical companies spend millions of dollars annually for screening soil and litter in search of beneficial microorganisms (Gadgil, 2000).

Plant diseases are responsible for enormous annual crop losses. Among the plant diseases, soil borne plant pathogens either fungi or bacteria are notoriously causing complete loss to crop plants and very difficult to control. These microbes reside in soil and survive on plant residues and penetrate the roots of crop plants. Although several microbes can act as plant pathogens, fungi are predominant in soil microflora which causes majority of soil borne diseases. Examples of economically important soil borne fungal plant pathogens include *Fusarium* spp., *Verticillium* spp, *Pythium* spp, *Phytophthora* spp, *Macrophomina* spp, *Sclerotium* spp, and *Rhizoctonia* spp. etc. Despite low initial inoculum in soil, these pathogens can cause complete destruction of plants and occasionally cause total loss of yield (Pal and Mc Spadden Gardener, 2006).

These soil borne plant pathogens especially fungi have been proven difficult to manage practically. Because most of the management practices for soil borne pathogens relied upon crop rotation for 4-5 year with non host crops or by modification of certain cultural practices. Application of fungicides and soil fumigants suitable as well as applicable for high value crops only. Moreover the 'one chemical kills all' approach for management of plant diseases is detrimental to the microbial diversity in agro-ecosystems. In most cases, plant resistance has not been available for all soil borne pathogens and therefore it is no longer an acceptable method of disease control.

Among the microbial community in soil, many antagonistic microbes are naturally present in soil and exert a certain degree of biological control over soil borne plant pathogens. However, this level of natural control is often insufficient for consistent and reliable disease free cropping. Since the earliest observations of antagonistic disease suppressing soil microorganisms more than 70 years ago, plant pathologists have been fascinated by the idea that such micro organisms could be used as eco-friendly method of managing the plant diseases. The use of microbial diversity to manage disease of crop plants falls into the category of biological control. Researchers are therefore attempting to enhance the effectiveness of antagonists in the cropping field, thus increasing suppressive-ness. Once introduced these disease suppressive effects of beneficial soil microbes are often long lasting because their populations continue to grow and will establish new generations in an environment with a diverse rhizosphere. After colonization around the roots of the crop plants, these biocontrol organisms offer preventive disease control for the life of the plant. Several studies have documented the relationship between the degree of soil suppressiveness to plant diseases and abundance of soil microbial communities (Nitta, 1991; Abawi and Widmer, 2000).

Plants are surrounded by diverse types of microbes, some of which can contribute to biological control of plant diseases. Due to the difficulty with which they can be cultured, most biocontrol research has focused on a limited number of bacterial (*Pseudomonas, Bacillus, Burkholderia, Lysobacter, Pantoea, Streptomyces* and *Serratia*) and fungal (*Trichoderma, Ampelomyces, Coniothyrium, Dactylella, Paecilomyces, Trichomycetes, Penicillium* and non pathogenic *Fusarium* and *Pythium*) genera. Still, other microbes that are more recalcitrant to *in vitro* culturing have been intensively studied and those include mycorhizal fungi, *e.g. Pisolithus* and *Glomus* spp. In this review, the exploitation and potential of soil microbial diversity to manage soil borne plant pathogens and their role in enhancing crop productivity have discussed in detail.

Diversity of Plant Pathogens in Soil

Among the soil borne microbes causing plant diseases, predominantly fungi, followed by bacteria and nematodes causes majority of soil borne plant diseases. Fungi are complex group of organisms with polyphyletic origins. Their unique pattern of mycelial growth and parasexual modes of reproduction have brought together enabling them to respond quickly to selection pressures. Changes in cropping sequences (cultivation of crop plants throughout the year), tillage practices, the deployment of disease-resistant cultivars, introduction of new crop varieties and use of selective systemic fungicides provides intense selection pressures that result in changes in soil fungal populations (Brasier, 1995; Ahmad *et al.*, 1995). These processes promote variability within pathogen populations and pathogens themselves contribute to soil biodiversity (Hawksworth and Rossman, 1997). This diversity within pathogens is important not only in terms of causing plant diseases but also with respect to other roles that they have. As colonizers of soil organic matter, many plant pathogens contribute to degradation processes in soil and recycling of nutrients (Durall and Parkinson, 1991). Some pathogens and soil fungi have industrial value as producers of useful secondary metabolites and valuable by products (Bills, 1995). Some soil fungi may cause human and animal diseases (Sternberg, 1994) and some others can be employed in management of weeds (Briere, *et al.*, 2000) and insects.

Microbial Diversity in Relation to Suppression of Soil Borne Plant Pathogens

Soil is considered to be the storehouse of microbial activity, though the space occupied by living microorganisms is estimated to be less than 5 per cent of the total space. However, major microbial activity is confined to the 'hot-spot' rhizosphere (Pinton *et al.*, 2001). The diversity of soil microbial communities can be key to the capacity of soils to suppress soil-borne plant diseases (Van Elsas *et al.*, 2002). Mechanisms within the microbial activity of soil are responsible for the suppression of plant pathogens (Van Os and Van Ginkel, 2001). The mechanisms by which these microorganisms make soil suppressive can be divided into several categories like nutrient competition, microbial antagonism, parasitism, and induced systemic resistance. The most important groups containing antagonistic microorganisms is the group of Fluorescent pseudomonads, *Bacillus* and *Serratia* sp. and number of rhizobacteria belonging to the genera *Azospirillum, Alcaligenes, Arthrobacter, Acinetobacter, Burkholderia, Enterobacter, Erwinia, Flavobacterium* and *Rhizobium* are able to exert a beneficial effect on plant growth. Spontaneous control of plant diseases by bacteria in some fields was discovered at several places around the world. Some soils, called suppressive soils, contain bacteria that protect plants against fungal diseases despite the presence of disease causing pathogens in soil (Haas and Defago, 2005).

Root colonizing plant beneficial fungi *Trichoderma* spp. and non-pathogenic *Fusarium* and *Pythium* spp. are also important in protecting plants from root pathogens and it is reviewed by Harman *et al.* (2004); Fravel *et al.* (2003) respectively and are not discussed in this review. These are well recognized antagonists of established root pathogens. The various mechanisms of disease suppression by beneficial microbes includes (i) microbe (pathogen)-microbe (biocontrol agent) interaction, (ii) plant-microbe (biocontrol agent) interaction, (iii) metabolites produced by biocontrol agents and (iv)induced systemic resistance. Multiple interactions by beneficial fungi and bacteria in the rhizosphere region are shown to provide enhanced level of disease suppression then when they acted / used individually.

It is a complex process involving not only the bio-control microbe, the pathogen and the plant, but also the indigenous microflora, macrobiota such as nematodes and the plant growth substrate such as soil. To act efficiently, microbes should remain active under wide range of natural conditions such as varying pH, temperature and different concentrations of ions. These requirements are not that much easy to satisfy. Therefore, it is not surprising that the efficacy of several bio-control products is not always sufficiently aimed to control soil borne plant pathogens. However, as our understanding and selection procedures for active strains increases and released, the biocontrol products will improve in future.

Diversity of Rhizosphere Microflora and their Importance

Earlier, Hiltner (1904) discovered that the rhizosphere ie., the layer of soil influenced by the root, is much richer in bacteria than the surrounding soil. Rhizosphere is an area encircling the plant root system, which is characterized by enhanced biomass productivity. The exudation of nutrients such as aminoacids, organic acids, enzymes, carbohydrates by plant roots creates nutrient rich environment in which microbial activity is enhanced. To exert their beneficial effects in the root environment, bacteria have to be rhizosphere competent *i.e.*, able to compete well with other rhizosphere microbes for nutrients secreted by the roots. The rhizosphere soils can be a good source of plant growth promoting rhizobacteria (PGPR) and fungi, although ordinary soil also contains these disease suppressing organisms. The dominant rhizobacteria that prefer to live in close vicinity of the roots or on their surface play a crucial role in soil health and plant growth. Although the concentration of bacteria in the rhizoplane is 10 to 1000 times higher than that of in bulk soil, it is still 100-fold lower than that in the average laboratory medium (Lugtenberg and Kamilova, 2009). The desirable trait of good root colonization can be selected by isolating bacteria, or fungi that remain attached to the root surface or have been penetrated into the intercellular spaces between the root epidermis and the cortex after extensive washing of the roots (Hallman *et al.*, 1997).

Both free living (*Pseudomonas* spp.) and symbiotic (VAM and *Rhizobium*) rhizobacteria are involved in such specific ecological niches that help in organic matter degradation, solubilization and mobilization of nutrients and thereby making them available to the plant for their growth. Plant roots secrete metabolites that can be utilized as nutrients, and among the substantial amount of carbon fixed by the plant, 5-21 per cent is secreted mainly as root exudates. In recent years, it has been proven that root colonization indeed required for some biocontrol mechanisms, such as antiobiosis and competition for nutrients and niches (Chin-A-Woneng *et al.*, 2000; Kamilova *et al.*, 2005). Hence, microbial interaction that takes place in the rhizosphere are very important in plant disease suppression and sustainable crop productivity.

The use of beneficial soil microorganisms as agricultural inputs for improved crop production requires the selection of rhizosphere competent microorganisms with plant growth- promoting attributes (Hynes *et al.*, 2008). The best known exudates composition known is probably in tomato, its major portion of root exudates contains citrate and glucose, allows the spores of the tomato root pathogen (*Fusarium oxysporum* sp. *radicis lycopersici*) to germinate. However the additional presence of *P. fluorescens* (WCS 365) in the rhizosphere region delays the germination process of pathogenic *Fusarium* (Kamilova *et al.*, 2008).

Role of Growth Promoting Rhizobacteria in Crop Plants

Free living, root colonizing bacteria (rhizobacteria) have been studied in the past century as possible source for increasing crop productivity. Beneficial rhizobacteria besides inhibiting the pathogenic growth, they also enhance the plant growth and yield, hence they referred as PGPR (plant growth promoting rhizobacteria). PGPR are non pathogenic, strongly root colonizing bacteria which increase the plant yield by one or more mechanisms (Glick, 1995). It enhances the availability and uptake of plant nutrients, improvement of soil structure, production of substances promoting plant growth and suppression of soil borne plant pathogens. This growth promoting rhizobacteria can be divided into two major groups according to their relationship with host plants (i) Symbiotic rhizobacteria (ii) free living rhizobacteria (Khan, 2005). The rhizobacteria which invades the interior of host cell and survive inside called intracellular PGPR Eg. nodule bacteria (*Rhizobium)* or remain outside the host plant cells called extracellular PGPR Eg. *Pseudomonas, Bacillus* and *Azotobacter*.

More specially the soil borne fluorescent pseudomonades have received particular attention because of their excellent root colonizing ability, catabolic versatility and their capacity to produce a wide range of antifungal metabolites (Olivain, *et al.*, 2004). In several instances, inoculations with plant growth promoting rhizobacteria were effective in controlling multiple diseases caused by different pathogens (Ongena *et al.*, 2004; Ryu *et al.*, 2004). Sangeetha *et al.* (2010) examined nine native bacterial strains and among all, the native strains *viz.*, non-fluorescent Pseudomonas (NFP6) followed by *Pseudomonas fluorescens* (Pf3a) have recorded maximum inhibition of mycelial growth (up to 65.1 per cent) of crown rot pathogens *viz.*, *Lasiodiplodia theobromae* and *Colletotrichum musae* under *in vitro* condition. They also reported that apart from antifungal activity, induction of defense-related enzymes plays a major role in reduction of crown rot severity of harvested banana fruits.

The beneficial effect of rhizobacteria on plant growth may be direct or indirect. The direct plant growth promotion includes bio-fertilization, stimulation of root growth, rhizo remediation and plant stress control. The indirect effects includes the reduction of level of disease *i.e.*, through antibiosis, induction of systemic resistance and competition for nutrients and niches (Lugtenberg and Kamilova, 2009). PGPR compensate the reduction in plant growth caused by salt stress (Kaymak, *et al.*, 2009); drought stress (Zahir *et al.*, 2009); heavy metals (Kumar *et al.*, 2009) and some other unfavourable environmental conditions.

Generally, PGPR traits associated with the biocontrol of plant pathogens include (i) Fixing atmospheric nitrogen and supply it to plants (ii) synthesizing precursors of various phytohormones (Ahmad *et al.*, 2008) (iii) Solubilization of inorganic phosphorus (Khan and Zaidi, 2007) and mineralizing organic 'P' (Khan *et al.*, 2007 (iv) antibiotic synthesis (Haas and Defago, 2005) (v) secretion of iron binding siderophores and vitamins (Glick, 2001) (vi) production of low molecular weight metabolites such as hydrogen cyanide with antifungal activity (Dowling and O'Gara, 1994) (vii) production of enzymes chitinase, β-1,3-glucanase, protease and lipase which cause lysis of some fungal cells (Chet and Inbar, 1994) (viii) production of oxidative stress enzymes such as catalases, superoxide dismutases and peroxidases for scavenging active oxygen species (ix) competition for nutrients and niches on the root surface (Kamilova *et al.*,.2005) and (x) lowering the level of ethylene in stressed plants with the enzyme ACC deaminase (Saravanakumar and Samiyappan, 2007). The growth promoting substances produced by various rhizobacteria are summarized in Table 3.1.

Table 3.1: Growth Promoting Substances Produced by Plant Growth Promoting Rhizobacteria

Rhizobacteria	*Growth Promoting Substances*	*References*
Pseudomonas fluorescens	IAA, siderophore and P-solubilization	Gupta *et al.* (2005)
P. putida	Siderophore	Tripathi *et al.* (2005)
Bacillus subtilis	IAA and P-solubilization	Zaidi *et al.* (2006)
Bacillus spp.	IAA, P- solubilization, siderophores, HCN and ammonia	Wani *et al.* (2007b)
Azotobacter sp.	Gibberellin, kinetin, IAA and P-solubilization	Kumar *et al.* (2001)
Bacillus and *Azospirillum* sp.	IAA and P-solubilization	Yasmin *et al.* (2004)
Rhizobium	HCN and siderophore	Deshwal *et al.* (2003)
Bradyrhizobium	Siderophore, IAA and P-solubilization	Deshwal *et al.* (2003)
Bradyrhizobium, Rhizobium	IAA and P-solubilization	Antoun *et al.* (2004)

Mechanisms of Biocontrol by Beneficial Rhizobacteria

Antibiotics

Antibiotics are microbial toxins that can, at low concentrations, poison or kill other microorganisms. Most microbes produce and secrete one or more compounds with antibiotic activity. In some instances, antibiotics produced by microorganisms have been shown to be particularly effective at suppressing plant pathogens and the diseases they cause. Antibiotics identified in antagonistic rhizobacteria, *Pseudomaonas* include the classical compounds HCN (Haas and Keel, 2003); phenazines (Mavrodi *et al.*, 2006), of which the major ones are phenazine-1-carboxylic acid and phenazine-1- carboxamide; 2,4-diacetyl phloroglucinol (Phl) (Dunne *et al.*, 1996); pyoluteorin (Nowak-Thompson, 1999) and pyrrolnitrin. Likewise, Zwittermycin A and kanosamine (Emmert *et al.*, 2004) are produced by *Bacillus cereus*. Lipopeptide biosurfactants like rhamnolipid and phenazine act synergistically against soilborne diseases caused by *Pythium* spp. (Perneel *et al.*, 2008). To be effective, antibiotics must be produced in sufficient quantities near the pathogen to get a better result.

Predation and Parasitism

This mechanism has not been thoroughly known in beneficial rhizobacteria as like well established mechanism in antagonistic fungi. Predation and parasitism, the major biocontrol mechanism used by some biocontrol fungi like *Trichoderma* species, is based on enzymatic destruction of the fungal cell wall (Harman *et al.*, 2004).

Induced Systemic Resistance

A number of strains of root-colonizing microbes have been identified as potential elicitors of plant host defenses. Some biocontrol strains of *Pseudomonas* sp. and *Trichoderma* sp. are known to strongly induce plant host defenses (Haas and Defago 2005, Harman *et al.*, 2004). Interaction of some bacteria with plant roots can result in resistance in plants to pathogenic microbes. This phenomenon is called induced systemic resistance (ISR). ISR differs from SAR (systemic acquired resistance). ISR is dependent on jasmonic acid and ethylene signalling in the plant (van Loon 2007) and SAR is mediated by salicylic acid. Many bacterial chemical elicitors of SAR and ISR may be produced by the PGPR strains upon inoculation includes salicylic acid, flagella, siderophores, cyclic lipopeptides, lipopolysaccharides, the antifungal factor 2,4-diacetyl phloroglucinol (Phl), the signal molecule N- acyl homoserine Lactone (AHL), 2,3-butanediol and other volatile substances (Ongena *et al.*, 2004, Ryu *et al.*, 2004). This information suggests that plants would detect the composition of their plant-associated microbial diversities and also respond to changes in the types and localization of many different signals.

Competition for Nutrients and Niches (CNN)

Soil and living plant surfaces are frequently nutrient limited environments from microbial perception. Competition of biocontrol bacteria with the pathogen for nutrients and niches in the rhizosphere has been suggested for decades as a possible mechanism of biocontrol, but experimental proof was lacking. Mutant studies confirmed the proposed mechanism (Kamilova *et al.*, 2005). In general, soil borne pathogens, such as *Fusarium* and *Pythium* that infect through mycelial contact are more susceptible to competition with other microbes than those pathogens that germinate directly (by forming appressoria and infection pegs) on plant surfaces. For example, effective catabolism of nutrients in the spermosphere by *Enterobacter cloacae* has been identified as a mechanism behind the suppression of *Pythium ultimum* (van Dijk and Nelson 2000).

Competition for Ferric Ions

Biocontrol based on competition are rare but essential micronutrients, such as iron, plays a major role in iron limiting environment. Iron is extremely limited in the rhizosphere, depending on soil pH. To survive in such an environment, organisms were found to secrete iron-binding ligands called siderophores having high affinity to sequester/seize iron from the micro-environment. Rhizobacteria producing high concentrations of high-affinity siderophores in the rhizosphere can inhibit the growth of fungal pathogens when the Fe^{3+} concentration is low, *e.g.*, in acid soils (Schippers *et al.*, 1987). And, a direct correlation was established *in vitro* between siderophore synthesis in fluorescent pseudomonads and their capacity to inhibit germination of chlamydospores of *F. oxysporum* (Sneh *et al.*, 1984). The increased efficiency in iron uptake by the benefical rhizobacteria is thought to be a contributing factor for their ability to aggressively colonize plant roots which helps in displacement of pathogenic organisms infecting roots of crop plants.

Exploitation of Fluorescent Pseudomonads and *Bacillus* spp. in Managing Soil Borne Plant Diseases and Enhancing Crop Productivity

The term pseudomonad (*Pseudomonas* like bacteria) is often used to describe strains for which the taxonomic affiliation has not been established in detail. The fluorescent pesuedomonads produce the fluorescent pigment pyoverdines (Pvd). They are Gram (–)ve, rod shaped bacteria with large heterogenous group comprises of *P. fluorescens, P. syringae, P. aeruginosa, P. putida* etc. Apart from *P. fluorescens, P. aureofaciens, P. chlororaphis, P. brassicacearum, P. thevervalensis* are also reported to have biocontrol activity against a wide range of plant pathogens. In case of *Pseudomonas* spp. the crucial colonization level that must be reached has been estimated at 10^5–10^6 CFU (colony forming units) g^{-1} of root, to protect plants from *Pythium* spp. Artificially introduce *Pseudomonas* can initially colonize roots at 10^7–10^8 CFU g^{-1}, but these levels always decline in a few weeks (Landa, 2003). The use of fluorescent pseudomonads for controlling soil borne plant diseases has been well documented (Radjacommare *et al.*, 2002; Ramamoorthy *et al.*, 2001; 2002).

Jonathan *et al.* (2009) reported that the combined application of *Pseudomonas fluorescens* (Pfbv22) and *Bacillus subtilis* (Babv57) significantly reduced the nematode infestation in banana. Naturally occurring root- associated fluorescent pseudomonas producing the antibiotic 2, 4-DAPG were highly enriched in take all suppressive soil and are key components of specific suppression of take all pathogen in wheat (Raaijmakers *et al.*, 1997; 1998). This suppression was lost when 2, 4-DAPG producing *Pseudomonas* spp. was eliminated and conversely, conducive soil gained supressiveness to take all when 2, 4-DAPG producing pseudomonas strains were introduced. The modes of action with specific mechanisms exerted by *Pseudomonas* and *Bacillus* against major soil borne plant pathogens are given in Table 3.2.

Table 3.2. Modes of Action with Specific Mechanisms Exerted by *Pseudomonas* and *Bacillus* Against Major Soil Borne Plant Pathogens

Soil Borne Pathogens	*Rhizobacteria*	*Host Plant*	*Mode of Action with Specific Mechanism*	*References*
Pythium spp.	*P. fluorescens*	Sugar beet	Antibiotics-2,4-diacetylphloroglucinol	Shanahan *et al.* (1992)
Phytophthora infestans	*Pseudomonas* sp.	Tomato	Antibiotics-Cyclic lipopeptides	Raaijmakers *et al.* (2006)
Phytophthora infestans	*P. putida*	Potato	Phytohormone-mediated induction	van Loon (2007)
Fusarium oxysporum	*Bacillus amyloliquefaciens*	Maize	Antibiotics-Bacillomycin, fengycin	Koumoutsi *et al.* (2004)
Fusarium oxysporum	*P. chlororaphis*	Tomato	Antibiotics-Phenazines	Chin-A-Woeng *et al.* (2001)
Fusarium oxysporum	*P. fluorescens*	Wheat	Physical/chemical interference-Molecular cross-talk confused	Duffy *et al.* (2003)
Fusarium oxysporum	*P. fluorescens* (WCS 417)	Carnation	Siderophore mediated competition and antibiosis	Ben *et al.* (1993)
Fusarium oxysporum f. sp. dianthi	*Pseudomonas* spp.	Carnation	Induced systemic resistance	Duijiff *et al.* (1998)
Pythium sp. and *R. solani*	*B. subtilis* QST713		Iturin A	Kloepper *et al.* (2004)
Pythium aphanidermatum	*B. subtilis* BBG100		Mycosubtilin	Leclere *et al.* (2005)

Recently in India, DAPG production by *P. fluorescens* was reported to suppress rice bacterial blight (Velusamy and Gnanamanickam, 2003). Seeds treated with fluorescent *Pseudomonas* resulted in increased number of tillers and grain yield in addition to control of sheath blight disease in rice (Nandakumar *et al.*, 2001). However, some pathogenic fungi can inactivate biocontrol factors, for example, using enzymes that are able to metabolize Phl (2, 4-biacetyl phlorohicinol) on HCN (Duffy, 2003). It shows that complex interactions between pathogens and biocontrol agents can determine the balance between plant disease and its health. Plant growth benefits due to the addition of PGPR include increases in seed germination rates, root growth, leaf area, chlorophyll content, magnesium, nitrogen and protein content, tolerance to drought and salt stress, shoot and root weights, yield and delayed leaf senescence (Lucy *et al.*, 2004). Although the strains of fluorescent pseudomonads have contributed greatly to disease suppression caused soil borne pathogens, these strains have a disadvantage from application point of view, they generally lose viability when stored for a period of several weeks. However the spore forming biocontrol strains of *Bacillus* spp. have much better shelf life.

The Gram (+), rod shaped *Bacillus* strains were appealing candidates for biocontrol because they produce endospores that are tolerant to heat and desiccation. Apart from *Bacillus subtilis*, there are several *Bacillus* species such *B. mycoides*, *B. pumilis*, *B. amyloliquefaciens*, *B. cereus* have good control over various plant pathogens (Jetiyanon *et al.*, 2003). The *Bacillus* spp. has modes of action that includes antibiosis, parasitism, and induced systemic resistance (Bargabus *et al.*, 2002). Strains of *B. subtilis* and *B. amyloliquefaciens* promote plant growth by releasing volatiles (Ryu *et al.*, 2003). El-Hassan and Gowen (2006) reported that formulation and delivery of the bacterial antagonist *Bacillus subtilis* for management of lentil vascular wilt caused by *Fusarium oxysporum* f. sp. *lentis*. More recently Zhang *et al.* (2008) found that *B. subtilis* (GB03) increases photosynthetic efficiency in *Arabidiopsis thaliana* and concluded that the bacterium plays a regulatory role in the acquisition of energy by the plant.

Exploitation of Rhizobacteria in Alleviating Biotic and Abiotic Stress in Plants

Several reports showed that rhizobacteria which contain the enzyme, ACC deaminase (1-aminocyclopropane-1- carboxylate deaminase) which markedly lowered the level of ACC in the stressed plants, thereby limiting the amount of stress ethylene synthesis and hence the damage to the plant. The ethylene is synthesized in plant tissues from the precursor of 1-aminocyclopropane-carboxylic acid (ACC) during biotic and abiotic stress conditions which in turn retarded plant growth (Ma *et al.*, 2003). Interestingly several PGPR strains processes the enzyme ACC deaminase (Shah *et al.*, 1998, Saravakumar *et al.*, 2006) and this enzyme cleave the plant ethylene precursor ACC and thereby lower the level of ethylene in a developing seedling and stressed plant (Mayak *et al.*, 2000). The fluorescent pseudomonads strains that posses ACC-deaminase activity have the selective advantage over other rhizobacteria during biotic and abiotic stresses (Wang *et al.*, 2000; Mayak *et al.*, 2004). Saravanakumar *et al.*, (2007) reported that the *P. fluorescences* strain TDK possessing ACC deaminase activity enhanced the saline resistance in groundnut plants which in turn resulted in increased yield compared with untreated plants.

Exploitation of Microbial Diversity as Biofertilizers/Symbiotic Nitrogen Fixers

Some rhizobacteria promote plant growth in the absence of pathogenic microbes. These play a key role in maintaining soil fertility by fixing atmospheric nitrogen. Microbial biofertilizers accounting for about 65 per cent of the nitrogen supply to crops worldwide. These are known to fix atmospheric molecular nitrogen through symbiotic and asymbiotic or associative nitrogen fixing processes (Anjum *et al.*, 2007). With more and more emphasis being placed on organic farming, PGPR are finding increased application today as biofertilizers.

The spectrum of microbes including members of family Rhizobiaceae, certain Actinomycetes, Cynobacteria, free living or loosely associated bacteria such as *Azotobacter*, *Azospirillum* and certain phosphate solubilizing fungi helps in making available of important nutrients such as nitrogen, phosphate to soil that in turn increases growth and yield of crop plants (Peoples *et al.*, 1995). N_2 fixing bacteria such as *Rhizobium* and *Bradyrhizobium* can form nodules on roots of leguminous plants such as soybean, pea and peanut and the nitrogen can be used by the plant as nitrogen source (Van Rhijn and Vander Leyden, 1995). *Azospirillum* is a free living N_2-fixer that can fertilize sorghum, wheat, maize and it also increase the yield by increased root development and thus to increased rates of water and mineral uptake (Okon *et al.*, 1998). Some plant growth promoting bacteria solubilize phosphate from organic or inorganic bound phosphates thereby facilitating plant growth (Vassilev *et al.*, 2006).

The symbiotic nitrogen fixing rhizobacteria enhances the growth of legumes by the following ways. (*i*) Providing N to the plants through nitrogen fixation (N_2) (Zaidi *et al.*, 2004), (*ii*) increasing the availability of nutrients in the rhizosphere (*iii*) increasing the surface area of root (*iv*) suppressing the deleterious effects of plant pathogenic microbes. Many rhizohacteria are known to produce Indole acetic acid (IAA), phytoharmone of the auxin series hence, act as plant growth promoters (Wani *et al.*, 2007b). Another growth promoting substance produced by these bacteria is siderophores, which is specific iron-chelating substance that makes the chelated iron unavailable to pathogenic micro organisms and leads to increased plant health (Wang *et al.*, 1993).

Exploitation of Rhizobacteria in Remediation of Heavy Metal Contaminated Soils

Heavy metals in general, cannot be biologically degraded to more or less toxic products in a cost effective manner. As heavy metals are common contaminants worldwide and are a threat to soil quality and sustainability agriculture (Mueller *et al.*, 2002), rescuing the heavy metal contaminated soils by microbes is a low cost and effective tool to minimize environment pollution (Mueller *et al.*, 2001). Microorganisms have diverse capacities to biotransform and in some cases completely destroy toxic chemicals in the environment called bioremediation. The elevated concentration of metals in soils and their uptake by plants adversely affect the growth, symbiosis and consequently the yield of crops (Wani *et al.*, 2008). Samanta *et al.* (2002) have detailed the role of *Ralstonia* sp. for the polycylic aromatic hydrocarbon degradation. Pointing (2001) explained the feasibility of bioremediation by white rot fungi.

In this context, numerous strains of rhizobacteria possessing metal reducing ability have been identified (Faisal and Hasnain, 2005). For example hexavalent chromium is more toxic and carcinogenic and numerous chromium reducing PGPR such as *Pseudomonas* sp. (Rahman *et al.*, 2007), *Bacillus* spp. (Wani *et al.*, 2007a), *Ochrobacterium intermedium* (Faisal and Hasnain, 2005) have been reported which helps in bioremediation. Besides their role in detoxification, rhizobacteria also promotes the plant growth by production of growth promoting substances and iron chelating agents siderophores. Hence, utilization of PGPR in heavy metal contaminated soil with view of restoring them, consequently promote crop productivity also. Some examples of plant growth promoting rhizobacteria are given in Table 3.3 which are found to be involved in detoxification of heavy metals. Rhizosphere with high range of nutrients exerted from roots of crop plants and attracts more bacteria compared to normal soil (Penrose and Glick, 2001). This in turn facilitates the growth of host plant and this also has been proved to be effective in minimizing the availability and toxicity of heavy metals (Khan, 2005).

Table 3.3: Plant Growth Promoting Rhizobacteria Used in Detoxification of Heavy Metals

Bacteria	*Plant*	*Heavy Metals*	*Role of PGPR*	*References*
Pseudomas fluorescens	Soybean	Hg	Increased plant growth	Gupta *et al.* (2005)
Pseudomonas sp.	Soybean, mungbean, wheat	Ni, Cd, Cr	Promotes growth of plants	Gupta *et al.* (2002)
Bacillus subtilis SJ-101	*Brassica juncea*	Ni	Facilitated Ni accumulation	Zaidi *et al.* (2006)
Pseudomonas sp., *Bacillus* sp.	Mustard	Cr (VI)	Stimulated plant growth and decreased Cr (VI) content	Rajkumar *et al.* (2006)
Azotobacter chroococcum HKN-5	*Brassica juncea*	Pb, Zn	Stimulated plant growth	Wu *et al.* (2006)

Exploitation of Diversity of Mycorrhizae for Plant Disease Suppression

The ubiquity of mycorrhizae deserves much special attention apart from various epiphytes and endophytes which contributes to biological control. Mycorrhizae are formed as the result of mutual symbiotic relationship between fungi and plants and assist the plants in uptake of nutrients especially phosphorus and some micronutrients Those endomycorrhizae/ Vesicular Arbuscular Mycorrhizal fungi (VAM) produce the characteristic structures like arbuscles and vesicles in the root cortex. Arbuscules are formed by repeated dichotomous branching of fungal hyphae but vesicles are basically hyphal swellings in the root cortex act as storage organ that contain lipids and other nutrients. These structures can often develop thick walls in older roots. During colonization, VAM fungi form the intricate network of fungal hyphae and can prevent root infections by other soil microbes and reducing the access sites and stimulating host defense. The other mechanisms employed by VAM fungi to indirectly suppress plant pathogens include enhanced nutrition to plants; increased root lignification; changes in the chemical composition of the plant tissues such as antifungal chitinase and isoflavonoids (Morris and Ward 1992). In tomato, the infecction due to *Pseudomonas syringae* significantly reduced when the plants are well colonized by mycorrhizae (Garcia-Garrido and Ocampo, 1989).

In contrast to endomycorrhizae/VAM fungi, ectomycorrhizae proliferate outside the root surface and form a sheath around the root by the combination of mass of roots and hyphae called a mantle. Disease protection by ectomycorrhizal fungi may involve a multiple mechanisms including physical barrier of the fungal mantle around the root zone, antibiosis and synthesis of fungistatic compounds by plant roots

(Duchesne 1994). However, difficulty in culturing of mycorrhizae under laboratory condition makes them lesser exploited organism in disease control.

Conclusion

The knowledge on microbial diversity and major groups of microorganisms involved in disease suppressiveness of soil is fundamental to better understanding of the relevance. The era of molecular microbial ecology has uncovered only a part of novel microbiota, most of which is based on rRNA and rDNA analysis (Torsvik and Ovreas, 2002). The quantitative description of microbial communities in terms of gene expression of particular function is now possible through the development of DNA microarray technology and its applications in the study of microbial community structure of an agroecosystem (Dennis *et al.*, 2003; Peplies *et al.*, 2003). In future it will be of important to exploit molecular techniques like specialized DNA micro arrays for studying the genome expression of plant-beneficial and plant pathogenic microbes *in situ* and to know the complete picture of rhizosphere biodiversity, since most of the bacteria are unculturable.

References

Abawi, G.S. and Widmer, T.L. (2000). Impact of soil health managment practices on soil borne pathogens, nematodes and root diseases of vegetable crops. *Appl. Soil Ecol.* **15:**37–47

Ahmad, I., Bissett, J. and Malloch, D. (1995). Influence of the bioherbicide phosphinothricin on interactions between phytopathogens and their antagonists. *Can J. Botany.* **73:** 1759-1760

Antoun, H., Beauchamp, C.J., Goussard, N., Chabot, R. and Llande, R. (2004). Potential of *Rhizobium* and *Bradyrhizobioum* species as plant growth promoting rhizobacteria on non-legumes: effect on radishes. *Plant Soil.* **204:** 57–67

Anjum, M.A., Sajjad, M.R., Akhtar, N., Qureshi, M.A., Iqbal, A., Jami, A.R., Hasan, M. (2007). Response of cotton to plant growth promoting rhizobacteria (PGPR) inoculation under different levels of nitrogen. *J Agric Res.* 45**(2):** 135

Bargabus, R.L., Zidack, N. Sherwood, J.K. and Jacobson, B.J. (2002). Characterization of systemic resistance in sugarbeet elicited by a non- pathogenic, phyllosphere-colonizing *Bacillus mycoides,* biological control agent. *PMPP.* **61:** 289-298

Ben J. Duijff, Meijer, W.J., Bakker, P.A.H.M. and Schippers, B. (1993) Siderophore-mediated competition for iron and induced resistance in the suppression of *Fusarium* wilt of carnation by fluorescent *Pseudomonas* spp. *Neth. J. P1. Path.* **99:** 277-289

Bills, G.F. (1995). Analyses of microfungal diversity from a user's perspective. *Can. J. Bot.* **73:** 33-41.

Brasier, C.M. (1995). Episode selection as a force in fungal microevolution with special reference to clonal speciation and hybrid introgression. *Can J. Bot.* **73:** 1213-1221.

Briere, S.C. Wastson, A.K. and Hallett, S.G. (2000).Oxalic acid production and mycelial biomass yield of *Sclerotinia minor* for the formulation of a granular turf bioherbicide. *Biocontrol Sci. Technol.* **10:** 281–289.

Chet, I. and Inbar, J. (1994). Biological control of fungal pathogens. Applied control of fungal pathogens. *Applied Biochem. Biotechnol.* **48:** 37-43.

Chin-A-Woeng, T.F.C, Bloemberg, G.V., Mulders, I.H.M, Dekkers, L.C. and Lugtenberg, B.J.J. (2000). Root colonization is essential for biocontrol of tomato foot and root rot by the phenazine-1-carboxamide-producing bacterium *Pseudomonas chlororaphis* PCL1391. *Mol. Plant-Microbe Interact.* **13:**1340–45

Chin-A-Woeng, T.F.C, van den Broek, D., de Voer, G., van der Drift, K.M.G.M., Tuinman, S., Thomas-Oates, J.E., Lugtenberg, B.J.J. and Bloemberg, G.V. (2001). Phenazine-1-carboxamide production in the biocontrol strain Pseudomonas chlororaphis PCL1391 is regulated by multiple factors secreted into the growth medium. *Mol Plant-Microbe Interact.* **14:** 969–979

Dennis, P., Edwards, E.A., Liss, S.N. and Fulthorp, R. (2003). Monitoring gene expression in mixed microbial communities by using DNA microarrays. *Appl. Environ. Microbiol.* **69:** 769–778.

Deshwal,V.K., Pandey, P., Kang, S.C. and Maheshwari, D.K. (2003). Rhizobia as a biological control agent against soil borne plant pathogenic fungi. *Ind. J. Expt. Biol.* **41:** 1160–1164

Dowling, D. and O'Gara, F. (1994). Metabolites of *Pseudomonas* involved in the biocontrol of plant disease. *Trends Biotechnol.* **12:** 133–141.

Duchesne, L.C. (1994). Role of ectomycorrhizal fungi in biocontrol. Mycorrhizae and Plant Health. F. L. Pfleger and R. G. Linderman, eds. APS Press, St. Paul, MN. Pp 27-45

Duffy, B., Schouten, A. and Raaijmakers, J.M. (2003). Pathogen self-defense: mechanisms to counteract microbial antagonism. *Annu. Rev. Phytopathol.* **41:** 501–538

Duijiff, B.J., Pouhair, D., Olivain, C., Alabouvette, C. and Lemanceau, P. (1998). Implication of systemic induced resistance in the suppression of Fusarium wilt of tomato by *Pseudomonas fluorescens* WSC417r and by non-pathogenic *Fusarium.oxysporum* Fo47. *European J. of Pl. Pathol.* **104:** 903–10.

Dunne, C, Mo¨enne-Loccoz, Y., McCarthy, J., Higgins, P., Powell, J., Dowling, D.N. and O'Gara, F. (1998). Combining proteolytic and phloroglucinol-producing bacteria for improved control of *Pythium*-mediated dampingoff of sugar beet. *Plant Pathol.* **47:** 299–307

Durall, D.M. and Parkinson, D. (1991). Initial fungal community development on decomposing timothy litter from a reclaimed coal mine spoil in Alberta, *Canada. Mycol. Res.* **95:** 14-18.

El-Hassan, S.A. and Gowen, S. (2006). Formulation and delivery of the antagonist *Bacillus subtilis* for management of lentil vascular wilt caused by *Fusarium oxysporum* f. sp. *lentis. J. of Phytopathol.*, **154:** 148-155.

Emmert, E.A., Klimowicz, A.K., Thomas, M.G. and Handelsman J. (2004). Genetics of zwittermicin A production by *Bacillus cereus. Appl. Environ. Microbiol.* **70:** 104–113

Faisal, M. and Hasnain, S. (2005). Bacterial Cr (VI) reduction concurrently improves sunflower (*Helianthus annuus* L.) growth. *Biotechnol. Lett.* **27:** 943–947

Fravel, D., Olivain, C. and Alabouvette, C. (2003). *Fusaruim oxyporum* and its biocontrol. *New Phytologist.* **157:** 492-502

Gadgil, M. (2000). Defining biodiversity resources. *Curr Sci.* 79 **(3):** 280–282.

Garcia-Garrido, J.M., and Ocampo, J.A. (1989). Effect of VA mycorrhizal infection of tomato on damage caused by *Pseudomonas syringae. Soil Biol. Biochem.* **21:** 165-167.

Glick, B.R. (1995). The enhancement of plant growth by free living bacteria. *Can. J. Microbiol.* **41:** 109–117

Glick, B.R. (2001). Phytoremediation: synergistic use of plants and bacteria to clean up the environment. *Biotechnol. Adv.* **21:** 383–393

Gupta, A., Meyer, J.M. and Goel, R. (2002). Development of heavy metal resistant mutants of phosphate solubilizing *Pseudomonas* sp. NBRI4014 and their characterization. *Curr Microbiol.* **45:** 323–332

Gupta, A., Rai, V., Bagdwal, N. and Goel, R. (2005). *In situ* characterization of mercury resistant growth promoting fluorescent pseudomonads. *Microbiol Res.* **160:**385–388

Haas, D. and Defago, G. (2005). Biological control of soil-borne pathogens by fluorescent pseudomonads. *Nat. Rev. Microbiol.* **3:** 307–19

Haas, D. and Keel, C. (2003). Regulation of antibiotic production in root-colonizing *Pseudomonas* spp. and relevance for biological control of plant disease. *Annu. Rev. Phytopathol.* **41:** 117–53

Harman, G.E., Howel, C.H., Viterbo, A., Chet, I. and Lorito, M. (2004). *Trichoderma* species–opportunistic, avirulent plant symbionts. *Nat. Rev. Microbiol.* **2:** 43–56

Hallman, J., Quadit-Hallman, A., Mahaffee, W.F. and Kloepper, J.W. (1997). Bacterial endophytes in agriculrural crops. *Can. J. Microbiolgy.* **43:** 895-914

Hawksworth, D.L. and Rossman, A.Y. (1997). Where are all the undescribed fungi? *Phytopathology.* **87:** 888-891.

Hiltner, L. (1904). U¨ ber neuere Erfahrungen und Probleme auf dem Gebiete der Bodenbakteriologie unter bessonderer Ber¨ ucksichtigung der Gr¨undung und Brache. *Arb. Dtsch. Landwirtsch. Ges. Berl.* **98:** 59–78

Hynes, R.K., Leung, G.C.Y, Hirkala, D.L.M. and Nelson, L.M. (2008) Isolation, selection, and characterization of beneficial rhizobacteria from pea, lentil, and chickpea grown in western Canada. *Can. J. Microbiol.* 54 **(4):** 248–258

Jetiyanon, K., Fowler, W.D. and Kloepper, J.W. (2003). Broadspectrum protection against several pathogens by PGPR mixtures under field conditions in Thailand. *Plant Dis.* **87:** 1390-1394.

Jonathan, E.I., Raguchander, T., Zareena Bagam, M. and Sundaramoorthy, S. (2009). Field evaluation of talc based bioformulations of Biocontrol agents for the management of *Radopholus similis* and *Helicotylenchus dihystera* in banana. *Nematol. Medit.* **37:** 139-144.

Kamilova, F., Validov, S., Azarova, T., Mulders, I. and Lugtenberg, B. (2005). Enrichment for enhanced competitive plant root tip colonizers selects for a new class of biocontrol bacteria. *Environ. Microbiol.* **7:** 1809–17

Kamilova, F., Lamers, G. and Lugtenberg, B. (2008). Biocontrol strain *Pseudomonas fluorescens* WCS365 inhibits germination of *Fusarium oxysporum* spores in tomato root exudate as well as subsequent formation of new spores. *Environ. Microbiol.* **10:** 2455–61

Kaymak, H.C., Guvenc, I., Yarali, F. and Donmez, M.F. (2009) The effects of bio-priming with PGPR on germination of radish (Raphanus sativus L.) seeds under saline conditions. *Turk. J. Agric. For.* 33 **(2):** 173–179

Khan, A.G. (2005). Role of soil microbes in the rhizospheres of plants growing on trace metal contaminated soils in phytoremediation. *J. Trace Elem. Med. Biol.* **18:** 355–364

Khan, M.S. and Zaidi, A. (2007). Synergistic effects of the inoculation with plant growth promoting rhizobacteria and arbuscular mycorrhizal fungus on the performance of wheat. *Turk. J. Agric.* **31:** 355–362

Khan, M.S., Zaidi, A. and Wani, P.A. (2007) Role of phosphate solubilizing microorganisms in sustainable agriculture: a review. *Agron. Sustain. Dev.* **27:** 29–43

Kloepper, J.W., Ryu, C.M. and Zhang, S. (2004). Induce systemic resistance and promotion of plant growth by *Bacillus* spp. *Phytopathology.* **94:**1259-1266.

Koumoutsi, A., Chen, X.H., Henne, A., Liesegang, H., Gabriele, H., Franke, P., Vater, J. and Borris, R. (2004). Structural and functional characterization of gene clusters directing nonribosomal synthesis of bioactive lipopeptides in *Bacillus amyloliquefaciens* strain FZB42. *J. Bacteriol.***186:** 1084–1096

Kumar, V., Behl, R.K. and Narula, N. (2001). Establishment of phosphate solubilizing strains of *Azotobacter chroococcum* in the rhizosphere and their effect on wheat cultivars under greenhouse conditions. *Microbiol Res.* **156:** 87–93

Kumar, A., Sharma, I.K., Sharma, A., Varshneg, S. and Verma. P.S. (2009) Heavy metal contamination of vegetable foodstuffs in Jaipur (India). *Electronic J. Environ. Agri. Food Chem.* 8 **(2):** 96–101

Landa, B.B., Mavrodi, D.M., Thomoshow, L.S. and Weller, D.M. (2003). Interactions between strains of 2,4- diacetylphloroglucinol-producing Pseudomonas fluorescens in the rhizosphere of wheat. *Phytopathology.* **93:** 982-994.

Leclere, V., Bechet, M., Adam, A., Guez, J. S., Wathelet, B., Ongena, M., Thonart, P., Gancel, F., Chollet-Imbert, M., and Jacques, P. (2005). Mycosubtilin over production by *Bacillus subtilis* BBG100 enhances the organism's antagonistic and biocontrol activities. *Appl. Environ. Microbiol.* **71:** 4577-4584.

Lucy, M., E. Reed and B.R. Glick. (2004). Applications of free living plant growth-promoting rhizobacteria. *A Van Leeuw J. Microb.* **86:** 1-25.

Lugtenberg, B. and Kamilova, F. (2009). Plant-Growth-Promoting Rhizobacteria. *Annu. Rev. Microbiol.* **63:** 541–56

Ma, W., Guinel, F.C. and Glick, G.R. (2003). Rhizobium leguminosarum biovar viciae 1-aminocyclopropane-1-carboxylate deaminase promotes nodulation of pea plants. *Appl. Environ. Microbiol.* **69:** 4396–4402.

Mavrodi, D.V., Blankenfeldt, W. and Thomashow, L.S. (2006). Phenazine compounds in fluorescent *Pseudomonas* spp.: biosynthesis and regulation. *Annu. Rev. Phytopathol.* **44:** 417–45

Mayak, S., Tirosh, T. and Glick, B.R. (2004). Plant growth promoting bacteria that confer resistance in tomato and pepper to salt stress. *Plant Physiol. Biochem.***167:** 650–656.

Morin, P.J. (2000). Biodiversity -ups and downs. *Nature,* **405:** 463-464.

Mueller, A.K., Westergaard, K., Christensen, S. and Surensen, S.J. (2001). The effect of long term mercury pollution on the soil microbial community. *FEMS Microbiol. Ecol.* **36:** 11-19.

Mueller, A.K., Westergaard, K., Christensen, S. and Surensen, S.J. (2002). The diversity and function of soil microbial communities exposed to different disturbances. *Microb. Ecol.* **44:** 49-58.

Nandakumar, R., Babu, S., Viswanathan, R., Sheela, J., Raghuchander, T. and Samiyappan, R. (2001). A new bioformulation of containing plant growth promoting rhizobacterial mixture for the management of sheath blight and enhanced grain yield in rice. *Biocontrol,* **46:** 493-510.

Nee, S. (2002). Thinking big in ecology. *Nature* **417:** 229-230.

Nitta, T. (1991). Diversity of root fungal floras: its implications for soil-borne diseases and crop growth. *Jpn. Agric. Res.* **25:** 6–11

Nowak-Thompson, B., Chaney, N., Wing, J.S., Gould, S.J. and Loper, J.E. (1999). Characterization of the pyoluteorin biosynthetic gene cluster of *Pseudomonas fluorescens* Pf-5. *J. Bacteriol.* **181:** 2166–74

Okon, Y., Bloemberg, G.V. and Lugtenberg, B.J.J. (1998). Biotechnology of biofertilization and phytostimulation. In *Agricultural Biotechnology*, ed. A Altman, pp. 327–49. New York: Marcel Dekker.

Olivain, C., Alabouvette, C. and Steinberg, C. (2004). Production of a mixed inoculums of *Fusarium oxysporum* Fo47 and *Pseudomonas fluorescens* C7 to control *Fusarium* diseases. *Bio. Sci. Technol.* **14:** 227–238.

Ongena, M., Duby, F., Rossignol, F., Fouconnier, M. L., Dommes, J., and Thonart, P. (2004). Stimulation of the lipoxygenase pathway is associated with systemic resistance induced in bean by a nonpathogenic *Pseudomonas* strain. *Mol. Plant-Microbe Interact.* **17:** 1009-1018.

Pal, K.K. and McSpadden Gardener, B. (2006) Biological control of plant pathogens. The Plant Health Instructor. doi:10.1094/PHI-A-2006-1117-02. 25 pp

Peoples, M.B., Herridge, D.F. and Ladha, J.K. (1995) Biological nitrogen fixation: An efficient source of nitrogen for sustainable agricultural production. *Plant Soil,* **174:** 3-28.

Peplies, J., Glõckner, F. O. and Amann, R. (2003). Optimization strategies for DNA microarray-based detection of bacteria with 16S rDNA-targeting oligonucleotide probes. *Appl. Environ. Microbiol.,* **69:** 1397–1407.

Perneel, M., D'Hondt, L., De Maeyer, K., Adiobo, A., Rabaey, K. and Hofte, M. (2008). Phenazines and biosurfactants interact in the biological control of soil-borne diseases caused by *Pythium* spp. *Environ. Microbiol.* **10:** 778–88

Pinton, R., Varanini, Z. and Nannipieri, P. (2001). *The Rhizosphere: Biodiversity and Organic Substances at the Soil–Plant Interface,* Marcel Dekker, New York.

Pointing, S.B. (2001). Feasibility of bioremediation using white-rot fungi. *Appl Microbiol. Biotechnol.* **57:** 20-33.

Raaijmakers, J.M., Weller, D.M. and Thomashow, L.S. (1997). Frequency of antibiotic producing *Pseudomonas* spp. in natural environments. *Appl. Environ. Microbiol.* **63:** 881–87

Raaijmakers, J.M. and Weller, D.M. (1998). Natural plant protection by 2, 4-dicetylphloroglucinol- producing *Pseudomonas* spp. in take-all decline soils. *Mol. Plant Microbe Interact.* **11:** 144–52

Raaijmakers, J.M, de Bruijn, I. and de Kock, M.J.D. (2006). Cyclic lipopeptide production by plant-associated *Pseudomonas* spp.: diversity, activity, biosynthesis, and regulation. *Mol Plant-Microb Interact* **19:** 699–710

Radjacommare, R., Nandakumar, R., Kandan, A., Suresh, S., Bharathi, M., Raguchander, T. and Samiyappan, R. (2002). *Psedomonas fluorescens* based bioformulation for the management of sheath blight and leaf folder in rice. *Crop Prot.* **21:** 671-677.

Rahman, M., Gul, S., and Haq, M.Z. (2007). Reduction of chromium (vi) by locally isolated pseudomonassp. C-171. *Turk J Biol* **31:**161–166

Rajkumar, M., Nagendran, R, Kui Jae, L., Wang Hyu, L. and Sung Zoo, K. (2006). Influence of plant growth promoting bacteria and Cr (vi) on the growth of Indian mustard. *Chemosphere* **62:** 741–748

Ramamoorthy, V., Viswanathan, R., Raguchander, T., Prakasam, V. and Samiyappan, R. (2001). Induction of systemic resistance by plant growth promoting rhizobacteria in crop plants against pest and diseases. *Crop Protect.* **20:** 1–11.

Ramamoorthy, V., Raghuchander, T. and Samiyappan, R. (2002) Enhancing resistance of tomato and hot pepper to *Pythium* diseases by seed treatment with fluorescent pseudomonads. *European J. Pl. Pathol.,* **108:** 429-441.

Ryu, C.M., Farag, M.A., Hu, C.-H., Reddy, M.S. and Wie, H.-X. (2003). Bacterial volatiles promote growth of *Arabidopsis*. *Proc. Natl. Acad. Sci. USA*. **100:** 4927–32

Ryu, C.M., Farag, M.A., Hu, C.H., Reddy, M.S., Kloepper, J.W., and Pare, P.W. (2004). Bacterial volatiles induce systemic resistance in *Arabidopsis*. *Plant Physiol.* **134:**1017-1026.

Samanta, S.K., Singh, O.V. and Jain, R.K. (2002). Polycyclic aromatic hydrocarbons: Environmental pollution and bioremediation. *Trends Biotechnol.* **20:** 243-248.

Sangeetha, G., Thangavelu, R., Usha Rani, S., Muthukumar, A. and Udayakumar, R. (2010). Induction of systemic resistance by mixtures of antagonist bacteria for the management of crown rot complex on banana. *Acta Physiol Plant.* **32:** 1177–1187

Saravanakumar, D. and Samiyappan, R. (2007). ACC deaminase from Pseudomonas fluorescens mediated saline resistance in groundnut (*Arachis hypogea*) plants. *J. of App. Microbiol.* **102:** 1283–1292

Saravanakumar, D., Harish, S., Loganathan, M., Vivekananthan, R. Rajendran, L., Raguchander, T. and Samiyappan, R. (2007). Rhizobacterial bioformulation for the effective management of *Macrophomina* root rot in mungbean. *Arch. of Phytopathol. and* Plant *Protect.* 40 **(5):** 323-337

Schippers, B., Bakker, A.W. and Bakker, P.A.H.M. (1987). Interactions of deleterious and beneficial microorganisms and the effect on cropping practices. *Annu. Rev. Phytopathol.* **25:** 339–58

Shah, S., Li, J., Moffatt, B.A. and Glick, B.R. (1998). Isolation and characterization of ACC deaminase genes from two different plant growth-promoting rhizobacteria. *Can. J. Microbiol.* **44:** 833–843.

Shanahan, P., O'Sullivan, D.J., Simpson, P., Glennon, J.D. and O'Gara, F. (1992). Isolation of 2, 4-Diacetylphloroglucinol from a fluorescent pseudomonad and investigation of physiological parameters influencing its production. *Appl. Environ. Microbiol.* **58:** 353–358

Sneh, B., Dupler, M., Elad, Y., and Baker, R. (1984). Chlamydospore germination of *Fusarium oxysporum* f. sp. *cucumerinum* as affected by fluorescent and lytic bacteria from *Fusarium* suppressive soils. *Phytopathology* **74:** 1115–1124.

Sternberg, S. (1994). The emerging fungal threat. *Science* **266:** 1632–1634

Subba Rao, N.S. and Gaur, D.Y. (2000). Microbial diversity-Management and Exploitation for Sustainable Agriculture, In: Extended summaries: International Conference on Managing Natural Resources for Sustainable Agricultural Production in the 21 st century (ed. Yadav, J.S.P. *et al.*), New Delhi, India. Pp. 48-58.

Torsvik, V, Goksoyr, J. and Daae, F.D. (1990). High diversity in DNA of soil bacteria. *Appl. Environ. Microbiol.* **56:** 782-87

Torsvik, V. and Ovreas, L. (2002). Microbial diversity and function in soil: from genes to ecosystems. *Curr. Opin. Microbiol.* **5:** 240–245

Tripathi, M., Munot, H.P., Shouche, Y., Meyer, J. M. and Goel, R. (2005). Isolation and functional characterization of siderophore producing lead and cadmium resistant *Pseudomonas putida* KNP9. *Curr. Microbiol.***50:** 233–237

van Dijk, K., and Nelson, E.B. (2000). Fatty acid competition as a mechanism by which *Enterobacter cloacae* suppresses *Pythium ultimum* sporangium germination and dampingoff. *Appl. Environ. Microbiol.* **66:** 5340-5347.

Van Elsas, J.D., Garbeva, P. and Salles, J. (2002). Effects of agronomical measures on the microbial diversity of soils as related to the suppression of soil-borne plant pathogens. *Biodegradation,* **13:** 29-40.

van Loon, L.C. (2007). Plant responses to plant growth-promoting bacteria. *Eur. J. Plant Pathol.* **119:** 243–254

Van Os, G.J and van Ginkel, J.H. (2001). Suppression of *Pythium* root rot in bulbous *Iris* in relation to biomass and activity of the soil microflora. *Soil Biol. Biochem.* **32:** 1447–54

Van Rhijn, P. and Vanderleyden, J. (1995). The *Rhizobium*-plant symbiosis. *Microbiol. Rev.* **59:** 124–42

Vassilev, N., Vassileva, M. and Nicolaeva, I. (2006). Simultaneous P-solubilizing and biocontrol activity of microorganisms: potentials and future trends. *Appl. Microbiol. Biotechnol.* **71:** 137–44.

Velusamy, P. and Gnanamanickam, S.S. (2003). Identification of 2,4 diacetylphloroglucinol production by plant associated bacteria and its role in suppression of rice bacterial blight. *Curr. Sci.* **85:** 1270-1273.

Wang, Y., Brown, H.N., Crowley, D.E. and Szaniszlo, P.J. (1993). Evidence for direct utilization of a siderophore, ferroxamine B, in axenically grown cucumber. *Plant Cell Env.* **16:** 579–585

Wang, C., Knill, E., Glick, B.R. and De'fago, G. (2000). Effect of transferring 1-aminocyclopropane-1-carboxylic acid (ACC) deaminase genes into *Pseudomonas fluorescens* strain CHA0 and its gacA derivative CHA96 on their growth-promoting and disease-suppressive capacities. *Can. J. Microbiol.* **46:** 898–907.

Wani, P.A., Khan, M.S., Zaidi, A. (2007a) Chromium reduction, plant growth promoting potentials and metal solubilization by *Bacillus* sp. isolated from alluvial soil. *Curr Microbiol.* **54:** 237–243

Wani, P.A., Khan, M.S. and Zaidi, A. (2007b). Effect of metal tolerant plant growth promoting *Bradyrhizobium* sp. (Vigna) on growth, symbiosis, seed yield and metal uptake by greengram plants. *Chemosphere* **70:** 36–45

Wani, P.A., Khan, M.S. and Zaidi, A. (2008). Effect of heavy metal toxicity on growth, symbiosis, seed yield and metal uptake in pea grown in metal amended soil. *Bull. Environ. Contam. Toxicol.* doi: 10.1007/s00128-008-9383-z

Wu, C.H., Wood, T.K., Mulchandani, A. and Chen, W. (2006). Engineering plant-microbe symbiosis for rhizoremediation of heavy metals. *Appl. Environ. Microbiol.* **72:** 1129–1134

Yasmin, S., Rahman, M. and Hafeez, F.Y. (2004). Isolation, characterization and beneficial effects of rice associated plant growth promoting bacteria from Zanzibar soils. *J. Basic. Microbiol.* **44:** 241–252

Zahir, Z.A., Munir, A., Asghar, H.N., Shaharoona, B. and Arshad, M. (2008). Effectiveness of rhizobacteria containing ACC deaminase for growth promotion of peas (*Pisum sativum*) under drought conditions. *J. Microbiol. Biotechnol.* 18 **(5):** 958–963

Zaidi, A. and Khan, M.S. (2005). Interactive effect of rhizospheric microorganisms on growth, yield and nutrient uptake of wheat. *J. Plant. Nutr.* **28:** 2079–2092

Zaidi, A., Khan, M.S. and Aamil, M. (2004) Bioassociative effect of rhizospheric microorganisms on growth, yield and nutrient uptake of greengram. *J. Plant Nutr.* **27:** 599–610

Zaidi, S., Usmani, S., Singh, B.R. and Musarrat, J. (2006). Significance of *Bacillus subtilis* strain SJ 101 as a bioinoculant for concurrent plant growth promotion and nickel accumulation in *Brassica juncea*. *Chemosphere* **64:** 991–997

Zhang, H., Xie, X., Kim, M.-S., Kornyeyev, D.A., Holaday, S. and Par´e, P.W. (2008). Soil bacteria augment *Arabidopsis* photosynthesis by decreasing glucose sensing and abscisic acid levels in plants. *Plant J.* **56:** 264–73.

Chapter 4

Soil Microbial Diversity in Relation to Maize Cultivation for Sustainable Agriculture

Mushtaq Ahmed, Dilshad Ahmed and R.S. Upadhyay*

Laboratory of Mycopathology and Microbial Technology, Centre of Advanced Study in Botany, Banaras Hindu University, Varanasi – 221 005, U.P.

Maize (*Zea mays* L) is one of the oldest cultivated food grains and one of the most productive crop species with a global average yield of more than 4 tones per hectare (Farnaham *et al.*, 2003). It is widely cultivated throughout the world and the production of maize is greater than any other cereal (http://fao.org). The value added products of maize that are of commercial use include maize starch, liquid dextrose, dextrose monohydrates, anhydrous dextrose, sorbitol, and corn gluten etc. (http://www.niir.org). A large proportion of maize produced is used as stock feed *i.e.* nearly 40 per cent in tropical areas and up to 85 per cent in developed countries (Farnaham *et al.*, 2003). It can be processed for a range of uses such as ingredients in food or drinks or for industrial purposes (White, 1994). Maize is the major source of starch in the world (Johnson and May, 2003). The paper and/or textile industry is the biggest non food buyer of maize starch (Boyer and Hannah, 1994). Maize can be grown under varied environment conditions (Farnaham *et al.*, 2003). Comparatively maize is a less water stress tolerant crop (Bachingham, 2007). The maize plant grows best in

* Corresponding Author E-mail: upadhyay_bhu@yahoo.co.uk

well drained, nutrient rich soils with a pH range of 5.5 to 7.0 (Farrell and O'keeffe, 2007). The nutrient status of the soil is very important for the productivity of this crop and substantial amount of nutrients is removed from the soil by harvesting maize plants (Farrell and O' Keeffe, 2007).

Maize is an important staple food in many countries of the world and the area of maize production in the world has been increasing continuously. The United States of America (USA) produces 40 per cent of the world's harvest; other top maize producing countries include China, Brazil, Mexico, India, Indonesia, France and Argentina (http://faostat.org). The world production of maize was 817 million tones in 2009 which is more than rice (678 million tones) or wheat (682 million tonnes) (http://faostat.fao.org). In 2009 over 159 million hectares of land was sown with this crop in the world and yield increased approximately over 5 tones / hectare (http://gazetteonline.com). The climatic changes at the global level have led to drastic reduction in agricultural soil fertility and crop productivity. This poses a serious threat to food security of man on this planet. Because of this, modern agricultural practices require the use of chemical fertilizers and pesticides that are of particular concern because of their high cost, limited supply and negative impact on human health. Therefore, to explore the possibility of supplementing the inorganic chemical fertilizers with organic ones such as the biofertilizers of microbial origin is the need of the day. A better alternative to the chemicals are the rhizosphere plant growth promoting microorganisms that have been reported to promote the growth and yield of crop plants. The soil microorganisms associated with maize rhizosphere establish positive interactions with the crop and play a key role in agriculture practices and are promising for their potential use in sustainable agriculture (Di Cello *et al.*, 1997). Weller (1998) reported that a beneficial soil microorganism that colonizes the maize roots, is ideal for use as biological control agent for soil born diseases and consequently, in improving plant growth (Youssef *et al.*, 2001). The beneficial soil microbes benefit the crop plants through their ability to produce growth regulators, siderophores, phosphate solubilisation, nutrient uptake and availability (Hofflich and Kuhn, 1996; Gupta *et al.*, 1998).

Importance and Cultivation of Maize

Maize is one of the oldest cultivated food grains and one of the most productive crop species with a global average yield of more than 4 tones per hectare (Paliwal, 2000b; Farnaham *et al.*, 2003). It is widely cultivated throughout the world and the production of maize is greater than any other cereal (http://fao.org). Maize is of immense commercial use. The value added products of maize that are of commercial use include maize starch, liquid dextrose, dextrose monohydrates, anhydrous dextrose, sorbitol, and corn gluten etc (http://www.niir.org). A large proportion of maize produced is used as stock feed *i.e.* nearly 40 per cent in tropical areas and up to 85 per cent in developed countries (Paliwal, 2000b; Farnaham *et al.*, 2003). It can be processed for a range of uses such as ingredients in food or drinks or for industrial purposes (White, 1994). Maize is the major source of starch in the world and is used as either in its native form such as baby foods, snack foods, salad dressing, paper products, insulating materials, paints, tablet binders or chemically modified forms like bakery products, sauces and gravies, kings, and glazes, pastes and glues, ceramics, dye and sandpaper (Johnson and May, 2003). Corn starch is also fermented to produce alcoholic beverages, flavor enhancers, engine fuel and solvents. The paper industry and textile industry is the biggest non food buyer of maize starch (Paliwal, 2000h; Hobbs, 2003).

Maize can be grown under varied environment conditions (Paliwal, 2000b; Farnaham *et al.*, 2003). Compratively maize is a less water stress tolerant crop (Farrell and O'keefffe, 2007; Bachingham, 2007). The maize plant grows best in well drained, nutrient rich soils with a pH range of 5.5 to 7.0 (Farrell and O'keeffe, 2007). The nutrient status of the soil is very important for the productivity of this crop and substantial amount of nutrients is removed from the soil by harvesting maize plants (Farrell and O' Keeffe, 2007). Therefore, soil test for various nutrients should be carried out before sowing this crop (Farrell and O'keeffe, 2007). Nutrient availability varies with soil types (Farrell and O'Keeffe, 2007). Nitrogen is yield limiting nutrient in maize production so that the amount of nitrogen that needs to be added to the soil depends on various factors like cropping history and yield target (Farrell and O'Keeffe, 2007). Small amount of nitrogen and phosphorus may be applied at the time of sowing to reduce the chance of damage of seedling by fertilizer burn and to make these nutrients available during the early stages of development (Farrell and O'Keeffe, 2007). Maize crop is also salinity sensitive or moderately salt tolerant plants and therefore, shows rapid productivity decline with increasing salinity (Yensen and Beil, 2006). Maize is an important staple food in many countries of the world and the area of maize production in the world has been increasing continuously (http://www.krvycomtrade.com). The United States of America (USA) produces 40 per cent of the world's harvest; other top maize producing countries include China, Brazil, Mexico, India, Indonesia, France and Argentina (http://faostat.org). The world production of maize was 817 million tones in 2009 which is more than rice (678 million tones) or wheat (682 million tonnes) (http://faostat.fao.org). In 2009 over 159 million hectares of land was sown with this crop in the world and yield increased approximately over 5 tones / hectare (http://gazetteonline.com). The present day agriculture requires the use of chemical fertilizers. But, these chemical fertilizers are expensive, hazardous to the environment and also are available in limited supply. Therefore, to explore the possibility of supplementing the inorganic chemical fertilizers with organic ones such as the biofertilizers of microbial origin is the need of the day.

Plant Growth Promoting Soil Microbes

There are many species of bacteria, actinomycetes, blue green algae (BGA), fungi including VAM (vesicular arbuscular mycorrhizae) and viruses that flourish in diverse habitats, have immense potential to enhance plant growth by a plethora of mechanisms including biological nitrogen fixation (Hashem, 2001), phytohormone production (Mehnaz *et al.*, 2001), complex substrate degradation and/or siderophore production (Masalha *et al.*, 2000) etc. and are therefore, of agricultural importance (http://www.jansamachar.net). Twenty first century agriculture needs to be more productive, environmentally benign, robust in the face of climate change and socially beneficial. A reliable approach is the sustainable agricultural system which maintains and improves human health, benefits producers and consumers both economically and spiritually, protects the environment, and produces enough food for an increasing world population (Higa, 1991). Sustainable agriculture that integrates environmental health, economic profitability and social or/and economic equity, is based on substantial use of beneficial soil microorganisms that hold tremendous potential for use to enhance plant growth and yield at a low cost (Higa, 1991). Utilization of growth promoting microorganisms for sustainable agriculture requires their isolation, identification and strain selection followed by plant or target pest test, pot culture test and field trials. (http://faperta.ugm.ac.id). There are techniques for the production of inoculum of the desired microbial strain and its application in soil to cause positive effect on plant growth and yield (Flores and O' Hara, 2006). The most reliable approach is to select microorganisms that are physiologically and ecologically compatible with each other, introduce them into the soil as part of a mixed culture at a sufficiently high inoculum density to have desirable positive effect on plant growth and yield (Parr *et al.*, 1994). Microbial inoculants are cheaper as compared to the chemical fertilizers, and are not hazardous to the environment.

Soil Microbes Associated with Maize Rhizosphere

The study of microorganisms associated with rhizosphere of plants is important for understanding their ecological role in natural environments (Di Cello *et al.*, 1997). The microorganisms that establish positive interactions with plant roots play a key role in agricultural practices and are promising for their potential use in sustainable agriculture (Di Cello *et al.*, 1997). Plant roots have been reported to lose up to 31 per cent of their assimilates in the form of amino acids, organic acids, and sugars (Rovira, 1959; Whipps, 1990). These compounds are known to support considerable bacterial activity (Guckert, 1992, Martin, 1975). The rhizosphere microbial population may show temporary changes in its diversity due to environmental variations over time (Wise *et al.*, 1996). It has been reported that the production and diffusion of root exudates is affected by plant growth and development (Hamlen *et al.*, 1972). These exudates are known to act as selective microbial stimulants that is vary in function with time due to plant age and related factors (Buir and Caesar, 1984; miller *et al.*, 1989). It is observed that the population of *Burkholderia cepacia* associated with maize roots decreased significantly during plant development (Di Cello *et al.*, 1997). Several studies on the rhizosphere microbial community structure of maize revealed that fast growing soil microbes predominantly colonise the immature roots of this crop plant whereas slow growing ones are found to be predominant on mature roots (Di Leij *et al.*, 1995; Nacamuli *et al.*, 1997). The endophytic rhizobacteria and fungi promoting plant growth and protecting the plant against pathogens have been reported (Benhamou *et at.*, 2000; Stirz and Nowak, 2000). It is found that *Enterobacter cloaceae*, endophytically associated with corn inhibited the systemic pathogens of corn (Hinton and Bacon, 1995). Among the microorganisms inhabiting the rhizosphere of maize, *Fusarium species* are the major soil born fungal pathogens (Marasas *et al.*, 1984). A current emphasis is developed on rhizosphere microporganisms by seed inoculation for biological control of soil borne plant pathogens or for improvement of plant growth (Lee *et al.*, 1990; *Crowford et al.*, 1993; Yuan and Crowford, 1995; Valois *et al.*, 1996; Baker *et al.*, 1998; Youssef *et al.*, 2001). Weller (1998) reported that a beneficial microorganism that colonizes plant roots, is ideal for use as biological control agent for soil borne diseases and consequently, in improving plant growth (Lee *et al.*, 1990; crowford *et al.*, 1993; Yuan and Crowford, 1995; Valois *et al.*, 1996; Baker *et al.*, 1998; Youssef *et al.*, 2001). The beneficial plant microbe interactions benefit the plant through different mechanism such as production of growth regulators, siderophores, phosphate solubilisation, nutrient uptake and availability (Hoflich and Kuhn, 1996; Gupta *et al.*, 1998; Bowen and Rovira, 1999).

Rhizosphere Bacterial Community of *Zea mays*

It is observed that the inoculation of soil, seeds or plants with specific *Pseudomonas spp.* significantly increases the yield of maize (Lemanceau, 1992). This increase is due to increase in plant growth and protection against pathogenic soil microorganisms (Lemanceau, 1992). The *pseudomonas spp.* Produce plant growth hormones that promote the root growth (Brown 1972; Vancura, 1970; Lemanceau, 1992). Under Fe^{3+} limitation stress conditions some *Pseudomonas spp.* produce siderophores that are able to scavenge the trace of Fe^{3+} and deprive other microbes of these ions (Benziri, 1995; Leong, 1986; Coutrade and Guckert, 1995; Neilands and Fravel, 1986; Schorth and Hancock, 1982). Some *Pseudomonas spp.* Secrete some volatile and antimicrobial compounds which reduce the density and activity of deleterious microorganisms (Benziri, Courtade and Guckert, 1995). Klopper and Schroth introduced the term plant growth promoting rhizobacteria (PGPR) to describe these beneficial bacteria that are capable of stimulating plant growth (KLoepper and Schroth, 1978). However very little is known about the mechanisms involved in root colonization and crop specificity of *Pseudomonas* strains (Benziri *et al.*, 1996). Other bacteria in (addition to *Pseudomonas sp*) colonizing the rhizosphere of maize are *Bacillus sp*, *Azotobacter sp*, *Arthrobacter sp*, *Listeria sp*, *Sporolactobacillus sp*, and *Micrococcus sp* (Lilia *et al.*, 2007). It has been reported that colonization of maize roots by some bacteria increases when the plant is inoculated with some fungal strains (Sarathchandra *et al.*, 1993). Approximately 85 spp. of actinomycetes that have been isolated from the rhizosphere of maize plants were screened for *in vitro* antagonism against *Cephalosporium maydis* the causal agent of late wilt disease of maize (Mehalawy *et al.*, 2004). Out of 85 spp. Six spp. of actinomycetes were found to be highly antagonistic to the pathogen (Hussain *et al.*, 2002). They were identified and tested for their ability to produce antifungal compounds against *Cephalosporium maydis* (Hussain *et al.*, 2002). This is the first record of this deleterious pathogen being controlled by the antagonistic activity of rhizosphere Streptomycetes actinomycetes (Yuan and Crowford, 1995; Takaki *et al.*, 1996; Benbow and Sufar, 1999; Buck and Adrews, 1999). The mechanism involved in disease control, mainly might be the antibiosis (Yuan and Crowford, 1995; Takaki *et al.*, 1996; Benbow and Sufar, 1999; Buck and Adrews, 1999). The actinomycetes isolate was capable of parasitizing *Cephalosporium maydis* hyphae in addition to producing antifungal metabolites (Yuan and Crowford, 1995; Takaki *et al.*, 1996; Benbow and Sufar, 1999; Buck and Adrews, 1999). Most workers have utilized actinomycetes as potential biological control agents against the plant pathogens (Yuan and Crowford, 1995; Takaki *et al.*, 1996; Benbow and Sufar, 1999; Buck and Adrews, 1999). Many actinomycetes species used as biocontrol against were found to produce antibiotics (Locci and Schofield, 1989; You *et al.*, 1996).tis activity mght be responsible for their ability to suppress pathogenic soil microbes (Locci and Schofield, 1989; You *et al.*, 1996).

Rhizosphere Fungal Community of *Zea mays*

It has been investigated that selected species of yeast fungi produce plant growth regulating factors and thus, significantly stimulated maize root and stem growth (Hoflish and Kuhn, 1996). The metabolites of these fungi enhance plant growth after being taken up by plants or indirectly by modifying the rhizosphere of maize (Hoflish and Kuhn, 1996). The yeast fungi also promoted plant growth by oxidizing ammonium to nitrate, oxidizing elemental sulphur to sulphate and solubilizing insoluble phosphate (Mehalawy *et al.*, 2004). About 40 yeast fungi were isolated from the rhizosphere of maize plant and were screened in vitro for their antagonistic activities against *Cephalosporium maydis* (Mehalawy *et al.*, 2004). Out of these, five yeast fungal isolates were found to be strongly antagonistic to the pathogen *in vitro* (Mehalawy *et al.*, 2004). These are *Candida glabrata, C. Maltose, C. Slooffii, Rhodotorula rubra* and *Trichosporon cutaneum* (Mehalawy *et al.*, 2004). They significantly reduced the growth of *Cephalosporium maydis in vitro* (Mehalawy *et al.*, 2004). In the absence of pathogen these yeast fungi significantly increase the maize plant growth as compared to control (Mehalawy *et al.*, 2004).

Recently, mineral phosphate solubilising (MPS) activity of the fungus *Penicillium rugulosum* was characterized in the wild type strain IR-94 MF1 and two UV- induced mutants derived from it with altered phenotypes for phosphate solubilisation (Reyes *et al.*, 1998). The nature and amount of organic acids secreted by *P.rugulosum* were found to strongly influence the solubilisation of rock phosphate deposits (Reyes *et al.*, 2001). Inoculation of soil with different *P. rugulosum* strains increased the maize yield (Kucey *et al.*, 1989). Indigenous soil phosphate solubilising arbuscular mycorhizal fungi could also have played an important role in phosphate up take by maize (Reyes *et al.*, 2002). It is observed that fungal community in rhizosphere of maize is more in seedling stage than silking stage (Windham, 1983). There are large number of fungal communities reported from growing maize plant at rhizoplane and endorhizosphere root levels (Lilia Cavagleiri *et al.*, 2007). Species of *Aspergillus, Fusarium, Cladosporium, Eurotium, Ulocladium* and *Trichoderma* are predominant in the rhizosphere of maize (Lilia Cavagleiri *et al.*, 2007). Some endophytes including *Beaveria bassiana, Trichoderma koningii, Alternaria alternate, Phoma spp* and *Acremonium strictum* isolated from maize roots have been found to be of immense benefit to the plant as a promoter of plant health improving growth potentials and act as biological control agents against fungal and bacterial disease of plants (Zinniel *et al.*, 2002; Compant *et al.*, 2005; Kirkpatrick and Wilhelm, 2006).

The different mechanisms through which these endophytes inhibit fungal pathogens in the rhizospheres include competition for available nutrients, oxygen and space, parasitism and physical distruction of fungal cell walls by the action of hydrolytic enzymes produced by the endophytes (Taechowisan *et al.*, 2009). The pathogens *Fusarium oxysporum, F. Pallidoroseum, F. Verticilloides* and *Cladosporium herbarum* were isolated from blighted maize plants (Orole and Adejumo, 2009). These pathogens can be reduced by the use of endophytes because the *in vitro* assay of the endophytes against these pathogens showed that *T. Koningii* and *A. Alternata* grew on the mycelia of all pathogens there by, reducing the radial growth by 25–75 per cent and 53–80 per cent respectively (Orole and Adejumo, 2009). The *in vivo* studies revealed that *Trichoderma koningii* and *Alternaria alternata* could be successfully formulated and applied as alternative fungicide in the management of maize wilt and seedling blight (Orole and Adejumo, 2009). Antagonism of wilt causing pathogens was found to be heighest by the action of *Alternaria alternata* and *Trichoderma koningii* (Orole and Adejumo, 2009).

The activities of endophytes in suppressing fungal pathogens were confirmed by polling *et al.*, 2008; Salehpour *et al.*, 2005; Muhammad and Amusa, 2003. It has been found that endophytic fungi isolated from healthy maize roots restricted the growth of wilt causing pathogens of maize seedlings (Ownley *et al.*, 2004; Posada and Vega, 2005). The biological control activity of *Beaveria bassiana* is less surprising but its dual purpose role against insect pests and plant pathogen has been highlited by Ownley *et al.*, 2004; Posada and Vega, 2005. The *Alternaria alternata* and *Trichoderma koningii* showed the highest potential as biocontrol agents requiring further testing and applications for use on the field (Orole and Adejumo, 2009).

Mechanisms of Plant Growth Promotion

The exact mechanism of plant growth promotion by inoculation with microorganisms is not well understood. However, plant growth promotion by the agriculturally important microorganisms may be because of their ability of nitrogen fixation, complex substrate degradation and/or siderophore production, phytohormone production and suppression of pathogenic microbes, etc.

Nitrogen Fixation

Numerous genera of Blue Green Algae (BGA) such as *Anabaena, Nostoc* etc. that reside in the rhizosphere of *Triticum aestivum* and *Oryza sativa* have been reported to enhance the growth of these crop plants due to their nitrogenase activity (Obreht *et al.*, 1993; Hashem, 2001). Many genera of rhizobacteria such as *Rhizobium, Mesorhizobium, Allorhizobium, Azorhizobium, Bradyrhizobium* and *Sinorhizobium* are known for their potential to fix nitrogen for their host plants (Gualtieri and Besseling, 2000; Sessitch *et al.*, 2002). Different rhizospheric or endophytic microorganisms which have been reported to promote plant growth due to their ability to fix N_2 are listed below (Table 4.1).

Complex Substrate Degradation and/or Siderophore Production

Plant growth promoting activity of microorganisms has been attributed by several workers to their ability to enhance the availability of nutrients in the rhizosphere by mineralizing complex substrates and/or producing siderophores which facilitate the transport of certain metal ions, notably Fe^{3+} ions (Wang *et al.*, 1993; Glick, 1995; Kim *et al.*, 1998; Rodriguez and Fraga, 1999; Hyakumachi, 2000). Some rhizospheric bacteria have been reported to produce siderophores which bind with ferric (Fe^{3+}) ions to form Fe^{3+}-siderophore complexes that can be easily absorbed by the root system of a number of plant species (Bar-Ness *et al.*, 1991). Masalha *et al.* (2000) suggested that uptake of ferric ions as bacterial Fe^{3+}-siderophore complex by the plant roots plays a vital role in the overall iron requirement of the plants especially, in calcareous soils. Phosphate-solubilizing bacteria which occur commonly in the rhizosphere of many crop plants (Vazquez *et al.*, 2000b; Nautiyal *et al.*, 2000), have been reported to secrete organic acids and phosphatases that aid in the conversion of insoluble forms of phosphorous to plant-available forms (Kim *et al.*, 1998) and enhance nutrient availability to host plants resulting in their better growth as well as yield (Richardson, 2001). *Azotobacter chroococcum, Enterobacter agglomerans, Pseudomonas chlororaphis* and *Pseudomonas putida, Rhizobium* sp., *Bradyrhizobium japonicum* etc., residing respectively, in the rhizospheres of wheat (kumar and Narula, 1999), tomato (Kim *et al.*, 1998), soybean (Cattelan *et al.*, 1999) and radish (Antoun *et al.*, 1998) have been reported to solubilize the phosphates and promote the growth of these crop plants.

Table 4.1: Microorganisms that Stimulate Plant Growth Due to their Mitrogenase Activity

Microorganism	*Host Plant*	*Reference*
Azospirillum sp.	Wheat	Boddey *et al.* (1986)
(Rhizospheric)	Maize	de Salamone *et al.* (1996)
	Rice	Malik *et al.* (1997)
Azotobacter sp	Maize	Pandey *et al.* (1998)
(Rhizospheric)	Wheat	Mrkovacki and Milic (2001)
Bacillus polymyxa	Wheat	Omar *et al.* (1996)
(Rhizospheric)		
Azoarcus sp.	Sorghum	Stein *et al.* (1997)
(Endophytic)	Rice	Egener *et al.* (1999)
	Kallar grass	Hurek *et al.* (2002)
Herbaspirillum sp.	Sugarcane	Pimentel *et al.* (1991)
(Endophytic)	Sorghum	James *et al.* (1997)
	Rice	James *et al.* (2002)
Gluconacetobacter	Sorghum	Isopi *et al.* (1995)
diazotrophicus	Sugarcane	Boddey *et al.* (2001);
(Endophytic)	Sevilla *et al.* (2001)	
Bulkholderia sp.	Rice	Baldani *et al.* (2001)
(Endophytic)		

Source: (Vessey, 2003).

Plant Growth Promoting Fungi (PGPF) including VAM fingi have been reported to mineralize the organic substrates and may, therefore, provide the plants with necessary mineral nutrients in an easily assimilating form (Hyakumachi, 2000). Altmore *et al.* (1999) investigated the capability of *Trichoderma harzianum* Rifai 1295-22 (T-22) to solubilize some insoluble or sparingly soluble minerals *in vitro* and reported that T-22 was able to solubilize MnO_2, metallic zinc and rock phosphate (mostly calcium phosphate) in a liquid sucrose-yeast extract medium. This phosphate solubilising activity of *T. harzianum* might be responsible for its plant growth promoting ability. Kang *et al.* (2002) reported the ability of *Fomitopsis* to solubilize tri-calcium phosphate. Gibson *et al.* (2004) studied the nutritional influences on solubilization of metal phosphates by ericoid mycorrhizal fungi and found that *Hymenoscyphus ericae* (an ericoid mycorrhizal mycobiont) had the ability to solubilize zinc and calcium phosphates ($CaHPO_4$) on solid agar plates. Richa *et al.* (2007) tested the efficacy of *Aspergillus tubingensis* and *A. niger* to solubilize rock phosphate and found that both these fungi had the ability to solubilize rock phosphate and also enhanced the growth and yield of maize in rock phosphate amended soil. El-Azouni (2008) tested the efficacy of *Aspergillus niger* and *Penicillium italicum* to solubilize tri-calcium phosphate (TCP) *in vitro* as well as their effect on the growth of soybean (*Glycine max*) *in vivo* and reported that both these fungi showed high TCP solubilising ability on agar plates and their dual inoculation in pot experiments significantly increased the yield and dry matter of soybean plants.

Phytohormone Production

Plant growth promoting activity of microorganisms may be related to their ability to produce phytohormones including auxins, gibberellins, cytokinins and ethylene (Brazani and Friedman, 1999; Unyayar *et al.*, 2000) which in most cases are believed to change assimilate partitioning patterns in plants, affect growth patterns in roots resulting in bigger roots system with greater surface area enabling the plant to access more nutrients from the soil (Salisbury, 1994). Brazani and Friedman (1999) observed that Indole-3-acetic acid (IAA) which induces root initiation, cell division and cell enlargement in plants, is very commonly produced by PGPR such as *Agrobacterium* sp., *Alcaligenes piechaudii*, *Comamonas acidovorans* associated with the rhizosphere of lettuce. Mehnaz *et al.* (2001) reported IAA production by the PGPR *Aeromonas veronii* and *Enterobacter cloacae* which are associated with the root system of *Oryza sativa*. The other plant growth promoting rhizobacteria which have been reported to produce IAA include *Azospirillum brasilense*, *Enterobacter* sp. residing respectively in the rhizosphere of wheat (Kaushik *et al.*, 2000) and sugarcane (Mirza *et al.*, 2001). *Rhizobium leguminosarum* and *Pseudomonas fluorescens* which stimulate the growth respectively, of rape or lettuce (Noel *et al.*, 1996) and soybean (de Salamone *et al.*, 2001) have been found to produce cytokinins which are known to induce cell division and cell enlargement (Salisbury, 1994). *Bacillus* sp. that stimulate the growth of alder, have been reported to produce gibberellins (Gutierrez-Manero *et al.*, 2001) which induce modification of plant morphology by the extension of plant tissue (Salisbury, 1994).

Unyayar *et al.* (2000) reported the production of auxin and abscisic acid by the fungus *Phanerochaete chrysosporium* ME446 immobilized on polyurethane foam. *Aspergillus niger*, *Aspergillus flavus*, *Penicillium corylophilum*, *Penicillium cyclopium*, *Penicillium funiculosum* and *Rhizopus stolonifer* have been reported to produce gibberellin (Hasan, 2002). Cristescu *et al.* (2002) studied ethylene production by phytopathogenic fungus *Botrytis cinerea* causing post harvest rot of perishable plant products including tomato and found that this fungus has the ability to produce ethylene *in vitro*. Ethylene (C_2H_2) is known as fruit ripening hormone in higher plants. Plant growth stimulating effect of some rhizobacteria such as *Alcaligenes* sp., *Bacillus pumilus, Enterobacter cloacae, Pseudomonas* sp., *Variovorax paradoxus* has been attributed to their ability to produce 1-aminocyclopropane-1-carboxylate (ACC) deaminase, an enzyme that breaks down ACC, the immediate precursor for ethylene in the biosynthetic pathway in plants thereby decreasing ethylene production in the roots of host plants resulting in root lengthening and better pant growth (Glick *et al.*, 1998; Belimov *et al.*, 2001; Saleh and Glick, 2001). Ankit *et al.* (2008) observed auxin-like activity of the extract from hypertrophied tissue of *Acacia eburnea* infected with the rust pathogen *Ravenelia esculenta* and reported that the hypertrophy of the host plant tissue might be due to indole-3-acetic acid (IAA) produced by this pathogen.

Suppression of Pathogenic Microbes

Hyakumachi (1994) observed plant growth promotion effect in cucumber which was due to the suppression of *Pythium* spp. by PGPF which were indigenous in the soil. Elad (2000) studied the biological control of foliar pathogens of cucumber by means of *T. harzianum* and found that four foliar pathogens namely *Botrytis cinerea, Pseuperonospora cubensis, Sclerotinia sclerotiorum* and *Sphaerotheca fusca* causing grey mould, downy mildew, white mould and powdery mildew diseases of cucumber, respectively, were suppressed by *T. harzianum* under greenhouse conditions. Narisawa *et al.* (2002) reported that *Verticillium dahliae* causing wilt disease of eggplant was suppressed by *Heterconium chaetospira, Phialocephala fortinii, Penicillium* sp. and *Trichoderma* sp. *T. harzianum, T. viride* and *T. virens* have been found to suppress the mycelial growth of *Fusarium oxysporum* f. sp. *ciceris* and enhance the growth and yield of this crop plant (Dubey *et al.*, 2007).

References

Adel A.E.L. Mehalawy, Naziha M. Husseinein, Hend M. Khater, EL-Zahira A., Karam E.L.-Din and Youssef (2004). Influence of maize root colonization by the Rhizosphere Actinomycetes snd Yeast fungi on plant growth and on the Biological control of Late wilt disease. International Journal of Agriculture and Biology. **6:** 599–605.

Altmore, C., Norvell W.A., Bjorkman, T. and Harman G.E. (1999). Solubilization of phosphates and micronutrients by the plant growth promoting and biocontrol fungus *Trichoderma harzianum* Rifai 1295-22. *Applied and Environmental Microbiology.* **65**: 2926-2933.

Aniket, K., Pankaj, P. and Ankita, K. (2008). Auxin-like activity of extract from hypertrophied tissue of *Acacia eburnea* infected with *Ravenelia esculenta. Plant Cell, Tissue and Organ Culture.* **94**: 101-104.

Antoun, H., Beauchamp, C.J., Goussard, N., Chabot, R. and Lalande, R. (1998). Potential of *Rhizobium* and *Bradyrhizobium* species as plant growth promoting rhizobacteria on non-legumes: effect on radishes (*Raphanus sativus* L.). *Plant and Soil.* **204:** 57-67.

Baker, R. (1991). Diversity in biological control. *Crop Protection.* **10:** 85–94

Baldani, V.L.D., Baldani, J.I. and Dobereiner, J. (2001). Inoculation of rice plants with the endophytic diazotrophs *Herbaspirillum seropedicae* and *Burkholderia* spp. *Biology and Fertility of Soils.* **30:** 485-491.

Barazani, O. and Friedman, J. (1999). Is IAA the major root growth factor secreted from plant-growth-mediating bacteria? *Journal of Chemical Ecology.* **25:** 2397-2406.

Bar-Ness, E., Chen, Y., Hadar, H., Marschner, H. and Romheld, V. (1991). Siderophores of *Pseudomonas putida* as an iron source for dicot and monocot plants. *Plant and Soil.* **130:** 231-241.

Bechingham, C. (2007). Commodity growing guide–sweet corn.NSW Department of primary industries. Available online at http://www.dpi.nsw.gov.

Belimov, A.A., Safronova, V.I., Sergeyeva, T.A., Egorova, T.N., Matveyeva, V.A., Tsyganov, V.E., Borisov, A.Y., Tikhonovich, I.A., Kluge, C., Preisfeld, A., Dietz, K. J. and Stepanok, V.V. (2001). Characterization of plant growth promoting rhizobacteria isolated from polluted soils and containing 1-aminocyclopropane-1-carboxylate deaminase. *Canadian Journal of Microbiology.* **47:** 642-652.

Benbow, J.M. and Sugar, D. (1999). Fruit surface colonization and biologicalcontrol of post-harvest diseases of pear by pre-harvest yeastapplication. *Plant Disease.* **3:** 839–44.

Benhamou, N., Gagne, S., Lequere, D. and Debbi, L. (2000). Bacteral mediated induced resistance in cucumber: beneficial effect of the endophytic bacterium Serratia plymuthica on the protection against infection by Phytium ultimum. Phytopathology. **90:** 45–56.

Beniziri, E., Coutrade, A., and Guckert, A. (1995). Fate of two microorganisms in maize stimulated rhizosphere under hydroponic and sterile conditions. Soil Boil Biochem. **27:** 71- 77.

Boddey, R.M., Baldani, V.L.D., Baldani, J.I. and Dobereiner, J. (1986). Effect of inoculation of Azospirillum spp. on nitrogen accumulation by field-grown wheat. *Plant and Soil.* **5:** 109-121.

Boddey, R.M., Polidoro, J.C., Resende, A.S., Alves, B.J.R. and Urquiaga, S. (2001). Use of the ^{15}N natural abundance technique for the quantification of the contribution of N_2 fixation to sugarcane and other grasses. *Australian Journal of Plant Physiology.* **28:** 889-895.

Bown, G.D. and Rovira, A.D. (1999). The rhizospher and its management toimprove plant growth. *Adv. Agron.* **66:** 1–102.

Boyer, C.D. and Hannah, L.C. (1994). Kernel mutants of corn.Chapter1. In: AR Hallauer, ed. *Specialty corns* CRC Press Inc BOCA Roton, USA. Pp 1-28.

Brown, M.E. (1972). plant growth substances produced by microorganisms of soil and rhizosphere. *J. Appl Bacteriol.* **35:** 443- 45.

Buck, J.W. and Andrews, J.H. (1999). Role of adhesion in the colonization ofleaves by yeast *Rhodosporidium toruloides. Candian J. Microbiol.* **45:** 433–40.

Buir T. J. and Caesar A. (1984). Beneficial plant bacteria. *Crit Rev Plant Sci.* **2:** 1–20.

Cattelan, A.J., Hartel, P.G. and Fuhrmann, J.J. (1999). Screening for plant growth promoting rhizobacteria to promote early soybean growth. *Soil Science Society of America Journal.* **63:** 1670-1680.

Compant, S., Reiter, B., Sessitsch, A., Nowak, J., Clément, C. and Barka, E.A. (2005). Endophytic Colonization of *Vitis vinifera* L. by Plant Growth-Promoting Bacterium Burkholderia sp. Strain PsJN. *Appl. And Environ. Microbiol.* **71(4):** 1685-1693.

Crawford, D.L., Lynch, J.M., Whipps, J.M. and Ousley, M.A. (1993). Isolationand characterization of actinomycetes antagonists of a fungal rootpathogen. *Appl. Environ. Microbiol.* **59:** 3899–3905

Cristescu, S.M., Martinis, D.D., Hekkert, S.T.L., Parker, D.H. and Harren, F.J.M. (2002). Ethylene production by *Botrytis cinerea in vitro* and in tomatoes. *Applied and Environmental Microbiology.* **68** (11): 5342-5350.

de Salammone, *I.E.G.*, Hynes, R.K. and NIlson, L.M. (2001). Cytokinin production by plant growth promoting rhizobacteria and selected mutants. *Canadian Journal of Microbiology.* **47:** 404-411.

de Salamone, *I.E.G.*, Dobereiner, J., Urquiaga, S. and Boddey, R.M. (1996). Biological nitrogen fixation in Azospirillum strain-maize genotype associations as evaluated by the ^{15}N isotope dilution technique. *Biology and Fertility of Soils.* **23:** 249-256.

DeLeij, F.A.A.M., Sutton, M.M., Ehipps, J.M. and Linch, J.M. (1999). Impact of field release of genetically modified Pseudomonas fluorescens on indigenous microbial populations of wheat. *Appl Environ Microbiol.* **61:** 3442–53.

Di Cello, F., Bevivino, A., Chiarini, L., Fani, R., Paffetti, D. and Tabacchioni, S. (1997). Biodiversity of a Burkholderia cepacia population isolated from maize rhizosphere at different plant growth stages. *Appl Environ Microbiol.* **63:** 4485–93.

Di Cello, F., Bevivino, A., Chiarini, L., Fani, R., Paffetti, D. and Tabacchioni, S. (1997). Biodiversity of a Burkholderia cepacia population isolated from maize rhizosphere at different plant growth stages. *Appl Environ Microbiol.* **63:** 4485–93.

Dubey, S.C., Suresh, M. and Singh, B. (2007). Evaluation of *Trichoderma* species against *Fusarium oxysporum* f. sp. *ciceris* for integrated management of chickpea wilt. *Sciencs Direct Biological Control.* **40:** 118-127.

Egener, T., Hurek, T. and Reinhold-Hurek, B. (1999). Endophytic expression of nif genes of *Azoarcus* sp. strain BH72 in rice roots. *Molecular Plant- Microbe Interactions.* **12:** 813-819.

Elad, Y. (2000). Biological control of foliar pathogens by means of *T. harzianum* and potential modes of action. *Crop Protection.* **19:** 709-714.

El-Azouni, I.M. (2008). Effect of phosphate solubilising fungi on growth and nutrient uptake of soybean (*Glycine max* L.) plants. *Journal of Applied Sciences Research.* **4 (6):** 592–598.

El-Tarabily, K.A., Hardy GEStJ, K. Sivasithamparam, A.M. Hussein and Kurtböke D.I. (1997). The potential for the biological control of cavity-spot disease of carrot, caused by *Pythium coloratum,* by streptomycete and non-streptomycete actinomycetes. *New Phytopathologist.* **137:** 459–507.

Farnaham, D.E., Benson, G.O., and Pearce, R.B. (2003). Corn perspective and culture. Chapter 1. In: PJ White. LA. Johanson.eds.corn: chemistry and technology, edition 2nd. American Association of cerial chemicals, Inc. st.Paul,Minesota, USA. Pp 1-33

Farrell, T. and O'Keeffe, K. (2007). Maize.NSW Department of primary industries available online at http://www.dpi.nsw.gov.au

Flores-Vargas, R.D. and O'Hara, G.W. (2006). Isolation and characterization of rhizosphere bacteria with potential for biological control of weeds in vineyyards. *Journal of Applied Microbiology.* **100:** 946-954.

Fravel, D.R. (1988). Role of antibiosis in Biocontrol of plant diseases. *Annu Rev Phytopathol* **26:** 75-91.

Frommel, M.I., Novak, J. and Lazarovits, G. (1993). Treatment of potato tubers with a growth promoting Pseudomonas sp. Plant growth responses and bacterium distribution in the rhizosphere. *Plant Soil.* **1(50):** 51–60.

Germida, J.J., Siciliano, S.D., de Freitas, J.R. and Seib, A.M. (1998). Diversity of root-associated bacteria associated wit held–grown canola (*Brassica napus* L.) and wheat *(Triticum aestivum L.)* FEMS *Microbiol. Ecol.* **26:** 43–50.

Gibson, B.R. and Mitchell, D.T. (2004). Nutritional influences on the solubilization of metal phosphate by ericoid mycorrhizal fungi. *Mycological Research.* **108 (8):** 947-954.

Glick, B.R. (1995). The enhancement of plant growth by free-living bacteria. *Canadian Journal of Microbiology.* **4:** 109-117.

Glick, B.R., Penrose, D.M. and Li, J.P. (1998). A model for the lowering of plant ethylene concentrations by plant growth promoting bacteria. *Journal of Theoretical Biology.* **190:** 63-68.

Gualtieri, G. and Bisseling, T. (2000). The evolution of nodulation. *Plant Molecular Biology.* **42:** 181-194.

Guckert, A. (1992). Bedeutung der pflanzenwurzeln und ihrer Ausscheidungen als Quellen organischer Stoffe im Boden. *Der Landwirtsch* **206:** 97-113.

Gupta, A., Gopal, M., Pandey, A., Soccol, C.R., Josk, V.K., Saxena, A.K. and Tilak, K.V.B.R. (1998). Bacterisation of green gram with rhizosphere bacteria for enhanced plant growth. *J. Sci. Indust. Res.* **57:** 126–31

Gupta, A., Gopal, M., Pandey, A., Soccol, C.R., Josk, V.K., Saxena, A.K. and Tilak, K.V.B.R. (1998). Bacterisation of green gram with rhizosphere bacteria for enhanced plant growth. *J. Sci. Indust. Res.* **57:** 126–31

Gutierrez-Manero, F.J., Ramos-Solano, B., Probanza, A., Mehouachi, J., Jadeo, F.R. and Talon, M. (2001). The plant growth promoting rhizobacteria *Bacillus pumilus* and *Bacillus licheniformis* produce high amounts of physiologically active gibberellins. *Physiology of Plants.* **111:** 206-211.

Hallmann, J., Quadt-Hallmann, A., Mahaffee, W.F. and Kloepper, J.W. (1997). Bacterial endophytes in agricultural crops. *Can J Microbiol.* **43:** 895–914.

Hamlen, R.A., Lukezie, F.L. and Bloom, J.R. (1972). Influence of age and state of development of the neutral carbohydrates components in root exudates from alfalfa plants grown in a gnotobiotic environment. *Can J Plant Sci* **52:** 633–42.

Hasan, H.A.H. (2002). Gibberellin and auxin-indole production by plant root-fung and their biosynthesis under salinity-calcium interaction. *Rostlinna Vyroba.* **48 (3):** 101-106.

Hashem, M.A. (2001). Problems and prospects of cyanobacterial biofertilizer for rice cultivation. *Australian Journal of Plant Physiology.* **28:** 881-888.

Higa, T. (1991). Effective microorganisms: A biotechnology for mankind. P. 8-14. In J.F. Parr, S.B. Hornick, and C.E. Whitman (ed.) Proceedings of the First International Conference on Kyusei Nature Farming U.S Department of Agriculture, Washington, D.C., USA.

Hobbs, L. (2003). Corn sweetners. Chapter 17. In: PJ White, LA Johanson, eds. Corn: chemistry and technology, Edition 2. American Association of cereal Chemicals, Inc. St. Mnesota, USA. Pp 635-669.

Hoflich, G. and Kühn, G. (1996). Promotion of plant growth and nutrient up take of cruciferous oil and inter crops by inoculated rhizosphere microorganisms. *Zeitscrift fur Pflanzenernährung und Bodenkunde.* **159:** 575–81

Hoflich, G. and Kühn, G. (1996). Promotion of plant growth and nutrient up take of cruciferous oil and inter crops by inoculated rhizosphere microorganisms. *Zeitscrift fur Pflanzenernährung und Bodenkunde.* **159:** 575–81

http://fao.org

http://faostat.fao.org

http://faostat.org

http://faperta.ugm.ac.id

http://gazetteonline.com

http://www.jansamachar.net

http://www.krvycomtrade.com

http://www.niir.org

Hurek, T., Handley, L.L., Reinhold-Hurek, B. and Piche, Y. (2002). *Azoarcus* grass endophytes contribute fixed nitrogen to the plant in an unculturable state. *Molecular Plant- Microbe Interactions.* **15:** 233-242.

Hussain, A. and Vancura, V. (1970). Formation o biologically active substances by rhizosphere bacteria and their effect on plant growth. *Folia microbial* **15:** 468-478.

Hyakumachi, M. (1994). Plant–growth–promoting fungi from turfgrass hizosphere with potential for disease suppression. *Soil Microorganisms.* **44:** 53-68.

Hyakumachi, M. (2000). Mehanisms of plant growth promotion by PGPF. In: *Fungal Biotechnology in Agricultural, Food, and Environmental Applications, Mycology*, Arora D. K. **21:** 103.

Isopi, R., Fabbri, P., Del-Gallo, M. and Puppi, G. (1995). Dual inoculation of *Sorghum bicolor* (L.) Moench ssp. *bicolor* with vesicular arbuscular mycorrhizas and *Acetobacter diazotrophicus*. Symbiosis. **18:** 43-55.

James, E.K., Gyaneshwar, P., Mathan, N., Barraquio, Q.L., Reddy, P.M., Iannetta, P.P. M., Olivares, F.L. and Ladha, J.K. (2002). Infection and colonization of rice seedlings by the plant growth promoting bacterium *Herbaspirillum seropedicae* Z67. *Molecular Plant- Microbe Interactions.* **15:** 894-906.

James, E.K., Olivares, F.L., Baldani, J.I. and Dobereiner, J. (1997). *Herbaspirillum*, an endophytic diazotroph colonizing vascular tissue in leaves of Sorghum bicolor L. Moench. *Journal of Experimental Botany.* **48:** 785-797.

Johanson, L.A. and May, J.B. (2003). Wet milling: the basis for corn biorefineries. Chapter 12. In : PJWhite, LA Johanson,eds. *Corn chemistry and technology.* American associan of cereal chemicals, Inc, st. paul, Minesota, USA. Pp 449-494.

Kang, S.C., Ha, C.G., Lee, T.G. and Maheshwari, D.K. (2002). Solubilization of insoluble inorganic phosphates by a soil inhabiting fungus *Fomitopsis* sp. Ps 102. *Current Science.* **82:** No. 4.

Kaushik, R., Saxena, A.K. and Tilak, K.V.B.R. (2000). Selection of Tn5: lacZ mutants isogenic to wild type *Azospirillum brasilense* strains capable of growing at sub-optimal temperature. *World Journal of Microbiology and Biotechnology.* **16:** 567-570.

Kim, K.Y., Jordan, D. and McDonald, G.A. (1998). Effect of phosphate solubilizing bacteria and vesicular-arbuscular mycorrhizae on tomato growth and soil microbial activity. *Biology and Fertility of Soils.* **26:** 79-87.

Kirkpatrick, B. and Wilhelm, M. (2006). Evaluation of grapevine endophytic bacteria for control of Pierce's disease. *Department of Plant Pathology University of California Davis CA 95616.*

Klopper, J.W. and Schroth, M.N. (1978). Plant growth promoting rhizobacteria on radishes. In proceeding of the 4th International conferences on plant pathogenic bacteria, Angers, France, 27 August–2 September, 1978, Vol. 2, Gilbert–Clary, Tours, pp 879- 882.

Kucey, R.M.N., Janzen, H.H. and Legget, M.E. (1989). Microbially mediated increases in plant available phosphorus. *Adv Agron.* **42:**199- 22

Kumar, V. and Narula, N. (1999). Solubilization of inorganic phosphates and growth emergence of wheat as affected by *Azotobacter chroococcum* mutants. *Biology and Fertility of Soils.* **28:** 301-305.

Lee, I.K., Kim, C.J., Kim, C.D. and Yoo, I. D. (1990). Antifungal antibiotic against fruit rot disease and red pepper from *Streptomyces parvullus.Korean J. Appl. Microbiol. Biotech.* **18:** 142–7

Lemanceau, P. (1992). Effcts benefiques de rhizobacteries sur les plantes: example des *pseudomonas spp* fluorescents. *Agronomie.* **12:** 413-437.

Leong, J. (1986). Siderophores : their chemistry and possible role in the biocontrol of plant pathogens. *Annu Rev Phytopathol.* **24:** 187- 208.

Locci, R. and Schofield, G.M. (1989). Genus Williams, S.T., M.E. Sharpe and J.G. Holt (eds.). *Bergey's Manual ofSystematic Bacteriology* (Vol. 4) pp: 2492–504. Williams and Wilkins, Baltimore, USA.

Malik, K.A., Bilal, R., Mehnaz, S., Rasul, G., Mirza, M.S. and Ali, S. (1997). Association of nitrogen-fixing, plant growth promoting rhizobacteria (PGPR) with kallar grass and rice. *Plant and Soil.* **194:** 37-44.

Martin, J.K. (1975). ^{14}C- labled material leached from the rhizosphere of plants supplied continuously with $^{14}CO_2$. *Soil Biol Biochem.* **7:** 395- 399.

Masalha, J., Kosegarten, H., Elmaci, O. and Mengel, K. (2000). The central role of microbial activity for iron acquisition in maize and sunflower. *Biology and Fertility of Soils.* **30:** 433-439.

Mehnaz, S., Mirza, M.S., Haurat, J., Bally, R., Normand, P., Bano, A. and Malik, K.A. (2001). Isolation and 16S rRNA sequence analysis of the beneficial bacteria from the rhizosphere of rice. *Canadian Journal of Microbiology.* **472:** 110-117.

Mirza, M.S., Ahmad, W., Latif, F., Haurat, J., Bally, R., Normand, P. and Malik, K.A. (2001). Isolation, partial characterization and the effect of plant growth promoting bacteria (PGPB) on micro-propagated sugarcane *in vitro. Plant and Soil.* **237:** 47-54.

Mrkovacki, N. and Milic, V. (2001). Use of Azotobacter chroococcum as potentially useful in agricultural application. *Annals of Microbiology.* **51:** 145-158.

Muhammed, S. and Amusa, N.A. (2003). *In vitro* inhibition of growth of someseedling blight inducing pathogens by compost–inhibiting microbes. *Afr. J. Biotechnol.* **2(6):** 161–164.

Nacamulli, C., Bevivino, A., Dalmastri, C., Tabacchioni, S. and Chiarini, L. (1997). Perturbation of maize rhizosphere microflora following gseed bacterization with Burkholderia cepacia. MCI 7. *FEMS Microbiol Ecol.* **23:**183–93.

Narisawa, K., Kawamata, H., Currah, R.S. and Hashiba, T. (2002). Suppression of *Verticillium* wilt in egg plant by some fungal endophytes. *European Journal of Plant Pathology.* **108:** 103-109.

Nautiyal, C. S., Bhadauri, S., Kumar, P., Lal, H., Mondal, R. and Verma, D. (2000). Stress induced phosphate solubilization in bacteria isolated from alkaline soils. *FEMS Microbiology Letters.* **182:** 291-296.

Neilands, J.B. and Leong, J. (1986). Siderophore in relation to plant growth and disease. *Annu Rev Plant physiol.* **37:**187- 208.

Noel, T.C., Sheng, C., Yost, C.K., Pharis, R.P. and Hynes, M.F. (1996). *Rhizobium leguminosarum* as a plant growth promoting rhizobacterium: direct growth promotion of canola and lettuce. *Canadian Journal of Microbiology.* **42:** 279-283.

Obreht, Z., Kerby, N.W., Gantar, M. and Rowell, P. (1993). Effects of root- associated N_2-fixing cyanobacteria on the growth and nitrogen content of wheat (*Triticum vulgare* L.) seedlings. *Biology and Fertility of Soils.* **15:** 68-72.

Omar, M.N.A., Mahrous, N.M. and Hamouda, A.M. (1996). Evaluating the efficiency of inoculating some diazotrophs on yield and protein content of 3 wheat cultivars under graded levels of nitrogen fertilization. *Annals of Agricultural Science.* **41:** 579-590.

Orole, O.O. and Adejumo, T.O. (2009). Activity of fungal endophytes against four maize wilt pathogens. *African journal of microbiology Research.* **3 (12):** 969- 973.

Ownley, B.H., Pereira, R.M., Klingeman, W.E., Quigley, N.B. and Leckie, B.M. (2004). *Beaveria bassiana,* a dual purpose biocontrol organism with activity against insect pests and plant pathogens. *Emerging Concepts in Plant Health Management.* 255-269.

Paliwal, R.L. (2006b). Introduction to maize and its importacce. Chapter 1. In : *Tropical maize : Improvement and production,* Food and Agriculture Organisation of United Nations Rome. pp 1-3

Pandey, A., Sharma, E. and Palni, L.M.S. (1998). Influence of bacterial inoculation on maize in upland farming systems of the Sikkim Himalaya. *Soil Biology and Biochemistry.* **30:** 379-384.

Parr, J.F., Hornick, S.B. and Kaufman, D.D. (1994). Use of microbial inoculants and organic fertilizers in agricultural production. In Proceedings of the International Seminar on the use of Microbial and Organic Fertilizers in Agricultural Production. Published by the Food and Fertilizer Technology Centre, Taipei, Taiwan.

Pimentel, J.P., Olivares, F., Pitard, R.M., Urquiaga, S., Akiba, F. and Dobereiner, J. (1991). Dinitrogen fixation and infection of grass leaves by *Pseudomonas rubrisubalbicans* and *Herbaspirillum seropedicae. Developments in Plant and Soil Sciences.* **48:** 225-229.

Poling, S.M., Wicklow, D.T., Rogers, K.D. and Gloer, J.B. (2008). *Acremoniumzeae* a protective Endophyte of maize, produces Dihydroresorcylideand 7–hydroxydihydroresorcylides. *J. Agric. Food Chem.* **56(9):** 3006-3009.

Posada, F. and Vega, F.E. (2005). Establishment of the fungal entomopathogen*Beauveria bassiana* (Ascomycota: Hypocreales) as anendophyte in cocoa seedlings (*Theobroma cacao*). *Mycol. Soc. Am.* **97(6):** 1195-1200.

Reyes, I., Berneir, L., Simard, R.R., Tanguay, P. and Antoun, H. (1998). Characteristicsnof phosphate solibilisation by an isolate of a tropical Penicillium rugulosum and two UV induced mutants. *FEMS Microbiol Ecol.* **28:** 291- 295.

Reyes, I., Baziramakenga, R., Bernier, L. and Antoun, H. (2001). Solubilization of phosphate rocks and minerals by a wild type strain and two UV induced mutants of Penicillium rugulosum. *Soil Biol. Biochem.* **33:** 1741- 1746.

Richa, G., Khosla, B., and Reddy, M.S. (2007). Improvement of maize plant growth by phosphate solubilising fungi in rock phosphate amended soils. *World Journal of Agricultural Sciences.* **3 (4):** 481-484.

Richardson, A.E. (2001). Prospects for using microorganisms to improve the acquisition of phosphorous by plants. *Australian Journal of Plant Physiology.* **28:** 897-906.

Rodriguez, H. and Fraga, R. (1999). Phosphate solubilizing bacteria and their role in plant growth promotion. *Biotechnology Advances.* **17:** 319-339.

Rovira A.D. (1959). Plant root excretion in relation to rhizosphere effect. *Plant soil.* **11:** 53- 63.

Saleh, S.S. and Glick, B.R. (2001). Involvement of gacS and rpoS in enhancement of the plant growth promoting capabilities of *Enterobacter cloacae* CAL2 and UW4. *Canadian Journal of Microbiology.* **47:** 698-705.

Salehpour, M., Etebarian, H., Roustaei, A., Khodakaramian, G. and Aminian, H. (2005). Biological Control of Common Root Rot of Wheat *(Bipolaris sorokiniand)* by *Trichoderma* Isolates. *Plant Pathol. J.* **4(1):** 85-90.

Salisbury, F.B. (1994). The role of plant hormones. *In* Plant-Environment Interactions. Ed. R. E. Wilkinson. Pp. 39-81. Marcel Dekker, New York, USA.

Sarathchandra, S.U., Lee, A., Perrott, K.W., Rajan, S.S.S. and Oliver, E.H.A. (1993). Effect of fertilizer application on microorganisms in pastoral soil. *Aust J Soil. Res.* **31:** 299- 309.

Schroth, M.N. and Hancock, J.G. (1982). Disease suppressive Soil and root colonizing bacteria. *Science.* **216:**1376- 1381.

Sevilla, M., Burris, R.H., Gunapala, N. and Kennedy, C. (2001). Comparison of benefit to sugarcane plant growth and 15N2 incorporation following inoculation of sterile plants with *Acetobacter diazotrophicus* wild-type and Nif-mutant strains. *Molecular Plant- Microbe Interactions.* **14:** 358-366.

Stein, T., Hayen-Schneg, N. and Fendrik, I. (1997). Contribution of BNF by *Azoarcus* sp. BH72 in *Sorghum vulgare. Soil Biology and Biochemistry.* **29:** 969-971.

Stirz, A.V. and Nowak, J. (2000). Endophytic communities of rhizobacteria and the strategies required to create yield enhancing association with crops. *Appl Soil Ecol.* **15:**183–90

Taechowisan, J.N., Chuaychot, S., Chanaphat, A. Wanbanjob and Tantiwachwutikul, Z.P. (2009). Antagonistic Effects of *Streptomyces* sp.SRM1 on

Colletotrichum musae. Biotechnology Asian Network for Scientific Information] T'Department of Microbiology, Silpakorn University, Nakorn Pathom 73000, Thailand.

Takaki, S.A., Kitamura, T., Marumota, D., Ishida and Tanaka, S. (1996). Control of *Fusarium* disease using antagonistic actinomycetes. III. Effect of a microbial inoculum (Material A) on microbial flora under some soil environmental conditions and development of effective method of application to radish yellows. *Soil Microorganisms*. **47:** 51–8

Unyayar, S., Unyayar, A. and Unal, E. (2000). Production of auxin and abscisic acid by *Phanerochaete chrysosporium* ME446 immobilized on polyurethane foam. *Turkish Journal of Biology*. **24:** 769-774.

Valois, D.K., Fayed, T., Barasubye, M., Garon, C., Dery, R., Brezezinski and Eaulieu, C. (1996). Glucanolytic actinomycetes antagonistic to *Phytophthora fragariae* var. *rubi*, the causal agent of raspberry root.*Appl. Environ. Microbiol.* **62:** 1630–5

Vazquez, P., Holguin, G., Puente, M.E., Lopez-Cortez, A. and Bashan, Y. (2000b). Phosphate solubilizing microorganisms associated with the rhizosphere of mangroves in semiarid coastal lagoon. *Biology and Fertility of Soils*. **30:** 460-468.

Wang, Y., Brown, H.N., Crowley, D.E. and Szaniszlo, P.J. (1993). Evidence for direct utilization of a siderophore, ferrioxamine B, in axenically grown cucumber. *Plant, Cell and Environment*. **16:** 579-585.

Weller, D.M. (1988). Biological control of soil-borne plant pathogens in therhizosphere with bacteria. *Ann. Rev. Phytopathol.* **26:** 379–407

Weller, D.M. (1988). Biological control of soil-borne plant pathogens in therhizosphere with bacteria. *Ann. Rev. Phytopathol.* **26:** 379–407

Whipps, J.M. (1990). Carbon Economy. In: Lynch JM 9ed) The rhizosphere. Chichester, England: John Willey 7 sons, pp59- 99

White, P.J. (1994). Propertioes of corn starch. Chapter 2. In; AR Hallauer, ed. *Specialty corns*. CRC Press Inc Boca Raton, USA. PP 29- 54.

Yensen, N.P. and Beil, K.Y. (2006). Soil remediation via salt conduction and the hypothesis and photoprotection. Chapter 21. In : MA Khan, DJ Weber, eds *Ecophysiology of high salinity tolerant plants*. Springer pp 313- 344.

You, M.P., Sivasithamparam, K. and Kurtböke, D.L. (1996). Actinomycetes in organic mulch used in avocado plantations and their ability to suppress *Phytophthora cinnamomi*. *Biol. Fertil. Soil*. **22:** 237–42

Young, T.R. and Kucharek, T.A. (1997). Succession of fungal communities in roots and stalks of hybrid field corn grown in Florida. *Plant Dis.* **61:** 76–80.

Youssef, Y.A., El-Tarabily, K.A. and Hussein, A.M. (2001). *Plectosporiu mtabacinum* root rot disease of white lupine (*Lupinus termis* Forsk.) and its biological control by *Streptomyces* species. *J. Phytopathol.* **1490:** 29–33

Youssef, Y.A., El-Tarabily, K.A. and Hussein, A.M. (2001). *Plectosporiu mtabacinum* root rot disease of white lupine (*Lupinus termis* Forsk.) and its biological control by *Streptomyces* species. *J. Phytopathol.* **1490:** 29–33

Yuan, W.M. and Crawford, D.L. (1995). Characterization of *Stretomyces lydicus* WYEC 108 as a potential l biocontrol agent against fungal root and seed rots. *Appl. Environ. Microbiol.* **61:** 3119–28

Zinniel, D.K., Lambrecht, P., Harris, N.B., Zhengyu, F., Kuczmarski, D., Higley, P., Ishimaru, C.A., Arunakumari, A., Barletta, R.G. and Vidaver, A.K. (2002). Isolation and Characterization of Endophytic Colonizing Bacteria from Agronomic Crops and Prairie Plants. *Appl. and Environ. Microbiol.* **68(5):** 2198–2208.

Chapter 5

Soil Beneficial Bacterial Diversity and its Influence on Plant Health

Neerja Asthana and Ranjana Kumari

Department of Botany, R.D.S. College, Muzaffarpur, Bihar

Soil biological communities are an important natural recourses like sun energy, air & water and an integral part of agriculture ecosystem by maintaining ecological functions, soil health and crop production. These organisms have specific role in complex web of life in the soil. Rhizosphere, or the zone of influence around the plant roots, harbors a multitude of micro-organisms that are affected by both a -biotic and biotic stresses. So, the plant bacterial interactions in the rhizosphere are the determinants of plant health and soil fertility (Burr &Caesar, 1984). While the rhizosphere as a domine of fierce microbial activity has been studied for over a century, the availability of modern tools in microbial ecology has now permitted the study of bacterial communities associated with plant growth and development, in situ. In an interesting study of microbial population and diversity reveals that bacteria influence soil processes through nutrient acquisition and release, interconvert ion through enzymatic processes mobilization and immobilization, influencing physical structure through aggregation and by producing dead cells mass. Thus their influence on soil processes is both direct and indirect.

The main functions of these bacteria are:

- ✫ To supply nutrients to crops by synthesizing particular compounds and facilitating their uptake from the soil.
- ✫ To stimulate plant growth through the production of plant hormones.

- ☆ To control diseases by inhibiting the activity of plant pathogens.
- ☆ To improve soil structure by bioaccumulation of in-organics and mineralization of organic pollutants, *i.e.* bioremediation of polluted soils (Middledrop *et al.*, 1990, Burd *et al.*, 2000, Zhuang *et al.*, 2007 Zaidi *et al.*, 2008).
- ☆ To enhance resistance to drought,(Alvarez *et al.*, 1996), salinity and water logging (Saleem *et al.*, 2007) and oxidative stress (Stajner *et al.*, 1995, 1997)
- ☆ Solubilizing and mineralizing of nutrients, particularly mineral phosphates.

So, soil bacteria are very important in biogeochemical cycle and in sustainable agriculture production. This review focuses on soil beneficial bacteria and their role in plant growth promotion both directly and indirectly.

Nitrogen Fixing Bacteria

Nitrogen is an essential plant nutrient. It is the nutrient that required to biosynthesize the basic building blocks of life, i;e nucleotides for DNA and RNA and aminoacids for proteins. Biological nitrogen fixation (BNF) accounts for 65 per cent of the nitrogen utilized in agriculture, so, it is very important in sustainable crop production system. Dinitrogen (N_2) is metabolically unavailable directly to higher plants, though it makes up four fifth of the atmosphere, but it is available to some micro-organisms through biological nitrogen fixation (BNF) in which atmospheric nitrogen is converted into ammonia. According to their relationship with the plants, and their residing sites these nitrogen fixing bacteria can be divided into two groups: symbiotic bacteria, which live inside the plant cells, produce nodules, and are localized inside the specialized structure, non-symbiotic nitrogen fixing bacteria (free living rhizobacteria), which live outside the plant cells and do not produce nodules, but still promotes plant growth (Gary and Smith, 2005).

Sybiotic Nitrogen Fixing Rhizobacteria

The best known symbiotic nitrogen fixing bacteria are commonly called rhizobia, all belongs to Rhizobiaceae family which produce nodules to leguminous plants. Rhizobiaceae is divided into six genera currently accepted as Rhizobium, Mesorhizobium, Bradyrhizobium, Azorhizobium, Allorhizobium and Sinorhizobium have been successfully used worldwide to permit an effective establishment of the nitrogen fixing symbiosis with leguminous crop plants (Bottomley and Maggard 1990, Bottomley and Dughri 1989).These bacteria formed intimate symbiotic association with some higher plants especially legumes by responding chemotactically to flavonoid molecules released, as signals by the legume as a host. Compounds secreted by these plants induce the expression of nodulation (nod) genes in rhizobia. Nodules, the site for symbiotic nitrogen fixation are formed as a result of series of interactions between rhizobia and leguminous plants. However, there are number of factors which affect the nodulation on legume roots including host microbe symbiotic compatibility, physio-chemical conditions of the soil and the presence of both known and unknown bio-molecules such as flavonoids, polysaccharides and hormones (Tisdale *et al.*, 1990; Zafar-ul-Hye *et al.*, 2007). The great majority of legumes have this type of association with Rhizobia, except few. Although by far the majority of nitrogen fixing plants is from the legume family, there are few non-leguminous plants, such as Alder, Casuarina and Baybeery etc, that can also fix nitrogen by forming symbiotic association with Frankia bacteria. These plants referred as "actinorhizal plants" consists of twenty four genera of woody shrubs and trees as well as two non woody species of Datisea, distributed among eight plant families.

Researchers have studied that Nitrogen fixed by rhizobium-legume symbiosis has not only benefit legumes but also benefit associated non-legumes by direct transfer of biologically fixed nitrogen to cereals growing in intercrop(Snapp *et al.*, 1998) or subsequent crops rotated with symbiotic legumes(Shah *et al.*, 2003; Hayat 2005; Hayat *et al.*, 2008). A number of researchers have experimentally demonstrated the ability of rhizobia to colonize roots of non-legumes and localize themselves internally in tissues, including the xylem (Spencer *et al.*, 1994). Yanni *et al.* (1997) isolated Rhizobium leguminosarum bv. trifolii as a natural endophyte from roots of rice, this is because rice has been grown in rotation with berseem clover for about seven countries in the Nile delta, promoted closer rhizobial affinity to this cereal as a "host plant". Similarly photosynthetic Bradyrhizobia isolated from the roots of African brown rice (Chaintreuil *et al.*, 2000). Besides rice, rhizobia have also been isolated as a natural endophyets from the roots of othere non- leguminous plants species such as cotton, sweet corn (Mclnroy and Kloepper, 1995), maize (Martinez-Romero *et al.*, 2000), wheat (Biederbeck *et al.*, 2000), cornola (Lupwayi *et al.*, 2000). Laboratery studies on the rhizobial infection in the cereals roots lead to the conclusion that during legume cereal rotation or mixed inter-cropping rhizobia are brought into closer contact with cereal roots, and this probably results in non-legume root infection by native Rhizobial population in the soil via cracks made by emerging lateral roots.

Leelahawonge *et al.* (2010) isolated root nodules bacteria from the medicinal legume Indigofera tinctoria and reported a new legume symbiont related to Pseudoalteromonas sp. from the gamma class of proteobacteria. The plant growth promoting ability of rhizobial inoculation varies with soil properties, crop rotation, soil moisture, and nitrogen availability, yield potential of the crop and the abundance and effectiveness of native rhizobial population (Venketeswarlu *et al.*, 1997).

Non-Symbiotic Nitrogen Fixing Rhizobacteria

Dinitrogen is fixed in the rhizosphere of a variety of non-nodulated angiosperms. Such associative fixation can be a significant source of nitrogen for plants (Boddey *et al.*, 1983, Rennie *et al.*, 1983, Biesboer,1984).Some of the major free living and associative rhizobacterial genera includes species belongs to Acetobacter, Bacillus, Flavobacter, Pseudomonas, Proteus, Serrate, Xanthomonas, Klebsiella, Entrobacter, Citrpbacter, and others. The rhizosphere population and community structure of such associative and free- living bacteria depends upon the quantity and quality of root exudates released by the plants. (Kennedy and Tehan 1992) reported that Azotobacter species (A. vinelandii and A. chroococcum) are free-living heterotrophic dizotrops that depends on adequate supply of reduced carbon compounds, such as sugars for their energy sourses. Clostridia isolates from rice soils, shows that, their activity increased after returning straw to field raised C to N ratio in the soil (Elbadry *et al.*, 1999). It has been reported that application of Azotobacter increases the yield of crops like rice (Yanni & El Fattah 1999), cotton (Iruthayaraj 1981, Patil & Patil 1984, Anjum *et al.*, 2007) and wheat (Barassi *et al.*, 2000). Azospirillum species are an aerobic hetrotrotrophs and an associative symbiont can fix nitrogen under microaerobic condition and grow extensively in the rhizosphere of gramineous plants (Kennedy and Tehan 1992, Kennedy *et al.*, 2004). Green house study by Mirza *et al.* (2000) showed that Azospirilum lipoferum inoculation to rice increased the yield up to 6.7 g plant^{-1}. Radioactive nitrogen tracer technique showed that Azospirillum bresilense & Azospirrilum lipoferum contributed 7-12 per cent of wheat plant nitrogen by biological nitrogen fixation (Malik *et al.*, 2002). The genus Burkholderia comprises sixty seven (67) validly published species are all being capable of fixing nitrogen (Estrada–delos Station *et al.*, 2001, Vandamme *et al.*, 2002).Rice seedlings when inoculated with some strains of Burkholderia species isolated from rice plants contributed a relatively higher level nitrogen to rice via biological nitrogen fixation(BNF) (Baldani *et al.*, 2000) and increased rice grain yields significantly up to 8 t ha^{-1} (Tran Van *et al.*, 2000), and are capable of saving 25-30 kg N ha^{-1} of fertilizer. Herbispirillum is an endophytic diazotroph which colonizes internally between the cells of the roots of sugarcane, rice, maize, sorghum and other cereals (James *et al.*, 2000) can fix nitrogen and increase yield sigificantly. Some other endophytes that can fix atmospheric nitrogen are Acetobacter diazotrophicus found associated with sugarcane and Azoarcus sp. BH72, originally isolated from kallar grass.

Phosphorus-Solubilizing Rhizobacteria

Phosphorus is a second important plant nutrient after nitrogen. Phosphorous availability is low (normally at levels of 1ppm or less, Goldstein 1994) in soil because of its fixation as an insoluble phosphates of iron, aluminum and calcium. Phosphorous solubilizing bacteria (PSB) is a group of beneficial bacteria capable of hydrolysing organic and in-organic phosphorous from insoluble compounds and it is considered to be one of the important traits associated with plant phosphate nutrition. It is generally accepted that the mechanism of mineral phosphate (tricalcium phosphate, dicalcium phosphate, hydroxyl apatite and rock phosphate) solubilization by PSB strains belonging to genera Pseudomonas, Bacillus, Flavobacterium, Erwina and others are associated with the release of low molecular weight organic acids (Goldstein, 1995; Kim *et al.*, 1997) through which their hydroxyl and carboxyl group chelate the cation bounds to phosphate, there by converting it into soluble forms(Kpomblekow & Tabatabai, 1994). Production of low molecular weight acid like gluconic acid seems to be most frequent agent of mineral phosphate solubilization by bacteria such as Pseudomonas cepacia, Erwinia herbicola and Burkholderia cepacia (Rodriguez and Fraga, 1999), and 2-ketogluconic acid in Rhizobium leguminosarum (Halder *et al.*, 1990), Rhzobium mililoti (Halder and Chakrabarty, 1993) and Bacillus firmus (Banik and Dey, 1982). Some Bacillus species also found to produce mixture of lactic acid, isovaleric acid, isobutyric acid and acetic acid. Other organic acids isolated from phosphate solubilizing bacteria are glycolic acid, oxalic acid, malonic acid, succinic acid citric acid and propionic acid (Illmer and Schinne,1992).

Some PSB also hydrolyze organic form of phosphate compounds efficiently and it is carried out by means of enzyme like phosphatase (Rodriguez and Fraga, 1999), phytase (Richardson and Hadobas, 1997), phosphonoacetate hydrolase (McGrath *et al.*, 1998), D-α-glycerophosphatase (Skrary and Cameron, 1998) and C-P lyase (Ohtake *et al.*, 1996). Bacteria expressing a significant level of acid phosphatases are from the genus Rhizobium, (Abd- Alla, 1994) Enterobacter, Serratia, Citrobacter, Proteus and Klebsiella (Thaller *et al.*, 1995), Pseudomonas (Gugi *et al.*, 1991) and Bacillus (Skrary and Cameron, 1998). Four new strains namely Phyllobacterium myrsinacearum, Rhodococcus erythropolis, Arthrobacter ureafaciens and Delftia sp. are also added in this list by Chen *et al.* (2006). One or both types of PSB have been introduced to agriculture community as phosphate bio-fertilizer and can save 50 per cent of the crop requirement of phosphate fertilizer but its effect on plant growth varies greatly with bacterial strain, host plant compatibility, types and pH of soil and other environmental factors.

Plant Growth Regulating Rhizobacteria

The important plant growth promoting mechanism of rhizobacteria besides biological nitrogen fixation (BNF) and phosphate solubilization is synthesis of phytohormones or plant growth regulating compounds (PGRs). There are five classes of well known PGRs, namely auxines, gibberllines, cytokinins, ethylene and abscisic acid (Zahir *et al.*, 2004) may play regulatory role in growth and developmentof plants. The physiologically most active auxine in plant is indole-3-acetic acid (IAA), which is known to stimulate both rapid and long term responses to plants (Cieiand 1990,Hagen 1990). About 80 per cent of bacteria isolated from rhizosphere can produce IAA as a plant growth regulator (Patten and Grhizoblick 1996), common among them are *Agrobacterium sp., Azospirillum brasilense, Aeromonas veronii, Alcaligenes plechaudii, Bordyrhizobium sp., Comamonas acidovorans, Entrobacter spp. and Rhizobium spp.* (Vessey, 2003), *Acetobacter diazotropnicus* and *Herbaspirillum seropedicae* (Batianet *et al.*, 1998). In addition to IAA some rhizosphere bacteria such as *Paenibacillus polymyxa* and *Azospirilla* release compounds like indole-3-butyric acid, Tryptophan and indole-3-ethanol that can indirectly contribute plant growth promotion (Lebuhn *et al.*, 9197); (El- Khawas and Adachi 1999). Cytokinines are other important phytohormones present in very small amount helps in cell division, root development and root hair formation (Frankenberger and Arshad, 1995). Dobbelacre *et al.* (2003) reported the presence of 89 types of gibberllins (GAs) and are numbered as GA1 through GA89 in approximate order of their discovery (Frankenberger and Arshad 1995). Among 89 gibberllines the GA1 is most widely recognized and GA3 is most active GA of plants responsible for stem elongation. Presence of abscisic acid is also detected from Azospirillum and Rhizobium sp. cultures (Dangar and Basu, 1987; Dobbelacre *et al.*, 2003). Its presence in the rhizosphere helps plants to grow under water stress environment such as arid and semi arid climates (Frankenberger and Arshad, 1995).Ethylene is a potent plant growth regulator that affects many aspects of plant growth particularly stimulates germination and breaks dormancy of seeds (Esashi, 1991). After germination the high level of ethylene is potentially deleterious to plant, it inhibits growth and root elongation and root nodulation, but the presence of PGPR induceses the synthesis of enzyme 1-aminocyclopropane-1-carboxylate (ACC) deaminase that cleaved 1-aminocyclopropane-1-caboxylate(ACC) an immidiate precursor to ethylene during ethylene biosynthesis into ammonia and α-Ketobutyrate which is readily metabolized by PGPR like Alcaligenes sp., Bacillus pumilus, Entrobacter cloacae, Pseudomonas sp. (Vessey 2003). In his way rhizobacteria act as a sink for ACC and promotes plant growth and yield. Some of the important PGRs produced by rhizobacteria is given in the Table 5.1.

Table 5.1: Some of the Plant Growth Regulator Produced by Rhizosbacteria and its Response to Crops

PGPR	*PGRs*	*Crops*	*Responses*	*Reference*
Azotobacter sp.	Indole-3-acitic acid	Maize	Strain efficient in IAA production had growth promoting effects on seedling	Zahir *et al.* (2000)
Azospirillum lipoferum	Indole-3-acitic acid	Wheat	Promote development of root system strains 15 even under crude oil contamination	Muratova *et al.*, 2005
Azospirillum brasilense *Bacillus circulans* P2, *Bacillus* sp. P3, *Bacillus magaterium* P5 *Bacillus sp.* Psd7 *Pseudomonas aeruginosa* Pds5 *Pseudomonas piekeyti* Pds6 *Pseudomonas fluorescens* MTCC103	Indole-3-acitic acid	Rice	All the bacterial strain increased rice A3, A4, A7, A10, CDJA	Thakuria *et al.*, 2004
Bacillus cereus RC 18 *Bacillus licheniformis* RC08	Indole-3-acitic acid	Wheat, spinach	All bacterial strains significantly increased growth of wheat and spinach	Cakmakci *et al.*, 2007
Bacillus subtilis RC11 *Pseudomonas putida* RC06 *Paenibacillus polymixa* RC05 and RC14 *Khuyvera ascorbata* SUD 165	Sidophores	Canola	Decreased inhibitory effect of heavy metals (nickel, lead, zinc) on plant growth	Burd *et al.*, 2000
Rhizobium leguminosum	Indole-3-acitic acid		Rice growth promoting effect on rice seedling	Biswas *et al.*, 2000
Rhizobacterial isolates	Auxins	Wheat	Significant growth promoting effect on wheat and rice	Khalid *et al.*, 2001
Streptomyces acidiscabies	Hydroxamate Sidophores	Cowpea	Promotes growth under nickel stress	Dimkpa *et al.*, 2008

Rhizobacteria as a Biocontrol Agent

Bacteria that reduce the incidence of plant diseases are often referred to as bio-control agent and are one of the potential alternatives for plant disease management. Biological control of soil born plant pathogen and the synthesis of antibiotics have been reported in is the production of several bacterial species (O' Sullivan and O'Gara, 1992). Bacillus subtilis are the most widely used PGPR due to their disease reducing and antibiotic producing capabilities (Brannen and Backman 1994).Some of the common antibiotics produced by PGPR are 2,4-diacelylphlorog lucinol (DAPG), Pyoluteorin, Pyrinolnitrin and Phenazines. Another mechanism by which rhizobacteria can inhibit phytopathogens is hydrogen cyanide (HCN). It (HCN) inhibits the electron transport, disturbing the energy supply to the cells, ultimately leading to the death of the pathogen (Knowels, 1976). Production of fungal cell wall degrading enzymes chitinase and β-1, 3-glucanase (Bloemberg and Lugtenberg, 2001; Persello-Cartieaux *et al.*, 2003) have been reported from some of the phytopathogenic rhizobacteria such as Pseudomonas sp., Bacillus sp., Streptomyces griseoviridies (Haas and Defago, 2005), Agrobacterium, Burknolderia cepacia, Paenibacillus polymyxa, Entrobacter cloacae Streptomyces sp. Serratia marcescens (Van Loonetal, 1998). Some strains of Klebsiella pneumoniae and Yersinia (Chatterjee *et al.*, 1978), and Frankia (Seguin and Lalonde, 1989) are also able to degrade pectin. Inoculation of Pseudomonas fluorescens isolates PGPR1, PGPR2 and PGPR4 reduced the seedling mortality caused by Aspergillus nigger (Dey *et al.*, 2004).

Specific mechanism involved in pathogen suppression by PGPR varies and include antibiotic production, substrate competition and induced systemic resistence in the host plant (Van Loon *et al.*, 1998).Some of the important studies of biological control by PGPR against certain diseases, pathogen in different crops cited in Table 5.2.

Sidophores

In soil, iron is found predominantly as ferric iron, a form that cannot be directly assimilated by micro-organisms. So, majority of micro-organism produced siderophores (Barton *et al.*, 1993, Winkelmann, 1991), a small iron binding molecules that produced in response to iron limitation and combined with iron to form ferric-siderophore complexes. This complex, binds with membrane receptor proteins and get transported to the cell, where it becomes available for metabolic functions, otherwise these plants would develop iron chlorosis and become susceptible to pathogen (Kong, 1980). Siderophores production is very common among various strains of Pseudomonas (O'Sullivan and O'Gara 1992), Frankia (Boyer *et al.*, 1999) and Streptomyces sp.(Loper and Buyer, 1991). Siderophore producing PGPR suppress plant diseases either by siderophore-mediated competition with pathogen by removing iron from the environment (O'Sullivan and O'Gara, 1992;Persello-Cartieaux *et al.*, 2003) or inducing resistance response to the plant through induced systemic resistance (ISR) process.

Table 5.2: Biological Control by some Strains of PGPR against Diseases

PGPR Strains	Crops	Disease/Pathogen	References
Azotobacter sp. *Pseudomonas* sp.	Wheat	Fungal biocontrol	Wachowaska, 2000
Bacillus amyloliquefaciens strain 1 N 937a, *Bacillus subtilis* 1N 937b	Tomato	Tomato motile virus	Murphy *et al.*, 2000
Bacillus sp. *Aphiz gossypii*	Cucumber	Cotton aphids	Glover Stout *et al.*, 2002
Bacillus amyloliquefaciens Strain 1 N 937a	Pepper	Myzus persicae	Kokalis-Burelle *et al.*, 2002
Bacillus subtilis G803			
Bacillus cereus MJ-1	Red pepper	Myzus persicae	Joo *et al.*, 2005.
Bacillus cereus BS 03	Pigeonpea	Fusarial wilt, Fuserium udum	Dutta *et al.*, 2008
Pseudomonas aeruginosa RRIJ04			
Rhizobia			
Pseudomonas aeruginosa	Mungbean	Root rot, root knot	Siddiqui *et al.*, 2000
Bacillus subtilis			
Pseudomonas sp.	Groundnut	Charcoal rot caused by *Rhizoctonia botaticola*	Gupta *et al.*, 2002.
Streptomyces marcescens 90-116	Tobacco	Blue mold	Zhang *et al.*, 2003
Bacillus pumilus SE 34			
Bacillus pumilus T4			
Bacillus pasteurii C-9			
Pseudomonas fluorescencs 89B-61			
Mesorhizobium loti MP6	Mustard	White rot *Sclerotinia sclerotiorum*	Chandra *et al.*, 2007

Conclusion

Studies have shown that plant growth promoting rhizobacteria (PGPR), not only have important roles in nutrient cycling but also protect crops against diseases. According to statistics by FAO about 42 million tons of nitrogen fertilizers is being used annually on a global scale for the production of major cereal crops that is wheat, rice and maize. Crop plants are able to use only 50 per cent of the applied nitrogen fertilizers and remaining 50 per cent lost due to many other factors caused not only the annual loss of US $3 billion but also cause pollution. They are beneficial not only from economical, but also from ecological point of view. They affect growth via N? fixation, P solubilization, phytohormones production that enhances root hair formation, root elongation, respiration, nodulation, nutrient uptake and also help plant in efficient use of water, increases dry weight of plants and render the plant more tolerant to salt stress. They promote plant growth by suppressing the growth of plant pathogens like fungi, bacteria, viruses and nematodes by producing siderophores, HCN, chitinase, ACC deaminase, antibiotics and other volatile metabolites or by induce systemic resistance or competing with the pathogens for nutrient or for colonizing space. Capturing the benefits of soil biological diversity of these rhizobacteria and its functioning for sustainable, value productive agriculture system requires, better understanding of linkage among soil life, biological process and the ecosystem functions.

References

Abd-Alla, M.H. (1994). Solubilization of rock phosphates by *Rhizobium* and *Bordyrhizobium. Microbiol.* **39:** 53-56.

Arihara, Ae.N., Okada, J., Yoshihara. K.T. and Johansen, C. (1990). Phosphorus up take by pigeon pea and its role in cropping Systems of the Indian subcontinent. *Science.* **248:** 477-480.

Ahmad, F., Ahmad, I. and Khan, M.S. (2005). Indole acetic acid production by the indigenous isolates of *Azotobacter* and *Flourescent Pseudomonas* in the presence and absence of tryptophan. *Turk J Biol.* **29:** 29-34.

Al-Niemi, T.S., Khan, M.I. and Mc Dermott, T.R. (1997). P. metabolism in the bean *Rhizobium tropici* symbiosis, *Plant Physiol* **113:** 1233-1242.

Angle, J.S. (1986) Pectic and proteolytic enzymes produced by fast and slow growing soybean *Rhizobia. Soil Biol Biochem.***18:** 115-116.

Anjum, M.A., Sajjid, M.R., Akahtar, N., Qureshi, M.A., Iqbal, A., Jami, A.R. and Hassan, M. (2007). Respons of cotton to plant growth Promoting rhizobacteria (PGPR) inoculation under different levels of nitrogen. *J Agric Res.* **45(2):** 135-143.

Arshad, M., and Frankenberger, W.T.Jr. (1998). Plant growth regulating substances in the rhizosphere. *Microbial production and Function Adv Agron* **62:** 46-51.

Balandreau, J. (2002). The spermosphere model to select for plant growth promoting rhizobacteria. In; Kenned IR, Choudhary ATMA (eds) Biofertilizer in action. Rural Industries Research and Development Corporation, Canbera, pp 55-63.

Banerjee, M.R. and Yasmin, L. (2002). Sulfer oxidizing rhizobacteria an innovative environment friendly soil biotechnological. Tool for better canola production. *Proceedings of AGROENVIRON Cairo,* Egypt, 2002, October 26-29, pp 1-7.

Banik, S. and Dey, B.K. (1982). Available phosphate contents of an alluvial soil is influenced by inoculation of some isolated phosphate solbilizing micro-organisms. *Plant Soil.* **69:** 353-364.

Bashan, Y. (1998). *Azospirillum* Plant growth promoting strains are nonpathogenic on tomato, pepper, cotton and wheat. *Can J Microbiol* **44:** 168-174.

Baldani, V.L.D., Balani, J.I. and Dobereiner, J. (2000). Inoculation of rice plants with the endophytic dizotrophs *Herbaspirillum seropediaceae* and *burkholderia sp. Biol Fertile Soils* 30: 485-491.

Chaintreuil., C., Girand, E., Prin, Y., Lorqin, J., Ba, A., Gillis, M., de Lajudie, P., and Dreyfus, B. (2000), Photosynthetic *Bradyrhizobia* are natural endophytes of the African wild rice *Oryza breviligualata Appl. Environ. Microbial,* **66:** 5437-5447.

Chandler, M.R. (1978). Some observation on the infection of *Arachis hypogaea* L. b Rhizobium. *J. Exp Bot.* **29:** 749-755.

Charon, C., Sousa, C., Crespi, M. and Kondorosi, A. (1999). Alteration of enod 40 expression modifies *Medicago truncatula* root nodule development indused by *Sinorhizobium meliloti, Plant cell* **11:** 1953-1966.

Cameron, R.K, Dixon, R. and Lamb, C. (1994). Biologically indused systemic acquired resistence in *Arabidopsis thaliana.* Plant J.

Chester, K. (1933). The problem of acquired physiological immunity in plants. *Quart. Rev. Biol.* **8:** 275-324.

Chopra, C.L. and SubbaRao, N.S. (1967). Mutual relationship among bacteroids leghaemoglobin and nitrogen content Of Egyptian clover and gram. *Arch. Microbiol.,* **58:** 71-76 (Arch Fur Mickbiol).

Dalton, D.A., Baird, L.M., Langeberg, I., Taugher, C.Y., Anyan, W.R, Vance, C.P. and Sarath, G. (1993) Subcellular Localization of oxygen defense enzymes in soybean.Glycine max (L) Merr. Root nodules. *Plant Phycol.* **102:** 481-489.

Doyle, J.J. (1998). Phylogenetic perspective on nodulation: evolveing views of plants and symbiotic bacteria. Trends. *Plant Sci.* **3:** 473-478.

Dalisay, R.F. and Kuc, J.A. (1995). Persistence of reduced penetration by *Collectotrichum lagenarium* into cucumber leaves with induced systemic resistance and its relation to enhanced peroxidase and chitinase activities. *Physiol. Mol. Plant Pathol.* **47:** 329-338.

De Meyer, G., Audenaert, K. and Hofte, M. (1999). *Pseudomonas aeruginosa* 7NSK2-induced systemic resistance in tobacco depends on in planta salicylic acid but is not associated with PR1 a gene expression. *Eur. J Plant Pathol,* **105:** 513-517.

De Wit, P.J.G.M. (1997). Pathogen a virulence and plant resistance : A key role for recognition. Trends Plant Sci., to : 452-458.

Ferris, H., Venette, R.C., Vander Meulen, H.R. and Lau, S.S. (1998). Nitrogen mineralization by bacterial feeding nimatodes: Verification and measurement. *Plant and Soil.* **203:** 159-171.

Frink, C.R., Waggoner, P.E. and Ausubel, S.H. (1999). Nitrogen fertilizer: retrospect and prospect. *Proc. Natl. Acad. Sci.* USA, **96:** 1175-1180.

Giller, K.E. (2001). Nitrogen Fixation in Tropical Croping Systems. CABI Publishing, Walingford, UK.

Gaffney, T., Firedrich, L., Vernnoij, B., Negrotto, D., Nye, G., Uknes, S., Ward, E., Kessmann, H. and Ryals, J. (1993). Requirement of salicylic acid for the induction of systematic acquired resistance. *Science.* **261:** 754-756.

Geels, F.P. and Schippers, B. (1983a). Selection of antagonistic fluorescent *Pseudomonas spp.* And their root colonization and persistence following treatment of seed potatoes. *J. Phytopathol.* **108:** 193-206.

Hemantaranjan, A. and Garg, O.K. (1986) Introduction of nitrogen fixing nodules through iron and zinc fertilization in non-nodule forming French beans (*phaseolus vulgaris* L). *J. Plant Nutrition* **9:** 281-288.

Hammerschmidt, R. and KuC, J. (1995). Induced resistance to disease in for the induction of systemic acquired resistance. *Science* **261:** 754-756.

Hoffland, E., Bakker, P.A.H.M. and Van loon, L.C. (1997). Multiple disease protection by rhizobacteria that induce systemic resistance reply. *Phytopathology,* **87:**138.

Iswaran, V., Sundara Rao, W.V.B., Mangu, S.P. and Jauhri, K.S. (1969). Indian peat as a carrier of *Rhizobium. Curr. Sci.,***38:** 468-469.

Jeewa, Ram, Jain, K.L. and Lodha, P.C. (2004). Compatibility of seed protectant with *Rhizobium leguminosarum* and its biological nitrogen fixation ability in mungbean. *J. Mycol. Pl. Pathol.,* 34 **(2):** 644-666.

Johnson, J.F., Vance, C.P. and Allan, D.L. (1996). Phosphorus deficiency in *Lupinus albus* altered lateral root development and enhanced expression of phosphoenolpyruvate carboxylase. *Plant Physiol.* **112:** 31-41.

Kandaswamy, R. and Prasad, N.N. (1971). Lignite as a carrier of rhizobia., *Curr. Sci.* **40:** 496.

Kassanis, B., Gianinazzi, S. and Whit, R.F. (1974). A possible explanation of the resistance of virus infected tobacco to second infection. *Journan of General Virology,* **23:** 11-16.

Kloepper, J.W. and Tuzun, S. (1996). Induced systemic to diseases and increased plant growth by growth promoting rhizobacteria under field conditions. *Phytopathology* **81:** 1508-1516.

Kloepper, J.W. (1991). Plant growth promoting rhizobacteria as biological control agents of soil born diseases. In: The Biological Control of Plant Diseases, (Ed.) Bay-Petersen. J. FFTC Book Series No. 42. Food and Fertilizer Technology Center, Taipei, Taiwan.

Knoester, M., Linthorst, H.J.M., Bol, J.F. and Van Loon, L.C. (2001). Involvement of ethylene in lesion development and systemic acquired resistance in tobacco during the hypersensitive reaction to tobacco mosaic virus. *Physiol. Mol. Plant Pathol.,* **59:** 45-57.

Kurz, N.G.W. and La Rue, T.A. (1975). Nitrogenase activity in rhizobia in the absence of host plant. *Nature* London, **256:** 407-409.

Mantelin, S. and Touraine, B. (2004). Plant growth-promoting bacteria and nitrate availability: impacts on root development and nitrate uptake. *J Exp Bot.* **55:** 27-34.

Matiru, V.N. and Dakora, F.D. (2004) Potential use of rhizobial bacteria as promoters of plant growth for increased yield in land races of African cereal crops. *Afr J Biotechnol* 3 **(1):** 1-7.

McGrath, J.W., Hammerschmidt, F., and Quinn, J.P. (1998) Biodegradation of phosphonomycin by *Rhizobium huakuii* PMYI. *Appl Environ Microbiol* **64:** 356-358.

McInroy, J.A. and Kloepper, J.W. (1995). Survey of indigenous endophytes from cotton and sweet corn. *Plant Soil* **173:** 337- 342.

Milton, H.S.J. (2007). Benificial bacteria and bioremediation. *Water Air Soil Pollut* **184:** 1-3.

Mirza, M.S., Rasul, G., Mehmaz, S., Ladha, J.K., So, R.B., Ali, S. and Malik, K.A. (2000) Benificial effects of inoculated nitrogen- fixing bacteria on rice. In: Ladha J K, Reddy P M. (eds) The quest for nitrogen fixation in rice. International Rice Research Institute, Los Banos, pp. 191-204.

Nieto, K.F. and Frankenberger, W.T.Jr. (1990a) Influence of adenine isopentyl alcohol and *Azotobacter chroococcum* on the growth of *Raphanus sativus. Plant Soil* **127:** 147-156.

O'Sullivan, D.J. and O'Gara, F. (1992) Traits of fluorescent *Pseudomonas* spp. Involved in suppression of plant root pathogens. *Micrcbiol Rev.* **56:** 662-676.

Pandey, A. and Kumar, S. (1989). Potential of *Azotobacter* and *Azospirilla* as biofertilizer for upland agriculture: a review. *J Sci. Ind Res.* **48:** 134-144.

Park, K.S. and Kloepper, J.W. (2000) Activation of PR-1a promoter by rhizobacteria which induces systemic resistance in tobacco against *Pseudomonas syringae pv. Tabaco Biol Cont.* **18:** 2-9.

Rao, D.L.N. (2001). B N F. research progress 1996-2000: all India coordinated research project on biological nitrogen fixation. IISS, Bhopal.

Reis, V.M., Baldani, J.I., Baldani, V.L.D. and Doberener, J. (2000) Biological dinitrogen fixation in the graminae and palm trees. *Crit Rev Plant Sci* **19:** 227-247.

Richardson, A.E. (2001) Prospects for using soil microorganisms to improve the acquisition of phosphorus by plants. *Aust J Plant Physiol* **28:** 897-906.

Rodriguez, H., Rossolini, G.M., Gonzalez, T., Jiping, L. and Glick, B.R. (2000) Isolation of a gene from *Burkholderi cepacia* IS-16 encoding a protein that facilitates phosphatase activity. *Curr Microbiol* **40:** 362-366.

Roper, M.M. and Ladha, J.K. (1995). Biological nitrogen fixation by heterotrophic and phototrophic bacteria in association with straw, *Plant Soil* **174:** 211-224.

Saleem, M., Arshad, M., Hussain, S. and Bhatti, A.S. (2007). Perspective of plant growth promoting Rhizobacteria (PGPR) Containing ACC deaminase in stress agriculture. *J Ind Microbial Biotechnol* **34:** 635-648.

Saravanakumar, D., Lavanya, N., Muthumeena, B., Raguchander, T., Suresh, S. and Samiyappan, R. (2008) *Pseudomonas fluorescens* enhances resistance and natural enemy population in rice plants against leaf folder past. *J Appl Entomol* 132 **(6): 469-479.**

Saubidet, M.I., Fatta, N. and Barneix, A.J. (2000) The effects of inoculation with *Azospirillum brasilense* on growth and nitrogen utilization by wheat plant. *Plant Soil* 245 **(2):** 15-222.

Shah, Z., Shah, S.H., People, N.B., Schwenke, G.D. and Hridge, D.F. (2003) Crop residue and fertilizer N effects on nitrogen fixation and yields of legume-cereal rotations and soil organic fertility, *Field Crops Res* **83:** 1-11.

Shaharoona, B., Arshad, M., Zahir, Z.A. and Khalid, A. (2006b). Performance of *pseudomonas spp.* containing ACC-deaminase for improving growth and yield of maize (*Zea mays L*) in the presence of nitrogenous fertilizers. *Soil Biol Biochem* **38:** 2971-2975.

Skrary, F.A. and Cameron, D.C. (1998) Purification and characterization of a *Bacillus licheniformis* phosphatase specific for D- alphaglycerphosphate, *Arch Biochem Biophys* **349:** 27-35.

Spencer, D., James, E.K., Ellis, G.J., Show, J.E. and Sperent, J.I. (1994) Interaction between *Rhizobia* and potato tissue. *J. Exp Bot* **45:** 1475-1482.

Thakuria, D., Taleekdar, N.C., Goswami, C., Hazarika, S., Boro, R.C. and Khan, M.R. (2004) Characterization and screening of bacteria from rhizosphere of rice grown in acidic soils of Assam. *Curr Sci* 86**(7):** 978-985.

Tye, A.Z., Sci, F.K., Leung, T.Y. and Linn, B.L. (2002). Molecular cloning and the biochemical characterization of two novel phytases from *Bacillus subitilis* 168 and *Bacillus licheniformis. Appl Microbiol Biotechnol* **59:** 190-197.

Vndamme, P., Goris, J., Chem, W.M., de Vos P. and Willems, A. (2002) *Burkholderia tuberum sp.* nov and *Burkholderia phymatum sp.* nov, nodulate the roots of tropical legumes. *Syst Appl Microbiol* **25:** 507-512.

Venkateswarlu, B., Hari, K. and Katyl, J.C. (1997) Influence of soil and crop factors on the rhizobia populations in soils under dry land farming. *Appl Soil Ecol* **7:** 1-10.

Webster, G., Gough, C., Basse, J., Batchlor, C.A., O'Callaghan, K.J., Kothari, S.L., Dabey, M.R., Denarie, J. and Cockaing, E.C. (1997). Interactions of *Rhizobia* was rice and wheat plant soil **194:** 115-122.

Yanni, Y.G. and El-Fattah, F.K.A. (1999). Towards integrated biofertilization management with free living and associative dinitrogen fixers for enhancing rice performance in the Nile delta. *Symbiosis* **27:** 319-331.

Yanni, Y.G., Rizk, R.Y., Corich, V., Squartini, A., Ninke, K., Philip, Hollingsworth, S. Orgabide, G., de Bruijn F. Stoltzfus, J. Buckley, D., Schmidt, T.M., Manteos, P.F., Ladha, J.K. and Dazzo, F.B. (1997). Natural entophytic association between *rhizobium leguminosarum bv trifolii* and rice roots and assessment of its potential to promote rice growth. *Plant Soil* **194:** 99-114.

Zahir, A.Z., Arshad, M. and Frankenverger, W.T.Jr. (2004). Plant growth promoting rhizobacteria: application and perspectives in *Agriculture Adv Agron* **81:** 97-168.

Zaidi, S., Usmani, S., Sinha, B.R. and Musarrat, J. (2008). Significance of *Bacillus subtulis* strains SJ-101 as a bioinoculant for concurrent plant growth promotion and nickel accumulation in *Brassica junceu. Chemosphere* **64:** 991-997.

Zhang, S., Moyne, A.L., Reddy, M.S. and Kloepp, J.W. (2002). The role of salicylic acid in induced systemic resistance elicited by plant growth promoting rhizobacteria against blue mold of tobacco *Biol Control* **25:** 288-296.

Zhuang, XL, Chen, Z., Shim, H. and Bai, Z. (2007). New advances in plant growth promoting rhizobacteria for bioremediation. *Environ Int* **33:** 406-413.

Chapter 6

Morphological, Pathogenic and Molecular Diversity in *Rhizoctonia solani* Kühn Causing Sheath Blight of Rice

Vineeta Singh and Prabhat Kumar*

Department of Mycology and Plant Pathology, Institute of Agricultural Sciences, Banaras Hindu University, Varanasi – 221 005, U.P.

Rhizoctonia solani Kühn [teleomorph: *Thanatephorus cucumeris* (Frank) Donk] is a destructive and widespread fungal plant pathogen of rice causing sheath blight (Parmeter, 1970; Ou, 1985). The first description of rice sheath blight was reported from Japan at the beginning of the 20th century (Miyaki, 1910). The disease has been subsequently reported from most rice-growing areas of the world like Bangladesh, China, Colombia, Cuba, Germany, India, Indonesia, Iran, Japan, Korea, Malaysia, Moscow, Netherlands, Nigeria, Philippines, Senegal, Sri Lanka, Taiwan, Thailand, Trinidad and Tobago, UK (Manchester), US (Louisiana, Mississippi) and Vietnam (Jones and Belmar, 1989; Dasgupta, 1992; Linde *et al.*, 2005). It is a polyphagous fungus having wide host range and capable of causing various symptoms like seedling damping off, collor rot, stem cancer, crown rot, bud and fruit rot as well as foliage blight on susceptible variety of agriculturally important crops (Baker, 1970;

* Corresponding Author E-mail: prabhatbhu@gmail.com

Anderson, 1990) like soyabean (Liu and Sinclair, 1992; Doupnik, 1993), cotton (Brown and McCatter, 1976), wheat (Wiseman *et al.*, 1995), beet (Carling *et al.*, 1987), potato (Escande and Echandi, 1991), cabbage (Keinath, 1995) and in turf grass species (Burpee and Martin, 1992; Couch, 1995). The emergence of *R. solani* as an economically important rice pathogen has been attributed to intensification of rice cropping system with the development of a new short stature, high tillering, high yielding varieties, high plant densities and increase in nitrogen fertilization (Gangopadhay and Chakrabarti, 1982; Ou, 1985; Savary *et al.*, 1995). Sheath blight starts during the maximum growth stage of the rice crop. In Japan, this disease causes a yield loss of 20 per cent and in the United States, a yield loss of 50 per cent were reported in susceptible cultivars (Prasad and Eizenga, 2008). In China, sheath blight disease affects about 15–20 million hectares and causes a yield loss of 6 million tons of rice grains per year (Ou, 1985; Ren *et al.*, 2001). The studies at IRRI showed that sheath blight causes a yield loss of 6 per cent in Tropical Asia (IRRI updates, 2008). In India grain yield loss varied from 4.9 per cent in variety Phalguna to 69 per cent in variety Mahsuri (Naidu, 1989). Rice sheath blight caused by *R. solani* belonging to anastomosis group, AG1-IA has become increasingly important in most rice production regions because of the use of high yielding cultivars with large number of tillers and more frequent application of nitrogen fertilizer (Cu *et al.*, 1996; Savary *et al.*, 1995). Many workers have been made attempt to classify *R. solani* isolates into various groups on the basis of morphological, physiological, pathologic characterization (Sherwood, 1969; Parmeter and Whitney, 1970; Vijayan and Nair, 1985; Jones and Belmar. 1989; Banniza *et al.*, 1996 ; Singh, 2009), anastomosis behaviour (Parmeter *et al.*, 1969; Ogoshi, 1987), intra- and extra-cellular enzyme, proteins (Matsuyama *et al.*, 1978; Liu and Sinclair, 1993), and cellular fatty acids (Jobaji- Hare *et al.*, 1990; Vilgalys and Gonzalez, 1990; Liu and Sinclair, 1993; Duncan *et al.*, 1993; Banniza *et al.*, 1996; Yang *et al.*, 1996 and Bounou *et al.*, 1999). Despite the high yield losses caused by *R. solani* AG1-IA on rice crops throughout the world, the population biology of the fungus is poorly understood. Information about the genetic structure of populations of *R. solani* AG1-IA, its reproductive mode, and the patterns of migration in natural populations of the pathogen has been reviewed by several workers (Rosewich *et al.*, 1999; Ciampi *et al.*, 2005; Linde *et al.*, 2005; Bernardes-de-Assis *et al.*, 2008, Ciampi *et al.*, 2008). Knowledge about the population genetic structure of a pathogen provides insights into the evolutionary processes that have shaped population and also can elucidate the evolutionary potential of population subjected to different control strategies such as pesticide applications, resistance gene deployment, and cultural practices (McDermott and McDonald, 1993; McDonald, 1997; Stukenbrock and McDonald, 2008). In present chapter the work done on characterization of *R. solani* causing sheath blight of rice by using morphological or cultural, pathological and molecular approaches has been reviewed.

Morphological Characterization

The cultural or morphological characteristics of *R. solani* isolates have been studied by several workers (Thakur *et al.*, 1992; Singh *et al.*, 1999; Meena *et al.*, 2001; Guleria *et al.*, 2007; Singh, 2009) on the potato dexrose agar (PDA) medium with different mycelial and sclerotial characters have been observed like abundance of mycelium (abundant, moderate and slight), colony colour (white to brown), growth on medium (fast, moderate and slow), sclerotial colour (Dark brown, medium brown, light brown and off white), size of sclerotia (micro sclerotia powdery and poorly melanised or micro sclerotia unmelanized and diffused myceloids), distribution pattern of sclerotia on petri-plate (uniformly distributed throughout the plate, near inoculation point form the ring and present on the periphery of the plate), location of sclerotia (arial, present on surface of medium and embedded inside the medium), sclerotial aggregation or separation (aggregate, separate and semi-aggregate), anatomically differentiation of sclerotia (differentiated or undifferentiated), sclerotial weight (micro and medium) which can be used for characterization of *R. solani* Isolates. Teheri *et al.* (2007) also characterized *R. solani* isolates in two groups as either binucleated or multinucleated by using trypan blue staining techniques described by Matrin and Lucas (1984).

Anastomosis Behaviour

The anastomosis grouping based on hyphal fusion is considered as an easiest method of characterizing variability in *R. solani* (Richter and Schneider, 1953). Anderson (1982) stated that hyphal anastomosis in *R. solani* is a manifestation of somatic or vegetative incompatibility between hyphae. Yokoyama and Ogoshi (1984) conducted an experiment to show that when isolates of *R. solani* are paired 2-3 cm apart on 2 per cent water agar medium, their mycelia grow and overlap, which can be observed under a light microscope at low magnification. If hyphal fusion occurs, these isolates belong to the same anastomosis group and often, attraction of hyphae and death do not occur, the isolates belong to different anastomosis group (AGs). Among the 14 anastomosis groups (AGs) that have been described in *R. solani* to date (Ogoshi, 1976; Ogoshi, 1987; Carling, 1996), isolates of AG1-1A have been associated with rice sheath blight pathogen (Gangopadhay and Chakrabarti, 1982; Banniza *et al.*, 1999). Several anastomosis groups are further subdivided into intra-specific groups (ISGs). Grouping isolates of *R. solani* by anastomosis reaction or by anastomosis reaction in combination with other methods of grouping produces subunits of *R. solani* that are sometimes called intra-specific groups (ISGs). Ogoshi (1987) characterized anastomosis and intraspecific groups of *R. solani* on the basis of pathological, morphological, vitamin requirement, serological, isoenzyme and DNA-base differences. He discussed the ecology and pathogenicity of individual groups. Isolates belonging to AG1 has been divided into three intra-specific groups, including 1A, 1B, and 1C based on host origin, symptoms and cultural characteristics (Ogoshi, 1987; Sneh *et al.*, 1991; Liu and Sinclair, 1993). The subgroup AG1- IA is one of the most important plant pathogens worldwide, causing foliar diseases on maize, rice, and soybean (Jones and Belmar, 1989). Although many attempts were made to characterize *R. solani* in several AG types, a report of hyphal anastomosis involving more than one AG and the occasional loss of anastomosis ability complicates the identification and characterization of AG types (Kuninaga *et al.*, 1979; Neate and Warcup, 1985). Anastomosis grouping is a convenient but not the ideal method for classification of *R. solani* as misidentification is frequent because of the varying frequency of hyphal fusion (Neate and Warcup, 1985). This method also does not offer reliable information on genetic variation or taxonomic relationships within and between AGs and sometimes isolates within an AG are found to be more similar than between AGs and even genetically distinct subgroups exist within AGs (Kuninaga and Yokosawa, 1985; Sneh *et al.*, 1991; Keijer *et al.*, 1996).Within the species complex, classification has been based on hyphal anastomosis grouping (Carling *et al.*, 2002), each AG is considered an independent evolutionary entity or phylospecies (Cubeta and Vilgalys, 1997).

Pathological Characterization

Rhizoctonia solani Kühn pathogenicity is complex; it has heterogeneous strains and diversity in host range. It can damage any part or all of a plant (Mubarak, 2003). *R. solani* penetrates plants in various ways: through the intact plant surface by means of complex infection structures (infection cushions), which are characteristic of different isolates, through natural openings and through wounds. R. solani may also penetrate the host mechanically or by means of enzymes or toxins (Dodman and Flentje, 1985). *R. solani* produces cutinolytic enzymes (Linskens and Haage, 1993), which could degrade the cuticle-like pectinases and cellulases, and then helps the pathogen to penetrate the host (Baker and Walker, 1962). The formation of infection cushion and lobate appressorium are also known to lead to host infection (Matsuura, 1986; Murray, 1982). Induction of infection cushions by R. *solani* as a response to root exudates has been reported by many researchers (Doman *et al.*, 1968; Kamara *et al.*, 1980; El-Samra *et al.*, 1981; Stockwell & Hanchey, 1984; El- Faham and Aboshosha, 1987; and El-Farnawany, 1991). Kamara *et al.* (1980) suggested 5 types of infection cushions and a correlation was established between the types of infection cushions developed and host resistance (El-Samra *et al.*, 1981). Hyakumachi *et al.* (1987) showed that, in some AGs of *R. solani*, melanin accumulated in hyphae by incubating fungal culture for 6 weeks. Particularly in AG 2-2 of *R. solani*, melanin biosynthesis was correlated with the anastomosing ability of hyphae and the ability to grow in soil (Hyakumachi and Ui, 1987). The loss of ability to grow in soil, which resulted from non-production of melanin, was correlated with the decrease of pathogenicity in seedlings and mature roots of sugarbeet. However, in some plant pathogenic fungi such as *Magnaporthe grisea* and *Colletotrichum lagenarium*, melanization of appresorium was also reported as an important factor to pathogenicity (Chida and Sisler, 1987a&b; Kubo *et al.*, 1984; Suzuki *et al.*, 1982; Woloshuk *et al.*, 1982; Yamaguchi *et al.*, 1983). Studies with melanin-deficient mutants of *M. grisea* or *C. lagenarium* showed that those mutants were not able to cause typical disease by failure in penetration to host epidermis. Melanization of mycelia of *R. solani* is an important pathogenicity factor in rice (Kim *et al.*, 2001). The isolates producing melanin (M+) on PDA medium with dark melanin showed strong pathogenicity to rice and self anastomosis. Meanwhile, M-type with white or less-melanized mycelia showed very weak pathogenicity and non-self anastomosis. Melanin production of *R. solani* was affected by incubation temperature in both M+ and M– types, but not by light treatment.

Molecular Characterization

In the absence of stable morphological, physiological characteristics (Mordue *et al.*, 1989) and anastomosis groups (Kuninaga *et al.*, 1979; Neate and Warcup, 1985), the identification and genetic characterization of *R. solani* isolates has proved to be very difficult and tedious. Molecular markers are used as important tools for the characterization of genetic diversity in the pathogens where morphological characteristics are either absent or not able to differentiate isolates properly (Vilgalys and Cubeta, 1994; Sharma *et al.*, 1999). Moreover, morphological characters are also influenced by the environmental and cultural conditions. Therefore, problems associated with studying different levels of genetic diversity in *R. solani* have been suggested to be best addressed by the use of molecular techniques (Toda *et al.*, 1999). The simplified molecular techniques like PCR-RFLP of the rDNA internal transcribed spacer (ITS) region, random amplified polymorphic DNA (RAPD), enterobacterial repetitive consensus (ERIC) analysis, simple sequence repeats (ISSR) and universal rice primers (URPs) have been utilized for rapid detection of variation in *Rhizoctonia* spp. and other fungi (Williams *et al.*, 1990; Liu *et al.*, 1993; Manaut *et al.*, 1998; Fenille *et al.*, 2002 & 2003; Neeraja *et al.*, 2002; Singh *et al.*, 2002; Guleria *et al.*, 2007; and Khodayari *et al.*, 2009). The use of RFLP in DNA analysis, which focuses on rDNA, has also been successfully utilized. Ribosomal RNA genes in fungi are conserved and contain sequence components reflecting different evolutionary rates which are phylogenetically and taxonomically important (Bruns *et al.*, 1991; Liu & Sinclair, 1993). Furthermore, the variation observed in the length of PCR fragments of the rDNA of ITS region was not random and could be used as a grouping characteristic (Liu *et al.*, 1993). The use of RAPD markers has great potential in population analysis (Welsh & McClelland, 1990; Williams *et al.*, 1990). The RAPD technique was successfully utilized by Duncan *et al.* (1993) in the analysis of variation of several AGs of *R. solani*. Aye and Matsumoto (2010) reported that 44 isolates of *Rhizoctonia solani*, 30 isolates of *R. oryzae* and 29 isolates of *R. oryzae-sativae* were collected from different regions of Myanamar were differentiated into two types of *R. solani* AG1, two types of *R. oryzae* and three types of *R. oryzae-sativae* by Repetitive- element Polymerase Chain Reaction (Rep-PCR) using the BOXA1R (5'-CTACGGCAAGGCGACGCTGACG-3') and ERIC2 primers (5'-CTACGGCAAGGCGACGCTGACG-3'). Yang *et al.* (1996) who have reported the considerable variation within *R. solani* AG-9 anastomosis group based on RAPD markers. Singh *et al.* (2002) have also reported that most of the microsclerotia- forming isolates were grouped together by using RAPD analysis. While testing the utility of three marker systems used in this study in terms of their efficiency in detecting polymorphisms among the isolates found that URP markers were more efficient when compared with the other two, *i.e.* RAPD and ISSR markers. These markers are more robust and specific because of long primers and high annealing temperature compared with the RAPD and ISSR markers (Kang *et al.*, 2002). These markers can be used for further analysis of more number of *R. solani* isolates collected from different epidemiological regions. UPGMA-SAHN clustering analysis for RAPD finger printing data of 30 haplotypes of *R.solani* AG1-IA isolates from the Philippines and Japan resolved seven groups of AG1-IA at 75 per cent similarity level. In PCR-RFLP analysis of the rDNA ITS region, no polymorphism was observed among the AG1-IA isolates but they were differentiated from subgroups AG1-IB and AG1-IC using endonucleases EcoRI, MboI and Hinf I (Pascual *et al.*, 2000).

Conclusion

Many problems associated with studying different levels of diversity in *Rhizoctonia* are best addressed through the use of molecular genetic markers. Molecular approaches have been very fruitful for answering basis question about genetic variation and systematic relationship in many grounds of phytopathogenic fungi. At higher taxonomic levels, molecular data can provide an inexhaustible source of information for determining phylogenetic and establishing predictive classification systems; at the species level, molecular markers aid in the development of species concepts by providing information about the limits of genetically isolated groups in relation to patterns of morphological variation and mating behaviours; at the population level, molecular markers provide a basis for identifying patterns of inheritance, dispersal and colonization in relation to spatial and temporal distribution. Considerable progress has been possible by recognizing ISG as fundamental taxonomic units within different groups in *Rhizoctonia.* Numerous studies have demonstrated the utility of AG and ISG concepts as indicators of genetic groupings. The most convincing validation of AG and ISG concepts has come from molecular studies. The variety of tools being applied towards *Rhizoctonia* includes isozymes, analysis of restriction fragment length polymorphisms (RFLP) the polymerase chain reaction (PCR) and DNA sequencing (Vigalys and Cubeta, 1994).

References

Anderson, T.F. (1990). A study of hyphal morphology in the form genus *Rhizoctonia. Mycotoxon,* **37:** 26-46.

Aye, S.S. and Matsumoto, M. (2010). Genetic characterization by rep-pcr of Myanmar isolates of *Rhizoctonia* spp., causal agents of rice sheath diseases. *Journal of Plant Pathology,* **92 (1):** 255-260.

Baker, K.F. (1970). Type of *Rhizoctonia,* diseases and their occurrence. In: *Rhizoctonia solani Biology and Pathology* (Ed. J.R. Parmeter), University of California Press, Berkley, USA, pp. 124–148.

Banniza, S., Rutherford, M.A., Bridge, P.D., Holderness, M. and Mordue, J.E.M. (1996). Biological characterization of *Rhizoctonia solani* in rice based cropping system. *In: Proc. Brighton Crop Protection Conference: Pest and Diseases,* Vol.1. The British Crop Protection Council, Fornham, UK, pp 399-404.

Banniza, S., Sy, A.A., Bridge, P.D., Simons, S.A., and Holderness, M. (1999). Characterization of population of *Rhizoctonia solani* in paddy rice field in Cote d' Ivoire. *Phytopathology,* **89:** 414-420.

Baker, K.R. and Walker, J.C. (1962). Relationship of pectolytic and cellulytic enzyme production by strains *Pellicularia filamentosa* to their pathogenicity. *Phytopathology,* **52:** 1119-1121.

Bernardes-de-Assis, J., Peyer, P., Zala, M., Rush, M., McDonald, B.A. and Ceresini, P.C. (2008). Divergence between sympatric rice- and soybean-infecting populations of *Rhizoctonia solani* AG-1 IA. *Phytopathology,* **98:** 1326-1333.

Bounou, S., Jabaji Hare, S.H., Hougue, R., and Charest, P.M. (1999). Polymerase chain reaction based assay for specific detection of Rhizoctonia solani AG-3 isolates. *Mycological Research*, **103:** 1–8.

Brown, E.A. and McCarter, S.M. (1976). Effect of seedling disease caused by *Rhizoctonia solani* on subsequent growth and yield of cotton. *Phytopathology*, **66:** 111-115.

Bruns, T.D., White, T.J. and Taylor, J.W. (1991). Fungal molecular systematics. *Annual Review of Ecology and Systematics*, **22:** 525–564.

Burpee, L. and Martin, B. (1992). Biology of *Rhizoctonia* species associated with turfgrasses. *Plant Dis.*, **76:** 112-117.

Carling, D.E. (1996). Grouping in Rhizoctonia solani by hyphal anastomosis. In: *Rhizoctonia Species: Taxonomy, Molecular Biology, Ecology, Pathology and Disease Control* (Eds. B. Sneh, S. Jabaji-Hare, S. Neate, and G. Dijst), Kluwer Academic Publishers, Dordrecht, Netherlands. pp. 35–48.

Carling, D.E, Leiner, R.H. and Kebler, K.M. (1987). Characterization of a new anastomosis group (AG-9) of *Rhizoctonia solani*. *Phytopathology*, **77:** 1609–1612.

Carling, D.E., Kuninaga, S. and Brainard, K.A. (2002). Hyphal anastomosis reactions, rDNA-internal transcribed spacer sequences, and virulence levels among subsets of *Rhizoctonia solani* anastomosis group-2 (AG-2) and AG-BI. *Phytopathology*, **92:** 43-50.

Chida, T. and Sisler, H.D. (1987a). Restoration of appressorial penetration ability by melanin precursors in *Pyricularia oryzae* treated with antipenetrants and in melanin deficient mutants. *J. Pestic. Sci.*, **12:** 49-55.

Chida, T. and Sisler, H.D. (1987b). Effect of inhibitors of melanin biosynthesis on appressorial penetration and reductive reactions in *Pyricularia oryzae*. *Pestic. Biochem. Physiol.*, **29:** 244-251.

Ciampi, M.B., Kuramae, E.E., Fenille, R.C., Meyer, M.C., Souza, N.L. and Ceresini, P.C. (2005). Intraspecific evolution of *Rhizoctonia solani* AG-1 IA associated with soybean and rice in Brazil based on polymorphisms at the ITS-5.8S rDNA operon. *Eur. J. Plant Pathol.*, **113:** 183-196.

Ciampi, M.B., Meyer, M.C., Costa, M.J.N., Zala, M., McDonald, B.A. and Ceresini, P.C. (2008). Genetic structure of populations of *Rhizoctonia solani* anastomosis group-1 IA from soybean in Brazil. *Phytopathology*, **98:** 932-941.

Couch, H.B. (1995). *Diseases of turf grass* (3rd ed.), Kreiger Publishing, Malabar, Florida.

Cubeta, M.A., and Vilgalys, R. (1997). Population biology of the *Rhizoctonia solani* complex. *Phytopathology*, **87:** 480-484.

Cu, R.M., Mew, T.W., Cassman K.G. and Teng, P.S. (1996). Effect of sheath blight on yield in tropical, intensive rice production system. *Plant Dis.*, **80:** 1103-1108.

Dasgupta, M.K. (1992). Rice sheath blight. The challenge continues. In: *Plant Diseases of International Importance: Disease of Cereals and Pulses.* Vol. I. (Eds. U.S. Singh, A.N. Mukhopadhyay, J. Kumar, and H.S. Chaube), Prentice Hall, Englewood Cliffs, New Jersey pp. 130-150.

Dodman, R.L. and Flentije, N.T. (1985). The mechanism and physiology of plant penetration by *Rhizoctonia solani*. In: *Rhizoctonia solani, Biology and Pathology* (Ed. J.R. Parameter Jr.), Univ. California Press, Berkeley, pp. 147-160.

Doman, R.L., Baker, K.R. and Walker, J.C. (1968). A detailed study of the different mode of penetration by *Rhizoctonia solani*. *Phytopathol.*, **58:** 1271-1276.

Doupnik, B. (1993). Soybean production and disease loss estimates for north central United States from 1989 to 1991. *Plant Dis.*, **77:** 1170-1171.

Duncan, S., Barton, J.E. and O'Brien P.A. (1993). Analysis of variation in isolates of *Rhizoctonia solani* by random amplified polymorphic DNA assay. *Mycological Research*, **97:** 1075–82.

El-Faham, Y.M. and Aboshosha, S.S. (1987). Formation of infection cushion by *Rhizoctoni solani* Kohn in relation to host isolate compatibility. *Com. In. Sci. & Dev. Res.*, **19:** 225-245.

El-Farnawany, M.A. (1991). Studies on host parasite interactions during infection of some important vegetable crops with different isolates of *Rhizoctoni solani*. PhD Thesis Alexandria Univ. pp. 156.

El-Samra, I.A., El-Faham, Y.M. and Kamara, A.M. (1981). Selective induction of infection cushions by *Rhizoctonia solani* in relation to host responses. *Phytopathol.*, **102:** 122-126.

Escande, A.R. and Echandi, E. (1991). Protection of potato from Rhizoctonia cancker with binucleated *Rhizoctonia* fungi. *Plant pathol.*, **40:** 197-202.

Fenille, R.C, Ciampi, M.B., Kuramae, E.E. and Souza, N.L. (2003). Identification of *Rhizoctonia solani* associated with soybean in Brazil by rDNA-ITS sequences. *Fitopatologia Brasileira*, **28:** 413–419.

Fenille, R.C, Souza, N.L. and Kuramae, E.E. (2002). Characterization of *Rhizoctonia solani* associated with soybean in Brazil. *European Journal of Plant Pathology*, **108:** 783–792.

Gangopadhyay, S. and Chakrabarti, N.K. (1982). Sheath blight of rice. *Rev. Pl. Pathol.*, **61:** 451-460.

Guleria, S., Aggarwal, R., Thind, T.S. and Sharma, T.R. (2007). Morphological and pathological variability in rice isolates of *Rhizoctonia solani* and molecular analysis of their genetic variability. *Journal of Phytopathology*, **155:** 654-661.

Hyakumachi, M. and Ui, T. (1987). Non-self-anastomosing isolates of *Rhizoctonia solani* obtained from fields of sugarbeet monoculture. *Trans. Br. Mycol. Soc.*, **89:** 155-159.

Hyakumachi, M., Yokoyama, K. and Ui, T. (1987). Role of melanin in susceptibility and resistance of *Rhizoctonia solani* to microbial lysis. *Trans. Br. Mycol. Soc.*, **89:** 27-33.

IRRI updates (2008). International Rice Research Institute, *www.irri.org.in*

Jabaji-Hare, S.H., Meller, Y., Gill, S. and Charest, P.M. (1990). Investigation of genetic relatedness among anastomosis groups of *Rhizoctonia solani* using cloned DNA probes. *Canadian Journal of Plant Pathology*, **12:** 393–404.

Jones, R.K. and Belmar, S.B. (1989). Characterization and pathogenicity of *Rhizoctonia* spp. isolated from rice, soybean, and other crops grown in rotation with rice in Texas. *Plant Dis.*, **73:** 1004-1010.

Kamara, A.M., El-Samra, I.A. and El-Faham, Y.M. (1980). Host pathogen interaction of *Rhizoctonia solani* Kuhn as a function of chemical stimulants in host root exudates. *Phytopath. Medit.* **70:** 1167-1172.

Kang, H.W., Go, S.J. and Eun, M.Y. (2002). Fingerprinting of diverse genomes using PCR with universal rice primers (URPs) generated from repetitive sequence of Korean weedy rice. Mol. Cells, 13**(2):** 281–287.

Keijer, J., Houterman, P.M., Dullemans, A.M. and Korsman, M.G. (1996). Heterogeneity in electrophoretic karyotype within and between anatomosis groups of *Rhizoctonia solani. Mycol. Res.*, **100:** 789–797.

Keinath, A.P. (1995). Relationships between inoculum density of *Rhizoctonia solani*, wirestem incidence and severity, and growth of cabbage. *Phytopathology*, **85:** 1487-1492.

Khodayari, M., Safaie, N. and Shamsbakhsh, M. (2009). Genetic diversity of Iranian AG1-IA isolates of *Rhizoctonia solani*, the cause of rice sheath blight, using morphological and molecular markers. *Journal of Phytopathology*, **157:** 708-714.

Kim, H.T., Chung, Y.R. and Cho, K.Y. (2001). Mycelial melanisation of *Rhizoctonia solani* AG1 affecting pathogenicity in rice. *Plant Pathol. J.*, 17 **(9):** 210-215.

Kubo, Y., Furusawa, I. and Yamamoto, M. (1984). Regulation of melanin biosynthesis during appressorium formation in *Colletotrichum lagenarium. Exp. Mycol.*, **8:** 364-369.

Kuninaga, S. and Yokosawa, R. (1985). DNA base sequence homology in *Rhizoctonia solani* Kuhn. VI. Genetic relatedness among seven anastomosis groups. *Ann. Phytopathol. Soc. Jpn.*, **51:** 127–132.

Kuninaga, S., Yokosawa, R. and Ogoshi, A. (1979). Some properties of anastomosis groups 6 and BI in *Rhizoctonia solani* Kuhn. *Ann. Phytopathol. Soc. Jpn.*, **45:** 207–217.

Linde, C.C., Zala, M., Paulraj, R.S.D., McDonald, B.A. and Gnanamanickam, S.S. (2005). Population structure of the rice sheath blight pathogen *Rhizoctonia solani* AG-1 IA from India. *Eur. J. Plant Pathol.*, **112:** 113-121.

Linskens, H.F. and Haage, P. (1993). Cutinase–Nachweis in phytopathogenen. *Pilzen. Phytopathol. Z.*, **48:** 306-311.

Liu, Z.I. and Sinclair, J.B. (1992). Genetic diversity of *Rhizoctonia solani* anastomosis group 2. *Phytopathology*, **82:** 778-787.

Liu, Z.L. and Sinclair, J.B. (1993). Differentiation of intraspecific groups within anastomosis group-1 of Rhizoctonia solani using ribosomal DNA internal transcribed spacer and isozyme comparisons. *Canadian Journal of Plant Pathology,* **15:** 272–280.

Liu, Z.L., Domier, L.L. and Sinclair, J.B. (1993). ISG-specific ribosomal DNA polymorphism of the *Rhizoctonia solani* species complex. *Mycologia,* **85:** 795–800.

Manaut, F., Hamaide, N., Vander Stappen, J. and Maraite, H. (1998). Genetic relationships among isolates of *Colletotrichum gloeosporioides* from *Stylosanthes* spp. in Africa and Australia using RAPD and ribosomal DNA markers. *Plant Pathology,* **47:** 641–648.

Martin, S.B. and Lucas, L.T. (1984). Characterization and pathogenicity of *Rhizoctonia* spp. and binucleate *Rhizoctonia*-like fungi from turfgrasses in North Carolina. *Phytopathology,* **74:** 170–175.

Matsuyaman., Moromizatoz., Ogoshi, A. and Wakimotos (1978). Grouping *Rhizoctonia solani* Kiihn with non-specific esterase zymogram. *Annals of the Phytopathological Society of Japan,* **44:** 652-658.

Matsuura, K. (1986). Scanning electron microscopy of the infection process of *Rhizoctonia solani* in leaf sheaths of rice plants. *Phytopathology,* **76:** 811-814.

McDermott, J.M., and McDonald, B.A. (1993). Gene flow in plant pathosystems. *Annu. Rev. Phytopathol.,* **31:** 353-373.

McDonald, B.A. (1997). The population genetics of fungi: tools and techniques. *Phytopathology,* **87:** 448-453.

Menna, B., Ramamoorthy, V. and Muthusamy, M. (2001). Morphological and pathological variations in isolates of *Rhizoctonia solani* causing sheath blight of rice. *Plant Dis. Res.,* **16:** 166-172.

Miyaki, I. (1910). Studies uber die reisptlance in Japan. *Jl. Cell. Agric. Imp. Univ.,* **2:** 237-276.

Mordue, J.E.M., Currah, R.S. and Bridge, P.D. (1989). An integrated approach to *Rhizoctonia* taxonomy cultural, biochemical and numerical techniques. *Mycol. Res.,* **92:** 78-90.

Mubarak, H.M. (2003). Control of root rot and damping-off disease of bean. PhD Thesis Tanta Univ. Botany Depart. Egypt.

Murray, D.I.L. (1982). Penetration of barley root and coleoptiles surfaces by *Rhizoctonia solani. Trans. Br. Mycol. Soc.,* **79:** 354-361.

Naidu, V.D. (1989). Influences of sheath blight of rice on grain and straw yield in some popular local varieties. *Journal of Research, Assam Agricultural University,* **10:** 78-80.

Neate, S.M. and Warcup, J.H. (1985). Anastomosis grouping of some isolates of *Thanatephorus cucumeris* from agricultural soils in South Australia. *Transactions of the British Mycological Society,* **85:** 615–20.

Neeraja, C.N., Vijayabhanu, N., Shenoy, V.V., Reddy, C.S. and Sarma, N.P. (2002). RAPD analysis of Indian isolates of rice sheath blight fungus *Rhizoctonia solani. Journal of Plant Biochemistry and Biotechnology,* **11:** 43-48.

Ogoshai, A. (1976). Studies on the grouping of *Rhizoctonia solani* Kuhn with hyphal anastomosis and on the perfect stages of groups. *Bulletin of the National Institute of Agricultural Sciences, Series C*, **30:** 1-63.

Ogoshi, A. (1987). Ecology and pathogenicity of anastomosis and intraspecific groups of *Rhizoctonia solani* Kuhn. *Annu. Rev. Phytopathol.*, **25:** 125–143.

Ou, S.H. (1985). *Rice disease* (2nd ed.), Common Wealth Mycological Institute, Kew, pp. 280-282.

Parmeter, J.R. and Whitney, H.S.Jr. (1970). Taxonomy and nomenclature of the perfect state. In: *Rhizoctonia solani Biology and Pathology* (Ed. J.R. Parmeter), University of California Press, Berkley, USA, pp. 6–19.

Parmeter, J.R. Jr., Sherwood, R.T. and Platt, W.D. (1969). Anastomosis grouping among isolates of *Thanatephorus cucumeris*. *Phytopathology*, **59:** 1270–1278.

Pascuala, C.B. Toda, T., Raymondo, A.D. and Hyakumachi, M. (2000). Characterization by conventional techniques and PCR of *Rhizoctonia solani* isolates causing banded leaf sheath blight in maize. *Plant Pathology*, **49:** 108–118.

Prasad, B. and Eizenga, G.C. (2008). Rice sheath blight disease resistance identified in *Oryza* spp. accessions. *Plant Dis.*, **92:** 1503–1509.

Ren, C.M., Gao, B.D., He, Y.C. (2001). Progress of research in rice sheath blight. *Plant Protection*, **27:** 32-36.

Richter, H. and Schneider, R. (1953). Untersuchungen zur morphologischen und biologischen von *Rhizoctonia solani* Kuhn. *Phytopathology*, **20:** 117-226.

Rosewich, U.L., Pettway, R.E., McDonald, B.A. and Kistler, H.C. (1999). High levels of gene flow and heterozygote excess characterize *Rhizoctonia solani* AG-1 IA (*Thanatephorus cucumeris*) from Texas. *Fungal Genet. Biol.*, **28:** 148-159.

Savary, S., Madden, L.V., Zadoks, J.C. and Klein-Gebbinck, H.W. (1995). Use of categorical information and correspondence analysis in plant disease epidemiology. In: *Advances in Botanical Research incorporating Advances in Plant Pathology*, Vol 21, Academic Press Ltd. London, pp. 213-240.

Sharma, T.R., Prachi and Singh, B.M. (1999). Application of polymerase chain reaction in phytopathogenic microbes. *Indian J. Microbial.*, **59:** 79-91.

Sherwood, R.T. (1969). Morphology and physiology in four anastomosis groups of *Thanatephorus cucumeris*. *Phytopathology*, **59:** 1924–1929.

Singh, V., Singh, M., Singh, U.S. and Kumar, A. (2002). Genetic diversity of *Rhizoctonia solani* isolates from rice: differentiation by morphological characteristics, pathogenicity, anastomosis behaviour and RAPD fingerprinting. *Journal of Mycology and Plant Pathology*, **32:** 332-344.

Singh, A., Singh, U.S., Willocquet L. and Savary, S. (1999). Relationship among cultural/morphological characteristics, anastomosis behaviour and pathogenicity of *Rhizoctinia solani* Kuhn on rice. *J. Mycol. Plant Pathol.*, **29:** 306-316.

Singh, V. (2009). Genetic diversity in Rhizoctonia solani isolates causing sheath blight of rice. *Journal of Scientific research,* **53:** 103-115.

Sneh, B., Bupree, L. and Ogoshi, A. (1991). *Identification of Rhizoctonia species.* The American Phytopathological society, St Paul, MN, USA.

Stockwell, V. and Hanchey, P. (1984). The role of the cuticle in the resistance of bean to *Rhizoctonia solani. Phytopathol.,* **74:** 1640- 1642.

Stukenbrock, E.H. and McDonald, B.A. (2008). The origins of plant pathogens in agro-ecosystems. *Annu. Rev. Phytopathol.,* **46:** 75-100.

Suzuki, K., Kubo, Y., Furusawa, I., Ishida, N. and Yamamoto, M. (1982). Behavior of colorless appressoria in an albino mutant of *Colletotrichum lagenarium. Can. J. Microbiol.,* **28:** 1210-1213.

Taheri, P., Gnanamanickam, S. and Hofte, M. (2007). Characterization, genetic structure, and pathogenicity of *Rhizoctonia* spp. associated with rice sheath diseases in India. *Phytopathology,* **97:** 373-383.

Thakur, R.S., Sugha, S.K. and Sharma, B.M. (1992). Morphological grouping of different isolates of *Rhizoctonia solani* Kuhn. *Plant Dis. Res.,* **7:** 58-59.

Toda, T., Hyakumachi, P. and Arora D.K. (1999). Genetic relatedness among and within different *Rhizoctonia solani* anastomosis groups as assessed by RAPD, ERIC and REP-PCR. *Microbiological Research,* **154:** 247-258.

Vijayans, R. and Nair, C.M. (1985). Anastomosis grouping of isolates of *Rhizoctonia solani* Kuhn [*Thanatophorus cucumeris* (Frank) Doonk] causing sheath blight of rice. *Current Sciences,* **54:** 289-291.

Vilgalys, R. and Cubeta, M.A. (1994). Molecular systematics and population biology of *Rhizoctonia. Annu. Rev. Phytopathol.,* **32:** 135-155.

Vilgalys, R. and Gonzalez, D. (1990). Ribosomal DNA restriction fragment length polymorphisms in *Rhizoctonia solani. Phytopathology,* **80:** 151–158.

Welsh, J.H. and McClelland, M. (1990). Fingerprinting genomes using PCR with arbitrary primers. *Nucleic Acid Research,* **18:** 7213–7218.

Williams, J.G.K., Kubelik, A.R., Rafalski, J.A. and Tingey, S.V. (1990). DNA polymorphisms ainplified by arbitrary primers are useful as genetic markers. *Nucleic Acids Research,* **18:** 6531–6535.

Wiseman, B.M., Neate, S.M., Keller, K.O. and Smith, S.E. (1995). Suppression of *Rhizoctonia solani* anastomosis group-8 in Australia and its biological nature. *Soil Biol. Biochem.,* **28:** 727-732.

Woloshuk, C.P. and Sisler, H.D. (1982). Tricyclazole, pyroquilon, tetrachlorophthalide, PCBA, coumarin and related compounds inhibit melanization and epidermal penetration by *Pyricularia oryzae. J. Pestic. Sci.,* **7:** 161-166.

Yamaguchi, I., Sekido, S. and Misato, T. (1983). Inhibition of appressorial melanization in *Pyricularia oryzae* by non-fungicidal anti-blast chemicals. *J. Pestic. Sci.*, **8:** 229-231.

Yang, J., Kharbanda, P.D., Wang, H. and McAndrew, D.W. (1996). Characterization, virulence, and genetic variation of *Rhizoctonia solani* AG-9 in Alberta. *Plant Dis.*, **80:** 513-518.

Yokoyama, K. and Ogoshi, A. (1984). Studies on hyphal anastomosis of *Rhizoctonia solani*, IV. Observation of imperfect fusion. *Ann. Phytopathol. Soc. Jpn.*, **50:** 398.

Chapter 7

Climate Change and Plant Diseases: Changing Responses of Plant Pathogenic Microbes

Ravindra Kumar[1] *and Asha Sinha*[2]

[1]*Indian Agricultural Research Institute, Regional Station, Kalimpong – 734 301, West Bengal*

[2]*Department of Mycology and Plant Pathology, Institute of Agricultural Sciences, Banaras Hindu University, Varanasi – 221 005, U.P.*

Since two decade back the topic of climate change and its potential impacts have consistently remained in the headlines of the scientific and popular press. This has been due to the mounting evidences for greenhouse gas induced changes in global and regional climate. Climate change represents one of the biggest scientific and political challenges of the 21st century. The greenhouse gas concentrations in the atmosphere are being altered by human activities thus causing global climate change. These activities have been intensified worldwide after the industrial revolution at the end of the 18th century result from the use of natural resources such as fossil fuel burning, deforestation and other land use changes. The atmospheric concentration of carbon dioxide (CO_2) has reached levels significantly higher than in the last 650 thousand years (Siegenthaler *et al.*, 2005). Since 2000, the growth rate of CO_2 concentration is increasing more rapidly than the previous decades (Canadell *et al.*, 2007). Similar trends have been observed for methane (CH_4), nitrous oxide (N_2O) and other green house gases (Spahni *et al.*, 2005; IPCC, 2007). Consequently, several changes in the climate have been recorded. The average global surface temperature has increased by 0.2°C per decade in the past 30 years (Hansen *et al.*, 2006). Alterations in the water cycle have also been observed. Changes will probably continue to happen even if greenhouse gas concentrations stabilize, due to the system's thermal inertia and to the long period necessary for returning to a lower equilibrium (IPCC, 2007).

The Intergovernmental Panel on Climate Change (IPCC), which was jointly established by the World Meteorological Organization (WMO) and the United Nations Environment Programme (UNEP) in 1988, has responsibility for assessing information relevant to climate change and summarizing this information for policy makers and the public. The reports of IPCC (2001 and 2007) are major landmarks in the global change history in that these reports provide strong evidences that human induced climate changes are already a reality. For example the instrumental record shows that global mean surface temperature has increased by 0.6±2 during 20th century, the rate of temperature increase observed during the past 25 years is unprecedented during the last millennium. The global climate is predicted to change drastically over the next century and various parameters will be affected in this changing environment (Houghton *et al.*, 2001). This is the case for atmospheric CO_2 concentrations that increase continuously (IPCC, 2007). Additionally, global surface temperatures are predicted to increase between 1.8 and 3.6°C by the year 2100, driven by increased atmospheric CO_2 levels derived from natural and/or anthropogenic sources (IPCC, 2007; Compant *et al.*, 2010).

Although there is uncertainty surrounding the projection of future warming and other changes in climate, since 1990, when the first IPCC report was published, actual increases in the global temperature per decade have been within the range of projected increase of between 0.15 and 0.3°C per decade (Chakraborty *et al.*, 2008). The fourth IPCC assessment report projects a 0.2°C warming per decade for the next two decades for a range of IPCC emission scenarios originally outlined in IPCC special report (2000). Other changes in climate include rising sea level, shrinking of glaciers and increased rainfall in the middle and high latitudes of the Northern Hemisphere but a decrease over the sub-tropics. The global atmospheric CO_2 concentration has increased from about 280 ppm since the pre-industrial times (till 1750) to 379 ppm in 2005. This concentration exceeds the 180–300 ppm range observed from air pockets trapped within ice cores for over the last 650,000 years. Atmospheric concentration of other greenhouse gases including N_2O and CH_4 has also increased since pre-industrial times. The combined radiative forcing resulting from increases in greenhouse gases has led to a warming of the globe. For example, CO_2 radiative forcing has increased by 20 per cent from 1995 to 2005 (IPCC, 2007).

In a comprehensive view, global change encompasses all changes in climate, land, oceans, atmospheric composition and chemistry, and ecological systems that influence the global environment. The interactions between atmosphere, hydrosphere, cryosphere and biosphere as driven by solar radiation make our earth's climate. A part of the radiation reaching the earth is absorbed to heat up the earth's surface and some is radiated back to space. The oceans, covering over 70 per cent of the earth's surface, absorb solar energy; while snow and ice reflect 60–90 per cent of the solar energy. The reflected radiation is trapped by radiatively active water vapour, CO_2, CH_4, N_2O and O_3 in the atmosphere, acting like the glass of a greenhouse that warms the earth's surface, a natural phenomenon known as "green house effect". Based on a range of emission scenarios for greenhouse gases and aerosol precursors, global mean temperature is projected to rise between 0.9 and 3.5°C by 2100, but the actual decadal changes would include considerable variability. The longevity and radiative efficiency of these greenhouse gases determine their global-warming potential. Human activities are increasingly influencing the atmosphere, oceans, cryosphere and the terrestrial and marine biospheres, which together constitute the global climate system. Increased emissions of CO_2 and other radiatively active gases from industrial and agricultural development are changing the atmospheric composition. There is a strong interactive link between the large-scale clearing of forests in the humid tropics for logging and intensive agriculture, which alters global carbon balance and climate (IPCC, 1996; Chakraborty *et al.*, 2000). Global change, including a changing climate, is one of the most critical issues facing our future today as terrestrial and aquatic ecosystems which sustain life on earth are being increasingly affected by it. While the global population continues to rise, productive land resource, necessary for food production, shrinks. Uncertainties of climate change only magnify the challenge of increasing agricultural production to feed the ever increasing/ expanding population. Climates continually change and due to climate change recent accelerated warming affect many biological systems (IPCC, 2007). The effects on the geographic distributions of pests and pathogens (Woods *et al.*, 2005; Admassu *et al.*, 2008; Elphinstone and Toth, 2008; Kudela, 2009), with potentially serious implications for food security is one of the most important issues (Newton *et al.*, 2010). However, cropping systems will also change in response to climate, with consequent impacts on their interactions with pests and pathogens.

The importance of the environment for the development of plant diseases has been known for over two thousand years. Theophrastus (370-286 B.C.) observed that cereals cultivated in higher altitude regions exposed to the wind had lower disease incidence than cereals cultivated in lower altitude areas. It is an established fact that the environment can potentially influence host plant growth and susceptibility, pathogen survival, reproduction, dispersal and activity as well as plant-microbe interaction or host- pathogen interaction (Garrett *et al.*, 2006). The classic disease triangle establishes the conditions for disease development, *i.e.*, the interaction of a susceptible host, a virulent pathogen and a favourable environment. This relationship is evidenced in the definition of plant disease itself (Gaumann, 1950). The classic disease triangle recognizes the role of physical environment in plant disease as no virulent pathogen can induce disease on a highly susceptible host if weather conditions are not favourable. Weather influences all stages of host and pathogen life cycles as well as the development of disease. Relationships between weather and disease are routinely used for forecasting and managing epidemics, and disease severity over a number of years can fluctuate according to climatic variation (Coakley, 1979; Scherm and Yang, 1995).

Plant pathogens are ubiquitous in natural and managed systems. They are among the first to demonstrate the effects of climate change due to the numerous populations, ease of reproduction and dispersal, and short time between generations. Therefore, they constitute a fundamental group of biological indicators that needs to be evaluated regarding climate change impacts. The plant pathogen groups include fungi, prokaryotes (bacteria and mycoplasmas), oomycetes, viruses and viroids, nematodes, parasitic plants and protozoa. The very different life histories of this diverse group of organisms and their different interactions with host plants produce a wide range of responses to environmental and climatic drivers. For example, viruses may be present in hosts while symptom expression is dependent on temperature thus, even the difficulty of detection of these pathogens varies with climate. Fungal pathogens are often strongly dependent on humidity or dew for plant infection (Huber and Gillespie, 1992), so changes in these environmental factors are likely to shift disease risk. Pathogen populations may explode when weather conditions are favourable for disease development. Despite the threat posed by climate change to plant protection in the near future, there are few reports about this subject (Newton *et al.*, 2010).

Effect of Climate Change on Plants

The geographical distribution and growth of plant species are influenced due to climate change around the world. These impacts on plants vary depending upon the species involved and their growth patterns *e.g.* annual vs perennials, type of plants *e.g.* agricultural or domestic plant vs natural vegetation, their competition ability, migration and ability of recovery from different stresses. The options for managing annual crops from the effects of climate changes are more because there is always opportunity to change annually the location, cultivar, time of sowing or planting and acerage of the crops. The direct effects of climate change on individual plants and plant communities may occur in the absence of pathogens, but may also bring about changes in plants that will affect their interactions with pathogens. Changes in plant architecture may affect microclimate and thus risks of infection (Burdon, 1987). In general, increased plant density will tend to increase leaf surface wetness and leaf surface wetness duration and so make infection by foliar pathogens more likely (Huber and Gillespie, 1992). Abiotic stress such as heat and drought may contribute to plant susceptibility to pathogens or it may induce general defense pathways which increase resistance (Garrett *et al.*, 2006).

Elevated CO_2 levels tend to result in changed plant structure. At multiple scales, plant organs may increase in size: Increased leaf area, increased leaf thickness, higher numbers of leaves, higher total leaf area per plant, and stems and branches with greater diameter have been observed under elevated CO_2 (Pritchard *et al.*, 1999). Enhanced photosynthesis, increased water use efficiency and reduced damage from ozone are also reported under elevated CO_2 (Seem, 2004). Since many foliar pathogens benefit from denser plant growth and the resulting more humid microclimate (Burdon, 1987), there is the potential for these changes in plant architecture to increase infection rates. The effects of elevated temperature on plants vary greatly throughout the year. During colder season, warming may relieve plant stress, whereas during hotter parts of the year it may increase stress. When high-temperature stress is exacerbated, plant responses may be similar to those induced by water stress, with symptoms including wilting, leaf burn, leaf folding, and abscission, and physiological responses including changes in RNA metabolism and protein synthesis, enzymes, isoenzymes, and plant growth hormones (Garrett *et al.*, 2006). These changes will certainly affect susceptibility to pathogens, though the wide range of changes can make interactions difficult to predict. The potential effects of temperature on crop plants can be understood by the fact that rice yield decline upto 10 per cent for each 1°C increase in the minimum temperature during the dry season (Peng *et al.*, 2004). The elevated ozone enhance the susceptibility of plant to several foliar pathogens by changing structure of leaf surface by altering physical topography as well composition of surfaces, including the structure of epicuticular wax (Karnosky *et al.*, 2002). The Ozone exposure has been proposed to enhance susceptibility of plants to necrotrophic fungi, root-rot fungi, and bark beetles (Sandermann, 2000; Garrett *et al.*, 2006).

Impacts of Climate Change on Host-Pathogens System

Climate is most frequently the primary driving force for successful host-pathogen interactions and more is known about how climate affects disease development than how various atmospheric chemicals do. The early studies of air pollutants focussed on direct damage to the host, *i.e.*, the air pollutant behaving as the pathogen. With increased evidences for climate change, the emphasis of research has shifted from pollutants as pathogens to the effect of these climate change components on host, pathogen and host-pathogen interactions. Climate change and its elements affect the host, pathogen and their interactions both directly and indirectly and at different levels. For example, the direct effect of ozone on a mycorrhizal fungus results in an indirect effect on its host. Similarly environmental factors that influence insect vector activities or weed competition have indirect effects on the host and therefore host-

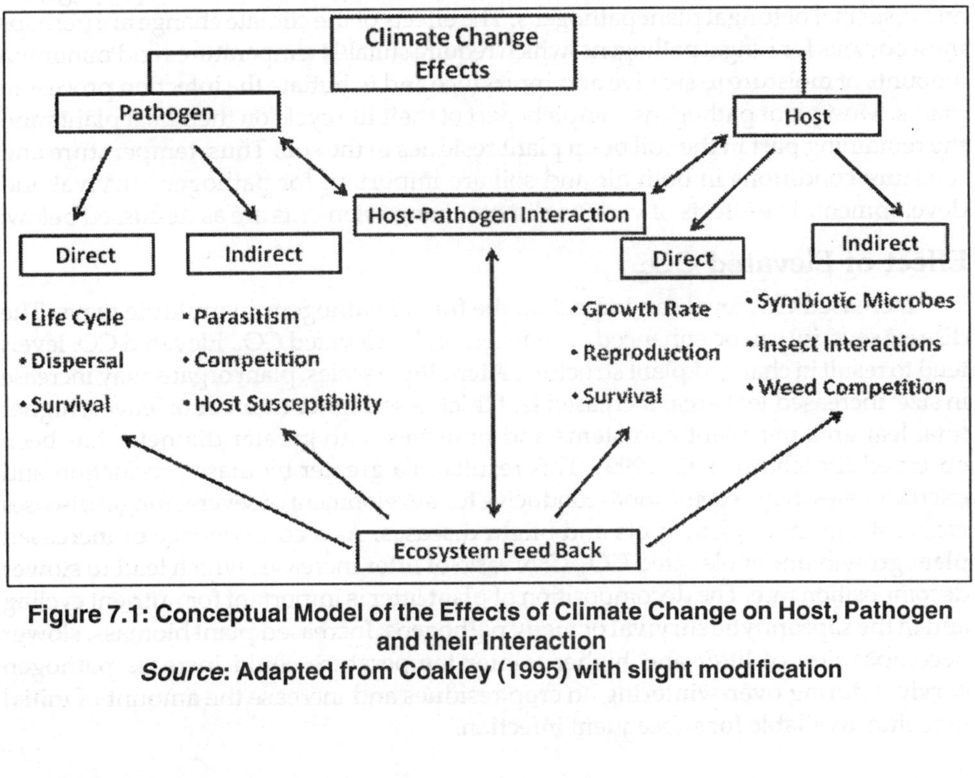

Figure 7.1: Conceptual Model of the Effects of Climate Change on Host, Pathogen and their Interaction

***Source*: Adapted from Coakley (1995) with slight modification**

pathogen interaction. Global climate change that increase or decrease biocontrol of a pathogen, or the competition among pathogens, indirectly affect the pathogens.

In case of direct effect of climate change elements on pathogens, it may influence its generation time, or its dispersal or survival. Equally as important as direct effect is the ecosystem feedback from the host-pathogen interaction. Interactions between different climate change factors may occur that enhance or inhibit their impact on the host or pathogen. Also, a particular climate change element may have a positive effect on one part of the disease triangle but a negative on another. For example elevated CO_2 may increase host growth, but also weed competition, increased temperature may increase evapotranspiration and increase water use or loss and ultimately results in water stress for the plants (Coakley, 1995). The range and severity of plant diseases increased by global warming have also been reported (Evans *et al.*, 2008). Due to comprehensive impacts of climate changes on host-pathogens system more aggressive strains of pathogen with broad host range, such as *Rhizoctonia*, *Sclerotinia*, *Sclerotium* and other necrotrophic pathogens can migrate from agroecosystems to natural vegetation, and less aggressive pathogens from natural plant communities can start causing damage in monocultures of nearby regions. Regarding unspecialized necrotrophs, the range of hosts can be extended due to crop migration (Chakraborty *et al.*, 2000; Ghini *et al.*, 2008).

Effects of Climate Change and its Various Elements on Fungal Pathogens

Environmental conditions have a major influence on the survival, propagation and dispersal of fungal plant pathogens. The effects of the climate change are perhaps most obvious for fungal pathogens, which require suitable temperatures and minimum amounts of moisture to survive and reproduce and to initiate the infection process in plants. Most plant pathogens complete part of their life cycle on their host plants and the remaining part in the soil or on plant residues in the soil. Thus, temperature and moisture conditions in both air and soil are important for pathogen survival and development. The effects of various climate change elements are as discussed below:

Effect of Elevated CO_2

Increased CO_2 level can impact on the fungal pathogens in multiple ways. The disease severity can be enhanced or reduced under elevated CO_2. Elevated CO_2 levels tend to result in changed plant structure. At multiple scales, plant organs may increase in size: increased leaf area, increased leaf thickness, higher number of leaves, higher total leaf area per plant and stems and branches with greater diameter has been observed (Pritchard *et al.*, 1999). This results in a greater biomass production and microclimates may become more conducive for development of several fungal diseases *viz.*, rusts, mildews, leaf spots and blight diseases. As a consequence of increased plant growth under elevated CO_2, C: N ratio of litter increases which lead to slower decomposition rate. The decomposition of plant litter is important for nutrient cycling and in the saprophytic survival of many pathogens. Increased plant biomass, slower decomposition of litter and higher winter temperature could increase pathogen survival during over-wintering on crop residues and increase the amount of initial inoculum available for subsequent infection.

Regarding fungal pathogens, important alterations can occur as a consequence of an increase in CO_2 concentration. There can also be direct effects on pathogen growth; for example the enhanced growth of *Colletotrichum gloeosporioides* infecting *Stylosanthes scabra* has been reported at high CO_2 (Chakraborty and Datta, 2003). CO_2 can also affect pathogen fecundity which was shown to increase under elevated CO_2 levels leading to enhanced rates of pathogen evolution (Chakraborty and Datta, 2003). They reported that the aggressiveness increased for the resistant cultivar, but not for the susceptible cultivar. Similar results were reported by Osswald *et al.* (2006), while working with host-pathogen system of potato and *Phytophthora infestans*. These results are extremely important for the study of epidemiology of the disease demonstrating that the pathogen can adapt to a new environment. The incidence of rice plants naturally infected with sheath blight (*Rhizoctonia solani*) was generally higher at elevated CO_2 concentrations under high nitrogen levels, but this trend was not apparent for sheath blight severity (Chakraborty *et al.*, 2008). *Arabidopsis thaliana* was found more susceptible to *Erysiphe cichoracearum* under high CO_2 concentration, correlated with increased stomatal density and guard cell length, but there were inherent differences between ecotypes in this response (Lake and Wade, 2009; Newton *et al.*, 2010). In rice, enhanced susceptibility to *Magnaporthe oryzae* under elevated CO_2 was attributed to lower leaf silicon content (Kobayashi *et al.*, 2006; Newton *et al.*, 2010).

Effect of Temperature

Temperature is one of the most important climate change factors that influence the disease severity and establishment of infection by fungal pathogens. Both temperature and the length of exposure are important in determining the effect of climate change on disease severity. The higher temperature in a particular area may lead appearances of invasive alien species of fungal pathogens that can cause severe epidemic on important crops. Change in temperature will directly influence infection, reproduction, dispersal, and survival between seasons and other critical stages in the life cycle of a fungal pathogen. Higher temperature can modify host physiology and resistance (Garrett *et al.*, 2006) for example at higher temperature, lignification of cell walls increased in forage species and enhanced resistance to fungal pathogens. Impact would, therefore, depend on the nature of the host- pathogen interactions and mechanism of resistance. A rise in temperature above 20°C can inactivate temperature sensitive resistance to stem rust in oat cultivars. Increase in temperature with sufficient soil moisture may increase evapo-transpiration resulting in humid microclimate in crop canopy and may lead to incidence of diseases favoured under warm and humid condition. Some of the soil-borne diseases may increase at the rise of soil temperature (Compant *et al.*, 2010). Bergot *et al.* (2004) predicted the geographic range expansion of *Phytophthora cinnamomi* in Europe in response to increased temperatures that would allow for overwintering of this oomycete in new areas. Several workers suggested that under conditions of increased temperature due to global warming the survival and degree of root disease seems likely to be enhanced, while the host range of microorganisms might also be increased (Brasier and Scott, 1994; Brasier *et al.*, 1996; Garrett *et al.*, 2006; Compant *et al.*, 2010).

Effects of Elevated Levels of Atmospheric Pollutants (Ozone and Nitrous Oxide)

Ozone (O_3) is a secondary pollutant that is increasing downwind of major metropolitan areas around the world (IPCC, 2001; Vingarzan, 2004). It is a highly phytotoxic pollutant that decreases carbon assimilation of O_3-sensitive plants through direct effects on photosynthesis, leaf area and leaf area duration (Krupa *et al.*, 2000; Karnosky *et al.*, 2007). It has long been known that ozone can also alter plant responses to biotic diseases (Manning, 1975; Sandermann, 2000; Fuhrer, 2003; Ashmore *et al.*, 2006). Effects of O_3 on pathogen interactions are variable depending on the timing of the exposure to both plant and the pathogen, the O_3 concentration, the stage of plant development, predisposing factors and environmental conditions (Fuhrer, 2003). Elevated ozone concentrations can change the structure of leaf surfaces, altering the physical topography as well as the chemical composition of surfaces, including the structure of epicuticular wax (Karnosky *et al.*, 2002). These changes in leaf structure may alter leaf surface properties such as leaf wettability and the ability of leaves to retain solutes; all influencing the ability of pathogens to attach to leaf surfaces and to infect (Karnosky *et al.*, 2002). Plants appear to be less sensitive to nitrous oxide; however, higher concentrations can cause water-soaked lesions, which soon turn brown. Ozone and nitrous oxide injury on plants in turn may add new problem to pathologists in diagnosis. Current climate change scenarios predict a further increase of tropospheric ozone, which is well known to inhibit plant photosynthesis and growth process. Ozone can also predispose plants to enhanced biotic attack, as proposed in particular for necrotrophic fungi, root rot fungi and black beetles (Sandermann, 2000; Garrett *et al.*, 2006). Several root pathogens show a preference for stressed trees, although the direct role of ozone in disease development is not always evident.

The occurrence of co-occurring elevated atmospheric CO_2 can also alter the O_3 disease interactions (von Tiedemann and Firsching, 2000; Plessl *et al.*, 2007). Thus, it is not surprising that O_3 biotic disease interactions have ranged from significant enhancement for diseases such as powdery mildew (*Erysiphe graminis*), leaf spot disease (*Septoria nodorum*) and spot blotch (*Bipolaris sorokiniana*) on wheat flag leaves exposed to O_3 (von Tiedemann *et al.*, 1991), tan spot fungus (*Pyrenophora tritici-repentis*) on wheat (Sah *et al.*, 1993) and other root diseases experiments (Fenn *et al.*, 1990; Pritsch *et al.*, 2005) to decreases or no impact with other diseases. The occurrence of mycorrhizal and non-mycorrhizal root-infecting fungi (Bonello *et al.*, 1993), powdery mildew (*Sphaerotheca fuliginea*) on cucumber (Khan and Khan, 1999), spot blotch on barley and fescue (Plazek *et al.*, 2001) and leaf rust (*Puccinia recondita*) on wheat (von Tiedemann and Firsching, 2000) have been shown to decrease with O_3. Finally, at least one study has shown no interaction of O_3 and pathogen occurrence in wheat (Pfleeger *et al.*, 1999).

Effect of Acid Rain

Acid rain is the result of human activities, primarily the combuston of fossil fuels (oils, coal and natural gas) and smelting of sulphide ores. These activities release large quantities of sulphur and nitrogen oxides in the atmosphere, which when in contact with atmospheric moisture are converted into two of the strongest acids known (sulphuric and nitric acid) and fall to the ground in the form of rain, snow and fog. The pH of acid rain over a large region of world ranges from 4.0 to 4.5 but the lowest rain pH values reported so far pH 1.5 in West Virginia and 1.7 in Los Angeles (Agrios, 2005). Most studies on the effect of acid rain were done with simulated acid rain since it is not easy to establish experiments under field conditions (Asai and Futai, 2005). Variable effects of acid rain on four different patho-systems: alfalfa leaf spot (ALS), peanut leaf spot (PLS), potato late blight (PLB), and soybean brown spot (SBS) have been observed by Campbell *et al.* (1988) during their two year's study.

Experiments conducted to determine the effect of acid rain on the initiation and development of plant diseases have shown that telia formation of *Cronartium fusiforme* rust of oak was same as at pH 6.0 under acid rain of pH 3.0, whereas similar nematode egg mass production was noticed in same host plant at acid rain of 3.2 pH as it was under rain of pH 6.0. However, a bacterial disease 'Halo blight' and rust disease of bean were some times more severe and other milder with acidic rain than with the pH 6.0 rain (Agrios, 2005).

Effect of Elevated Ultraviolet B

The effect of UV-B on the incidence and development of pathogen-induced diseases on crops is dependent upon the crop cultivar, age, pathogen inoculum level and the timing and duration of UV-B exposure (Krupa *et al.*, 2000). In some cases, a shift in one kind of atmospheric component may have a profound effect on another, for example the ozone hole thought to be caused by chlorofluoromethanes has led to an increase in UV-B radiation (Ashmore and Bell, 1991). The effect of UV-B radiation on fungal- host interactions is now an increasingly studied subject. One of the first studies about this subject was conducted by Luo *et al.* (1995). They carried out a risk analysis of rice blast epidemics and plant growth associated with climate change in several Asian countries due to the importance of this crop and to the losses related to this disease, caused by *Magnaporthe grisea*. Simulations were made to study the risk of blast epidemics under the effects of temperature change and enhanced UV-B radiation. The results demonstrated that changes in the amount of rainfall do not affect the occurrence of the epidemics since they have little effect on the leaf wetting period. In cool subtropical zones, higher temperatures caused increases in disease severity and in the area bellow the disease progress curve, because higher risk of epidemics occurs under higher temperatures. In humid tropical and humid warm subtropical zones, such as Southern China, Philippines and Thailand, the opposite effect was observed. Lower temperatures increased the risk of rice blast epidemics since the current temperatures in these regions are above favourable values for the occurrence of this disease. However, a larger area bellow the disease progress curve does not always result in lower rice yield, since the effect in plant growth also takes place. The effects of the increase in UV-B radiation were highly significant for the occurrence of epidemics.

Effect of Climate Change and its Elements on Bacterial Pathogens

At present about 400 bacterial plant pathogens are known (Kudela, 2009). Bacteria multiply with astonishing rapidity and their significance as pathogens originates primarily from the fact that they can produce tremendous numbers of cells in short period of time (Agrios, 2005). The climate change induced effects on geographic distribution of plant pathogenic bacteria have been reported (Kudela 2009). Due to climate change factors like climate warming and draught stress thermophillic bacteria can emerge potentially where as decrease in the cold tolerant species of these bacteria can be observed. Insect vector borne xylem limited bacteria can pose new emerging threat to agricultural crops in changing global climate, as these insect vectors can adjust effectively with the changing climate (Hamilton *et al.*, 2005). The consequences of climate change impacts on bacteria are as follows:

Emergence of Thermophillic (Heat-Loving) Bacteria

Temperature is undoubtedly one of the most important factors influencing the occurrence and development of many plant pathogenic bacteria. Climate model simulations using future emission scenarios of greenhouse gases and aerosols suggest an increase in global mean temperature between 1 and 3.5°C by the year 2100. Since, there is solid evidence that global warming is occurring and if such conditions continue, heat-loving plant pathogenic bacteria should be expected to increase. A common trait of these high-temperature bacteria is an optimum growth temperature of 32–36°C (most grow well up to 41°C) whereas most other plant pathogenic bacteria grow best at lower temperatures (Kudela, 2009). Among heat-loving plant pathogenic bacteria that have emerged as serious problem worldwide belong following bacterial plant pathogens: *Ralstonia solanacearum*, *Acidovorax avenae* subsp. *aveane*, and *Burkholderia glumea* (Schaad, 2008).

Why the increase in these heat-loving bacteria? The most likely explanation is the influence of global warming on the World's climate. The evidence for global warming is quite broad, including, major shifts in recorded temperature and precipitation, melting glaciers and reduced snow cover, and more frequent and severe storms and droughts. Studies have shown global warming is caused primarily by heat-trapping greenhouse gas (GHG) emissions (IPCC, 2007). Industrial activity, from the burning of fossil fuels such as coal, oil, and gas, generates CO_2 and other gases which trap the sun's rays in the atmosphere and enhance the natural "greenhouse effect" (Gore, 2006). Although automobiles and industry are considered the major contributors of GHG, the increase in air traffic is emerging as a major factor. "Concentrations of CO_2, the prime GHG, have increased 30 per cent during the past 100 years" (Fallon, 1997). These comprehensive impacts of climate change element favour the emergence of thermophillic bacteria.

Changes in the Spectrum of Pectolytic Bacteria

Some bacterial phytopathogens have strong pectolytic capacities that enable them to cause soft-rots (*i.e.*, tissue-macerating diseases and storage rots) in a wide variety of plants. They are economically important because of the crop loss they can cause both in the field and after harvest in transit and in storage (Perombelon, 1982; Starr, 1983; Schaad, 2008; Kudela, 2009). Among the species of pectolytic bacteria associated with crop loss, *Erwinia* spp. (also known as *Pectobacterium* spp.) and *Clostridium* are economically important in temperate area. In warmer climates species of the other genera may play an important role (*e.g., Bacillus* spp., whose pathogenicity is often greater at high temperatures–Perombelon, 1982). Ecology of soft rot erwinia reviewed by Perombelon and Kelman (1980). Soft rot erwinia differ in temperature optima and requirements. Strains of *Erwinia carotovora* subsp. *carotovora* (*Ecc*) but not *Erwinia carotovora* subsp. *atroseptica* (*Eca*) will grow at 37°C and, whereas most strains of the former are inhibited at 39°C, those of *Erwinia chrysanthemi* (*Echr*) grow relatively well at > 39°C. These temperature characteristics are reflected in their host range as affected by geographical distribution (Kudela, 2009).

A Decrease in the Frequency of Occurrence of Cold Tolerant Pseudomonads and an Increase in more Thermophilic Xanthomonads Population

The most of plant pathogenic bacteria belong to the *Pseudomonas* genus or *Xanthomonas* genus (Kudela, 2009). In general, pseudomonads (namely *Pseudomonas syringae* group) and the most of xanthomonads produce necrotic lesions on foliage, stems, or fruit that develop into spots, streaks, or cankers. They affect plants worldwide, causing varying amounts of damage in crops of nearly every plant family (Starr, 1983). Minimal growth temperature of *Pseudomonas* spp. is 4–5°C, whereas it is 7–9°C in *Xanhomonas* spp. (Klement *et al.,* 1990) Therefore, cold tolerant pathogenic pseudomonads cause serious losses in cooler areas including Central Europe. In contrast to pseudomonads, xanthomonads are more commonly found in tropical and subtropical conditions. These comprehensive impacts of climate change elements mainly climate warming reduce the frequency of occurrence of cold tolerant pseudomonads and as a result of climate warming, population of thermophillic xanthomonads will increase.

Increased Risk of Xylem-Limited Bacteria which Overwinter in Insect Vectors

Some plant pathogenic bacteria overwinter within bodies of their insect vectors. Insect transmission of plant pathogenic bacteria is usually non-specific. Examples of the specific transmission are xylem limited bacteria *Xylella fastidiosa* subsp. *fastidiosa* causing diseases in grape (Pierce's disease), alfalfa, maple and almond and *X. f.* subsp. *multiplex* causing diseases in peach (phony peach disease), plum, almond, elm *etc.* Causal agent, *Xylella fastidiosa,* is vectored by xylem sucking insects, such as sharphooters (subfamily Cicadellinae in the leaf-hopper family Cicadellidae) and spittlebug (*Philaenus spumarius,* family Cercopidae) found also in Europe. *P. spumarius* is associated with Pierce's disease of grapevine and almond leaf scorch, however, its relative threat as invasive vector is reported low in USA (Redak *et al.,* 2004). As these insect vectors can adjust effectively with the changing climate (Hamilton *et al.,* 2005), they can pose new emerging threat to agricultural crops due to their associated xylem limited bacteria in changing global climate.

Effect of Climate Change and its Elements on Viral Plant Pathogens

Most of the studies related to climate change and its impacts have been conducted on fungal plant pathogens. Only few studies have reported the response of plants infected with viral diseases to various climate change components. It has been observed that oats infected with *Barley yellow dwarf virus* (BYDV) showed three fold greater biomass accumulation to CO_2 enrichment than the healthy plant (Malmstrom and Field, 1997). Tobacco plants grown at elevated CO_2 concentrations showed a markedly decreased spread of virus. It appears that CO_2 rise in the air may have some positive effects, which may likely offset the negative effects of virus infection. Gioria *et al.* (2008) during their prediction of important tomato diseases and influence of climate change showed that climate change will not alter the importance of tomato mosaic disease caused by *tomato mosaic virus* (ToMV). In contrast authors considered that the importance of *tomato spotted wilt virus* (TSWV), *tomato chloratic spot virus* (TCSV), *groundnut ring spot virus* (GRSV), *Chrysanthemum stem necrosis virus* (CSNV) and *yellow leaf curl virus* (Geminivirus) will be increased due to climate change. Because of the elevated temperature soil water content is expected to decrease (Compant *et al.*, 2010) which leads to draught condition. Drought stress and disease stress may have additive effects on plants, as observed for infection by *Beet yellows virus* (Clover *et al.*, 1999), and *Maize dwarf mosaic virus* (Olson *et al.*, 1990).

Effect of Climate Change on Nematode

Most of the plant pathogenic nematodes spend part of their lives in soil and therefore soil is the source of primary inoculum. Life cycle of a nematode can be completed within 2-4 weeks under favourable environmental conditions. Temperature is the most important factor influencing the population dynamics of plant pathogenic nematodes. The development of plant parasitic nematodes is slower with cooler soil temperatures. Warmer soil temperatures are expected to accelerate nematode development, perhaps resulting in additional generations per season. While drier temperatures are expected to increase symptoms of water stress in plants infected with nematodes such as soybean cyst nematode. Overwintering of nematodes is not expected to be significantly affected by changes in climate, although for some such as the soybean cyst nematode, egg viability may be reduced in mild winters. The effect of climate change on distribution of *Meloidogyne incognita* in coffee crop was evaluated by Ghini *et al.* (2008). The distribution map indicated that there could be an increase in infestation of this nematode due to the higher number of generations per month as compared to previous years. Similar results are reported earlier by Carter *et al.* (1996) during studies on distribution of the potato cyst nematode (*Globodera rostochiensis*) and Boag *et al.* (1991) also obtained similar results for the plant-parasitic nematodes *Xiphinema* and *Longidorus* during the study of the geographical distribution of these virus-vector nematodes.

Effects of Climate Change on Insect Vector

Insects are key factors in the transmission of several plant diseases. Besides numerous viral diseases they are potential vectors of other disease caused by several plant pathogens *viz.*, fungi, bacteria, phytoplasma, viriods etc. Hence, in these pathosystems, the effect of climate on vector survival, reproduction and efficiency of pathogen transmission is directly linked to disease development. Insects are cold-blooded organisms. The temperature of their body is approximately same that of the environment. Therefore, temperature is probably the single most environmental factor influencing each and every sphere of their life cycle *viz.* insect behaviour, distribution, development, survival and reproduction. Insect life stage predictions are most often calculated using accumulated degree days from a base temperature and biofix point. Some researchers believe that the effect of temperature on insects largely overwhelms the effect of other environmental factors (Bale *et al.*, 2002). It has been estimated that with a 2°C temperature increase, insect might experience one to five more life cycles per season (Yamamura and Kiritani, 1998). Several other researchers have found that moisture and CO_2 effects on insects can be potentially important considerations in a global climate change setting (Coviella and Trumble, 1999; Hunter, 2001; Hamilton *et al.*, 2005). Thus, the effect of climate change on vector-borne plant diseases can be complex and it is difficult to generalize potential future impacts due to climate change.

Climate Change and Alien Invasive Species

The alien invasive species are both a cause and a consequence of global change (Scherm and Coakley, 2003). Being one of the major contributors to global change, invasive non-indigenous organisms are already having serious adverse impacts on our ecosystem (Scherm and Coakley, 2003; Admassu *et al.*, 2008). Similar to other global change drivers such as climate warming and changes in land-use patterns, the magnitude of the problem has increased considerably during the second half of the 20th century mainly due to upsurge in global travel and trade during the past 25 years. At the same time, stressor such as rising temperatures and habitat degradation may predispose ecosystems to biological invasions and these invasions thus become a consequence of other global changes. According to an estimation 239 species of non-indigenous plant pathogens had become established by the early 1990s in United States (National Research Council, 2002). Most of these, including highly devastating pathogens such as Wheat rust (*Puccinia* spp.), White pine blister rust (*Cronatium ribicola*), chestnut blight (*Cryphonectria parasitica*) and Dutch elm disease (*Ophiostoma ulmi*) were introduced before the mid-1900s (Yarwood, 1983). Although many introductions of novel plant pathogens have already occurred in different parts of the world, climate change may facilitate their further establishment and spread. A new race of *Puccinia graminis* f. sp. *tritici*, Ug99 has been reported from Uganda in 1999 and since then it is causing severe losses to wheat production in many countries where it had not been reported earlier (Admassu *et al.*, 2008). In some cases, however, there is the possibility that the risk of the introduction of some plant pathogens may decrease due to changes in precipitation patterns predicted under climate change. However, the pathogen propagule pressure due to modern trade within and between continents, with plants moved around the world in both shipping and air networks, makes it possible that new plant health problems will arise with their potential threat.

Climate Change and Plant Disease Management

Diseases are responsible for losses of at least 10 per cent of global food production, representing a severe threat to food security (Strange and Scott, 2005). Agrios (2005) estimated that annual losses by diseases cost US$ 220 billion. Besides direct losses, the methods for disease control especially the chemical methods can result in environmental contamination and in residual chemicals in food, in addition to social and economic problems. The close relationship between the environment and diseases suggests that climate change will cause modifications in the current phytosanitary scenario. The impacts can be positive, negative or neutral, since there can be a decrease, an increase or no effect on the different pathosystems, in each region. The analysis of the potential impacts of climate change on plant diseases is essential for the adoption of effective management practices including development of resistant cultivars in order to avoid more serious crop losses (Chakraborty and Pangga, 2004; Ghini, 2005; Chakraborty *et al.*, 2008).

The well-known dependence of plant diseases on weather has long been exploited for predicting epidemics and to time applications of control measures for tactical disease management. Disease management strategies depend on climate conditions. Climate change will cause alterations in the disease geographical and temporal distributions and consequently the control methods will have to be adapted to this new reality. Changes in temperature and precipitation can alter fungicide residue dynamics in the foliage and the degradation of products can be modified. Alterations in plant morphology or physiology, resulting from growth in a CO_2-enriched atmosphere or from different temperature and precipitation conditions, can affect the penetration, translocation and mode of action of systemic fungicides. Besides, these changes in plant growth can alter the period of higher susceptibility to pathogens which can determine a new fungicide application calendar (Coakley, 1995; Chakraborty and Pangga, 2004; Pritchard and Amthor, 2005). The per acre pesticide usage average cost for corn, cotton, potatoes, soybeans and wheat were found to increase as precipitation increases. Similarly, the pesticide usage average cost for corn, cotton, soybean and potatoes also increase as temperature increases, while the pesticide usage cost for wheat decreases (Ghini *et al.*, 2008).

The physiological changes in host plants may result in higher disease resistance under climate change scenarios, host resistance to disease may overcome more quickly by more rapid disease cycles, resulting in a greater chance of pathogens evolving to overcome host plant resistance. Fungicide and bactericide efficacy may change with increased CO_2, moisture, and temperature. The more frequent rainfall events predicted by climate change models could result in farmers finding it difficult to keep residues of contact fungicides on plants, triggering more frequent applications. Systemic fungicides could be affected negatively by physiological changes that slow uptake rates, such as smaller stomatal opening or thicker epicuticular waxes in crop plants grown under higher temperatures. These same fungicides could be affected positively by increased plant metabolic rates that could increase fungicide uptake. Genetic variation in pathogen populations often makes plant disease management more complicated when pathogens overcome host disease resistance (Strange and Scott, 2005). Pathogen species may quickly develop resistance to pesticides or adapt to overcome plant disease resistance, and may also adapt to environmental changes, where the rate of adaptation depends on the type of pathogen (McDonald and Linde, 2002). The potentially rapid onset of disease makes it difficult to anticipate the best timing of management measures, especially in areas with high levels of interannual variability in climatic conditions.

There is relatively less information on the impacts of climate change on plant disease biological control. The few results obtained focus on climate change impacts on the composition and dynamics of the microbial community of the phyllosphere and the soil, which can be very important for plant health (Ghini *et al.*, 2008; Compant *et al.*, 2010). The prediction of the effects of climate change on plant disease biological control is complex and currently based on indirect observations. And one of the major problems with applications of biological control for plant disease management in the field has been the vulnerability of biocontrol agent populations to environmental variation and environmental extremes (Grevstad, 1999; Wong *et al.*, 2002; Garrett *et al.*, 2006; Compant *et al.*, 2010). If appropriate temperature and moisture are not consistently available, biocontrol agent populations may reach densities that are too small to have important effects, and may not recover as rapidly as pathogen populations when conducive conditions recur (Gibson *et al.*, 1999; Garrett *et al.*, 2006). The increased efficiency of *Chlonostachys rosea*, an important biological control agent of *Botrytis* spp. and other pathogens, and *Metarrhizium anisopliae,* one of the most important entomopathogens for insect pest control, has been reported (Rezacova *et al.*, 2005) strongly associated with the cover crop in a high CO_2 concentration environment. The authors suggested the abundance of these fungal species can indicate an increase in the soil suppressiveness to phytopathogenic fungi and other pests.

Climate Change and International Collaboration

The speed of climate change and unpredictability of its characteristics are of great concern to biologists as well as social activists. The number of studies on the effects of climate change has escalated in the last 10 years due to the increase availability of funding to examine climate change related questions. Numerous international and interagency efforts have been undertaken. One example is the International Geosphere-Biosphere Programme: A study of Global Change of the International Council of Scientific Unions; this group has produced an operational plan for a study of global change and terrestrial ecosystem. International, interdisciplinary collaboration on aspects of global change affecting plant disease in natural and managed system will play more important role in years ahead (Scherm and Coakley, 2003). This may range from small, specialist research networks with a relatively narrow focus to broad umbrella projects that encompass multiple science and policy themes and involve numerous intergovernmental agencies as well as national and international donors. A good example of this type is the potato late blight simulation network established jointly by Global Initiative for Late Blight (GILB) and Global Change and Terrestrial Ecosystem (GCTE) project. The overall goal of this particular group is to develop an operational platform for simulating the effects of selected global change drivers on late blight intensity and potato yields on global scale. The Global Environmental Change and Food System Project (GECAFS) represents a recent example of a broader umbrella network that aims to determine strategies to cope with the impact of global change on food provision systems and to analyse the environmental and socio-economic consequences of adoption. This project was launched officially in 2002; the project builds upon and adds value to the work of the International Geosphere-Biosphere Programme (IGBP), the international Human Dimensions Programme on Global Environmental Change (IHDP), and the World Climate Research Programme (WCRP). GECAFS includes three science theme and concentrate on (1) vulnerability and impacts of global change on food provision, (2) adaptation to global change and option for enhancing food provision, and (3) feedbacks in which environmental and socio-economic consequences of adaptation are evaluated. One of the first such projects examines the rice-wheat rotational cropping system that is central to the Indo-Gangetic Plain food system and where yields have been stagnating or declining in recent years. Millions of people depend on this cropping system for staple food grains. Production in this region is highly sensitive to climate variability and may be negatively affected by competitive demands for water. Several plant diseases can have catastrophic impacts on rice-wheat production in this environment and active participation by plant pathologists is, therefore critical to solve new emerging problems.

Conclusion

The climate change effects are challenging to study but of potentially great importance. The impact of climate change on disease for a given plant species will depend on the nature of the effects climate change has on both the host and its pathogens. Climate change could first affect disease directly by either decreasing or increasing the encounter rate between pathogens and host by changing ranges of the two species. Disease severity should be positively correlated with increases in virulence and aggressiveness of pathogens. However, both of these effects on disease will be mediated by host resistance and encounter rates, which in turn are potentially affected by climate change. Thus a positive effect of climate change on conduciveness to infection or pathogen aggressiveness or virulence could be offset by a concurrent increase in resistance, yielding no net change in disease impact. Species at highest risk for an increase in disease will be those with positive effects of climate change on encounter rates, environmental conduciveness to infection, aggressiveness, or virulence, but with neutral or negative effects on resistance. The effects of climate change on all these traits will ultimately be modified by the evolutionary potential of host and pathogen. Certainly, all agree on paucity of knowledge prompting a need to generate new empirical data on host–pathogen biology under a changing climate. Therefore, there is a need to encourage the investigations to study the effects of changing climate on host-pathogen biology to manage the plant diseases in their best effective ways. There is also a need to promote the research based on effect of climate change on biocontrol agents and their interaction with plant pathogens to make these biological control strategies more effective against the plant disease under changing climate.

References

Admassu, B., Lind, V., Freidt, W. and Ordon, F. (2008). Virulence analysis of *Puccinia graminis* f. sp. *tritici* populations in Ethiopia with special consideration of Ug99. *Plant Pathology*. **58**:362-369.

Agrios, G.N. (2005). *Plant pathology*. 5th edition. Elsevier Academic press, London.

Asai, E. and Futai, K. (2005). Effects of inoculum density of pinewood nematode on the development of pine wilt disease in Japanese black pine seedlings pretreated with simulated acid rain. *Forest Pathology*. **35**: 135–144.

Ashmore, M., Toet, S. and Emberson, L. (2006). Ozone–a significant threat to future world food production? *New Phytologist*. **170**: 201–204.

Ashmore, M.R. and Bell, J.N.B. (1991). The role of ozone in global change. *Annals of Botany*. **67**:39-48.

Bale, J.S., Masters, G.J., Hodkinson, I.D., Awmack, C., Bezemer, T.M., Brown, V.K. and Butterfield, J., *et al*. (2002). Herbivory in global climate change research: direct effects of rising temperatures on insect herbivores. *Global Change Biology*. **8**: 1-16.

Bergot, M., Cloppet, E., Perarnaud, V., Deque, M., Marcais, B. and Desprez-Loustau, M.L. (2004). Simulation of potential range expansion of oak disease caused by *Phytophthora cinnamomi* under climate change. *Global Change Biololgy*. **10**:1539–1552.

Boag, B., Crawford, J.W. and Neilson, R. (1991). The effect of potential climatic changes on the geographical distribution of the plant-parasitic nematodes *Xiphinema* and *Longidorus* in Europe. *Nematologica*. **37**:312-323.

Bonello, P., Heller, W. and Sandermann, Jr. H. (1993). Ozone effects on root-disease susceptibility and defence responses in mycorrhizal and non-mycorrhizal seedlings of Scots pine (*Pinus sylvestris* L.). *New Phytologist*. **124**:653–663.

Brasier, C.M. and Scott, J.K. (1994). European oak declines and global warming: a theoretical assessment with special reference to the activity of *Phytophthora cinnamomi*. *Bulletin OEPP/EPPO Bulletin*. **24**:221-232.

Brasier, C.M., Dreyer, E. and Aussenac, G. (1996). *Phytophthora cinnamomi* and oak decline in southern Europe: Environmental constraints including climate change. *Annales des Sciences Forestieres*. **53**:347-358.

Burdon, J.J. (1987). *Diseases and Population Biology*. Cambridge Univ. Press, New York.

Campbell, C.L., Bruck, R.I., Sinn, J.P. and Martin, S.B. (1988). Influence of acidity level in simulated rain on disease progress in four pathosystems. *Environmental Pollution*. **53**: 219-234.

Canadell, J.G., Le Quere, C., Raupach, M.R., Field, C.B., Buitenhuis, E.T., Ciais, P., Conway, T.J., Gillett, N.P., Houghton, R.A. and Marland, G. (2007). Contributions to accelerating atmospheric CO_2 growth from economic activity, carbon intensity, and efficiency of natural sinks. *Proceedings of the National Academy of Sciences*. **104**: 1866-1870.

Carter, T.R., Saarikko, R.A. and Niemi, K.J. (1996). Assessing the risks and uncertainties of regional crop potential under a changing climate in Finland. *Agricultural and Food Science in Finland*. **5**:329-350.

Chakraborty, S. and Datta, S. (2003). How will plant pathogens adapt to host plant resistance at elevated CO_2 under a changing climate? *New Pathologist*. **159**:733-742.

Chakraborty, S. and Pangga, I.B. (2004). Plant disease and climate change. In: Gillings, M. and Holmes, A. (eds.). *Plant microbiology*. BIOS Scientific, London, pp.163-180.

Chakraborty, S., Luck, J., Hollaway, G. Freeman, A., Norton, R., Garrett, K.A. Percy, K., Hopkins, A., Davis, C. and Karnosky, F. (2008). Impacts of global changes on diseases of agricultural crops and forest trees. CAB Reviews: Perspective in agriculture, veterinary science, nutrition and natural resources. **3**(54): 1-15.

Chakraborty, S., Tiedermann, A.V. and Teng, P.S. (2000). Climate change: potential impact on plant diseases. *Environment Pollution*. **108**:317-326.

Clover, G.R.G., Smith, H.G., Azam-Ali, S.N. and Jaggard, K.W. (1999). The effects of drought on sugar beet growth in isolation and in combination with beet yellows virus infection. *Journal of Agricultural Sciences*. **133**: 251–261.

Coakley, S.M. (1979). Climatic variability in the Pacific Northwest and its effect on stripe rust of winter wheat. *Climate Change*. **2**: 33-51.

Coakley, S.M. (1995). Biospheric change: will it matter in plant pathology? *Canadian Journal of Plant Pathology*. **17**:147-153.

Compant, S., van der Heijden, M.G.A. and Sessitsch, A. (2010). Climate change effects on beneficial plant-microorganism Interactions. *FEMS Microbiology Ecology*. **73**: 197–214.

Coviella, C. and Trumble, J. (1999). Effects of elevated atmospheric carbon dioxide on insect-plant interactions. *Conservation Biology*. **13**:700-712.

Elphinstone, J. and Toth, I.K. (2008). *Erwinia chrysanthemi* (*Dikeya* spp.)–the facts. Potato Council, Oxford, U.K.

Evans, N., Baierl, A., Semenov, M.A., Gladders, P. and Fitt, B.D.L. (2008). Range and severity of a plant disease increased by global warming. *Journal of Royal Society Interface*. **5**:525-531.

Fallon, S. (1997). A few degrees makes a world of difference. *The Planet Newsletter*. **4**:1–4.

Fenn, M.E., Dunn, P.H. and Wilborn, R. (1990). Black stain root disease in ozone-stressed ponderosa pine. *Plant Disease*. **74**:426–430.

Fuhrer, J. (2003). Agroecosystem responses to combinations of elevated CO_2, ozone, and global climate change. *Agriculture, Ecosystems and Environment*. **97**:1–20.

Garrett, K.A., Dendy, S.P., Frank, E.E., Rouse, M.N. and Travers, S.E. (2006). Climate change effects on plant disease: genomes to ecosystems. *Annual Review of Phytopathology*. **44**: 489-509.

Gaumann, E. (1950). *Principles of plant infection*. Crosby Lockwood, London, pp. 543.

Ghini, R. (2005). Mudanças climáticas globais e doenças de plantas. Jaguariúna: Embrapa Meio Ambiente, pp. 104.

Ghini, R., Hamada, E. and Bettiol, W. (2008). Climate change and plant diseases. *Scientia Agricola*. **65**: 98-107.

Gibson, G.J., Gilligan, C.A. and Kleczkowski, A. (1999). Predicting variability in biological control of a plant-pathogen system using stochastic models. *Proceedings of Royal Society of London Ser.* B **266**:1743–1753.

Gioria, R., Brunelli, K.R. and Kobori, R.F. (2008). Impacto potencial das mudanças climáticas sobre as doenças de hortaliças: tomate, um estudo de caso. *Summa Phytopathologica*. **34**:121-122.

Gore, A. (2006). An inconvenient truth: The planety emergency of global warming and what we can do about it. Rodale, Emmaus, PA.

Grevstad, F.S. (1999). Factors influencing the chance of population establishment: implications for release strategies in biocontrol. *Ecological Application*. **9**:1439–1447.

Hamilton, J.G., Dermody, O., Aldea, M., Zangerl, A.R., Rogers, A., Berenbaum, M.R. and Delucia, E. (2005). Anthropogenic Changes in Tropospheric Composition Increase Susceptibility of Soybean to Insect Herbivory. *Environmental Entomology*. **34**(2): 479-485.

Hansen, J., Sato, M., Ruedy, R., Lo, K., Lea, D.W. and Medinaelizade, M. (2006). Global temperature change. *Proceedings of the National Academy of Sciences*. **103**:14288-14293.

Houghton, J.T., Ding, Y., Griggs, D.J., Noguer, M., van der Linden, P.J. and Xiaosu, D. (2001). Climate change: The scientific basis (Houghton, J.T., Ding, Y., Griggs, D.J., Noguer, M., van der Linder, P.J., Dai, X., Maskell, K. and Johnson, C.A. (eds.), pp. 1–83. Cambridge University Press, Cambridge, UK.

Huber, L. and Gillespie, T.J. (1992). Modeling leaf wetness in relation to plant disease epidemiology. *Annual Review Phytopathology*. **30**:553–577.

Hunter, M.D. (2001). Effects of elevated atmospheric carbon dioxide on insect-plant interactions. *Agricultural and Forest Entomology*. **3**:153-159.

Intergovernmental Panel on Climate Change (2007). Summary for Policymakers. In: Solomon, S., Qin, D., Manning, M., Chen, Z., Marquis, M. and Averyt, K.B., *et al.* (eds.). Climate Change 2007: The Physical Science Basis. Contribution of Working Group I to the Fourth Assessment Report of the Intergovernmental Panel on Climate Change. Cambridge University Press, New York, USA.

Intergovernmental Panel on Climate Change (2000). Special Report. Emission Scenarios. Summary for Policymakers. Available from: URL: http://www.ipcc.ch/pdf/special-reports/ spm/sres-en.pdf.

IPCC. (1996). Climate Change 1995: The Science of Climate Change. Houghton, J.T., Meira Filho, L.G., Callander, B.A., Harris, N., Kattenberg, A. and Maskell, K. (eds.). Contribution of Working Group I to the Second Assessment Report of the Intergovernmental Panel on Climate Change. Cambridge University Press, Cambridge, UK.

IPCC. (2001). Climate change 2001: the scientific basis. Contribution of Working Group I to the Third Assessment Report of the Intergovernmental Panel on Climate Change. Cambridge University Press, Cambridge, UK, New York.

Karnosky, D.F., Percy, K.E., Xiang, B., Callan, B. and Noormets, A., *et al.* (2002). Interacting elevated CO_2 and tropospheric O_3 predisposes aspen (*Populus tremuloides* Michx.) to infection by rust (*Melampsora medusae* f. sp. *tremuloidae*). *Global Change Biology*. 8:329–338.

Karnosky, D.F., Skelly, J.M., Percy, K.E. and Chappelka, A.H. (2007). Perspectives regarding 50 years of research on effects of tropospheric ozone air pollution on US forests. *Environmental Pollution*. **147**:489–506.

Khan, M.R. and Khan, M.W. (1999). Effects of intermittent ozone exposures on powdery mildew of cucumber. *Environmental and Experimental Botany*. **42**:163–171.

Klement, Z., Rudolph, K. and Sands, D.C. (eds.). (1990). *Methods in Phytobacteriology*. Academia Kiado, Budapest.

Kobayashi, T., Ishiguro, K., Nakajima, T., Kim, H.Y., Okada, M. and Kobayashi, K. (2006). Effects of elevated atmospheric CO_2 concentration on the infection of rice blast and sheath blight. *Phytopathology*. **96**:425-431.

Krupa, S., McGrath, M.T., Andersen, C.P., Booker, F.L., Burkey, K.O. and Chappelka, B.I., *et al.* (2000). Ambient ozone and plant health. *Plant Disease*. **85**:4–12.

Kudela, V. (2009). Potential impact of climate change on geographic distribution of plant pathogenic bacteria in Central Europe. *Plant Protection Science*. **45**:27–32.

Lake, J.A. and Wade, R.N. (2009). Plant-pathogen interactions and elevated CO_2: morphological changes in favour of pathogens. *Journal of Experimental Botany*. **60**: 3123-3131.

Luo, Y., Tebeest, D.O., Teng, P.S. and Fabellar, N.G. (1995). Simulation studies on risk analysis of rice leaf blast epidemics associated with global climate change in several Asian countries. *Journal of Biogeography*. **22**:673-678.

Malmstrom, C.M. and Field, C.B. (1997). Virus-induced differences in the response of oat plants to elevated carbon dioxide. *Plant, Cell and Environment*. **20**:178-188.

Manning, W.J. (1975). Interactions between air pollutants and fungal, bacterial and viral plant pathogens. *Environmental Pollution*. **9**:87–90.

McDonald, B.A. and Linde, C. (2002). Pathogen population genetics, evolutionary potential and durable resistance. *Annual Review Phytopathology*. **40**:349–379.

National Research Council. (2002). Predicting invasions of non-indigenous plants and plant pests. National Academy Press, Washington, DC.

Newton, A.C., Johnson, S.N. and Gregory, P.J. (2010). Implication of climate change for diseases, crop yields and food security. BGRI 2010, Technical Workshop, 30-31st May, 2010, St. Petersburg, Russia, pp. 1-17.

Olson, A.J., Pataky, J.K., D'Arcy, C.J. and Ford, R.E. (1990). Effects of drought stress and infection by maize dwarf mosaic virus in sweet corn. *Plant Disease*. **74**:147–151.

Osswald, W.F., Fleischmann, F. and Heiser, I. (2006). Investigations on the effect of ozone, elevated CO_2 and nitrogen fertilization on host-parasite interactions. *Summa Phytopathologica*. **32**:111-113.

Peng, S., Huang, J., Sheehy, J.E., Laza, R.C. and Visperas, R.M., *et al.* (2004). Rice yields decline with higher night temperature from global warming. *Proceeding of National Academy of Sciences*. **101**:9971–9975.

Perombelon, M.C.M. (1982). The impaired host and soft rot bacteria. In: Mout, M.S. and Lacy, G.H. (eds.). *Phytopathogenic Prokaryotes*. Vol. 2. Academic Press, New York.

Perombelon, M.C.M. and Kelman, A. (1980). Ecology of the soft rot erwinias. *Annual Review of Phytopathology*. **18**: 361–387.

Pfleeger, T.G., da Luz, M.A. and Mundt, C.C. (1999). Lack of a synergistic interaction between ozone and wheat leaf rust in wheat swards. *Environmental and Experimental Botany*. **41**:195–207.

Plazek, A., Hura, K., Rapacz, M. and Zur, I. (2001). The influence of ozone fumigation on metabolic efficiency and plant resistance to fungal pathogens. *Journal of Applied Botany*. **75**: 8–13.

Plessl, M., Elstner, E.F., Rennenberg, H., Habermeyer, J. and Heiser, I. (2007). Influence of elevated CO_2 and ozone concentrations on late blight resistance and growth of potato plants. *Environmental and Experimental Botany*. **60**:447–457.

Pritchard, S.G. and Amthor, J.S. (2005). *Crops and environmental change*. Food Products Press, Binghamton, pp. 421.

Pritchard, S.G., Rogers, H.H., Prior, S.A. and Peterson, C.M. (1999). Elevated CO_2 and plant structure: a review. *Global Change Biology*. **5**:807–837.

Pritsch, K., Luedemann, G., Matyssek, R., Hartmann, A., Schloter, M. and Scherb, H., *et al.* (2005). Mycorrhizosphere responsiveness to atmospheric ozone and inoculation with *Phytopthora citricola* in a phytotron experiment with spruce/ beech mixed cultures. *Plant Biology*. **7**:718–727.

Redak, R.A., Purcell, A.H., Lopes, J.R.S., Blua, M.J. and Mizell, R.F. (2004). The biology of xylem fluid–feeding insect vectors of *Xylella fasidiosa* and their relation to disease epidemiology. *Annual Review of Entomology*. **49**: 243–270.

Rezacova, V., Blum, H., Hrselova, H., Gamper, H. and Gryndler, M. (2005). Saprobic microfungi under *Lolium perenne* and *Trifolium repens* at different fertilization intensities and elevated atmospheric CO_2 concentration. *Global Change Biology*. **11**: 224-230.

Sah, D.N., von Tiedemann, A. and Fehrmann, H. (1993). Effects of ozone exposure on resistance of wheat genotypes to *Pyrenophora tritici-repentis*. *Journal of Phytopathology*. **138**:249–256.

Sandermann, H. Jr. (2000). Ozone/biotic disease interactions: molecular biomarkers as a new experimental tool. *Environmental Pollution*. **108**:327–332.

Schaad, N.W. (2008). Emerging plant pathogenic bacteria and global warming. In: Fatmi, M.B., Collmer, A., Iacobellis, N.S., Masfield, J.W., Murillo, J., Schaad, N.W. and Ulrich, M. (eds.). *Pseudomonas syringae* Pathovars and Related Pathogens–Identification Epidemiology and Genomics. Springer, Dordrecht, pp. 369–379.

Scherm, H. and Coakley, S.M. (2003). Plant pathogens in changing world. *Australasian Plant Pathology*. **32**:157-165.

Scherm, H. and Yang, X.B. (1995). Interannual variations in wheat rust development in China and the United States in relation to the El Nino/Southern Oscillation. *Phytopathology*. **85**:970-976.

Seem, R.C. (2004). Forecasting plant disease in a changing climate: a question of scale. *Canadian Journal of Plant Pathology*. **26**:274–283.

Siegenthaler, U., Stocker, T.F., Monnin, E., Luthi, D., Schwander, J., Stauffer, B., Raynaud, D., Barnola, J.M., Fischer, H., Delmotte, V. M. and Jouzel, J. (2005). Stable carbon cycle-climate relationship during the late pleistocene. *Science*. **310**:1313-1317.

Spahni, R., Chappellaz, J., Stocker, T.J., Loulergue, L., Hausammann, G., Kawamura, K., Fluckiger, J., Schwander, J., Raynaud, D., Delmotte, V.M. and Jouzel, J. (2005). Atmospheric methane and nitrous oxide of the late pleistocene from Antarctic ice cores. *Science*. **310**:1317-1321.

Starr, M.P. (eds.) (1983). *Phytopathogenic Bacteria*. Springer-Verlag, New York.

Strange, R.N. and Scott, P.R. (2005). Plant disease: A threat to global food security. *Annual Review of Phytopathology*. **43**:83- 116.

Vingarzan, R. (2004). A review of surface ozone background levels and trends. *Atmospheric Environment*. **38**:3431–3442.

von Tiedemann, A. and Firsching, K.H. (2000). Interactive effects of elevated ozone and carbon dioxide on growth and yield of leaf rust-infected versus non-infected wheat. *Environmental Pollution*. **108**:357–363.

von Tiedemann, A., Weigel, H. and Jager, H.J. (1991). Effects of open-top chamber fumigations with ozone on 3 fungal leaf diseases of wheat and the mycoflora of the phyllosphere. *Environmental Pollution*. **72**:205–224.

Wong, P.T.W., Mead, J.A. and Croft, M.C. (2002). Effect of temperature, moisture, soil type and *Trichoderma* species on the survival of *Fusarium pseudograminearum* in wheat straw. *Australasian Plant Pathology*. **31**:253–257.

Woods, A., Coates, K.D. and Hamann, A. (2005). Is an unprecedented Dothistroma needle blight epidemic related to climate change? *Bioscience*. **55**:61-69.

Yamamura, K. and Kiritani, K. (1998). A simple method to estimate the potential increase in the number of generations under global warming in temperate zones. *Applied Entomology and Zoology*. **33**:289- 298.

Yarwood, C.E. (1983). History of plant pathogen introduction. In: Wilson, C.L. and Graham, C.L. (eds.). *Exotic plant pests and North American agriculture*. Academic press, New York, pp. 39-63.

Chapter 8

Fungal Diversity in Leguminous Crop Field Soil

S.K. Dwivedi and Sangeeta

Department of Environmental Science, Babasaheb Bhimrao Ambedkar (A Central) University, Lucknow – 226 025, U.P.

Biological diversity encompasses the variety of life forms occurring in nature, from the ecosystem to the genetic level, as a result of evolution (Wilson, 1992). According to Nannipieri *et al.* (2003) biodiversity studies include number and species richness of the concerned taxa and genetic diversity or in other words biodiversity refers to the variability of life on Earth in all the living species of animals, plants and microorganisms. According to Hawksworth (2002), fungi are a major component of biodiversity essential for the survival of other organisms and are vital in global ecological processes.

Soil is a very species-rich habitat containing all major groups of microorganisms like bacteria, algae, protists and fungi (Hagvar, 1998). Soil contains the most complex and dynamic microbial assembly in the biosphere (Curtis *et al.*, 2002).It is a complex ecosystem surrounded by physico-chemical parameters that grasp huge number of living organisms. In the agricultural field soil organisms provide benefits to crop growing in an ecosystem. The soil microbes decompose the dead remains of plants and animals entering the soil and convert them into organic matter which affects the physical, chemical and biological properties of the soil (Olsen *et al.*, 1965).

Soil biodiversity plays a very important role in the sustainability of agriculture systems and it also indicates the level of health of soil, particularly while considering the richness of microorganisms which are involved in the biological control of soil-borne diseases. Cultural practices may produce changes in soil micro flora (Vargas Gil *et al.*, 2009). Soil quality is a suitable indicator of the effectiveness of sustainable agro ecosystem management. Soil microorganisms may reflect the changes in soil quality since their population dynamics describes the status and trend of soil conditions in response to management practices (Johnson *et al.*, 1960; Doran and Parkin, 1994; Mazzola, 2004).

The term microbial diversity comprises the (i) genetic diversity *i.e.* the amount and distribution of genetic information in microbial species; (ii) diversity of bacterial and fungal species in microbial communities and (iii) ecological diversity *i.e.* variation in community structure, complexity of interactions, number of trophic levels, and number of guilds. Microbial diversity simply can be defined as the number of different fungal and bacterial species (richness) and their relative abundance (evenness) in soil micro flora. Equations used to calculate diversity indices and species richness and evenness have been discussed by Kennedy & Smith (1995).

Microbial communities can also be used as soil health parameters because they maintain the ecological equilibrium between pathogens and biological control agents. They naturally suppresses the incidence of diseases. Microbial diversity includes the number of different fungal and bacterial species and their relative abundance (Nannipieri *et al.*, 2003). Microbial activity indicate the huge range of activities being carried out by microorganisms in the soil whereas biological activity includes microbial activities as well as the activities of other organisms in the soil including plant roots (Nannipieri *et al.*, 1990).

Fungi play vital roles in ecosystem function (Christensen, 1989; Doran and Parkin, 1994, 1996; Hawksworth *et al.*, 1996). It mediates plant health and promotes growth through mycorrhizal and parasitic associations. Fungi and bacteria are the dominant microorganisms which are involved in *C/N* cycling and in the degradation of organic matter. The filamentous fungi are the major contributors to the soil biomass (Alexander, 1977). Fungi account for up to 90 per cent of total living biomass in forest soils (Frankland, 1982). Faegri *et al.*, 1977; Anderson and Domsch, 1973 found fungal respiration to be two to four times greater than that of bacteria.

Fungi are heterotrophic, eukaryotic organisms. They do not have the capacity to produce their own food and therefore they are completely dependent on preformed organic matters. They do not have photosynthetic or chemosynthetic pigments. Fungi colonize, multiply and survive in diversified habitats, *i.e.* water, soil, air, litter, dung, foam, etc. They are ubiquitous and cosmopolitan in distribution from tropics to poles and from mountain tops to the deep oceans Hawkswork *et al.* (1995). About 90 percent of all the biological processes going on in the soil are known to be carry out by fungi, along with some soil bacteria (Nannipieri *et al.*, 2003).

Fungi are among one of the few most diverse organism groups (Hammond, 1995). They represent the second largest group in the world after insects (Hawksworth, 1991, 2001). They are the major components of tropical ecosystems throughout the world and are involved in numerous interactions with plants, animals and man such as saprophytism, parasitism and symbiosis etc.

Soil fungi play an important role as major decomposers in the soil ecosystem. They also provide very useful pharmaceutical products, such as antibiotics and other valuable substances including organic acids, enzymes, pigments and secondary metabolites used in the food industry and fermentation. Beside these, many soil fungi are used as biological control agents against plant pathogens and insect pests (Manoch, 1998 ; Giller,1996).

Baath, 1981; Babjeva and Chernow, 1995; Babjeva and Reshetova, 1998; Benkova, 1999; Buckova *et al.*, 2000; Cabello and Arambarri, 2002; Kok *et al.*, 1984; Slavikova and Vadkertiova, 2000; Vishniac, 1996 carried out the studies worldwide on the diversity of micro fungi and yeasts in soil. A lot of studies on soil microorganisms have been carried out on forest and grassland soils (Mishra, 1966; Christensen, 1969; Lewis *et al.*, 1971; Widden, 1979) but microbial studies on agricultural soil have received a very less attention (Soderstrom *et al.*, 1983). The myxomycetes have also been studied traditionally by mycologists (Everhart & Keller, 2008; Rojas and Stephenson, 2008).

Micro fungi play a crucial role in nutrient cycling by regulating soil biological activity (Hao-quin *et al.*, 2008) and in soil formation, soil fertility, soil structure and soil improvement (Hao-quin *et al.*, 2008).They forms the major group of organotrophic organisms responsible for the decomposition of organic compounds. Their activity participates in the biodeterioration and biodegradation of toxic substances in the soil (Rangaswami and Bagyaraj, 1998). It has been found that maximum number of fungi exist in soil than in any other environment (Nagmani *et al.*, 2005).

The leguminous crops can reduce the amount of N-fertilization required for successive crops. However, this strongly depends on the type of leguminous crop, *e.g.* peas may negatively affect N-balance in the soil due to significant nitrogen removal by a rich grain harvest. Besides providing protein rich grain yields, the forage legumes are known to improve agricultural sustainability. They have ability to decrease soil erosion, to maintain soil organic matter, and to improve the soil structure. The use of legume cropping for improving the soil properties is a highly recognized strategy in agriculture. Soil is a complex and dynamic biological system, and still in 2003 it is difficult to determine the composition of microbial communities in soil.It is a complex microhabitat for the following distinctive properties (Nannipieri & Badalucco, 2003).

Kirk *et al.* (2008) reported 1039 species chromistan fungal analogues and 1165 as protozoan in which 1038 are regarded as protozoan fungal analogues: *Percolozoa (Acrasida), Amoebozoa (Dictyostelia, Myxogastria, Protostelia), Cercozoa (Plasmodiophorida)* which were previously treated as *Myxomycota* and *Plasmodiophoromycota*.

About 205 new genera have been described from India, of which 32 per cent have been discovered by (Subramanian). Out of these, approximately 27,000 species are reported to colonize diversified habitats (Sarbhoy *et al.*, 1982-1992). The kingdom of fungi contains 1.5 million fungal species, of which 74,000 species are named Hawkswork *et al.* (1995). Manoharachary and his co-workers (2001) have added 12 new genera, 60 new taxa and 500 new additions to fungi of India.

Extraordinary numbers of genera and species of fungi including many new taxa have been isolated from soils of varied habitats: from barren desert sands to forest soils rich in humus. There are many reports on biodiversity studies of soil fungi carried out in the dense forests of the Western Ghats (Satish *et al.*, 2007), deciduous forest soils (Vishwanathan, 2010) of India, hilly terrains of Israel (Grishkan and Nevo, 2008),damp and musty caves in Puerto-Rico (Nieves-Rivera, 2003) and icy plains of the Antarctic (Ruisi *et al.*, 2007).

About one -third of the fungal diversity of the globe is believed to exist in India, referred to as 'the cradle for diverse groups of fungi' (Manoharachary *et al.*, 2005). There are about 1.5 million species of fungi out of which only 5 percent are described and out of one million species of bacteria only about 5,000 have been described (Tilak, 2000). Hawksworth (1991) compared the estimated and the actually described number of species of bacteria, fungi, algae and viruses with the culture collections, and found that the described ones are few while those actually cultured in the laboratory are still fewer.

Fungi are important components of biodiversity in tropical forest. It is a major contributor for the maintenance of earth's ecosystem, biosphere and biogeochemical cycles, they perform unique and vital activities on which organisms including human depend. Bilgrami *et al.*, 1991 reported numerous species of fungi from the Western Ghats, there appears to have been no study related to the diversity or dynamics of fungal population in forest.The tropics are generally considered as storehouse for fungi, large number of new fungal species have been collected from the tropics from time to time by mycologists. New fungi have been analyzed from the different parts of world during 1981-90 in which around 50 per cent were only discovered in the tropics (Hawksworth, 1993). The number of species of fungi described from India is around 6900 (Bilgrami *et al.*, 1991). Cannon, 1997; Hyde *et al.*, 2007 estimated that global fungal numbers ranges from less than one million to more than nine million, even though 1.5 million is generally considered to be a reasonably accurate working figure (Hyde *et al.*, 2007).

Fungi at Different Soil Profile

Usually the top soil contains high organic matter which in the presence of adequate supply of moisture is acted upon by the microorganisms to decompose the complex organic substances into simpler inorganic forms; hence the microbial population is higher in the surface soil layer (Shamir and Steinberger, 2007 ; Classen *et al.*, 2007) as compared to the lower depths. Beside this the distribution of microbial population is soil is affected by a number of environmental factors like pH, Moisture content and soil organic matter (Kennedy *et al.*, 2005). Arunachalam *et al.*, 1997 reported the higher fungal population during rainy and autumn seasons. He observed that the litter and other dead plant parts decompose faster during rainy season and as a result sufficient amount of soil organic matter and humus accumulates which enhances the colonization of soil microbes. However, Shukla *et al.* (1989) found negligible difference in fungal population across depths.

Dominance of soil fungi in different ecosystem are as below:

Ecosystem Type	*Relative Dominance*		
	Dominant	*General*	*Rare*
Agriculture	*Aspergillus, Fusarium, Monodyctis*	*Alternaria, Cladosporiun, Doratomyces, Drechslera, Gliocladium, Gloeosporium, Humicola, Penicillium, Trichoderma, Verticicladiella*	*Alysidium. Bipolaris, Candida, Cephalosporium, Curvularia, Emonsia, Gliocephalis, Gliomastix, Nigrospora, Paecilomyces, Papulospora, Rhizoctonia, Sclerotium, Thermomyces, Verticillium*
Barren	*Alternaria, Aspergillus, Cladosporium, Sporobolomyces*	*Acremonium, Cylindrocarpon, Gliocladium, Gloeosporium, Myrothecium, Paecilomyces, Phialomyces, Staphylotrichum, Trichoderma*	*Allesscheriella, Auriobasidium, Bipolaris, Cephalosporium, Curvularia, Epicoccum, Fusarium, Gilmaniella, Gonatobotrium, Humocola, Papulospora, Penicillium, Periconia, Sclerotium, Scytilidium, Verticicladiella, Verticillium*
Garden	*Aspergillus, Cladosporium, Doratomyces, Penicillium, Trichocladium, Trichoderma, Verticicladiella*	*Alternaria, Curvularia, Fusarium, Geotrichum, Gliocladium, Gloeoesporium, Verticillium, Drechslera*	Nil

Source: Wahegaonkar *et al., Annals of Biological Research*, 2011, 2 (2):198-205.

Effect of Physical Parameters on the Population of Fungi

Fungal diversity of any soil depends on a large number of factors of the soil such as pH, organic contents, and moisture (Alexander 1977, Rangaswami and Bagyaraj 1998). Environmental factors controlls the distribution and abundance of soil microorganisms are still poorly understood. However, much of this studies has been conducted using various techniques they do not permit detailed and comprehensive phylogenetic or taxonomic surveys of microbial communities (for example, DNA fingerprinting-based approaches, (Fierer and Jackson, 2006), phospholipid fatty acid (PLFA) analyses, (Ba°a°th and Anderson, 2003).Much of this previous studies have also paid attention on the distribution patterns exhibited by single taxonomic groups, for example, Acidobacteria, (Jones *et al.*, 2009, fungi: Bennett *et al.* (2009) and Bue´e *et al.* (2009), or SR1 bacteria: Davis *et al.* (2009).

There are several workers who worked on the microbial distribution in the soil. Most of them studied on pH and some of them on microbial biomass (Lauber *et al.*,

2008 ; Rousk *et al.*, 2009 ; Aciego Pietri and Brookes, 2007b, 2009 ; Rousk *et al.*, 2010a; Fierer *et al.*, 2008; Hamady *et al.*, 2008 ; Lauber *et al.*, 2009 ; Nilsson *et al.*, 2007; Ba°a°th and Anderson, 2003 ; Wheeler *et al.*, 1991; Nevarez *et al.*, 2009 ; Domsch *et al.*, 1980 ; Beales, 2004).

Plant Growth Promoting Fungi

Like PGPR (plant growth promoting rhizobacteria), some rhizosphere fungi able to promote the plant growth upon root colonization and are functionally designated as 'plant-growth-promoting-fungi' (PGPF) (Hyakumachi, 1994).PGPF belong to genera *Penicillium, Trichoderma, Fusarium* and *Phoma*. The plant growth promoting fungi (PGPF) are associated with plant roots and they secrete a number of secondary metabolites including GAs in the rhizosphere (Hamayun *et al.*, 2009).

Evidences of naturally occurring rhizospheric phosphorus solubilizing microorganism (PSM) dates back to 1903 (Khan *et al.*, 2007). Alam *et al.*, 2002 reported that bacteria are more effective in phosphorus solubilization as compared to fungi. Among the entire microbial population present in soil, PSB constitute 1 to 50 per cent, while phosphorus solubilizing fungi (PSF) are only 0.1 to 0.5 per cent in P solubilization potential (Chen *et al.*, 2006; Kucey 1983).

Microorganisms which are involved in phosphorus acquisition include mycorrhizal fungi and PSMs (Fankem *et al.*, 2006). Among the soil fungal communities, *Penicillium* and *Aspergiilus* fungi have been described as effective phosphate solubilizers (Whitelaw, 2000).

Usually, the P-solubilizing fungi produce more acids than bacteria as a result exhibit greater P-solubilizing activity (Venkateswarlu *et al.*, 1984). After a study done by Srivastav *et al.* (2004) reported about the P solubilization and antifungal activity of *Aspergillus niger, Curvularia lunata, Rhizoctonia solani* and *Fusarium oxysporium*.He suggested about the beneficial effect of these microorganisms in increasing the crop productivity. Later on in another study, *Aspergillus tubingensis* and two isolates of *Aspergillus niger* have also shown the highest solubilization of RP under *in vitro* condition (Reddy *et al.*, 2002). Rudresh *et al.* (2005) studied the capability of nine isolates of *Trichoderma* spp. and found their ability to solubilize insoluble P.All nine *Trichoderma* isolates solubilized insoluble tri-calcium phosphate (TCP) to various extents.

The P-solubilizing microorganisms can belong to any of the microbial groups, are ubiquitous whose numbers vary from soil to soil and are influenced greatly by nutritional status of soils and environmental factors.The concentration of iron ore, temperature, C and N sources greatly effect the P-solubilizing potentials of these microbes. Among the various nutrients used by these microorganisms are ammonium salts that has been found to be the best N source followed by asparagine, sodium nitrate, potassium nitrate, urea and calcium nitrate (Ahuja *et al.*, 2007).

Recently, a total of 62 fungi have been obtained from heavy metal mines of Orissa (India) and were evaluated for their P-solubilizing potential (Gupta *et al.*, 2007). And it is found that of these 62 fungi, 12 fungi solubilized TCP; the highest P-solubilizing potential was being displayed by *Penicillium* sp. 21, which released 81.48 mg P ml71 of TCP, followed by *Penicillium* sp. 2 that produced 4.87 mg P ml 71 into the liquid culture.

Aspergillus and *Penicillium* are two important genera of phosphate solubilizing fungi (Omar, 1998; Seshardi *et al.*, 2004; Wakelin *et al.*, 2004).This ability is generally associated with the release of organic acids, decreasing the pH (Seshardi *et al.*, 2004).

The inoculation of P-solubilizing microorganisms is a promising technique because it can increase P availability in soils fertilized with rock phosphates (Reyes *et al.*, 2002). Several authors reported yield increasing on wheat (Whitelaw *et al.*, 1997), onion (Vassilev *et al.*, 1997), alfalfa (Rodríguez *et al.*, 1999) and soybean (Abd-Alla *et al.*, 2001) through inoculation of P-solubilizing fungi (PSF). Inoculation of phosphate solubilizing fungi and mycorrhizal fungi improves the physio-chemical, biochemical and biological properties of rock-P amended soil (Caravaca *et al.*, 2004). Whitelaw *et al.*, 1997 recommended the application of P solubilizing fungi is as a sustainable way for increasing cropyield. Many reports had shown the improvement in plant growth using P-solubilizing fungi.

Recent advances in DNA sequence technologies and analytical methods have changed fungal systematics. Molecular characters such as DNA sequence data are essentially beneficial as they offer a greater number of distinct characters, which can be analyzed statistically to conclude phylogenetic relationships (Shenoy *et al.*, 2007).

A novel DNA barcoding system for fungal identification at the species level using mitochondrial *cytochrome c oxidase 1 (CO1)* sequences (Seifert *et al.*, 2007); *COX1*, ITS and D1/D2 sequences (Letourneau *et al.*, 2010) and multiple loci (Roe *et al.*, 2010) have been reported. Phylogenetic analyses are then performed using different methods based on nuclear ribosomal DNA sequences, 5.8S gene region (Huang *et al.*, 2009); ITS sequences (Nilsson *et al.*, 2008) and protein-coding gene sequences (Tang *et al.*, 2009) followed by application of different sequence analysis methods (Peláez *et al.*, 2008).

High-resolution analyses allow the detection of microbial strains at the species and subspecies level. They usually give 'fingerprints' of non-coding DNA regions or involve the sequencing of both coding and non-coding regions. These techniques include rep-polymerase chain reaction (PCR) amplification of sequences between repetitive elements, ribosomal inter space analysis (RISA), which is based on the length polymorphism of the spacer region between 16S and 23S rRNA genes (Borneman & Triplett, 1997), and random amplified polymorphic DNA (RAPD), which does not require a preliminary knowledge of the genome (Nei & Li, 1979). In RISA, PCR products are separated by gel electrophoresis, and the separated bands can be sequenced. A limit of this technique is the number of spacer sequences in the database. Recently, Borneman (1999) monitored microorganisms responding to nutrient addition to soil with the thymidine analogue bromodeoxyuridine. Bromodeoxyuridinelabelled DNA extracted from soil DNA by immunocapture was subjected to RISA analysis.

Species identification and classification by morphological methods can be supplemented by terminal restriction fragment length polymorphism (T-RFLP) analysis of the Internal Transcribed Spacer (ITS) sequences (Ortega *et al.*, 2008); (Cui *et al.*, 2008); membrane-based ITS macroarrays coupled with community ITS probes (Izzo and Mazzola, 2009); fluorescein diacetate (FDA) hydrolysis, single carbon source substrate utilization (SU) profiles and fatty acid methyl ester (FAME) profiles (Larkin, 2003); 18S rRNA profiles; fungal-specific PCR of soil DNA coupled with denaturing gradient gel electrophoresis (DGGE) (Oros-Sichler *et al.*, 2006); EF1-α nucleo-tide sequences (Alves *et al.*, 2008); GC content analysis (Nusslein and Tiedje, 1999) and DNA-DNA hybridization (Greene and Voordouw, 2003).

The assessment of fungal diversity in soil by molecular techniques has not been as successful as the characterization of bacterial diversity because the concentration of fungal DNA is much less than that of bacterial DNA (Borneman & Hartin,2000). We can now use DGGE and TGGE to generate fingerprints for the fungal community of soil because specific primers are available for fungal 18S rRNA (Smit *et al.*, 1999; Borneman & Hartin, 2000; van Elsas *et al.*, 2000). Another approach used to overcome the problem of selective culturing for assessing the composition of soil microflora is phospholipid fatty acid (PLFA) analysis (Tunlid & White,1992; Frostega? rd & Ba? a? th, 1996; Bossio & Scow, 1998; Zelles, 1999; Pankhurst *et al.*, 2001). This technique is based on the extraction, fractionation, methylation and chromatography of the phospholipid component of soil lipids. Phospholipids are thought to be related to the viable component of soil microflora because they are present as important components of membranes of living cells and break down rapidly when the cells die, and they cannot survive long enough to interact with soil colloids (Zelles, 1999).

The observed diversity of soil fungi largely depends on the method of isolation used and the numbers of isolates obtained. Particle-plating usually yields higher numbers of taxa than dilution plating (Gams 1992) and the curve of numbers of species versus numbers of isolates is initially steeper with the former technique.

Comprehensive reviews on the microbial community profiling methods in vogue were recently written by Kennedy and Clipson (2004), Leckie (2005) and Nocker *et al.* (2007), where as Hyde and Soytong (2007) provide a critical evaluation of the advances in microfungal diversity.

References

Abd-Alla, M.H., Omar, S.A. and Omar, S.A. (2001). Survival of rhizobia/ bradyrhizobia and a rockphosphate- solubilizing fungus *Aspergillus niger* on various carriers from some agro-industrial wastes and their effects on nodulation and growth of faba bean and soybean. *Journal of Plant Nutrition*, 24 **(2):** 261-272.

Aciego Pietri, J.C. and Brookes, P.C. (2007b). Relationships between soil pH and microbial properties in a UK arable soil. *Soil Biol Biochem* **40:** 1856–1861.

Aciego Pietri, J.C. and Brookes, P.C. (2009). Substrate inputs and pH as factors controlling microbial biomass, activity and community structure in an arable soil. *Soil Biol Biochem* **41:** 1396–1405.

Ahuja, A., Ghosh, S.B. and D'Souza S.F. (2007). Isolation of a starch utilizing, phosphate solubilizing fungus on buffered medium and its characterization. *Bioresource Technol.* **98:** 3408–3411.

Alam, S., Khalil, S., Ayub, N. and Rashid, M. (2002). In *vitro* solubilization of inorganic phosphate by phosphate solubilizing microorganism (PSM) from maize rhizosphere. *Intl. J. Agric. Biol.* **4:** 454-458.

Alexander, M. (1977). Introduction to soil microbiology. Wiley, New York

Alves, A., Crous, P. Correia, A. and Phillips, A. (2008). Mor-phological and molecular data reveal cryptic speci-ation in Lasiodiplodia theobromae. *Fungal Diversity,* **28:** 1-13.

Anderson, J.P.E. and Domsch, K.H. (1973). Quantification of bacterial and fungal contribution to soil respiration. *Arch Microbiol* **93:** 113–27.

Arunachalam, K., Tripathi, R.S. and Pandey., H.N. (1997). *Trop.Ecol.,* **38:** 333-341.

Ba°a°th, E. and Anderson, T.H. (2003). Comparison of soil fungal/ bacterial ratios in a pH gradient using physiological and PLFA-based techniques. *Soil Biol Biochem* **35:** 955–963.

Baath, E. (1981): Microfungi in a clear-cut pine forest soil in central Sweden. *Can. J. Bot.* 59: 1331–1337.

Babjeva, I. and Chernow, I.Y. (1995): Geographic aspects of yeast ecology. In: TURPAEV, T. M. (ed.): Physiology and general biology reviews 9, Part. 3. Harwood Academic Publishers, New York, 1–54.

Babjeva, I. and Reshetova, I. (1998). Yeast resources in Natural Habitats at Polar Circle Latitude. *Food technol. biotechnol.* 36 **(1):** 1–5.

Beales, N. (2004). Adaptation of microorganisms to cold temperature, weak acid preservatives, low pH, and osmotic stress: a review. *Comp Rev Food Sci* F **3:** 1–20.

Bennett, L.T., Kasel, S., and Tibbits, J. (2009). Woodland trees modulate soil resources and conserve fungal diversity in fragmented landscapes. *Soil Biol Biochem* **41:** 2162–2169.

Benkova, S. (1999). The occurrence and abundance of microscopic fungi in floodplain forest soils. *Biologia* **54 (3):** 297–301.

Bilgrami, K.S., Jamaluddin, S. and Rizwi, M.A. (1991). *Fungi of India. List and References.* Today & Tomorrow's Printers and Publishers, New Delhi. pp.1-798.

Borneman, J. and Hartin, R.J. (2000). PCR primers that amplify fungal rRNA genes from environmental samples. *Applied and Environmental Microbiology* **66:** 4356–4360.

Borneman, J. and Triplett, E.W. (1997). Molecular microbial diversity in soils from eastern Amazonia: evidence for unusual microorganisms and microbial population shifts associated with deforestation. *Appl Environ Microbiol* **63:** 2647-2653.

Borneman, J. (1999). Culture-independent identification of microorganisms that respond to specific soil stimuli. *Appl. Environ. Microbiol.* **65:** 3398–400.

Bossio, D.A. and Scow, K.M. (1998). Impacts of carbon and flooding on soil microbial communities: phospholipid fatty acid profiles and substrate utilization patterns. *Microbial Ecology,* **35:** 265–278.

Buckovo, E.K., Bacigalova, A., Simonovicova, E.Frankova and Benkova, S. (2000): Occurence and abundance of microscopic fungi in floodplain forest soils. *Ekologia (Bratislava)* **19 (1):** 3–9.

Bue´e M., Reich, M., Murat, C., Morin, E., Nilsson, R.H. and Uroz, S. *et al.* (2009). 454 pyrosequencing analyses of forest soils reveal an unexpectedly high fungal diversity. *New Phytol* **184:** 449–456.

Cannon, P. (1997). Strategies for rapid assessment of fungal diversity. *Biodivers Conserv.* **6:** 669-680.

Cabello, M. and Arambarri, A. (2002). Diversity in soil fungi from undisturbed and disturbed *Celtis tala* and *Scutia bifolia* forests in the eastern Buenos Aires province (Argentina). *Microbiol. Res.* **157:** 115–125.

Caravaca, F., Alguacil, M.M., Azcón, R., Diaz, G. and Roldan, A. (2004). Comparing the effectiveness of mycorrhizal inoculation and amendment with sugar beet, rock-phosphate and *Aspergillus niger* to enhance field performance of the leguminous shrub *Dorycnium pentaphyllum* L. *Appl. Soil Ecol.,* **25:** 169-180.

Chen, Y.P., Rekha, P.D., Arunshen, A.B., Lai, W.A. and Young, C.C. (2006). Phosphate solubilizing bacteria from subtropical soil and their tricalcium phosphate solubilizing abilities. *Appl. Soil Ecol.* **34:**33-41.

Christensen, M. (1969). Soil microfungi of dry to mesic conifer-hardwood forests in northern Wisconsin. *Ecology* **50:** 9-27.

Christensen, M. (1989). *A view of fungal ecology. Mycologia* **81:** *1-19.*

Classen, A.T., Overby, S.T., Hart, S.C., Koch, G.W. and Witham, T.G. (2007). *Plant Soil,* **295:** 217-227.

Cui, B., Wang, Z. and Dai, Y. (2008). *Albatrellus piceiphilus sp. nov.* on the basis of morphological and molecular characters. *Fungal Diversity.* **28:** 41-48.

Curtis, T.P., Sloan, W.T., Scannel, J.W. (2002). Estimating prokaryotic diversity and its limits. *Proc. Natl. Acad. Sci.* U. S. A. **99:** 104914–110499.

Davis, J.P., Youssef, N.H. and Elshahed, M.S. (2009). Assessment of the diversity, abundance, and ecological distribution of candidate division SR1 reveals a high level of phylogenetic diversity but limited morphotypic diversity. *Appl Environ Microbiol* **75:** 4139–4148.

Domsch, K.H., Gams, W. and Anderson, T.H. (1980). Compendium of Soil Fungi Volume 1 Academic Press: London.

Doran, J.W. and Parkin, T.B. (1994). Defining and assessing soil quality. In: DORAN, J. W. (ed.): Defining Soil Quality for a Sustainable Environment. SSSA Special Publication 35, *Soil Science Society of America,* Madison, 3–12.

Doran, J.W. and Parkin, T.B. (1996). Quantitative indicators of soil quality: A minimum data set. In: DORAN, J. W. and A. J. JONES (eds.): Methods for assessing soil quality. SSSA Special Publication 49, Soil Science Society of America, Madison, 25–37.

Everhart, S.E. and Keller, H.W. (2008). Life history strategies of corticolous myxomycetes: the life cycle, fruiting bodies, plasmodial types, and taxonomic orders. *Fungal Diversity* **29:** 1–16.

Faegri, A., Torsvik, V.L. and Goksoyr, J. (1977). Bacterial and fungal activities in soil: separation of bacteria and fungi by a rapid fractionated centrifugation technique. Soil *Biology and Biochemistry,* **9:** 105-112.

Fankem, H.D., Nwaga, A., Deubel, L., Dieng, W. Merbach and Etoa, F.X. (2006). Occurrence and functioning of phosphate solubilizing microorganisms from oil palm tree (*Elaeis guineensis*) rhizosphere in Cameroon. *African J. Biotech.* **5:** 2450-2460.

Fierer, N. and Jackson, R.B. (2006). The diversity and biogeography of soil bacterial communities. *Proc Natl Acad Sci* USA **103:** 626–631.

Fierer, N., Hamady, M., Lauber, C.L., and Jackson, R.B. (2008). The influence of sex, handedness, and washing on the diversity of hand surface bacteria. *Proc Natl Acad Sci* USA **105:** 17994–17999.

Frankland, J.C. (1982). Biomass and nutrient cycling by decomposer basidiomycetes. In: Decomposer basidiomycetes: their biology and ecology. (Eds. Frankland, J.C., Hedger, J.N., Swift, M.J.), pp. 241-261, Cambridge University Press,Cambridge.

Frostega? R.D.A. and Ba? a?, T.H.E. (1996). The use of phospholipid fatty acid analysis to estimate bacterial and fungal biomass in soil. *Biology and Fertility of Soils,* **22:** 59–65.

Gams, W. (1992). The analysis of communities of saprophytic microfungi with special reference to soil fungi. In: Fungi in vegetation science. (Ed. Winterhoff, W.), pp. 183-224, Kluwer Academic Publishers, Netherlands.

Giller, P.S. (1996). The diversity of soil communities, the"poor man's tropical rainforest". *Biodiv. Cons.* **5:** 135–168.

Greene, E. and Voordouw, G. (2003). Analysis of environ-mental microbial communities by reverse sample genome probing. *J Microbiol Methods* **53:** 211-219.

Grishkan, I. and Nevo, E. (2008). Soil microfungal communities of 'Evolution Canyons' in Israel-extreme differences on a regional scale. *Biol J Linn Soc.* **93:**157-163.

Gupta, N., Sabat, J., Parida, R. and Kerkatta, D. (2007). Solubilization of tricalcium phosphate and rock phosphate by microbes isolated from chromite, iron and manganese mines. Acta Botanica Croatica. **66:** 197–204.

Hagvar, S. (1998). The relevance of the Rio-Convention on biodiversity to conserving the biodiversity of soils. *Applied Soil Ecology* **9:** 1–7.

Hamady, M., Walker, J.J., Harris, J.K., Gold, N.J. and Knight, R. (2008). Error-correcting barcoded primers for pyrosequencing hundreds of samples in multiplex. *Nat Methods* 5: 235–237.

Hamayun, M., Khan, S.A., Kim, H.Y., Chaudhary, M.F., Hwang, Y.H., Shin, D.H., Kim, I. K., Lee, B.H. and Lee, I.J. (2009). Gibberellin production and plant growth enhancement by newly isolated strain of Scolecobasidium tshawytschae. *J. Microbiol. Biotechnol.* **19:** 560-565.

Hammond, PM. (1995). Magnitude and distribution of biodiversity. In: Heywood VH. Global Biodiversity Assessment. Cambridge University Press, Cambridge, UK.

Hao-quin Pan, Jin- Feng Yu, Yue- ming Wu, Tian- Yu Zhang and Hong- Feng Wang. (2008). Area: 1., *J. Zhejiang Univ. Sci.*, B **9 (10):** 829-834.

Hawksworth, D.L. (1991). The fungal dimension of biodiversity: magnitude, significance and conservation. *Mycol. Res.* **95:** 641-655.

Hawksworth, D.L., Kirk, P.M., Sutton, B.C. and Pegler, D.N. (1996). Ainsworth & Bisby's Dictionary of the Fungi. 8th edition. CAB International, Wallingford.

Hawksworth, D.L. (1993). The tropical fungal biota. Census, pertinence, prophylaxis and prognosis. In: *Aspects of tropical Mycology*. Isacc, S., Frankland, J.C., Vatling, R. and Whalley, A.J.S. (eds.), Cambridge University Press, London. pp. 265-293.

Hawksworth D.L. (2002). The magnitude of fungal diversity: the 1·5 million species estimate revisited. *Mycol Res.* **105:**1422-1432.

Hawksworth D.L. (2001). The magnitude of fungal diversity: the 1.5 million species estimate revisited. *Mycological Research* **105:**1422–1432.

Hawksworth, D.L., Kirk, P.M., Sutton, B.C. and Pegler, D.N. (1995). *Dictionary of the Fungi*, CAB Intl., pp. 616.

Huang, W., Cai, Y., Surveswaran, S., Hyde, K., Corke, H. and Sun, M. (2009). Molecular phylogenetic identification of endophytic fungi isolated from three *Artemisia* species. *Fungal Diversity.* **36:** 69-88.

Hyakumachi, M. (1994). Plant growth-promoting fungi from turfgrass rhizosphere with potential for disease suppression.*Soil Microorganisms* **44:** 53-68.

Hyde, K. and Soytong, K. (2007). Understanding microfungal diversity: A critique. *Cryptogamie Mycologie.* **28:** 281-289.

Hyde, K.D., Bussaban, B., Paulus, B., Crous, P.W., Lee, S. Mckenzie, E.H.C., Photita, W. and Lumyong, S. (2007). Diversity of saprobic microfungi. *Biodiver. Conserv.* 16: (in press). Izzo A and Mazzola M, (2009). Hybridization of an ITS-based macroarray with ITS community probes for characterization of complex communities of fungi and fungal-like protists. *Mycol Res.* **113:** 802-812.

Johnson, L.F., Curl, E.A., Bond, J.H. and Fribourg, H.A. (1960). Methods for studying soil microflora–plant disease relationships, 2nd ed. Minneapolis: Burgess Publishing Company; p. 178.

Jones, R.T., Robeson, M.S., Lauber, C.L., Hamady, M., Knight, R. and Fierer, N. (2009). A comprehensive survey of soil acidobacterial diversity using pyrosequencing and clone library analyses. ISME J **3:** 442–453.

Kennedy, N. and Clipson, N. (2004). Fingerprinting the fungal community. *Mycologist.* **17:** 158-164.

Kennedy, A.C. and Smith, K.L. (1995). Soil microbial diversity and the sustainability of agricultural soil. *Plant Soil* **170:** 75–86.

Kennedy, N.M., Gleeson, D.E., Connolly, J. and Clipson, N.J.W. (2005). *FEMS Microb. Ecol.*, **53:** 329-337.

Khan, M.S., Zaidi, A. and Wani, P.A. (2007). Role of phosphate-solubilizing microorganisms in sustainable agriculture–A review. Agron. Sustain. Dev. **27:** 29-43.

Kirk, P., Cannon, P.F., Minter, D.W. and Stalpers, J.A. (2008). *Ainsworth &* Kirk P, Cannon PF, Minter DW, Stalpers JA. (2008). *Ainsworth & Bisby's Dictionary of the Fungi.* 10th edn. CAB International, Wallingford, UK.

Kok, W.I., Luizao, R.C.C., Do Socorro, B.D.S., Teixeira, M. F. S. and Munitz, *E.G.* (1984). Ecology of pathogenic yeasts in Amazonian soil. *Appl. Environ. Microbiol.* **47:** 390–394.

Kucey, R.M.N. (1983). Phosphate solubilizing bacteria and fungi in various cultivated and virgin Alberta soils. *Can J Soil Sci.* **63:** 671–678.

Larkin, R. (2003). Characterization of soil microbial communities under different potato cropping systems by microbial population dynamics, substrate utilization, and fatty acid profiles. *Soil Biol Biochem.* **35:** 1451-1466.

Lauber, C.L., Hamady, M., Knight, R. and Fierer, N. (2009). Pyrosequencing- based assessment of soil pH as a predictor of soil bacterial community composition at the continental scale. *Appl Environ Microbiol* **75:** 5111–5120.

Lauber, C.L., Strickland, M.S., Bradford, M.A. and Fierer, N. (2008). The influence of soil properties on the structure of bacterial and fungal communities across land-use types. *Soil Biol Biochem* **40:** 2407–2415.

Leckie, S. (2005). Methods of microbial community profiling and their application to forest soils. *For Ecol Manage.* **220:** 88-106.

Letourneau, A., Seena, S., Marvanová, L. and Bär-locher, F. (2010). Potential use of barcoding to identify aquatic hyphomycetes. *Fungal Diversity.* **40:** 51-64.

Lewis, J.K., Dodd, J.L., Hutchensen, H.L. and Hansen, C.L. (1971). Antibiotic and herbage dynamics studies on the cotton site. U.S. IBP Tech. Rep. No. 111, Colorado, Colorado State Univ.

Manoch, L. (1998). Biodiversity of soil fungi in Thailand, pp. 126-140. *In* Proceedings of the Asia-Pacific Mycological Conference on Biodiversity and Biotechnology, Hua Hin.

Manoharachary, C., Sridhar, K., Singh, R., Ad-holeya, A., Suryanarayanan, T., Rawat, S. and Johri, B. (2005). Fungal biodiversity: Distribution, conservation and prospecting of fungi from India. *Curr Sci.* **89:** 58.

Manoharachary, C., Sridhar, K., Singh, R., Adholeya, A., Suryanarayanan, T. (2001). Biodiversity, Conservation and Biotechnology of Fungi. Presidential Address, Section–Botany, 89th Session of Indian Science Congress, Lucknow.

Mazzola, M. (2004). Assessment and management of soil microbial community structure for disease suppression. *Ann Rev Phytopathol* **42:** 35–59.

Mishra, R.R. (1966). Influence of soil environment and surface vegetation on soil mycoflora. *Proc. Natl. Acad. Sci. India* **36 B:** 117-123.

Nagmani, A., Kunwar, I.K., Manoharachary, C. (2005). Handbook of Soil Fungi. Published by I. K.International Pvt, Ltd. New Delhi.

Nannipieri, P., Ascher, J., Ceccherini, M.T., Landi, L., Pietramellara, G. and Renella, G. (2003). Microbial diversity and soil functions. *Eur J Soil Sci* **54:** 655–70.

Nannipieri, P., Grego, S. and Ceccanti, B. (1990). Ecological significance of the biological activity in soil. In: Soil Biochemistry, Volume 6 (eds J.-M. Bollag & G. Stotzky), pp. 293–355. Marcel Dekker, New York.

Nannipieri, P. and Badalucco, L. (2003). Biological processes. In: Processes in the Soil–Plant System: Modelling Concepts and Applications (eds D.K. Bembi & R. Nieder). The Haworth Press, Binghamton, NY, in press.

Nei, M. and W.H. Li. (1979). Mathematical Model for Studying Genetic Variation in Terms of Restriction Endonucleases. Proceedings of the National Academy of Sciences of the United States of America **76:** 5269-5273.

Nevarez, L., Vasseur, V., Le Madec, L., Le Bras, L. Coroller, L. and Legue´rinel, I. ***et al.*** (2009). Physiological traits of Penicillium glabrum strain LCP 08.5568, a filamentous fungus isolated from bottled aromatised mineral water. *Int J Food Microbiol* **130:** 166–171.

Nieves-Rivera, A. (2003). Mycological survey of Río Camuy Caves Park, Puerto Rico. *J Cave Karst Stud.* **65:** 23-28.

Nilsson, L.O., Ba°a°th, E., Falkengren-Grerup, U. and Wallander, H. (2007). Growth of ectomycorrhizal mycelia and composition of soil microbial communities in oak forest soils along a nitrogen deposition gradient. *Oecologia* **153:** 375–384.

Nilsson, R., Kristiansson, E., Ryberg, M., Hallenberg, N. and Larsson, K. (2008). Intraspecific ITS variability in the kingdom fungi as expressed in the international sequence databases and its implications for molecular species identification. *Evolutionary bioinformatics online.* **4:**193.

Nocker, A. Burr, M. and Camper. A. (2007). Genotypic microbial community profiling: A critical technical review. *Microb Ecol.* **54:** 276-289.

Nusslein, K. and Tiedje, J. (1999). Soil bacterial community shift correlated with change from forest to pasture vegetation in a tropical soil. *Appl Environ Microbiol.***65:** 3622-3626.

Omar, S.A. (1998). The role of rock-phosphate-solubilizing fungi and vesicular-arbuscular mycorrhiza (VAM) in growth of wheat plants fertilized with rock phosphate. *World J. Microbiol. and Biotechnol.* **14:** 211-218.

Oros-Sichler, M., Gomes, N., Neuber, G. and Smalla, K. (2006). A new seminested PCR protocol to amplify large 18S rRNA gene fragments for PCR-DGGE analysis of soil fungal communities. *J Microbiol Methods.* **65:** 63-75.

Ortega, A., Suárez-Santiago, V. and Reyes, J. (2008). Morphological and ITS identification of *Cortinarius* species (section Calochroi) collected in Mediterranean Quercus woodlands. *Fungal Diversity.* **29:** 73-88.

Olsen, S.R. and Dean, L.A. (1965). Methods of soil Analysis, Agronomy Ser. No. 9, Part 2. American Society of Agronomy, Madison. Phosphorus, pp. 1035-1049 in C.A Black (Ed.).

Pankhurst, C.E., Yu, S., Hawke, B.G. and Harch, B.D. (2001). Capacity of fatty acid profiles and substrate utilization patterns to describe differences in soil microbial communities associated with increased salinity or alkalinity at three locations in south Australia. Biology and Fertility of Soils, **33:** 204–217.

Pelaez, F., Gonzalez, V., Platas, G., Sanchez-Ballesteros, J. and Rubio, V. (2008). Molecular phylogenetic studies within the Xylariaceae based on ribosomal DNA sequences. *Fungal Diversity.*

Rangaswami, G. and Bagyaraj, D.J. (1998). Agricultural Microbiology II edition published by Prentice Hall of India Pvt. Ltd. N. Delhi.

Reddy, M.S., Kumar, S.K. and Babita, M.S.R. (2002). Biosolubilization of poorly soluble rock hosphates by Aspergillus tubingensis and Aspergillus niger. *Bioresource Technol.* **84:**187–189.

Reyes, I., Bernier, L. and Antoun, H. (2002). Rock phosphate solubilization and colonization of maize rhizosphere by wild and genetically modified strains of *Penicillium rugulosum. Microbial Ecology,* **44:** 39-48.

Rodríguez, R., Vassilev, N. and Azcón, R. (1999). Increases in growth and nutrient uptake of alfalfa grown in soil amended with microbially-treated sugar beet waste. *Applied Soil Ecology,* **11:** 9-15.

Roe, A., Rice, A., Bromilow, S., Cooke, J. and Sprling, F. (2010). Multilocus species identification and fungal DNA barcoding: Insights from blue stain fungal symbionts of the mountain pine beetle. *Molecular Ecology Resources.* 10.1111/ j.1755-0998.2010.02844.

Rojas, C. and Stephenson, S.L. (2008). Myxomycete ecology along an elevation gradient on Cocos Island, Costa Rica. *Fungal Diversity* **29:** 117–127.

Rousk, J., Brookes, P.C. and Ba°a°th, E. (2010a). The microbial PLFA composition as affected by pH in an arable soil. *Soil Biol Biochem* **42:** 516–520.

Rousk J, Brookes, P.C. and Ba°a°th, E. (2009). Contrasting soil pH effects on fungal and bacterial growth suggests functional redundancy in carbon mineralisation. *Appl Environ Microbiol* **75:** 1589–1596.

Rudresh, D.L., Shivaprakash, M.K. and Prasad, R.D. (2005). Tricalcium phosphate solubilizing abilities of Trichoderma spp. in relation to P uptake and growth and yield parameters of chickpea (*Cicer arietinum* L.). *Can J Microbiol.* **51:** 217–222.

Ruisi, S., Barreca, D., Selbmann, L., Zucconi, L. and Onofri, S. (2007). Fungi in Antarctica. *Reviews in Environmental Science and Biotechnology.* **6:** 127-141.

Sarbhoy, A.K., Agarwal, D.K. and Varshney, J.L., Fungi of India (1982–1992). *CBS Publishers and Distributors, New Delhi, 1996, pp. 350.*

Satish, N., Sultana, S. and Nanjundiah, V. (2007). Diversity of soil fungi in a tropical deciduous forest in Mudu-malai, southern India. *Curr Sci.* **93:** 669-677.

Seifert, K., Samson, R., DeWaard, J., Houbraken, J., Lévesque, C., Moncalvo, J., Louis-Seize, G. and Hebert, P. (2007). Prospects for fungus identification using *CO1* DNA barcodes, with *Penicillium* as a test case. *Proceedings of the National Academy of Sciences.* **104:** 3901.

Seshadri, S.S., Ignacimuth, U. and Lakshminarasimhan, C. (2004). Effect of nitrogen and carbon sources on the inorganic phosphate solubilization by different *Aspergillus niger* strains. *Chemical Engineering Communications,* 191: 043-1052. Shamir, I. and Steinberger.I. (2007). *Microb. Ecol.,* **53:**340-347.

Shenoy, B., Jeewon, R. and Hyde, K. (2007). Impact of DNA sequencedata on the taxonomy of anamorphic fungi. *Fungal Diversity.* **26:** 1-54.

Shukla, A.K., Tiwari, B.K. and Mishra, R.R. (1989). *Revue a Ecologie et de Biologie du sol,* **26:** 249-265.

Slavikova, E. and Vadkertiova, R. (2000). The occurrence of yeasts in the forest soils. *J. Basic Microbiol.* **40:** 207–212.

Smit, E., Leeflang, P., Glandorf, B., van Elsas, J.D. and Wernars, K. (1999). Analysis of fungal diversity in the wheat rhizosphere by sequencing of cloned PCR-amplified genes encoding 18S rRNA and temperature gradient gel electrophoresis. *Applied and Environmental Microbiology,* **65:** 2614–2621.

Soderstrom, B.E., Baath, E. and Lundgren, B. (1983). Decrease in soil microbial activity and biomass owing to nitrogen amendments. *Canad. J. Microbiol.* **29:** 1500-1506.

Srivastav, S., Yadav, K.S. and Kundu, B.S. (2004). Prospects of using phosphate solubilizing Pseudomonas as biofungicide. *Indian J Microbiol.* **44:** 91–94.

Tang, A., Jeewon, R. and Hyde, K. (2009). A re-evaluation of the evolutionary relationships within the Xylariaceae based on ribosomal and protein-coding gene sequences. *Fungal Diversity.* **34:** 127-155.

Tilak, K.V.B.R. (2000). Conservation and utilization of microbial diversity for natural resource management, In: International Conference on Managing Natural Rescurces for Sustainable Agricultural Production in the 21 st century (ed. Yadav, J.S.P. *et al.*), New Delhi, India. pp. 76-78.

Tunlid, A. and White, D.C. (1992). Biochemical analysis of biomass, community structure, nutritional status, and metabolic activity of microbial communities in soil. In: Soil Biochemistry, Volume 7 (eds G. Stotzky & J.-M. Bollag), pp. 229–262. Marcel Dekker, New York.

Van Elsas, J.D., Duarte, G.F., Keijzer-Wolters, A. and Smit, E. (2000). Analysis of the dynamics of fungal communities in soil via fungal specific PCR of soil DNA followed by denaturing gradient gel electrophoresis. *Journal of Microbiological Methods,* **43:** 133–151.

Vargas, Gil S., Pastor, S. and March, G.J. (2009). Quantitative isolation of biocontrol agents *Trichoderma* spp., *Gliocladium* spp. and actinomycetes from soil with culture media. *MicrobiologicalResearch.* **164:** 196–205.

Vassilev, N., Toro, M., Vassileva, M., Azcón, R. and Barea, J.M. (1997). Rock phosphate solubilization by immobilized cells of *Enterobacter* sp. in fermentation and soil conditions. *Bioresource Technology,* **61:** 29-32.

Venkateswarlu, B., Rao, A.V., Raina, P. and Ahmad, N. (1984). Evaluation of phosphorus solubilization by microorganisms isolated from arid soil. *J Indian Soc Soil Sci.* **32:** 273–277.

Vishwanathan, A.S. (2010). Diversity of Hyphomycetes in deciduous forest soils: A survey of local soil Hy-phomycete mycoflora of the Puttaparthi mandal, Andhra Pradesh, India. LAP Lambert Academic Publishing, Saarbrucken, Germany.

Vishniac, H.S. (1996). Biodiversity of yeasts and filamentous microfungi in terrestrial Antarctic ecosystems. *Biodiversity Conservation* **5:** 1365–1378.

Wahegaonkar, N., Salunkhe, Palsingarkar, S.M., and S.Y., Shinde (2011). Diversity of fungi from soils of Aurangabad, M.S., India *Annals of Biological Research,* 2 **(2):** 198-205.

Wakelin, S.A., Warren, R.A., Harvey, P.R. and Ryder, M.H. (2004). Phosphate solubilization by Penicillium spp. closely associated with wheat roots. *Biol Fertility Soils.* **40:**36–43.

Wheeler, K.A., Hurdman, B.F. and Pitt, J.I. (1991). Influence of pH on the growth of some toxigenic species of *Aspergillus, Penicillium* and *Fusarium. Int J Food Microbiol* **12:** 141–150.

Whitelaw, M.A., Harden, T.J. and Bender, G.L. (1997). Plant growth promotion of wheat inoculated with *Penicillium radicum* sp. nov. *Australian Journal of Soil Research,* **35:** 291-300.

Whitelaw, M.A. (2000). Growth promotion of plants inoculated with phosphate-solubilizing fungi. *Adv. Agron.* **69:** 99–151.

Widden, P. (1979). Fungal population from forest soils in Southern Quebec. *Canad. J. Bot.* **57:** 1324-1331.

Wilson, E.O. (1992). *The Diversity of Life.* Harvard University Press, Cambridge, Massachusetts, USA.

Zelles, L. (1999). Fatty acid patterns of phospholipids and lipopolysaccharides in the characterization of microbial communities in soil: a review. *Biology and Fertility of Soils,* **29:** 111–129.

Chapter 9

Diversity and Potentiality of Actinomycetes in Biological Control

Smita Srivastava, B.K. Sarma and Asha Sinha

Department of Mycology and Plant Pathology, Institute of Agricultural Sciences, Banaras Hindu University, Varanasi – 221 005, Uttar Pradesh

Actinomycetes are gram-positive filamentous bacteria having high DNA G+C content (>55 mol per cent). The name actinomycetes derived from the greek word *aktis* (a ray beam) and *mykes* (fungus). Earlier they were thought to be fungus but now recognized as Prokaryotes. Majority of them are aerobic but few are anaerobic such as *Oerskovia* sp. In the microbial world including other bacteria, fungi, nematodes, etc, actinomycetes play a vital role in biological control of plant pathogens due to their antibiotic production. They are good organic matter decomposer in the soil and abundantly found in humus soil. Some genera and species of actinomycetes are reported human pathogens and very few are plant pathogens but majority of them having beneficial properties for plants such as ability to colonize plant surface, antibiosis against plant pathogens, synthesis of extracellular proteins and phytohormones and degradation of phytotoxins, etc. (Doumbou, *et al.*, 2002).

Different Habitats of Actinomycetes

Soil

Actinomycetes are efficiently degraded organic matter in the soil and provide nutrient for the plant growth. They are widely distributed and can be easily isolated

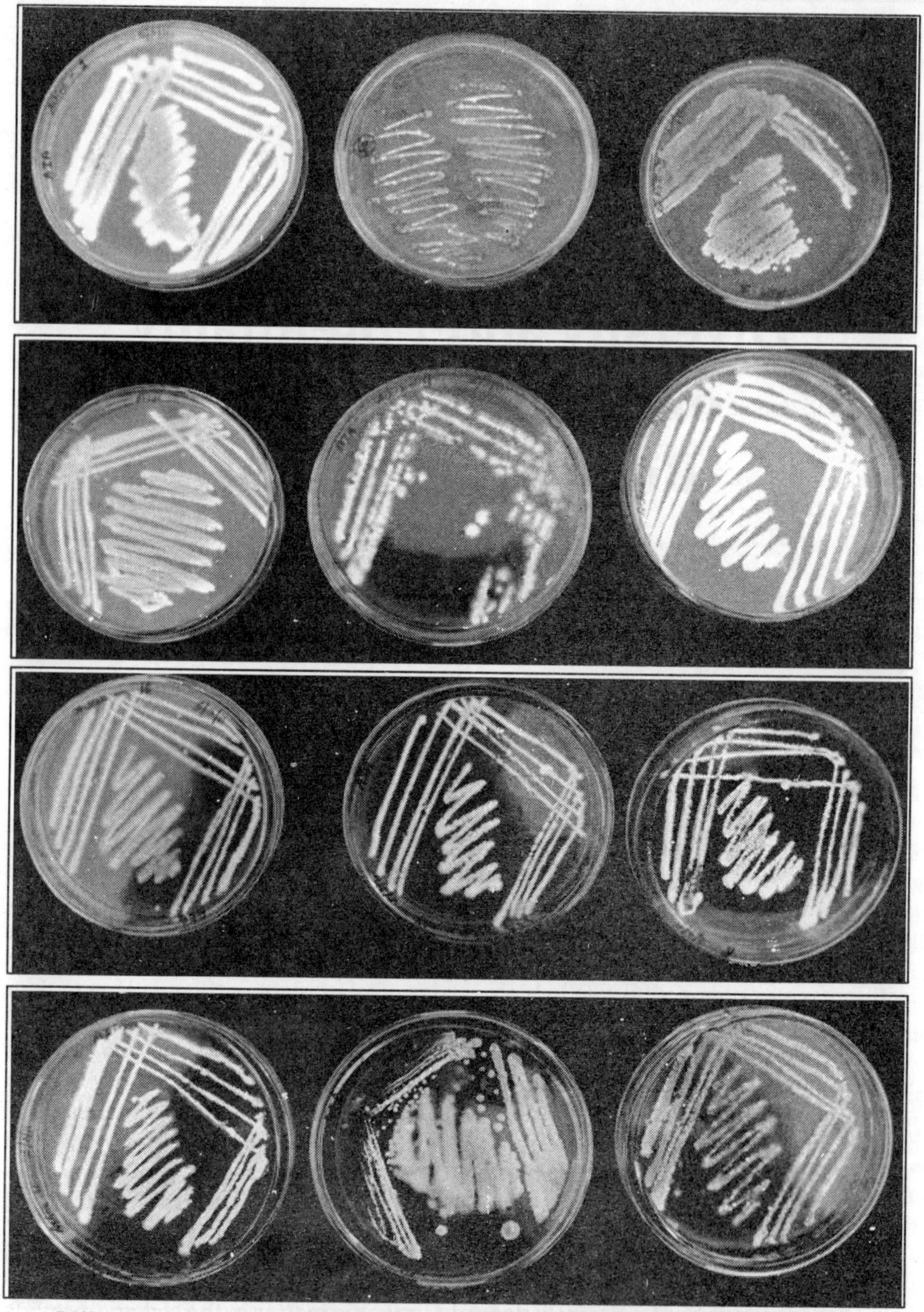

Different Isolates of Actinomycetes on AIA (Actinomycetes Isolation Agar) Media Plates Isolated from the Agricultural and Non-Agricultural Soil of Eastern Uttar Pradesh, India

from the humus rich soil. About one million of actinomycetes are present in one gram of soil. According to Berthelot and Andre (1891) earthy odor of soil is mainly produced by actinomycetes. Goodfellow and Williams (1983) describe the streptomycetes group of actinobacteria, on the basis of several microscopic studies, that they are able to exist in the soil for longer period, as resting arthrospores and germinate occasionally in the presence of exogenous nutrient such as root fragments, dead fungal hyphae, etc. whereas non-sporing genera such as *Arthrobacter*, etc. are existed in soil as resting cocci. According to them spores of streptomycetes are dispersed by wind, water, rain and arthropods and their distribution and activities in the soil are generally influence by some major factors such as temperature, pH, clay and humic colloids. There are some examples–*Streptomyces sodiiphilus*, a novel, alkaliphilic actinomycetes isolated from muddy sample of salty lake (Li, *et al.*, 2005), *Streptacidiphilus oryzae*, an acidiphilic bacteria, isolated from an acidic rice field soil (Wang, *et al.*, 2006). *Streptomyces beijiangensis*, a novel, psychrotolerant actinomycetes, it can grow well at 8°-20°c temperature (Li, *et al.*, 2002). Streptomycetes group of bacteria are able to degrade different polymers in the soil such as hemicelluloses, pectin, keratin and chitin (Iizuka and Kawaminami, 1965; Young and Smith, 1975; Hsu and Lockwood, 1975).

Compost

There are several mesoplilic and thermophilic actinomycetes are reported that can actively grow on manures, sewage sludge, animal faeces and badly stored fodder and grains, etc. For example–*Rhodococcus coprophilus* grow on sewage and animal faeces and a specific indicator of faecal pollution arising from farm animal wastes (AI-Diwany and Cross, 1978; Mara and Oragui, 1981). *Nocardia pinensis* isolated from activated sludge foams (Blackall, *et al.*, 1989). Activated sludge from a dairy contained *Corynebacterium*, *Microbacterium* and *Rhodococcus* strains whereas municipal sludge contain *Arthrobacter* at the place of *Rhodococcus* (Seiler, *et al.*, 1980). Streptomycetes can be predominately grown in silo-stored corn with 27 per cent moisture content (Lyons, *et al.*, 1975).

Aquatic Habitat

Actinomycetes are widely distributed in aquatic habitats. *Actinoplanes*, *Micromonospora*, *Rhodococcus*, *Streptomyces* and *Thermoactinomyces* can readily be isolated from freshwater (Cross, 1981). By some workers it has been found that earthy taste and odor produced by actinomycetes are sometimes occurred in drinking water. Gerber (1979) concluded that geosmin and methyl iso-boreol compounds are responsible for earthy odor and taste. Cross (1981) emphasized that such compounds are produced only during or after hyphal growth. Geosmin production occurred in the soil, so their runoff or seepage into reservoir quite possible, therefore, water also have some earthy taste.

Recent investigation suggests that certain actinobacterial clades identified in freshwater and estuaries are indigenous and not inoculated from terrestrial sources (Warnecke, *et al.*, 2004). Many new genera of actinomycetes are described which require seawater for growth and have marine chemotype signature such as *Salinospora* (Mincer, *et al.*, 2002; Jensen, *et al.*, 2004; Maldonado, *et al.*, 2004a), *Marimomyces* (Jensen, *et al.*, 2004), etc. In marine ecosystem, actinomycetes play a vital role in mobilization of DOC (Dissolved Organic Matter) and POM (Particulate Organic Matter) that settles at the sediment–water interface (Bull, *et al.*, 2005).

Endophytes

An endophyte is an endosymbiont, which lives inside the plant for at least part of its life without causing apparent disease. *Streptomyces* sp. strain EN27, an endophytic actinobacteria isolated from healthy wheat tissue, which are capable of suppressing a number of wheat fungal pathogens both in vitro and in planta, were investigated for the ability to activate key genes in the systemic acquired resistance (SAR) or the jasmonate/ethylene (JA/ET) pathways in *Arabidopsis thaliana* (Conn, *et al.*, 2008). Several workers isolated *Frankia* spp. from the root nodules of different plants, which are capable to fixes atmospheric nitrogen into the soil.

Selective Isolation

The selective isolation and enumeration of actinomycetes are generally influenced by pretreatment of the sample before plating and by condition under which the propagules of the sample are cultivated in the lab (Goodfellow and Williams, 1983). There are several pretreatment methods are used such as–heat treatment have been used to isolate *Actinomadura, Microbispora, Rhodococcus* and *Streptomyces*, (Athalye, *et al.*, 1981; Nonomura and Ohara, 1969; Rowbotham and Cross, 1977; Williams, *et al.*, 1972), various baits have been used to attract the motile spores of *Actinoplanetes* in water or soil suspension and chemical treatment for the selective isolation from soil (Palleroni, 1980), etc. Several media have been recommended for isolation of one or more actinomycete genera including, colloidal chitin-mineral salts (Lingappa and Lockwood, 1962; Hsu and Lockwood, 1975), Half-strength nutrient agar (Gregory and Lacey, 1963), Starch casein (Küster and William, 1964), Humic acid–vitamin agar medium (Hayakawa and Hideo, 1987) and Actinomycete Isolation Agar (*Link- http://www.bd.com/europe/regulatory/Assets/IFU/Difco_BBL/228220.pdf*), etc. Antifungal and antibacterial antibiotics, which do not inhibit actinomycetes have been widely used to improve media selectivity, for example, "Novobiocin" for *Thermactinomyces vulgaris* (Cross, 1968; Cross and Johnston, 1972) and *Micromonospora* sp. (Goofellow and Haynes, 1983), "Chlortetracycline" and "Methacycline" for *Nocardia* spp. (Orchard, *et al.*, 1977), "Kanamycin" for *Thermomonospora chromogena* (McCarthy and Cross, 1981) and "Rifampin" for *Actinomadura* strains (Athalye, *et al.*, 1981). Most of the colonies are develop within 14 days at 25°–30°C. Prolong incubation is needed for the isolation of root endophyte *Frankia* spp. Actinomycetes are widely distributed in soil, water, colonizing plants and various genera have been isolated from compost and related materials.

Role of Actinomycetes in Biological Control

The term "Biological Control" means total or partial inhibition or destruction of pathogen populations by other organism (Agrios, 2006). Several evidences indicates that actinomycetes play a significant role in rhizosphere where they influence plant growth and gives protection against soil and root borne plant pathogens (Lechevalier, 1988). The colonization by introduced bacteria is essential for the biological control of root pathogens (Suslow and Schroth, 1982).Within actinomycetes, *Streptomyces* spp. investigated predominantly because of their dominance, ease of isolation and commercial interest due to their antibiotic production. According to Doumbou, *et al.* (2002), *Streptomyces griseoviridis* strain K61 is isolated from light coloured *Sphagnum* peat and showed antagonistic against several plant pathogens including *Alternaria brassicola, Botrytis cinerea, Fusarium avenaceum, Fusarium culmorum, Fusarium oxysporum* f. sp. *dianthi, Pythium debaryanum, Phomopsis sclerotioides, Rhizoctonia solani* and *Sclerotinia sclerotiorum* (Tahvonen, 1982a, 1982b; Tahvonen and Avikainen, 1987). Similarly there are several examples of actinobacteria that are widely used in the biological control of different plant diseases such as- *Streptomyces griseus* for biological control of *Rhizoctonia solani* (Merriman, *et al.*, 1974); El-Abyad *et al.* (1993) described the use of three *Streptomyces* spp. (*S. pulcher, S. canescens* and *S. citreofluorescens*) against bacterial, *Fusarium* and *Verticillium* wilts, early blight and bacterial canker of tomato. The *Fusarium* wilt of carnation, Damping-off of *Brassica* and root rot of cucumber control by an antagonistic actinobacteria, *Streptomyces griseoviridis* which is reported by Tahvonen and Lahdenpera (1988). *Streptomyces lydicus* WYEC108 is a potentially potent biocontrol agent against *Pythium* seed and root rot (Yuan and Crawford, 1995).

Actinomycetes have the ability to produce a wide variety of extracellular enzymes like chitinases, glucanases, etc. These enzymes are directly targeted to chitin and b-1, 3-glucans which are major constituents of many fungal cell walls (Sietsma and Wessels, 1979). The chitinase producing strains of actinobacteria could be used directly in biocontrol of fungi or indirectly by using purified proteins or through gene manipulation (Doumbou, *et al.*, 2002). Vernekar *et al.* (1999) discovered an alkaline protease inhibitor (API) as a novel class of antifungal protein against phytopathogenic fungi such as *Alternaria, Fusarium* and *Rhizoctonia.* The characteristic feature of actinomycetes is the production of antibiotic that play a significant role in biological control of plant diseases. Antibiotics are generally considered to be organic compounds of low molecular weight produced by microbes. There are several species of *Streptomyces* have been reported that produces antibiotics against different phytopathogens, some are listed below:

Antibiotics	*Antibiotic Producing Actinobacteria*	*Plant Diseases*	*Target Phytopathogens*
Cycloheximide (A.J. Whiffen, 1950)	*Streptomyces griseus*	Leaf spots, Powdery mildews and Blister rust of Pine, etc.	Several Phytopathogenic Fungi (*Pythium debaryanum, Erysiphe lagerstroemia, Cronartium ribicola, Sclerotium rolfsii*, etc.)
Geldanamycin (DeBoyer, 1976)	*Streptomyces hygroscopicus* var. *geldanus*	Root rot of Pea	*Rhizoctonia solani*
Kasugamycin (Umezawa, *et al.*, 1965)	*Streptomyces kasugaensis*	Rice blast	*Pericularia oryzae*
Mildiomycin (Iwasa, *et al.*, 1978)	*Streptoverticillium rimofaciens*	Powdery mildew	*Leveillula taurica*

Antibiotics	*Antibiotic Producing Actinobacteria*	*Plant Diseases*	*Target Phytopathogens*
Nigericin, Geldanamycin and complex of polyketide like compound AFA (Antifungal Activity that included Guanidylfungin–A) (Trejo-Estrada, *et al.*, 1998)	*Streptomyces violaceusniger* YCED9	Damping-off of Lettuce	*Pythium ultimum*
Oxytetracycline (Finlay *et al.*, 1950)	*Streptomyces viridifaciens*	Citrus canker and Peach bacterial leaf spot	*Xanthomaonas* sp.
Polyoxins B & D (Suzuki, *et al.*, 1965 & 1966)	*Streptomyces cacaoi* var. *asoensis*	Rice Sheath Blight	*Rhizoctonia solani*
Streptomycin (Ryley, *et al.*, 1981)	*Streptomyces griseus*	Bacterial disease of fruits and vegetables	*Erwinia, Pseudomonas, Xanthomonas,* etc.

(Reference book: Modern crop protection compounds by Wolfgang Krämer and Ulrich Schirmer, 2007 and Natural Products in Crop Protection.

Review article: Actinomycetes, promising tools to control plant diseases and to promote plant growth. Doumbou, *et al.*, 2002. *Phytoprotection.* 82:85-102).

Among different biocontrol agents, actinomycetes are very less explored but several evidences indicates that actinomycetes group of bacteria, especially *Streptomyces* spp. are effective biocontrol agents against different phytopathogenic fungi and other bacteria and helpful in plant growth promotion. They are efficient root colonizers and can be explore in bioformulation also. Several advance researches are going on to understand genetic and molecular approaches of these microbes to overcome biotic and abiotic stress of plants and prepare them for future adverse climate change.

References

Agrios, G.N. (2006). "Plant Pathology", 5th Ed. Elsevier Academic Press, New York.

Al-Diwany, L.J. and Cross, T. (1978). Ecological studies on nocardioforms and other actinomycetes in aquatic habitats. *Zentralblattfuer, Bakteriologie Parasitenkunde Infektionskrnkheiten und Hygiene.* 6: 153-160.

Athalye, M., Lacey, J. and Goodfellow, M. (1981). Selective isolation and enumeration of actinomycetes using rifampicin. *Journal of Applied Bacteriology.* **51**:289-298.

Blackall, L.L., Parlett, J.H., Hayward, A.C., Minnikin, D.E., Greenfield, P.F. and Harbers, A.E. (1989). *Nocardia pinensis* sp. nov., an Actinomycete Found in Activated Sludge Foams in Australia. *Journal of General Microbiology.* **135**:1547-1558.

Berthelot, M., and Andre, G. (1891). Sur l'odeur propre de la terre. *Comptes Rendus Academy of Science Paris.* **112:** 598-599.

Bull, A.T., Stach, J.E.M., Ward, A.C. and Goodfellow, M. (2005). Marine actinobacteria: perspectives, challenges, future directions. *Antonie van Leeuwenhoek.* **87**: 65–79.

Conn, V.M.,Walker, A.R. and Franco, C.M.M. (2008). Endophytic Actinobacteria Induce Defense Pathways in *Arabidopsis thaliana. Molecular Plant-Microbe Interaction.* 21 **(2)**: 208–218.

Cross, T. (1968). Thermophilic actinomycete. *Journal of Applied Bacteriology.* **31**: 36-53.

Cross, T. (1981). Aquatic actinomycetes. A critical survey of the occurrence, growth and role of actinomycetes in aquatic habitats. *Journal of Applied Bacteriology.* **50**: 397-424.

Cross, T. and Johnston, D.W. (1972). "Spore Research". London:Academic.

Doumbou, C.L., Salove, M.K.H., Crawford, D.L. and Beaulieu, C. (2002). Actinomycetes, promising tools to control plant diseases and to promote plant growth. *Phytoprotection.* **82**: 85-102.

El-Abyad, M.S., El-Sayed, M.A., El-Shanshoury, A.L. and El-Sabbagh, S.M. (1993). Towards the biological control of fungal and bacterial diseases of tomato using antagonism *Streptomyces* spp. *Plant Soil.* **149**:185-195.

Finlay, D.B., Shull, G.M., Sobin, B.A., Solonons, I.A., Winson, J.W. and Kane. J.H. (1950). Terramycin, a new antibiotic. *Science.* **11**: 85.

Gregory, P.H. and Lacey, M.E. (1963). Mycological examination of dust from mouldy hay associated with farmer's lung disease. *Journal of General Microbiology.* **30**: 75-88.

Goodfellow, M. and Williams, S.T. (1983). Ecology of Actinomycetes. *Annual Review of Microbiology.* **37**:189-216.

Goodfellow, M. and Haynes, J.A. (1983). "The Biology of the Actinomycete". Academic press, New York.

Gerber, N.N. (1979). Odorous substances from actinomycetes. *Developments in industrial microbiology.* **20**: 225-238.

Hayakawa, M. and Nonomura, H. (1987). Humic acid-vitamin agar, a new medium for the selective isolation of soil actinomycetes. *Journal of Fermentation Technology.* 65 **(5)**: 501-509.

Hsu, S.C. and Lockwood, J.L. (1975). Powdered chitin agar as a selective medium for enumeration of actinomycetes in water and soil. *Applied Microbiology.* **29**: 422-426.

Iizuka, H. and Kawaminami, T. (1965). Studies on the xylanase from *Streptomyces.* Part I. Purification and some properties of xylanase from *Streptomyces xylophagus* nov. sp. *Agricultural and Biological Chemistry.* **29**:520-524.

Iwasa, T., Suetomi, K. and Kusuka, T. (1978). Taxonomic study and fermentation of producing organism and antimicrobial activity of mildiomycin. *Journal of Antibiotics.* **31**:511-518.

Jensen, P.R., Mincer, T.J., Williams, P.G. and Fenical, W. (2004). Marine actinomycete diversity and natural product discovery. *Antonie van Leeuwenhoek.* **87:** 43–48.

Krämer, W. and Schirmer, U. (2007). Modern crop protection compounds. Wiley-VCH, Germany.

Küster, E. and Williams, S.T. (1964). Selection of media for isolation of Streptomycetes. *Nature.* **202:** 92-29.

Link- http://geldanamycin.info/.

Link- http://www.bd.com/europe/regulatory/Assets/IFU/Difco_BBL/228220.pdf.

Lechevalier, M.P. (1988). Actinomycetes in agriculture and forestry. 327-358 *in* Goodfellow, M., Williams, S. T. and Mordarski, M. Actinomycetes in biotechnology. Academic Press, New York.

Lingappa, Y. and Lockwood, J.L. (1962). Chitin media for selective isolation and culture of actinomycetes. *Phytopathology.* **52:** 317-323.

Li, W.J., Zhang, Y.G., Zhang, Y.Q., Tang, S.K., Xu, P., Xu, L.H. and Jiang, C.L. (2005). *Streptomyces sodiiphilus* sp. nov., a novel alkaliphilic actinomycete. *International Journal of Systematic and Evolutionary Microbiology.* **55:** 1329–1333.

Li, W.J., Zhang, L.P., Xu, P., Cui, X.L., Lu, Z.T., Xu, L.H. and Jiang, C.L. (2002). *Streptomyces beijiangensis* sp. nov., a psychrotolerant actinomycete isolated from soil in china. *International Journal of Systematic and Evolutionary Microbiology.* **52:** 1695–1699.

Lyons, A.J., Pridham, T.G. and Rogers, R.F. (1975). Actinomycetales from corn. *Applied Microbiology.* **29:** 246-249.

Maldonado, L.A., Stach, J.E.M., Pathom-aree, W., Ward, A.C., Bull, A.T. and Goodfellow, M. (2004a). Diversity of cultivable actinobacteria in geographically widespread marine sediments. *Antonie van Leeuwenhoek.* **87:** 11–18.

McCarthy, A.J. and Cross, T. (1981). A note on a selective isolation medium for the thermophilic actinomycete *Thermonospora chromogena. Journal of Applied Bacteriology.* **51:** 299-302.

Mara, D.D. and Oragui, J.I. (1981). Occurrence of *Rhodococcus coprophilus* and associated actinomycetes in faeces, sewage and freshwater. *Applied and Environmental Microbiology.* **42:** 1037- 1042.

Merriman, P.R., Price, R.D., Kollmorgen, J.F., Piggott, T. and Ridge, E.H. (1974). Effect of seed inoculation with *Bacillus subtilis* and *Streptomyces griseus* on the growth of cereals and carrots. *Australian Journal Agricultural Research.* **25:** 219-226.

Mincer, T.J., Jensen, P.R., Kauffman, C.A. and Fenical, W. (2002). Widespread and persistent populations of a major new marine actinomycete taxon in ocean sediments. *Applied and Environmental Microbiology.* **68:** 5005–5011.

Nonomura, H. and Ohara, Y. (1969). Distribution of actinomycetes in soil. VI. A culture method effective for both preferential isolation of *Microbispora* and *Streptosporangium Journal of Fermentation Technology.* **47:** 463-469.

Orchard, V.A., Goodfellow, M. and Willianas, S.T. (1977). Selective isolation and occurrence of Nocardiae in soil. *Soil Biology and Biochemistry.* **9**: 233-238.

Palleroni, N.J. (1980). A chemotactic method for the isolation of Actinoplanaceae. *Archives of Mikrobiology.* **128**:53-55.

Rowbotham, T.J. and Cross, T. (1977). Ecology of *Rhodococcus coprophilus* and associated actinomycetes in fresh water and agricultural habitats. *Journal of General Microbiology.* **100**:231-240.

Ryley, J.F., Wilson, R.G., Gravestock, M.B. and Poyser, J.P. (1981). Experimental approaches to antifungal chemotherapy. *Advances in Pharmacology and Chemotherapy.* **18**: 49-176.

Seiler, H., Braatz, R. and Ohmayer, G. (1980). Numerical cluster analysis of the coryneform bacteria from activated sludge. *Zentralblattfuer, Bakteriologie Parasitenkunde Infektionskrnkheiten und Hygiene. Abt. 1* C **1**: 357-375.

Sietsma, J.A., and Wessels, J.G.H. (1979). Evidence for covalent linkages between chitin and b-glucan in a fungal wall. *Journal of General Microbiology.* **114**: 99-108.

Suslow, T.V., and Schroth, M.N. (1982). Rôle of deleterious rhizobacteria as minor pathogens in reducing crop growth. *Phytopathology.* **72**: 111-115 in Doumbou, C.L., Hamby Salove, M.K., Crawford, D.L. and Beaulieu, C. (2002). Actinomycetes, promising tools to control plant diseases and to promote plant growth. *Phytoprotection.* **82**: 85-102.

Suzuki, S., Isono, K., Nagatsu, J., Mizutani, T., Kawashima, K. and Mizuno, T. (1965). A new antibiotic, polyoxin A. *Journal of Antibiotics. (Tokyo) Ser. A.* **18**:131-132.

Suzuki, S., Isono, K., Nagatsu, J., Kawashima, Y., Yamagata, K., Sasaki, K. and Hashimoto, K. (1966). Studies on polyoxins, antifungal antibiotics. IV. Isolation of polyoxins C, D, E, F and G, new components of polyoxin complex. *Agricultural and Biological Chemistry.* **30**: 817-819.

Tahvonen, R. (1982a). The suppressiveness of Finnish light coloured *Sphagnum* peat. *Journal of Agricultural Sciences.* **54**: 345-356 in Doumbou, C.L., Hamby Salove, M.K., Crawford, D.L. and Beaulieu, C. 2002. Actinomycetes, promising tools to control plant diseases and to promote plant growth. *Phytoprotection.* **82**: 85-102.

Tahvonen, R. (1982b). Preleminary experiments into the use of *Streptomyces* spp. isolated from peat in the biological control of soil and seedborne disease in peat culture. *Journal of Agricultural Sciences.* **54**:357-369 in Doumbou, C. L., Hamby Salove, M. K., Crawford, D. L. and Beaulieu, C. 2002. Actinomycetes, promising tools to control plant diseases and to promote plant growth. *Phytoprotection.* **82**:85-102.

Tahvonen, R., and Avikainen, H. (1987). The biological control of seedborne *Alternaria brassicicola* of cruciferous plants with a powdery préparation of *Streptomyces* sp. *Journal of Agricultural Sciences.* **59**: 199-208.

Tahvonen, R., and Lahdenpera, M. L. (1988). Biological control of *Botrytis cinerea* and *Rhizoctonia solani* in lettuce by *Streptomyces* sp. *Annales Agriculturae Fenniae.* **27:** 107-116.

Trejo-Estrada, S.R., Paszczynski A. and Crawford, D.L. (1998). Antibiotics and enzymes produced by the biocontrol agent *Streptomyces violaceusniger* YCED-9. *Journal of Industrial Microbiology & Biotechnology.* **21:** 81–90.

Umezawa, H., Okami, T., Hashimoto, T., Suhara, Y., Hamada, M. and Takeuchi, T. (1965). A new antibiotic, kasugamycin. *Journal of Antibiotics Ser. A.* **18:** 101.

Vernekar, J.V., Ghatge, M.S. and Deshpande, V.V. (1999). Alkaline protease inhibitor: a novel class of antifungal proteins against phytopathogenic fungi. *Biochemical and Biophysical Research Communications.* **262:** 702-707.

Whiffen, A.J. (1950). The Activity in Vitro of Cycloheximide (Acti-Dione) against Fungi Pathogenic to Plants. *Mycologia.* **42:** 253-258.

Williams, S.T., Shameemullah, M., Watson, E.T. and Mayfield, C.I. (1972). Studies on the ecology of actinomycetes in soil VI. The influence of moisture tension on growth and survival. *Soil Biology and Biochemistry.* **4:** 215-225.

Wang, L., Huang, Y., Liu, Z., Goodfellow, M. and Rodriguez, C. (2006). *Streptacidiphilus oryzae* sp. nov., an actinomycete isolated from rice-field soil in Thailand. *International Journal of Systematic Evolutionary Microbiology.* **56:** 1257–1261.

Warnecke, F., Amann, R. and Pernthaler, J. (2004). Actinobacterial 16S rRNA genes from freshwater habitats cluster in four distinct lineages. *Environmental Microbiology.* **6:** 242–253.

Young, R.A. and Smith, R.E. (1975). Degradation of feather keratin by culture filtrates of *Streptomyces fradiae. Canadian Journal of Microbiology.* **21:**583.

Yuan, W.M., and Crawford, D.L. (1995). Characterization of *Streptomyces lydicus* WYEC108 as a potential biocontrol agent against fungal root and seed rots. *Applied and Environmental Microbiology.* **612:** 3119-3128.

Chapter 10
Environmental Factors and their Impact on Fungal Diversity in some Crop Field Soil

S.K. Dwivedi and Neetu Dwivedi

Department of Environmental Science, Babasaheb Bhimrao Ambedkar University, Raibareli Road, Lucknow – 226 025

Soils are very diverse and complex systems with full of life. The soil itself can be viewed as a living organism because it is a habitat for plants, animals and micro-organisms that are all interlinked. Soil consists mineral particles, organic matter and pores. The soil micro and mesoorganisms basically classified into two groups i.e flora and the fauna. Soil flora contains fungi, bacteria and algae and tends to be more static. The fauna on the other hand include nematode, protozoan, collembolan and acarids and tend to move about once the food source get exhausted Singh (2002). Soil organisms are important because they play essential role in the mineralization of nutrients into forms available for plant uptake, and a number of other advantages as well. The associations existing between different soil microorganisms, whether of a symbiotic or antagonistic nature, influence the activities of microorganisms in the soil. Microflora composition of any habitat is governed by the biological equilibrium created by the associations and interactions of all individuals found in the community. Some of the interactions or associations are mutually beneficial or mutually detrimental or neutral. The various types of possible interactions/associations occurring among the microorganisms in soil can be: (*a*) Beneficial; (*i*) Mutualism; (*ii*) Commensalisms and (*iii*) Proto-cooperation or (*b*) Detrimental/Harmful: (*i*) Amensalism;

(*ii*) Antagonism; (*iii*) Competition; (*iv*) Parasitism and (*v*) Predation. The physico-chemical factors of natural environment determine the rates of microbial growth and the nature and size of the indigenous population.The survivorship of microorganism is affected by environmental factors. These includes soil pH, moisture content, temperature, relative humidity and some other biological factors like antagonistic action, antibiosis, predation etc. Nannipieri *et al.* (2003).

Fungi are considered to be the next most abundant micro-flora after the bacteria and actinomycetes in the soil. Curl and Truelove (1986) revealed that the fungi are actually more difficult to assess than bacteria because of filamentous in nature. Fungi belong to a large and diverse group of micro organisms. These are actually a form of cells that are made of a membrane bound nucleus and are devoid of chlorophyll. They also have rigid cell walls. All fungi have a basic characteristic that they contain a vegetative body, of which some parts extend into the air and others penetrate the substrate of the organism that it grows on. Many fungi are also saprophytic in nature persist in soil and water and acquire their food by absorption. Characteristically they also produce sexual and asexual spores. Moulds are composed of numerous, microscopic, branching hyphae known collectively as a 'mycelium' Alexander (1961). Some soil fungi are listed in Table 10.1.

Table 10.1: List of some Soil Fungi

Acremonium roseogriseum (S.B. Saksena) W. Gams
A. strictum W. Gams
Acrophialophora fusispora (S.B. Saksena) Samson
Alternaria alternata (Fr.) Keissl.
Aspergillus candidus
A. clavatus Desm.
A. giganteus Wehmer
A. niger Tiegh. *A. ochraceus* K. Wilh.
Brachysporium sp.
Cercospora fusimaculans G.F. Atk.
Cladosporium cladosporioides (Fresen.) G.A. de Vries
C. oxysporum Berk. and M.A. Curtis
C. variabile (Cooke) G.A. de Vries
Colletotrichum gloeosporioides (Penz.) Sacc.
Curvularia pallescens Boedijn
Dark sterile mycelia
Dictyoarthrinium sp.
Fusariella obstipa (Pollack) S. Hughes
Fusarium incarnatum (Roberge) Sacc.
F. oxysporum Schltdl.
F. solani (Mart.) Sacc.
Gilmaniella humicola G.L. Barron
Gliocladium sp.
Lasiodiplodia theobromae (Pat.) Griffon and Maubl.

Monilia *sp.*

Monodictys lepraria (Berk.) M.B. Ellis

M. nigrosperma (Schwein.) W. Gams

Mucor hiemalis Wehmer

M. hiemalis f. *luteus* (Linnem.) Schipper

M. hiemalis f. *silvaticus* (Hagem) Schipper

M. racemosus f. *racemosus* Fresen.

Mycocladus corymbifer (Cohn) Vánová

Penicillium sp.

Penicillium sp.

P. capsulatum Raper and Fennell

P. chrysogenum Thom

P. citrinum Thom

P. diversum Raper and Fennell

P. expansum

P. funiculosum

P. glabrum (Wehmer) Westling

P. glandicola (Oudem.) Seifert and Samson

P. implicatum Biourge

P. indonesiae Pitt

P. klebahnii Pitt

P. lineatum Pitt

P. nalgiovense Laxa

P. rubrum Stoll

P. rugulosum Thom

P. spinulosum Thom

P. thomii Maire

P. turbatum Westling

Periconia hispidula (Pers.) E.W. Mason and M.B. Ellis

Pestalotiopsis disseminata (Thüm.) Steyaert

Phialophora fastigiata (Lagerb. and Melin) Conant

Phoma glomerata (Corda) Wollenw. and Hochapfel

Pithomyces chartarum (Berk. and M.A. Curtis) M.B. Ellis

Rhizopus oryzae Went and Prins. Geerl.

Thielavia terricola (J.C. Gilman and E.V. Abbott) C.W. Emmons

Trichoderma sp

T. koningii Oudem.

T. viride Pers. *Trichothecium roseum* (Pers)

Triposporium elegans Corda White sterile mycelia

Xylohypha ferruginosa (Corda) S. Hughes

X. nigrescens (Pers.) E.W. Mason ex Dieghton

Zygosporium gibbum (Sacc., M. Rousseau and E. Bommer) S. Hughes

Fungal Population Associated with Guava Fruit Crop Field Soil

Guava (*Psidium guajava*) is one of the most common fruit of India because of its high nutritional value and popularity of the processed products, hence considered as the "Poor men's apple". Besides that, the trees bear fruit twice a year and give generous returns involving very little inputs. Being very hardy it gives an assured crop even with very little care and has the ability to withstand adverse climate conditions and cultivated even in neglected problematic soils Mishra *et al.* (1994).

Guava occupies an important place in the horticultural wealth and economy of India being a hardy and tolerant to water logging. Guava belongs to the family *Myrtaceae* containing 80 genera and 3000 species. In India, 112 species of this family are present, out of which, *Psidium guajava L* is being commercially and economically harvested. Guava is cultivated in 148200 hectares (ha), with a production of 163 million tons all over the country, while Bihar is largest producer of guava producing 0.32 million tons followed by Maharashtra, Uttar Pradesh, Karnataka and Andhra Pradesh. Uttar Pradesh ranks second after Bihar in area under cultivation (12399 ha) and ranks third in guava production (145068 tons). The major guava producing areas in Uttar Pradesh are Allahabad, Varanasi, Lucknow, Kanpur, Aligarh and Agra of approximately 0.325 million ton followed by Andhra Pradesh and Uttar Pradesh Mathi and Pandey (2008). The popular varieties of guava grown in India are Sardar, Allahabad Safeda, Lalit, Pant Prabhat, Dhareedar, Arka Mridula, Khaja (Bengal Safeda), Chittidar, Harija etc. Hybrid varieties like Arka Amulya, Safed Jam and Kohir Safeda were also developed Mitra *et al.* (1929). Some important fungal species for guava and other crop fields and web link have been presented in Tables 10.2–10.4.

Table 10.2: List of Fungal Species Associated in Guava Crop Fields

Fungal Species	*References*
Rhizoctonia solani, R. bataticola	Dwivedi,(1991), Kataria and Verma (1992); Singh and Dwivedi (1987)
Macrophomina phaseolina,	Chattopadhyay and Sen gupta (1955)
Cephalosporium spp.	Dwivedi (1990) Tondon and Agrawal (1954)
Gloeosporium psidii	Tondon and Agrawal (1954)
Verticillium albo atrum	Mishra *et al.* (2004)
Fusarium solani, F. longipes,	
F. moniliforme, F. oxysporum f.sp. *psidii, F. coeruleum*	Chattopadhyay and bhattcharya (1968); Dwivedi and Dwivedi (1999); Edward (1960)
Gliocladium roseum, G.vermoesenii G.virens,G. penicilloides	Mishra and Panday(1999)
Cylindrocarpon lucidium	Mishra and Panday (1992)
Bartilinia robillardoides	Mishra and Panday (1992)
Botyodiplodia theobromae	Pandit and Samajpati (2002)

Contd...

Table 10.2–*Contd...*

Fungal Species	*References*
Clitocybe tabescens	Webber (1928)
Meloidogyne sp.	Rodrigues and Landa (1977)
Helictylenchus sp.	Rodrigues and Landa (1977)
Pratylenlenchus sp.	Rodrigues and Landa (1977)
Myxosporium psidii	Leu *et al.* (1979) and Schroers *et al.* (2005)
Pseudomanas sp.	Tokeshi *et al.* (1980)
Ervinia psidii	Rodrigues *et al.* (1987)
Steptofusidium sp.	Grech (1985)
Colletotrichum gloeosporiodes	Panday and Dwivedi (1985); Ansar *et al.* (1994)
Penicillium vermoensenii, Penicillium citrinum	Upadhyay and Rai (1987) Lim and Manicom (2003)
Aspergilius sp.	Dwivedi (1992)
Trichoderma sp.	Papavizas 1985; Ahmed and Baker (1988);
Curvularia sp.	Panday and Dwivedi(1985); Rajput *et al.* (2008)
Alterneria sp.	Panday and Dwivedi(1985)
Sclerotium rolfsii	Singh and Dwivedi (1987); Khan *et al.* (2001)

Table 10.3: Fungal Diversity in some Important Crop Field Soil

Crop	*Scientific Name*	*Fungal Species*	*Disease*	*Reference*
Cotton	*Gossypium hirsutum* L.	*Fusarium* sp. *Rhizoctonia solani, Pythium* spp., *Phoma exigua* (Ascochyta), and *Fusarium* spp.	Fusarium wilt Seedling diseases	Wang *et al.*, 2009 (College of agricultural and Life Sciences, Plant Pathology Extenstion North Carolina State University), http://www.ces.ncsu.edu/depts/pp/notes/Cotton/cdin1/cdin1.htm
Tomato	*Solanum lycopersicum*	*Didymella lycopersici, Fusarium oxysporum f. sp.conglutinans, Verticillium dahliae Phytophthora cinnamomi*	Stem-rot Wilt Corkey root	Knight, 1960; Erwin and Ribeiro 1996; Panthee and Chen 2010
Cucumber	*Cucumis sativus*	*Fusarium oxysporum f. sp.conglutinans*	Fusarium wilt	Gerlagh and Blok, 1988
		Corynespora cassicola	Leaf spot disease	Kwon *et al.*, 2003
Strawberry	*Fragaria* sp.	*Fusarium oxysporum f. sp.conglutinans*	Fusarium wilt	Hartz *et al.*, 1993
		Botrytis cinerea	Fruit rot	Hauke *et al.*, 2004
Onion	*Allium cepa*	*Plosmodiophoro brassicae, Scle rotium*	Pink root White rot	Champawat and Sharma, 2003

Contd...

Table 10.3–*Contd...*

Crop	*Scientific Name*	*Fungal Species*	*Disease*	*Reference*
Brinjal	*Leucinodes orbonalis*	*Fusarium solani f.sp. melongenae, sclerotium rolfsii*	Wilt Collar rot	Babu *et al.*, 2008 Vanitha and suresh, 2002
Potato	*Solanum tuberosum*	*Alternaria alternate, sclerotium rolfsii*	Leaf spot, wilt	Zaker and Mosallanejad, 2010
Mango	*Mangifera indica*	*Fusarium sterili-hyphosum, Fusarium proliferatum, Fusarium mangiferae, Fusarium pseudocircinatum, Fusarium mexicanum, Fusarium subglutinans*	Mango malformation disease (MMD)	Marasas *et al.*, 2006; Ploetz *et al.*, 2002; Freeman *et al.*, 2004; OteroColina *et al.*, 2010; (Freemam 1999) (Mango (Manako, *Mangifera indica*) Pest and Disease Image Gallery) http://www.ctahr.hawaii.edu/nelsons/mango/
Apple	*Malus domestica*	*Venturia inaequalis* *Helminthosporium papulosum* *Podosphaera leucotricha* *Botryosphaeria obtusa*	Apple scab Black pox Powdery mildew Black rot	Hirst and Stedman, 1962 Natalie (2005)
Wheat	*Triticum aestivum*	*Mycosphaerella graminicola*	Septoria tritici blotch disease of wheat	Marshall *et al.*, 2011
Banana	*Musa acuminata*	*Colletotrichum musae,* *Fusarium oxysporum* f. sp. *cubense* *Dieghtoniella torulosa* *Curvularia eragrostidis* *Fusarium oxysporum f. sp. cubense*	Anthracnose Wilt Leaf spot Fusarium Wilt (Panama disease)	Khan *et al.*, 2001 Ploetz and Pegg 1999; Tushemereirwe *et al.*, 2004 Ploetz, 2006 Raut and Suvarna Ranade, 2004
Sunflower	*Helianthus annuus*	*Puccinia helianthi*	Rust	Qi *et al.*, 2011
Pepper	*Capsicum annuum*	*Alternaria tenuissima*	Leaf spot and fruit rot	Li *et al.*, 2011
Sweet cherry fruit	*Prunus avium*	*Penicillium expansum*	blue mold rot	Xu and Tian, 2008
Citrus	*citrus*	*Alternaria citri*	Causes Alternaria black rot	Katoh *et al.*, 2006
Blackberry-lily	*Iris domestica*	*Alternaria iridicola*	leaf blight	Yu *et al.*, 2002
Cucumber	*Cucumis, Cucurbita*	*Alternaria cucumerina*	Leaf spot of pumpkin	Gannibal, 2011
Maize		*Pestalotiopsis neglecta*	chlorotic spot	Tagne and Mathur, 2001

Contd...

Table 10.3–*Contd...*

Crop	*Scientific Name*	*Fungal Species*	*Disease*	*Reference*
Mulberry	*(Morus* spp.)	*Phyllactinia guttata*	Powdery Mildew	Kurt and Soylu, 2001
Citrus	*Citrus sinensis*	*Xanthomonas campestris pv. citri*	Citrus canker	Braithwaite *et al.*, 2001
Nasturium	*Tropaeolum majus*	*Acroconidiella tropaeoli*	Leaf-spot	Vieira and Barreto, 2002
Rice	*Oryza sativa* L	*Sclerotium hydrophilum*	Leaf sheath	Lanoiselet *et al.*, 2002
Sweetcorn	*(Zea mays L.) cv. 'Golden Millennium'*	*Exserohilum pedicellatum*	cob-rot outbreak	Gilbert, 2002
Garlic	*Allium sativum*	*Fusarium proliferatum*	Causing rot	Dugan *et al.*, 2003.
Muskmelon	*Cucumis melo*	*Didymella bryoniae*	Gummy stem blight	Sudisha *et al.*, 2004
Tomato	*Solanum lycopersicum*	*Fusarium oxysporum f.sp. radicis-lycopersici*	Fusarium crown and root rot	Can *et al.*, 2004
Tomato	*Solanum lycopersicum*	*Alternaria alternata*	leaf blight	Akhtar *et al.*, 2004
Cherry	*Prunus serotina*	*Apiosporina morbosa*	Black Knot	(College of agricultural and Life Sciences, Plant Pathology Extenstion North Carolina State University) http://www.ces.ncsu.edu/depts/pp/notes/oldnotes/fd4.htm
Corn	*Zea mays convar. saccharata var. rugosa*	*Cercospora zeae-maydis*	Gray Leaf Spot	http://www.ces.ncsu.edu/depts/pp/notes/Corn/corn003.html
Blueberry	*Vaccinium* sp.	*Alternaria tenuissima* *Colletotrichum acutatum* *Phomopsis vaccinii*	Alternaria rot Ripe Rot Phomopsis soft rot	http://www.ces.ncsu.edu/depts/pp/notes/Fruit/blueberryinfo/erryrots.htm

Table 10.4: Internet Resources for Fungal Disease with Special Reference to Plants

Database	*Source*	*Web Address*
Tree Fruit database	Davis College of Agriculture, Natural Resources, and Design, Division of Plant and Soil Science, Kearneysvillel Tree Fruit Research and Education Cente,	http://www.caf.wvu.edu/kearneysville/wvufarm1.html
Literature database	AgEcon Search, Waite Library, Dept. of Applied Economics University of Minnesota, USA	http://ageconsearch.umn.edu/
Vegetable Disease	Cornell University Plant Pathology Department	http://vegetablemdonline.ppath.cornell.edu/

Contd...

Table 10.4–*Contd...*

Database	*Source*	*Web Address*
Professional Societies	American Phytopathological Society ASA-CSSA-SSSA Entomological Society of America	http://www.apsnet.org/ http://www.agronomy.org/ http://www.entsoc.org/
University of Wisconsin	Plant Disease and Diagnostic Clinic	http://pddc.wisc.edu/
Fungal database in the Pacific Northwest	.Pacific Northwest Fungi Database	http://pnwfungi.wsu.edu/programs/about Database.asp
AGIS, Genome Informatics Group. Get Information on soybean diseases	Soybean Pathology Database (SoyBase)	http://probe.nalusda.gov:8300/cgi-bin/dbrun/soybase?find+Pathology
AGIS is a cooperative effort between the University of Maryland, College Park, Department of Plant Biology and the National Agricultural Library	Agricultural Genome Information Server	http://probe.nalusda.gov:8000/index.html
Database for Wheat, Cron, Alfa alfa, Soyabean	Field Crop Plant Pathology	http://www.uwex.edu/ces/croppathology/
Fungal Database	Pest and Diseases Image Library (PaDIL)	http://www.mycology.net/ http://www.padil.gov.au/
Cornell University, Department of Pathology	Vegitable MD Online	http://vegetablemdonline.ppath.cornell.edu/
UC IPM Online	Disease models	http://www.ipm.ucdavis.edu/DISEASE/DATABASE/diseasemodeldatabase.html
Explanation of the Database of Plant Diseases in Japan	NAIS genebank	http://www.gene.affrc.go.jp/databases-micro_pl_diseases_en.php
Databases of the U.S. National Fungus Collection	Databases on fungi available through Telnet	http://nt.ars-grin.gov/NFCTEL.HTM
Center for Economic Entomology, Illinois Natural History Survey, USA	Ecological Database of the World's Insect Pathogens (EDWIP)	http://insectweb.inhs.uiuc.edu/pathogens/EDWIP/index.html

Soil-Borne Fungal Pathogens

An interaction among soil-born plant pathogens and fungi causes disease incidence and severity of many crop plants Pieezarka and Abawi (1978). *Macrophomina phaseolina* is reported to cause seedling blight, charcoal rot, root- rot, stem rot and wilt disease on more than 500 species of plants. (Sinclair, 1982 and shamama shameem, 2006). Another soil-borne fungus *Rhizotonia solani* persist as an active mycelium in soil and attacks a wide range of plants causing seed rot, damping-off of seedlings, wilt and root- rot (Baker 1970). Similarly *Fusarium spp.* are known to attack a wide variety of plants causing root- rot, stem rot and wilt diseases (Booth 1971).

Among the *Fusarium* species, *Fusarium oxysporum* and *Fusarium solani* are very common and most destructive pathogens in the agriculture fields and causes root-rot stem rot and wilt diseases on wide range of plants (Booth 1971; Etheshamul-Haque and Ghaffar 1994).

Disease severity varies with the year and the location, and is greatly influenced by farming practices and environmental conditions. Some fungi are the main pathogens responsible for plant disease and they may cause high yield losses (Park *et al.*, 2005; Pereira *et al.*, 2007; Shenoy *et al.*, 2007; Soares and Barreto, 2008; Than *et al.*, 2008). (Latiffah, 2007) isolated *Fusarium* species from 12 cultivated soil planted with different crops in Penang. A total of 42 *Fusarium* isolates were recovered in which four *Fusarium* species identified were *F. solani, F. semitectum, F. equiseti* and *F.oxysporum.* The most prevalent *Fusarium* species recovered was *F. solani* (84 per cent), followed by *F. semitectum* (7 per cent), *F.equiseti* (7 per cent) and *F. oxysporum* (2 per cent). *Sclerotium rolfsii Sacc.*, a facultative saprophyte has an ability to persist in soil for several years and causes disease in over 500 plant species throughout the world specially in tropics and subtropics. (Punja, 1996; Mukherjee & Rajhu, 1997; Harlton *et al.*, 1995; Cilliers *et al.*, 2000 Yaqub and Shahzad, 2009).

Nur (2011) stated that *Fusarium* species associated with maize are widely distributed in Malaysia. During his research work he isolated and identified 167 *Fusarium* isolates from seven locations throughout Malaysia in which eight *Fusarium* spp. were found dominant i.e *F. proliferatum, F. semitectum, F. verticillioides, F. subglutinan, F. oxysporum, F. solani, F. equiseti,* and *F. pseudograminearum.* The determination was based on micro and macro morphological features (growth rates, colony features, mode of production of microconidia, macroconidia, conidiophores, and chlamydospores).

Natalie (2005) reported that fungal species in apple orchards in New Mexico were found highly susceptible to seasonal and environmental factors.

Factors Affecting the Fungal Diversity in Soil Effect of Soil pH

There is not much more information about the effect of pH on fungal diversity but some ecologist and pathologist gave the information about the effect of pH on fungal pathogens associated with different crop field. Jaaffar *et al.* (2010) reported that *Rhizoctonia* spp. can easily survive at soil pH between (<pH 4 to >pH 7.5) in the fields of eastern Washington. Soil pH also affects the availability of nutrients to the plant. Some of these nutrients may be needed for strong cell wall and resistance to fungi. If the pH is too extreme, the plant will be stressed and may be less resistant to attack by the pathogen. Mehta (1951) reported that the pH ranging from 7.5 to 9.0 were found significant for the growth of fungal pathogens in alkaline soil of guava orchards. Most studies on *Rhizoctonia* root rots of broad leaf crops have shown that NO3-N results in less disease compared to NH4-N fertilizers Huber and Watson (1974). Brown patch on wheat caused by *R. solani.* Plots that received urea generally had less disease than the nitrate-treated plots Fidanza and Dernoeden (1996a).

The disease-producing activity of different inoculum levels of *P. ultimum*, in both pasteurized and natural soils. The optimum disease activity was from pH 5.0-5.5. There was a slight decline in disease from pH 5.5 to 6.5, but above pH 6.5, the disease activity dropped significantly. In one soil where the pH was decreased to 4.3, disease also declined significantly. Whereas Sen and Verma (1954) studied that fungal pathogens can best survive in lateritic soil at pH 6.5 in guava orchards.

Smiley *et al.* (1996) found *Pythium* root-rot was more prevalent in sites with inorganic N fertilizers, as opposed to those fertilized with cow manure or pea vines. The effect of the type of N on *Fusarium* foot rot or crown rot disease caused by *Fusarium pseudograminearum* and *F. culmorum* is well-known. Applications of NH4-N increase disease severity and incidence, while NO3-N fertilizers decrease the disease Smiley *et al.* (1972). This is similar to *Fusarium* wilt diseases which are suppressed by alkaline soil and nitrate fertilizers Nelson *et al.* (1981). Chattopadhyay and Bhattacharyya (1966) determined the effect of different factors on growth and reproduction of *Fusarium solani* and *Macrophomina phaseoli* incitants of guava wilt disease in West Bengal. Optimum pH of both fungi was found to be 6.0. Generally many types of fungi grow best at pH level of 4 to 7. Alexopoulous *et al.* (1996). Four *Fusarium* species *viz., F. solani, F. semitectum, F. equiseti* and *F. oxysporum* were isolated from silty loam soil having pH 3.9 Latiffah (2007).

Smiley *et al.* (1996) also found a strong correlation between crown rot and N application, and the disease was inversely proportional to soil pH, at least in the range measured (4.3 to 5.3). Liming the soil to increase soil pH from 5.1-5.3 to >6.0 decreased this disease 2 to four years and there was a significant correlation between soil pH and infected stems Murray *et al.* (1992). Kulkarni (2001) reported that variation due to change in pH level was evident in *Macrophomina phaseolina* isolates responsible for dry stalk rots of maize (*Zea mays* L.). Highest growth was observed at pH 7.0 closely followed by pH 6.5 indicating preferential range to be between pH 6.5 and 7.0. Whereas Shanmugam and Govindaswamy (1973) studied the physiological aspects of *Macrophomina phaseolina* causing groundnut root rot and found that *Macrophomina phaseolina* grew best at optimum pH of 5.0 in soil.

While in case of pigeon pea field soil Lokesha (2002) reported that *Macrophomina phaseolina* causing groundnut root rot best grow in neutral pH (7.0).

Effect of Temperature

Environmental temperature is one of the most important factor affecting the growth rate of microbes. There is a minimum temperature, below which growth does not occur. Most microorganisms have a growth optimum between 20°C and 40°C and are called mesophilic. Higgins (1927) emphasized that temperature was the limiting factor in the geographical distribution of the fungus.

In case of guava field soil (Das gupta and Rai, (1947); Edward (1960); Suhag (1976) reported that from August to October maximum population of fungal pathogens were observed whereas Sowmya (1993) reported that the pathogen *Fusarium oxysporum f. sp. cubense* causing panama disease in banana produced maximum growth at 35°C. But, at 40°C growth was drastically reduced. Clayton (1923) tested the behaviour of

Fusarium lycopersici causing wilt disease in tomato at different temperature ranges. He found that the minimum and maximum temperature required for growth of fungal pathogen ranged between 28°C to 37°C respectivly. Sahi *et al.* (1992) reported that 30°C was ideal for the growth of *Macrophomina phaseolina* causing dry root rot of mungbean. Hari *et al.* (1988) reported that 26°C was optimum temperature for growth of *Sclerotium rolfsii.* Further, they observed maximum growth of *Sclerotium rolfsii* at 30°C in groundnut field soil.

Bai *et al.* (1988) isolated and identified six *Fusarium* spp. from maize stalk rot in China including *Fusarium fusarioides* (*Fusarium chlamydosporum*) which can easily survive at temperature ranges from of 8-30°C. Singh and Mehrotra (1980) observed that rich mycelial growth *of Rhizoctonia bataticola* causing damping- off in gram seeds was observed at temperature 35°C. Sandhu *et al.* (1999) reported that 30°C favoured maximum disease development and growth of *M. phaseolina* in cowpea. Kulkarni (2001) reported variation in growth of *Macrophomina phaseolina* due to temperature ranges to be between 35°C and 40°C.

Natalie (2005) reported that in apple orchards diversity of fungal species *Podosphaera leucotricha* were noticed at warm temperature and high humidity (above 70 per cent) while *Phytophthora* spp. were found at relatively cool temperature at New Mexico.

Sulladmath *et al.* (1977) studied variation in requirement of temperature by different isolates and found that all isolates grew well between 23°C and 25°C. The optimum temperature for ground nut isolate was 25 °C and 30°C for tobacco and potato. Hari *et al.* (1988) reported that, 26°C was optimum temperature for growth of *Sclerotium rolfsii* and maximum growth was recorded at 30°C and maximum number of sclerotia formation was observed at 25°C in case of groundnut. Dalvi and Raut (1986) found that the optimum temperature and relative humidity for the growth of *Sclerotium rolfsii* causing groundnut wilt in culture were 28±1°C and 77 per cent respectively. Manjappa (1979) reported that diversity of *Sclerotium rolfsii* occur at temperature 25°C to 30°C in sunflower crop field. They concluded that least growth was noticed at 40°C.Weerapat and Schroeder (1966) observed that diversity of two strains of *Sclerotium rolfsii* was at temperature 30-35°C in rice seedlings.(Mishra and Panday (2000) revealed that maximum to minimum temperature around 33.5°C to 25°C supports the richness of fungal diversity in guava orchards.

The application of soil solarization, one of the most promising technique has been shown to be effective because it increases soil temperature which reduced or eradicate the fungal populations/soil microbiota from soil (Grinstein *et al.*, 1979; Elad *et al.*, 1980; Katan, 1981,1983; Pullman *et al.*, 1981 a,b; Stapleton and De Vay,1982, 1987, 1995; Katan and De Vay 1991; Dwivedi and Dubey, 1987; Abdel-Rahim *et al.*, 1988 Abu-Gharbieh *et al.*, 1990 a,b; Tjamos *et al.*, 1992; El-Zayat *et al.*, 1990; Dwivedi 1993; Gamliel *et al.*, 2000; Keinath, 1995, 1996 and Blok *et al.*, 2000; Botross *et al.*, 2000).

Dwivedi (1993b) reported that there was gradual reduction in the population densities of three fungal pathogens *viz., Fusarium oxysporurn f. sp. psidii, F. solani* and *Rhizoctonia solani* at temperature 55°C by the use of soil solarization in guava orchards.

Some authors (Gamliel and Katan, 1991; Chen *et al.*, 2000) reported that solarization practice is beneficial to soil microflora whereas others demonstrated that soil heating at temperature around 50 °C are detrimental for microbial biomass (Palese *et al.*, 2004; Scopa and Dumontet, 2007; Scopa *et al.*, 2008).

During soil solarization, temperature commonly reach up to 95°F to 140°F (35-60°C) depending on soil type, season, location, soil depth and other factors. These high temperatures induce changes in soil volatile compounds that are toxic to organisms already weakened by high temperature. Soil solarization is effective against fungal pathogens such as *Verticillium* spp. (wilt), *Fusarium* spp. (several diseases), and *Phytophthora cinnamomi* (*Phytophthora* root-rot), and bacterial pathogens such as *Streptomyces scabies* (potato scab), *Agrobacterium tumefaciens* (crown gall), and *Clavibacter michiganensis* (tomato canker). Pokharel and Hammon (2010).

Yaqub and Shahzad (2009) reported that soil solarization by polyethylene mulching significantly enhanced soil temperature in mulched soils. A difference of 16°C was recorded in mulched and non-mulched soils at 5cm depth. Maximum sclerotial mortality occurred at 5 cm depth followed by 10 cm in mulched soil. Maximum sclerotial mortality and minimum pathogen infection was noted after 15 days of mulching.

Effect of Soil Moisture

Webber (1931) emphasized that the soil moisture was most important environmental factor affecting the disease. Under optimum soil moisture, he reported that potato plants were affected near the soil surface, but infection occurred at greater depths as the soil dried out. Epps *et al.* (1951) observed no growth of fungus through sand from infested wheat seeds when the moisture content was 0.93 per cent or less, but good growth occurred at 1.02 per cent.

Fladns (1958) found the reduction in growth of the fungus with the increase in soil moisture and also reported that the organisms can grow from an inoculum source through air dry soils. Curl (1961) reported that low level of irrigation enhance the the population of *Sclerotium rolfsii* than high levels of irrigations. The widespread fungus *Glomerella psidii* (*Gloeosporium psidii*) caused dieback symptoms on young shoots and fruit, and also caused necrotic leaf spots and fruit canker. Disease development was enhanced by high humidity Cook (1975).

Lingaraju (1977) reported that the saprophytic activity of the fungus was more at 10 per cent soil moisture and the fungus did not survive well at 50 and 70 per cent moisture levels. Ramarao and Raju (1980) reported that fungal diversity in rhizosphere and non-rhizosphere soil of wheat plants increased at higher moisture levels. Population of actinomycetes was maximum with low moisture in rhizosphere soil and decreased gradually as the moisture level increased in non-rhizosphere soil they were favored by higher soil moisture content.

Khati *et al.* (1983) reported that the survival of *Sclerotium rolfsii* was highest at soil moisture levels between 30 and 50 per cent of water holding capacity. Palakshappa (1986) studied the effect of seven soil moisture levels on foot rot of betelvine caused by *Sclerotium rolfsii* and reported that *Sclerotium rolfsii* survived better at low soil moisture levels than at high soil moisture levels ranges between 20 and 40 per cent However, highest sapropohytic activity of the fungus was observed at 40 per cent moisture level (86.66 per cent). Least was found at 60 and 70 per cent soil moisture levels where the saprophytic activity of the fungus was found to be very less (43.33 and 30.00 per cent, respectively). Harlapur (1988) studied the effect of soil moisture on foot rot of wheat caused by *Sclerotium rolfsii* and reported that the fungus survived better at low soil moisture levels as compared to high level. Thirty percent soil moisture was found to be optimum for maximum saprophytic activity of the fungus. Dwivedi *et al.* (1990) also found more pathogenic fungi during rainy and winter seasons surviving better in association with root bits in guava plant.

Mishra and Panday (2000) revealed that higher rainfall during July- September, with high humidity of 76 per cent supports maximum fungal diversity in guava orchards associated with wilt disease. While in case of apple orchards in New Mexico fungal pathogens were found *viz., Podosphaera leucotricha* and *phytophthora spp* in which *Podosphaera leucotricha* best survived at low moisture while *phytophthora spp.* were found most abundant at high moisture condition soil. Natalie (2005)

Dolly *et al.* (2006) reported that August to October was the favonrable period for the development of fungal pathogens, while May to July was found least favonrable period. The optimum soil moisture *for Fusarium solani* was 60 per cent saturation, and 40 per cent was for *M. phaseoli*. Growth of both fungi was reduced with increase or decrease in the moisture content.

(Latiffah 2007) reported that highest percentage of moisture of 75 per cent was obtained from silty loam soil planted with oil palm. With high moisture content, all the four *Fusarium* species *viz., F. solani, F. semitectum, F. equiseti* and *F. oxysporum* were isolated. In contrast, lowest moisture content of 18 per cent–19 per cent, only *F. solani* was isolated.

Biological Interaction

Biological interactions in an ecosystem mainly comprised of mycorrhizal interaction, antagonism, antibiosis, mycoparasitism etc. enhance or suppress the growth of fungal populations in a particular area.

Mycorrhizal fungi are the main pathway through which most plants obtain mineral nutrients and, as such, are critical in terrestrial ecosystem functioning. In this mutualistic symbiosis plants exchange photosynthates not only for mineral nutrients but also for increased resistance to disease, drought and extreme temperatures (Smith and Read 1997). Some groups of mycorrhizal fungi may also mediate plant competition through the formation of mycelial linkages through which carbon is shared among different plant species (Simard and Durall, 2004). Some endo-mycorrhizae release volatile and non-volatile compounds through cortical cells inhibit pathogens and support the growth of antagonist in the rhizosphere Mukhopadhyay, (1994). Srivastava *et al.* (2001) reported that VA-Mychorrizae is useful for controlling fungal pathogens responsible for wilt disease of guava in alfisols. Multiple fungal interactions involving bacteria and fungi in the rhizosphere provide enhanced biocontrol in many cases in comparison with biocontrol agents used singly.

VAM technology is a beneficial tool for controlling fungal pathogens in guava.(Whipps, (2001) Davis *et al.* (1978); Baltruachart and Schoenbeck (1972); Duchense *et al.* (1987); Dehne and Schoenbeck (1979); Dehne (1982); Zambolim and Schenck (1983); Caron *et al.* (1986); Bagyaraj (1984); Bagyaraj and Menge (1978); Ratnayake *et al.* (1978); Gerdemann (1975). Plant community productivity may also influence mycorrhizal diversity. Zak *et al.* (2003) demonstrated that several indicators of microbial activity including fungal abundance increased with plant productivity.

An *Antagonist* is the microorganism that adversely affects another growing in association with it. Webster and Lomas (1964) reported that *Trichoderma viride* produces gliotoxin and viridin which easily inhibit the growth of pathogens.

Parasitic activity of *T. viridi* on sheath blight fungus of rice was reported by Roy (1977). The efficacy of *Streptosporangium pseudovulgare* in controlling rot of guava caused by *Lasiodiplodia theobromae* was reported by Neelima *et al.* (2003). Abada (1994) reported that among the biocontrol agents, fungi in the genus *Trichoderma* are considered to be the most important ones because they control various root diseases caused by a wide range of fungal pathogens. The bio-control by *Trichoderma* could be attributed to nutrient competition, release of toxic metabolites and extra-cellular hydrolytic enzyme activity (Elad, 2000). *Trichoderma* was reported to produce volatile and non volatile antibiotic compounds which inhibit the fungal growth at very low concentration (Weindling and Emerson (1936); Weindling 1941). Antifungal compounds released from *Trichoderma sp.* such as phenol like compound isolated from *Trichoderma harzianum* inhibited the uredospore germination of rust pathogen of groundnut, *Puccinia arachidis* (Govindasamy and Balasubramanian, 1989). *T. harzianum*, which produces an endochitinase and N-acetyl-β-glucosaminidase was reported to be a potent biocontrol agent against several phytopathogenic fungi like *Rhizoctonia solani, Sclerotium rolfsii* and *Penicillium ultimum* (Cook and Baker, 1983).

The application of *Trichoderma* species control a large number of foliar and soil-borne fungi *i.e. Fusarium* spp., *R. solani, Pythium* spp., *S. sclerotium, S. rolfsii* in vegetables, field, fruit and industrial crops (Tran 1998; Ngo *et al.*, 2006).

Dolly *et al.* (2006) indicated *in-vitro* efficacy of *T. viridi* and *Aspergillius niger* reduced the growth *Fusarium sp.* and *Verticillium sp.* by 81 to 90 per cent against guava wilt disease respectively. *Trichoderma* species were found as potential bioagent of several soil borne pathogen *viz. Rhizoctonia soloni, Sclerotium rolfsi, Penicillium ultimum, and B. theobromae.* (Cook and Baker(1983); Simon and Sivasithaparam, (1988); Chet and Inbar, (1994); Elad, (2000); Singh and Singh, (2004); Kidwai *et al.* (2006); Kumar *et al.* (2007) Gupta *et al.* (2003), Bokhari, (2008); Gupta and Mishra, (2009).

Trichoderma effectively control of plant diseases, several commercial biological products based on *Trichoderma* species are manufactured and marketed in Asia, Europe and USA for use on a wide range of crops. These can be efficiently used as conidia, mycelium and chlamydospores which are produced in either solid state or liquid fermentation Harman *et al.* (2004). Gupta and Mishra (2009) reported that treatment by volatile compounds against wilt pathogens of guava were found more effective than culture filtrate of *Penicillium citrinum.*

Antibiosis

Haas and keel (2003) describe the term antibiosis in its original meaning, substances produced by certain microorganism that inhibit or kill other microorganism at very low concentration. Rose (1979) and Fravel (1988) reported that there were an estimated 3000 known antibiotics, with approximately 50-100 new antibiotics being discovered each year. The antibiotic produced by *Bacillus subtilis* has been isolated (Vasudeve *et al.*, 1958) and named as 'bulbiformin' from soil able to suppress diversity of *Fusarium udum* in pigeon pea wilt affected soil. Singh *et al.* (1965) reported that seedling of pigeon pea gained resistance to *F. udum* when the seeds were bacterized with *Bacillus subtilis* befor sowing. Production of antibiotic by *T. harzianum*, inhibitory to growth of *S.rolfsii*, has been reported by Upadhyay and Mukhopadhyay (1983).

Rhizobium spp. secretes a chemical 'rizoblitoxin', inhibitory to fungal population found in field of *Pisium sativum* Chakraborty and Chakraborty (1988). *Pseudomonas fluroscens* were found effective against the fungal population of *Gaeumannomyces graminis* var. *tritici* causal agent of take-all of wheat field soil. (Weller 1985; Gutterson *et al.*, 1986; Thomashow *et al.*, 1986, 1987).

Howie and Suslow (1986) used an antifungal minus mutant of *P. fluroescens* to identify two components of suppression of *Pythium ultimum* in cotton. Wright (1956) reported that gliotoxin-producing strain of *T. viride* gave better control of *Pythium*-induced damping-off than did a viridin producing isolate.

Mutants also have been used to study antibiosis in other systems. Colyer *et al.* (1984) demonstrated that an antibiotic- negative mutant produced by treatment of *Pseudomonas putida* with nitrosoguanidine gave effective control of *Erwinia carotovora* on potato tubers. Kloepper *et al.* (1978, 1981,1981) studied that mutant also provide evidence that antibiosis is involved in the suppression of native rhizosphere microorganisms by plant growth promoting *Fluorescent Pseudomonas* spp.

Christensen (1981) reported that species diversity of soil fungi is a reflection of multiple factors and appears to be reduced by disturbances and manipulation activities. Wicklow *et al.* (1974) found positive correlations between species diversity of soil fungi and plant species diversity soils with rich organic matter, less physical disturbance and low nutrient inputs tend to be fungal dominated whereas frequently tilled soils receiving high inorganic fertilizer inputs are dominated by bacteria Ritz and Young (2004).

The effects of different *land uses* on the biological properties of soil have been considerably investigated, though little is known about the effects on soil microbial functional capacity Degens and Vojvodic-Vukovic (1999). Vinyoles (2008) suggested that microbial functional capacity varies according to the land use and may be restored by increased tree cover and active soil management practices such as agroforestry. The mangrove ecosystems around the world have suffered massive destruction due to activities such as *mining* Kongsangchai (1984) and mariculture Landseman (1994). Destruction of mangrove forests will decrease the amount of substrates available for colonization, thus reducing the number of species of manglicolous fungi and their evenness.

References

Abada, K.A. (1994). Fungi causing damping off and root rot on sugar beet and their biological control with *Trichoderma harzianum*. *Agri. Ecosys. Environ.*, **51:** 333-7.

Abu-Gharbieh, W.I, Saleh, H. and Abu-Blan, H. (1990a). Use of black plastic for soil solarization and post plant mulching. Proceeding of the first international conference on soil solarization, Amman, Jordan, 19-25 February 1990. 229-237.

Abu-Gharbieh, W.I., Saleh, H. and Al-Banna, L. (1990b): Solarization. Proceedings of the first international conference on soil solarization. Amman, Jordan, 19-25 February 1990. 69-77.

Abdel-Rahim, M.F., Satour, M.M., Mickail, K.Y., El-Eraki, S.A., Grinstein, A., Chen, Y., and Katan, J. (1988). Effectiveness of soil solarization in furrow-irrigated Egyptian soils. Plants diseases : **72:** 143-146.

Ahmad, J.S. and Baker, R. (1987). Rhizosphere competence of *Trichoderma harzianum*. *Phytopathology.*, 77 **(2):** 182-189

Allison, S.D., Hanson, C.A. and Treseder, K.K. (2007). Nitrogen fertilization reduces diversity and alters community structure of active fungi inboreal ecosystems. *Soil Biol. Biochem.*, **39:** 1878-1887

Alexopoulous, C.J., Mims, C.W. and Blackwell, M. (1996). Introductory Mycology. 4th Edition. John Wileyand Sons, Inc.

Akhtar, K.P, Saleem, M.Y., Asghar, M. and Haq, M.A. (2004). New report of *Alternaria alternata* causing leaf blight of tomato in Pakistan. *New Disease Reports* (2004) 9: 43.

Bai, J.K., Yin, Z. and Hu, J.C. (1988). A study on the pathogen of maize stalk rot in Northeast China. *Acta Phytophylactica-Sinica.*, 15 **(2):** 93-98.

Baker, K.F. (1970). Types of Rhizoctonia diseases and their occurrence. In: Parmeter, Jr., J.R. (Ed.), Rhizoctonia solani, Biology and Pathology. University of California Press, Berkeley, CA, pp. 125–148.

Bagyaraj, D.J. (1984). Biological interactions with VA mycorrhizal fungi. *In: VA mycorrhizae* (Ed. Powell, C.L & Bagyaraj, D.J.). CRC Press, Boca Raton, USA.

Bagyaraj, D.J. and Menge, J.A. (1978). Interaction between a VA mycorrhizae and azobactor and their effects on rhizosphere microflora and plant growth. *New Phytol.*, **80:** 567-73.

Blatruschat, H. and Schoenbeck, F. (1972). Influence of endotrophic mycorrhizae on chlamydospore production of *Thielaviopsis basicolaphytopathol.*, Z. **74:**172.

Blok, W.J., Lamers, J.G., Termorshuizen, A.J. and Bollen, G.J. (2000): Control of soilborne plant pathogens by incorporating fresh organic amendments followed by tarping. *Phytopathology* **90.** 253-259.

Botross, S.E., El-Assiuty, E.M., Zeinab, M. Fahmi and Abd El-Rahman, T.M. (2000): Long-term effects of soil solarization on density levels of soil-borne fungi and stalk rot incidence in sorghum. Egypt. J.Agric. Res., 78 **(2):** 275-283.

Bokhari, A.A., Sahi, T.S., Khan, A.M. and Ahmad, R. (2008). *In vivo* studies on the biological and chemical control of guava decline caused by different soil borne pathogens *Pak. J. Agri. Sci.*, 45 **(1):** 54-56.

Can,C., Yucel, S. Korolev and Katan, T. (2004). First report of Fusarium crown and root rot of tomato caused by *Fusarium oxysporum* f.sp. radicis-lycopersici in Turkey *New Disease Reports* **9:** 36.

Caron, M., Fortin, J.A. and Richard, C. (1986). Effect of *Glomus intraradices* on infection by *Fusarium oxysporum f. sp. radices- lycopersia* on tomato over a 12 week period. *Canad. J. bot.*, **64:** 552.

Christensen, M. (1981). Species diversity and dominance in fungal communities. In: Wicklow DT, Carroll GC (eds), The Fungal Community: Its Organization and Role in the Ecosystem. Marcel Dekker Inc, New York, pp. 201-232.

Chattopadhyay, S.B. and Bhattacharyya, S.K. (1966). Physiological studies on incitants of guava wilt *Fusarium solani* (Mart.) App. and Wr. emend Snyder and Hansen and *Macrophomina phaseoli* (Maubl.) Ashby. *Indian Journal of Mycological Research.*, 4 **(1&2):** 22-31.

Chattopadhyay, S.B. and Bhattacharjya, S.K. (1968). Formation of Tylosis in xylem vessels by xylem invading pathogens. *Indian J. Agri. Sci.*, **38:** 65-72.

Chet, I. and Inbar, J. (1994). Biological control of fungal pathogens, *Appl Biotechnol.*, **48:** 37-43.

Cilliers, A.J., Herselman, L. and Pretorius, Z.A. (2000). Genetic variability within and among mycelia compatibility groups of *Sclerotium rolfsii* in South Africa. *Phytopathology*, **90:** 1026-1031.

Chen, Y., Katan, J., Gamliel, A., Aviad, T. and Schnitzer, M. (2000). Involvement of soluble organic matter in increase plant growth in solarized soils. *Biol. Fertil. Soil* **32:** 28–34.

Clayton, E.E. (1923). The relation of temperature on the *Fusarium* wilt of tomato. *American Journal of Botany.*, **2:** 71-87.

Cook, A.A. (1975). Diseases of tropical and subtropical fruits and nuts. Guava: *Psidium guajava L.* Hafner Press, a division of Macmillan Publishing Co., New York Pp. 207-212.

Cook, R.J. and Baker, K.F. (1983). The nature and practice of biological control of plant pathogens. *American Phytopathological Society.*, St. Paul, MN, pp.539.

Colyer, P.D. and Mount, M.S. (1984). Bacterization of potatoes with *Pseudomonas putida* and its influence on postharvest soft rot disease. *Plant Dis.*, **68:** 703-6.

Curl, F.A. (1961). Influence of sprinkler irrigation and four forage crops; on populations of soil microorganisms including those antagonistic to *Sclerotium rolfsii* Sacc. *Plant Disease Reporter.*, **45 :** 517-519.

Davis, R.M., Menge, J.A and Erwin, D.C. (1979). Influence of *Glomus fassiculatum* and soil phosphorus on *verticillium* wilt of cotton. *Phytopathology.* **68:** 1614.

Das Gupta, S.N. and Rai, J.N. (1947). Wilt disease of guava (*P. guajava*). *Curr. Sci.* **16:** 256-258.

Dalvi, M.B. and Raut, S.P. (1986). Effect of temperature and humidity on growth and sclerotial formation of *Sclerotium rolfsii* causing groundnut wilt. *Indian Journal of Mycology and Plant Pathology.,* **16:** 175-176.

Degens, B.D. and Vojvodic-Vukovic, M. (1999). A Sampling strategy to assess the effects of land use on microbial functional diversity in soils. *Australian Journal of Soil Research.,* **37:** 593-601.

Dehne, H.W. and Schoenbeck, F. (1979). The influence of endotrophic mycorrhizae on plant disease. I. Colonization of tomato plant by *Fusarium oxysporum f. sp. lycopersicum. Phytopathol.,* Z **95:** 105.

Dehne, H.W. (1982). Interaction between vesicular arbuscular mycorrizal fungi and plant pathogens *Phytopatholog.,* **72:** 1115.

Deborah, R. Fravel. (1988). Role of antibiosis in the biocontrol of plant diseases. *Ann Rev. Phytopathol.,* **26:** 75-91.

Dugan, F.M., Hellie, R.B.C. and Lupien, S.L. (2003). First report of *Fusarium proliferatum* causing rot of garlic bulbs in North America *New Disease Reports* **6:** 25.

Dieter, H. and Christop, K. (2003). Regulation of antibiotic production in root colonizing *Pseudonomas* spp. and relevance for biological control of plant disease. *Annu.Rev.Phytopathol.,* 2003 **41:** 117-53.

Dolly, J. Gupta, P. and Gupta, V. (2006). Etiology and management of guava wilt disease in Allahabad and Kaushambi Districts of Uttar Pradesh, India. *J. Mycol. Pl. Pathol.,* **36:** 3, 365-72.

Duchesne, L.C., Peterson,.R.L. and Ellis, B.E. (1987). Pine roots exudates stimulate antibiosis by *Paxillus involut*us against *Fusarium oxysporum. In: Proc 7TH NACOM.* (Eds Sylvia, D.M., Hugg,L.L and Graham, J.H) PP 193. Gainesville, USA.

Dwivedi, S.K. (1991). Population dynamics of micro—fungi including pathogenic forms in the beds of completely healthy partially wilted and completely wilted guava trees grown on a line. *Intern. J. Trop. Plant Dis.,* **9:** 95-109.

Dwivedi, S.K. and Dwivedi, P. (1999). Wilt disease of guava: A national problem. *J. Applied Hort.,* **1:** 151-154.

Dwivedi, S.K. (1992). Effect of culture filtrate of soil microbes of pathogens inciting wilt disease of guava (*Psidium guajava* L.) under *in vitro* conditions. *Natl. Acad. Sci. Lett.,* **15:** 33-35.

Dwivedi, S.K. (1993a). Fungitoxicity of *Foeniculum vulgare* seed oil used against a guava wilt pathogen. *Natl. Acad. Sci. Lett.,* **16:** 207-207.

Dwivedi, S.K. (1993b). Soil solarization adversely affects some fungal pathogens causing wilt disease of guava (psidium *guajava* l.) Soil Eiol. *Biochem.* Vol. 25, No. 11, pp. 1635-1636, Dwivedi, S.K. 1990. Efficacy of some antibiotics on *Fusarium oxysporum* f.sp. *psidii* causing wilt disease of guava.*Hindustan Antibiotics Bulletin.,* 32 **(3-4):** 88-90.

Dwivedi, R.S. and Dubey, R.C. (1987). Effect of soil solarization on the population dynamics of *Macrophomina phaseolina* (Tassi) Goid. and general soil mycoflora. *International Journal of Tropical Plant Diseases* **5:** 67-73.

Etheshamul-Haque, S. and Ghaffar, A. (1994). New records of root infecting fungi from Pakistan. *Pak. J. Phytopath.,* **6:** 50-57.

Edward, J.C. (1960). Penetration and establishment of *Fusarium oxysporum* f.sp. *psidii* in Guava root. *Indian Phytopath.,* 13**(1)**:168-70.

Epps, W.M., Patterson, J.C. and Freeman, *I.E.* (1951). Physiology and parasitism of *Sclerotium rolfsii. Phytopathology.,* **41:** 245-255.

Elad, Y. (2000). Biological control of foliar pathogen by means of *Trichoderma harzianum* and potential modes of action. *Crop Prot.,* **19:** 709-14.

Elad, Y., Katan, J. and Chet, I. (1980). Physical, biological, and chemical control integrated for soil-borne diseases in potatoes. *Phytopathology,* **70:** 418-422.

El-Zayat, M.M., Ashour, W.E. and El-Shami-Mona, A. (1990): Residual effect of soil solarization for management of Fusarium wilt of tomato in the Nile Delta. Proceeding of the first international conference on soil solarization 19-25 February 1990. Amman, Jordan, p 35.

Freeman, S., Klein-Gueta, D., Korolev, N. and Sztejnberg, A. (2004). Epidemiology and survival of Fusarium mangiferae, the causal agent of mango malformation disease. *Acta Hort.,* **645:** 487-491.

Freeman, S., Maimon, M., and Pinkas, Y. (1999). Use of GUS transformants of Fusarium subglutinans for determining etiology of mango malformation disease. *Phytopathology.,* **89:** 456-461.

Fidanza, M.A. and Dornoeden, P.H. (1996). Brown patch severity in perennial ryegrass as influenced by irrigation, fungicide, and fertilizers. *Crop Sc.,* **36:** 1631-1638.

Flados, N.D. (1958). Ecological factors affecting the growth of *Sclerotium rolfsii,* (Abstract) *Phytopathology.,* **48:** 342.

Grech, N.M. (1985). First report of guava rapid death syndrome caused by *Septofusidium* sp. in South Africa. *Plant Dis.,* **69:** 726-726.

Grinstein, A., Katan, J., Abdul-Razik, A., Zeydan, O. and Elad, Y. (1979) Control of *Sclerotium rorfsii* and weeds in peanuts by solar heating of the soil. *Plunf Disease Reporter* **63:** 195-1959.

Gamliel, A. and Katan, J. (1991). Involvement of fluorescent Pseudomonads and other microorganisms in solarized soils. *Phytopathology* **81:** 494–502.

Gerlagh, M. and Blok, W.J. (1988). *Fusarium oxysporum f.sp. cucurbitacearum n.f.* embracing all formae speciales of F. oxysporum attacking Cucurbitaceous crops *European Journal of Plant Pathology* Volume 94 **(1):** 17-31.

Glòria Pallarès Vinyoles. (2008). Effects of land use on soil microbial community function in western Kenya highlands SLU, Department of Ecology, Uppsala,

Sweden Grafe, U.(1992) Biochemic der antibiotika: Struktur-Biosynthese-Wirkmechanismus. Spktrum Akademischer Verlag, Heideberg, p.389.

Gutterson, N.I., Layton, T.J., Ziegle, J.S. and Warren, G.J. (1986). Molecularcloning of genetic determinants for inhibition of fungal growth by. *fluorescent Pseudomonad. J.Bacteriol.,***165:** 696-703.

Gupta, R.P., Yadava, B.C., Singh, R.V. and Rai, O. (2003). Evaluation of biocontrol agents against *Fusarium* wilt of pigeon pea. Abstr. BS-57. National symposium of pulses for crop diversification and natural resource management, Kanpur, Dec., 20-22, 2003.

Govindasamyt, V. and Balasubramanian, R. (1989). Biological control of groundnut rust, *Puccinia arachidis* by *Trichoderma harzianum* Z. pflkrankh. Pflschutz, **96:** 337-345.

Gerdemann, J.W. (1975). Vesicular arbuscular mycorrhizae *In: Development and function of roots* (Ed. Torry, J.G. & Clarkson, D.T.) pp 575-91. Academic Press, New York

Gupta, V.K. and Misra, A.K. (2009). Efficacy of bioagents against *Fusarium* wilt of guava. *J. Mycol. Plant Pathol.,* **39:** 101-106.

Gamiel, A., Grinstein, A., Zilberg, V., Beniches, M., Katan, J. and Ucko,O. (2000). Control of soil-borne diseases by combining soil solarization and fumigants. *Acta Horticulturae* **532:** 157–164.

Harlapur, S.I. (1988). Studies on some aspects of foot rot of wheat caused by *S. rolfsii* Sacc. M.Sc. (Agri.) Thesis, University of Agricultural Sciences, Dharwad.

Higgins, B.B. (1927). Physiology and parasitism of Sclerotium rolfsii Sacc. *Phytopathology.,* **17:** 417-448.

Harman, G.E., Howell, C.R., Viterbo, A., Chet, I. and Lorito, M. (2004). Trichoderma species opportunistic, avirulent plant symbionts. *Nat Review Microbiol.,* 2 **(1):** 43-56.

Huber, D.M. and Watson, R.D. (1974). Nitrogen form and plant disease. *Annu. Rev. Phytopathology.,* **12:** 139-165.

Hari, B.V.S.C., Chiranjeevi, V., Sitaramaiah, K.A. and Subrhamaniyam, K. (1988). *In vitro* screening of fungicides against groundnut isolate of *S. rolfsii* Sacc. by soil vial technique. *Pesticides.,* **10:** 47-49.

Howie, W., Suslow, T. (1986). Effect of antifungal compound biosynthesis on cotton root colonization and *Pythium* suppression by a strain of *Pseudomonas fluorescens* and antifungal minus isogenic mutant. *Phytopathology.,* **76:** 1069 (Abstr.).

Harlton, C.E., C.A. Lévesque and Z.K. Punja. (1995). Genetic diversity in *Sclerotium* (*Athelia*) *rolfsii* and related species. *Phytopathology,* **85:** 1269-1281.

Jaaffar, A.M., Weller, D.M., Paulitz, T.C. and Thomashow, L.S. (2010). Effect of pH on the growth of Rhizoctonia spp. from cereal-based cropping systems in eastern Washington State. *Phytopathology.,* Vol. 100 **(6):** S85.

Katan, J. and De Vay, J.E. (1991). Soil Solarization. CRC Press, Inc., Boca Raton, Florida 33431 267pp.

Katan, J. and De Vay, J.E. (1991). Soil Solarization. CRC Press, Inc., Boca Raton, Florida 33431 267pp.

Katan, J. (1981). Solar heating (solarization) of soil for the control of diseases caused by soilborne pests. *Ann. Rev. Phytopathol.* **19:** 211-236.

Katan, J., Fisher, G. and Grinstein, A. (1983). Short- and long- term effects of soil solarization and crop sequence on *Fusarium* wilt and yield of cotton in Israel. Phytopathology, **73:** 1215-1219.

Keinath, A.P. (1995). Reduction in inoculum density of Rhizoctonia solani and control of belly rot on pickling cucumber with solarization. Plant Dis, **79:** 1213-1219.

Kumar, P., Mishra, AK. and Pandey, B.K. (2007). *In vitro* evaluation of *Trichoderma* spp. against vegetative mango malformation pathogen *Fusarium miniliformae* var. subglutinans. *J Eco- Friend Agri* **2:** 187-189.

Khan, R.M., Kumar, S. and Reddy, P.P. (2001). Role of plant parasitic nematode (s) and fungi in guava wilt. *Pest Manage. Horticult. Ecosyst.*, **7:** 152-161.

Kulkarni, S. (2001). Biology and management of dry stalk rots of maize (*Zea mays* L.) caused by *Fusarium moniliforme* Sheldon and *Macrophomina phaseolina* (Tassi) Goid. *Ph.D. Thesis,* University of Agricultural Sciences, Dharwad.

Kongsangchai, L. (1984). Impacts upon mangrove forests in Thailand. In: Proceedings of the Asia Symposium on Mangrove Environment–Research and Management. Thailand: 558-567.

Kloepper, J.W. and Schroth, M.N. (1978). Association of *in vitro* antibiosis with inducibility of inceased plant growth by *Pseudonomas* spp. *Phytopathol.*, News **12:** 136 (Abstr.).

Kwon, M.K., Kang, B.R, Cho, B.H. and Kim, Y.C. (2003). Occurrence of target leaf spot disease caused by *Corynespora cassicola* on cucumber in Korea. *New Disease Reports* **6:** 23.

Kloepper, J.W. and Schroth, M.N. (1981a). Plant growth promoting rhizobacteria and plant growth under genobiotic conditions *Phytopathology.*, **71:** 642-44.

Kloepper, J.W. and Schroth, M.N. (1981b). Relationship of *in vitro* antibiosis of Plant growth promoting rhizobacteria to plant growth and the displacement of root microflora. *Phytopathology.*, **71:**1020-24.

Khati, D.V.S. Bhargava, S.N. and Shukla, D.N. (1983). Some studies on the saprophytic survival of Sclerotium rolfsii in soil. *Z. pflalsrankh pflpath. Pllsohuts.*, **90:** 382–387.

Kataria, H.R. and Verma, P.R. (1992). Rhizoctonia solani damping-off and root rot in oilseed rape and canola. *Crop Prot.* **11:** 8–13.

Kidwai, M.K, Srivastava, Vikas, S. and Singh, HB. (2006). Compatibility of *Trichoderma harzianum* to selected fungicide. *J.Eco- Friend Agri.*, **1:**156-161.

Lokesha, N.M. (2002). Management of dry root rot of pigeonpea (Cajanus cajan (L.) Millsp) caused by Macrophomina phaseolina (Tassi.) Goid. M.Sc. (Agri.) Thesis,University of Agricultural Sciences, Dharwad.

Lanoiselet,V.M., Cother, E.J., Ash, G.J. and Harper, J.D.I. (2002). First report *of Sclerotium hydrophilum* on leaf sheath of rice (Oryza sativa L.) in south eastern Australia. *New Disease Reports,* **5:** 20.

Lodge, D.l. and Cantrell, S. (1995). Fungal communities in wet tropical forests: variation in time and space. *Canadian Journal of Botany* **73:** S1391-S1398.

Leu, L.S., Kao, C.W., Wang, C.C., Liang, W.J. and Hsieh, S.P.Y. (1979). *Myxosporium* wilt of guava and its control. *Plant Dis. Rep.,* **63:** 1075-1077.

Lim, T.K. and Manicom, B.Q. (2003). Diseases of Guava. In: Diseases of Tropical Fruit Crops, Ploetz, R.C. (Eds.). CABI Publication, Wallingford, UK., pp: 275-289.

Lingaraju, S. (1977). Studies on *Sclerotium rolfsii* Sacc. with respect to its survival in soil. *M.Sc. (Agari.) Thesis,* University of Agricultural Sciences, Bangalore, p. 87.

Landseman, L. (1994). Negative impacts of coastal aquaculture development.*WorldAquaculture.,* **25:** 12-17.

Latiffah, Z. Mohd Zariman, M. and Baharuddin, S. (2007). Malaysian Diversity of *Fusarium* species in cultivated soils in Penang. *Journal of Microbiology.,* Vol 3 **(1):** 27-30.

Marasas, W.F.O., Ploetz, R.C., Wingfield, M.J., Wingfield, B.D. and Steenkamp, E.T. (2006). Mango malformation disease and the associated Fusarium species. *Phytopathol.,* **96:** 667-672.

Misra, A.K. and Pandey, B.K. (1999). Natural wilting of guava plants during different months. *Indian Phytopathol.,* **52:** 312-312.

Misra, A.K. and Pandey, B.K. (1992). Wilt of guava and associated pathogen. *Indian J. Mycol & Pl. Path.,* **22:** 85-6.

Murray, T.D., Walter, C.C. and Anderegg, J.C. (1992). Control of Cephalosporium stripe of winter wheat by liming. *Plant Dis.,* **76:** 282-286.

Manjappa, B.H. (1979) Studies on the survival and variation in *S. rolfsii* Sacc. *M.Sc. (Agari. Thesis,* University of Agricultural Sciences, Bangalore, p. 86.

Mukhopadhyay, A.N. (1994). Biocontrol of soil borne fungal plant pathogens- current status, future prospect and potential limitations *Indian Phytopath.,* 47 **(2):** 119-26.

Misra, A.K. and Pandey, B.K. (2000). Progressive natural wilting of guava plants during different months. *Indian Phytopathol.,* **53:** 423-427.

Mehta, P.R. (1951). Observations on new and known diseases of crop plants in Uttar Pradesh. *Plant Prot. Bull.,* **3:** 712-712.

Mathi, M.K. and Pandey, A.P. (2008). marketing of guava in Allahabad district uttar pradesh. The IUP *Journal of agricultural Economics* **5:** 1 7-23.

Marshall, R, Kombrink, A., Motteram, J., Loza-Reyes, E., Lucas, J., Hammond-Kosack, K.E., Thomma, B.P. and Rudd, J.J. (2011) Analysis of two in planta expressed LysM effector homologs from the fungus *Mycosphaerella graminicola* reveals novel functional properties and varying contributions to virulence on wheat. Plant Physiol. 2011 Jun;156 **(2)**: 756-69.

Mukherjee, P.K. and Raghu, K. (1997). Effect of temperature on antagonistic and biocontrol potential of shape *Trichoderma* sp.,on *Sclerotium rolfsii. Mycopathologia,* **139:** 151-155.

Neelima, G., Prakash, O.M., Pathak, R.K., Garg, N. and Prakash, O. (2003). Biocontrol of rot of guava by *Streptosporangium pseudovulgare* of cow dung origin. *Farm Sci. J.,* 12 **(2):** 162.

Ngo, B.H., Vu, D.N. and Tran, D.Q. (2006). Analyze antagonist effects of *Trichoderma* spp. for controlling southern stem rot caused by *Sclerotium rolfsii* on peanut. *Plant Protection.,* **1:** 12-14.

Nelson, P.E., Toussoun, T.A. and Cook, R.J. (1981). *Fusarium*: Disease, Biology, and Taxonomy. Penn. State Univ., University Park, PA.

Natalie, P. Goldberg (2005). Apple disease control. New Mexico state university.

Nur Ain Izzati, M.Z., Azmi, A.R., Siti Nordahliawate, M.S. and Norazlina, J. (2011). Contribution to the knowledge of diversity of *Fusarium* associated with maize in Malaysia. Plant Protect. Sci., **47:** 20–24.

Qtero-Colina, G., Rodríguez-Alvarado G., Fernández-Pavía, S., Maymon, M., Ploetz, R.C., Aoki, T., O'Donnell, K. and Freeman, S. (2010). Identification and Characterization of a Novel Etiological Agent of Mango Malformation Disease in Mexico, Fusarium mexicanum sp. nov. *Phytopathology.,* 100 **(11):** 1176-1184.

Pandit, P.K. and Samajpati, N. (2002). Wilt disease of guava (*Psidium guajava* L.) by *Botryodiplodia theobromae Pat. J. Mycological Res.,* **40:** 71-73.

Pandey, R.R. and Dwivedi, R.S. (1985). *Fusarium oxysporum* f. sp. *psidii* as a pathogen causing wilt of guava in Varanasi district in India. *J. Phytopathol.,* **114:** 243-248.

Papavizas, G.C. (1985). Trichoderma and Gliocladium Biology, Ecology andPotential for Biocontrol. *Annual Review of Phytopathology.,* **23:** 23-54.

Ploetz, R.C., Zheng, Q., Vazquez, A. and Sattar, M.A.A. (2002). Current status and impact of mango malformation in Egypt. *Int. J. Pest Manag.,* **48:** 279-285.

Ploetz, R.C. (2006). Fusarium Wilt of Banana Is Caused by Several Pathogens Referred to as *Fusarium oxysporum f. sp. cubense. Phytopathology,* 96**(6):** 653-656.

Park, J.H., Choi, G.J., Jang, S.K., Lim, K.H., Kim, T.H., Cho, Y.K. and Kim, J.C. (2005). Antifungal activity against plant pathogenic fungi of chaetoviridins isolated from *Chaetomium globosum.* FEMS *Microbiology Letters.* **252:** 309-313.

Pereira, O.L., Barreto, R.W., Cavallazzi, J.R.P. and Braun, U. (2007). The mycobiota of the cactus weed Pereskia aculeata in Brazil, with comments on the life-cycle of Uromyces pereskiae. *Fungal Diversity.,* **25:** 127-140.

Pieczarka, D.J., and Abawi, G.S. (1978). Popualation and biology of Pythium spp. associated with snap been roots and soil in New York. *Phytopathology.,* **68:** 409-416.

Palakshappa, M.G. (1986). Studies on foot rot of betelvine caused by *Sclerotium rolfsii* Sacc. in Karnataka. *M. Sc. (Agri.) Thesis,* University of Agricultural Sciences,Bangalore.

Pokharel, R. and Hammon, R. (2010). Increased efficacy of biofumigation by soil solarization and integrating with Brassica meal cake and poultry manure to manage soil-borne problem in onion. Report Submitted to EPA, PESP program. Pp 23.

Punja, Z.K. (1988). *Sclerotium* (*Athelia*) *rolfsii,* a pathogen of many plant species. In: *Advances in Plant Pathology,* (Ed.): G.S. Sidhu. pp. 523-534. Vol. 6, Academic Press, London.

Pullman, G.S., Devay, J.E., Garber, R.H., and Weinhold, A.R. (1981a): Soil solarization : Effects on Verticillium wilt of cotton and soil-borne populations of V.dahliae, Pythium spp., Rhizoctonia solani and Thielaviopsis basicola. Phytopathology, **71:** 954-959.

Pullman, G.S., De Vay, J.E. and Garber, R.H. (1981 b): Soil solarization and thermal death: Alogarithmic relationship between time and temperature for four soil-borne plant pathogens. Phytopathology, 71, **9:** 959-964.

Panthee, D.R. and Chen, F. (2010). Genomics of fungal disease resistance in tomato. *Curr.Genomics,* 11 **(1):** 30-39.

Qi, L.L., Hulke, B.S., Vick, B.A., and Gulya, T.J. (2011). Molecular mapping of the rust resistance gene R (4) to a large NBS-LRR cluster on linkage group 13 of sunflower. *Theor.Appl.Genet.,* 123 **(2):** 351-358.

Roy, A.K. (1977). Parasitic activity of *Trichoderma viride* on sheeth blight fungus of rice. *J. Pl. Dis. Prot.* **84:** 675-683.

Raut, S.P. and Ranade, S. (2004). Diseases of Banana and their Management. Diseases of Fruits and Vegetables: Volume II 37-52, DOI: 10.1007/1-4020-2607-2_2.

Rajput, N.A., Pathan, M.A., Jiskani, M.M., Rajput, A.Q. and Arain, R.R. (2008). Pathogenicity and host range of Fusarium solani (Mart.) Sacc., causing dieback of Sisham (Dalbergia sissooRoxb.). *Pak. J. Bot.,* 40 **(6):** 2631-2639.

Ritz, K. and Young, I.M. (2004). Interactions between soil structure and fungi. *Mycologist,* **18:** 52-59.

Ratnayake, N.L., Leonard, R.T. and Menge, J.A. (1978). Root exudation in relation to supply to phosphorus and its possible relevance to mycorrhizal formation- *New Phytol.,* **81:** 543-52.

Ramarao, P. and Raju, U. (1980). Effect of soil moisture on development of foot and root rot of wheat and on other soil microflora. *Indian Journal of Mycology and Plant Pathology.,* **10:** 17-22.

Rodriguez, F.M.E. and Landa, J.B. (1977). Chemical soil disinfection against parasitic nematode in guava nurseries. *Central Agricola Facultod Cieneias Agricolos,* **4:** 57-77.

Rodriguez, N.J., Robbs, C.F. and Yamashino, T. (1987). A bacterial disease of guava (*Psidium guajava*) caused by *Erwinia psidii* sp. nov. Fitopathologia Brasileira., **12:** 345-350.

Rose, A.H., ed. (1979). *Economic microbiology. Secondary products of metabolism* Vol.3 New York: Academic 595 pp.

Scopa, A. and Dumontet, S. (2007). Soil solarization: effects on soil microbiological parameters. *J. Plant Nutr.* 30 **(4):** 537–547.

Scopa, A., Candido, V., Dumontet, S. and Miccolis, V. (2008). Greenhouse solarization: effects on soil microbiological parameters and agronomic aspects. *Scientia Horticulturae* 116 (2008) 98–103.

Smith, S. and Read, D.J. (1997). Mycorrhizal Symbiosis. Second Edition.Academic Press, London.,p. 605.

Simard, S.W. and Durall, D.M. (2004). 'Mycorrhizal networks: a review of their extent, function, and importance'. *Canadian Journal Of Botany-Revue Canadienne De Botanique.*, 82.8: 1140–1165.

Schroers, H.J., Geldenhuis, M.M., Wingfield, M.J., Schoeman, M.H., Yen, Y.F., Shen, W.C. and Wingfield, B.D. (2005). Classification of the guava wilt fungus *Myxosporium psidii*, the palm pathogen *Gliocladium vermoesenii* and the persimmon wilt fungus *Acremonium diospyri* in *Nalanthamala*. Mycologia., **97:** 375-395.

Singh, R.K. and Dwivedi, R.S. (1987). Effect of oils on Sclerotium rolfsii causing root rot of barley. *Ind. J. Phytopath.*, **40:** 531-533.

Smiley, R.W., Collins, H.P. and Rasmussen, P.E. (1996a). Diseases of wheat in long-term agronomic experiments at Pendleton, Oregon. *Plant Dis.*, **80:** 813-820.

Smiley, R.W., Cook, R.J. and Pappendick, R.I. (1972b). Fusarium foot rot of wheat and peas as influenced by soil application of anhydrous ammonia and ammonia potassium azide solutions. *Phytopath.*, **62:** 86-91.

Singh, P.J. and Mehrotra, R.S. (1980). The influence of cultivar and temperature on carbohydrate and amino acid exudation from gram seeds and on pre-emergence damping off by *Rhizoctonia bataticola. Plant and Soil, Netherland.*, **55:** 262-268.

Sandhu, A., Singh, R.D. and Sandhu, A. (1999). Factors influencing susceptibility of cowpea to *M. phaseolina. Journal of Mycology and Plant Pathology.*, **29:** 421-424.

Shanmugam, N. and Govindaswamy, C.V. (1973). Control of *Macrophomina phaseolina* root rot of groundnut. *Madras Agricultural Journal.*, **60:** 500-503.

Sulladmath, V.V., Hiremath, P.C. and Anilkumar, T.B. (1977). Studies on variation in *Sclerotium rolfsii. Mysore Journal of Agricultural Sciences.*, **11:** 374-380.

Sudisha, J., Kumar, T. Niranjana, S.R. and Shetty, S.H. (2004). First report of gummy stem blight caused by *Didymella bryoniae* on muskmelon (*Cucumis melo*) in India New Disease Reports (2004) **9:** 7.

Suhag, L.S. (1976). Observations in guava decline in Haryana and its control. *Pesticides.*, **10:** 42-44.

Sowmya, G.S. (1993). Studies on Panama disease of banana caused by *Fusarium oxysporum f.sp. cubense.* M.Sc. (Agri) Thesis submitted to University of Agricultural Sciences, Bangalore. pp. 11–27.

Sahi, S.T., Shakir, A.S., Bajwa, M.N. and Intizar-ul-Hassan, M. (1992). Physiological studies on *Macrophomina phaseolina* causing dry root rot of mungbean. *Journal of Agricultural Research.*, Lahore, **30:** 409-413.

Singh, A. and Singh, H.B. (2004). Control of color rot in mint (*Mentha* sp.) caused by *Sclerotium rolfsii* using biological mean. *Curr Sci.*, **87:** 362-66.

Simon, C. and Sivasithaparam, M. (1998). Interaction among G. *graminum* var. tritci, *Trichoderma koningii* and soil bacteria. *Can J Microbiol.*, **34:** 871-76.

Sen, P.K. and Verma, B.S. (1954). Studies on the die-back disease of guava (*Psidium guajava* L.): A survey of the incidence of the disease in Jhargram area (West Bengal). Proceedings of the 41st Indian Science Congress, (ISC'54), India, pp: 25-25.

Shenoy, B.D., Jeewon, R. and Hyde, K.D. (2007). Impact of DNA sequence-data on the taxonomy of anamorphic fungi. *Fungal Diversity.*, **26:** 1-54.

Soares, D.J. and Barreto, R.W. (2008). Fungal pathogens of the invasive riparian weed Hedychium coronarium from Brazil and their potential for biological control. *Fungal Diversity.*, **28:** 85-96.

Sinclair, J.B. (1982). Compendium of Soybean disease. 2nd Ed. by American Phytopathology Society, St. Paul, Minnesota, USA.

Singh, R.K. and Dwivedi, R.S. (1987). Effect of oils on Sclerotium rolfsii causing root rot of barley. *Indian Phytopathology.*, **40:** 531-533.

Srivastava, A.K., Ahmad, R., Kumar, S. and Mohandos, S. (2001). Role of VA - Mycorrhiza in the management of wilt disease of guava in the alfivols of Chotanagpur. *Indian Phytopathol.*, **54:** 78-81.

Singh, P., Vasudeva, R.S. and Bajaj, B.S. (1965). Seed bacterization and biological activity of bulbiformin. *Ann. Appl. Biol.* **55:** 89-97.

Stapleton, J.J. and DeVay, J.E. (1982). Effect of soil solarization on populations of selected soilborne microorganisms and growth of deciduous fruit tree seedlings. Phytopathology **72:** 323–326.

Stapleton, J.J., Lear, B. and Devay, J.E. (1987). Effect of combining soil solarization with certain nematicides on target and nontarget organisms and plant growth. *Ann. Appl. Nematology*, **1:** 107-112.

Stapleton, J.J. and DeVay, J.E. (1995). Soil solarization: A natural mechanism of integrated pest management. In: Reuveni, R. (Ed.), Novel Approaches to Integrated Pest Management. Lewis Publishers, Boca Raton, pp. 309}322.

Than, P.P. Prihastuti, H., Phoulivong, S. Taylor, P.W.J. and Hyde, K.D. (2008). Chilli anthracnose disease caused by *Colletotrichum* species. *Journal of Zhejiang University Science* B **9:** 764-778.

Thomashow, L.S., Weller, D.M. (1987). Role of a phenazine antibiotic in disease suppression by *Pseudomonas fluorescens* 2-79. *Phytopathology.*, **77:** 1724 (Abstr.).

Thomashow, L.S., Weller, D.M. and Cook, R.J. (1986). Molecular analysis of phenazine antibiotic synthesis by *Pseudomonas fluorescens* strain 2-79. *Third Int. Symp. Mol. Genet. Plant microbe Interact. McGill Univ.,Montreal, Canada.* (Abstr.).

Thomashow, L.S., Weller, D.M. (1988). Role of a phenazine antibiotic from *Pseudomonas fluorescens* in biological control of *Gaeumonnomyces graminis var. tritici. J.Bacteriol.*, **170:** 3499–508.

Tanden, R.N. and Agarwal, R.K. (1954). Reported wilt and dieback of guava due to *Rhizoctonia* sp. *Proc. Indian Acad. Sci.*, **40:** 102-109.

Tokeshi, H., Valdebenito, R.M. and Dias, A.S. (1980). Occurrence of a bacterial disease of guava in Sao Paulo State. Summa *Phytopathologica.*, **6:** 85-87.

Tran, T.T. (1998). Antagonistic effectiveness of *Trichoderma* against plant fungal pathogens. *Plant Protection.*, **4:** 35-38.

Tsui, K.M., Fryar, S.C., Hodgkiss, U., Hyde, K.D., Poonyth, A.D. and Taylor, lE. (1998). The effect of human disturbance on fungal diversity in the tropics. *Fungal Diversity.*, **1:** 19-26.

Tjamos, E.C., Papavizas, G.C. and Cook, R.J. (1992). Biological Control of Plant Diseases. Plenum,New York.

Upadhyay, R.S. and Rai, B. (1987). Studies on antagonism between *Fusarium udum*, Butler and root region microflora of pigeonpea. *Plant & Soil.*, **101:** 79-93.

Chakraborty, Usha and Chakraborty, B.N. (1989). Interaction of *Rhizobium leguminosarum* and *Fusarium solanif.*sp. *pisi* on pea affecting disease development and phytoalexin production *Canadian Journal of Botany*, 67: **(6)** 1698-1701.

Vasudeva, R.S., Subbaiah, T.V., Satry, M.L.N., Rangaswamy, G. and Iyengar, M.R.S. (1958)–Bulbiformi, an antibiotic produced by Bacillus subtilis. *Annals of Applied Biology*, **46:** 336-45.

Whipps, J.M. (2001). Microbial control and biocontrol in the rhizosphere. *J. Exp.Bot.* **52:** 487- 511.

Webber, G.F. (1928). Plant pathology. Annual Report of Florida Agricultural University Exp. Stat.

Weerapat, P. and Schroeder, H.W. (1966). Effect of soil temperature on resistance of rice to seedling blight caused by *Sclerotium rolfsii. Phytopathology.*, **56:** 640-644.

Weller, D.M., Howie, W.J. and Cook, R.J. (1985). Relationship of *in vitro* inhibition of *Gaeumonnomyces graminis var. tritici* and *In vivo* suppression of take all by *fluorescent Pseudomonad. Phytopathology.*, **75:** 1301 (Abstr.).

Weight, J.M. (1956). Biological control of soil- borne Pythium infection by seed inoculation. *Plant Soil.*, **8:** 132-40.

Webber, G.F. (1931). Blights of carrot caused by *Sclerotium rolfsii* with geographic distribution and host range of the fungi. *Phytopathology.*, **21:** 103-109.

Weindling, R. and Emerson, O.H. (1936). The isolation of toxic substance from the culture of a *Trichoderma. Phytopath.*, **31:** 991-1003.

Weindling, R. (1941). Experimental consideration of the mould toxins of *Gliocladium* and *Trichoderma. Phytopath.*, **31:** 991-1003.

Webster, J. and Lomas, N. (1964). Does *Trichoderma viride* produce gliotoxin andvridin? These transactions. *Trans. Brit. Mycol. Soc*, **47:** 536-40.

Wicklow, M., Bollin, W. and Denison, W.C. (1974). Comparison of soil microfungi in 40-year-old stands of pure alder, pure conifer and alder conifer mixtures. *Soil Biol. Biochem.*, **6:** 73-78.

Xu, X.B. and Tian, S.P. (2008). Reducing oxidative stress in sweet cherry fruit by Pichia membranaefaciens: a possible mode of action against *Penicillium expansum. J. Appl. Microbiol.*, 105 **(4):** 1170-1177.

Yaqub, F. and Shahzad, S. (2009). Effect of solar heating by polyethylene mulching on sclerotial viability and pathogenicity of sclerotium rolfsii on mungbean and sunflower Pak. J. Bot., 41**(6):** 3199-3205.

Zambofim, L. and Schenck, N.C. (1983). Reduction of the effect of pathogenic root infecting fungi on soybean by the mycorrhizae fungus *Glomus mosseae. Phytopathology.*, **73:**1402.

Zak, D.R., Holmes, W.E., White, D.C., Peacock, A.D. and Tilman, D. (2003). Plant diversity, soil microbial communities, and ecosystem function: Are there any links? *Ecology* **84:** 2042–2050.

Zhang, Li.Y.D., Xu, W., Wu, Z., Guo, M. and Cao, A. (2011). *Alternaria tenuissima* causing leaf spot and fruit rot on pepper (*Capsicum annuum*): first report in China. *New Disease Reports* **24:** 3. [http://dx.doi.org/10.5197/j.2044-0588.2011.024.003].

Chapter 11

Economic Valuation of Agro-Biodiversity

Jyothi Badri and S.K. Soam

National Academy of Agricultural Research Management, Hyderabad, A.P.

Any discussion of the value of a particular object requires an understanding of what exactly the object of value is. Biodiversity is innate in nature. Convention on Biological Diversity (CBD, 1993)[1] defines biodiversity as: "the variability among living organisms from all sources including, *inter alia*, terrestrial, marine and other aquatic ecosystems and the ecological complexes of which they are a part; this includes diversity within species, between species, and of ecosystems." Agro-biodiversity, a human engineered ecosystem is an important component of natural biodiversity. It is as a result of human intervention in the form of constant and careful selection of economically useful plants and animals coupled with natural selection. It encompasses different kinds of plants and animals that people raise for food. Domestication and artificial selection of plants and animals by humans has created innumerable wealth of biota with great evolutionary significance. This has resulted in the increase of variability, an essential prerequisite for any crop or animal improvement programmes adding to the naturally existing enormous wealth of biodiversity.

---

1 The Convention on Biological Diversity (CBD) entered into force on 29 December 1993. It has 3 main objectives: (*a*) The conservation of biological diversity, (*b*) The sustainable use of the components of biological diversity and (*c*) The fair and equitable sharing of the benefits arising out of the utilization of genetic resources.

Threat to Agro-biodiversity

It is estimated that there are between 50,000-60,000 crop wild relatives in the wild (Maxted & Kell, 2009). 10,739 of these are important plant genetic resources for food and agriculture (PGRFA) and 700 of these, representing less than 0.26 per cent of the world flora, are the most important in terms of global food security and the ones requiring urgent conservation measures. Unfortunately, depletion of valuable sources of genetic diversity due to genetic erosion as a consequence of continuous use of uniform and high yielding improved varieties in day to day farming activities is quite alarming. Over the course of domestication, traits from wild relatives perceived to be undesirable were selected against in the early domesticates (land races) and eventually linked traits to the targets of selection were left behind in the wild gene pool. The result was a narrowing of the domesticated gene pool (Tanksley and McCouch 1997)[3]. For instance *Oryza sativa* cultivars are estimated to retain only approximately 10-20 per cent of the genetic diversity present in its wild rice ancestor, *Oryza rufipogon* (Caicedo *et al.*, 2007; Zhu *et al.*, 2007).

Consequences of Agro-biodiversity Loss

Rapid strides in research and development in the course of modernising agriculture have led to marginalization of the bulk of the agro-biodiversity developed over centuries of patient famer's efforts. Human efforts to ensure food security resulted in high yielding; uniform crop varieties with narrow adaptation and people engaged in crop improvement programmes tend to restrict the amount of variation that they deal with as they work in controlled and constrained environments. These modern practices ultimately limit the choices available for future performance enhancements. Local/land races, traditional cultivars and wild and weedy relatives of cultivated crop species though yield comparatively less and are not uniform, are widely adaptable to different environmental conditions. Many valuable genes/traits selected against are harboured in these types. These have to be exploited to achieve another breakthrough in yield level and to have a second green revolution. Farmers cease to cultivate traditional varieties and land races that they nurtured over the years leading to extinction of such types. For future crop improvement programmes, exploitation of genetic variability available in the wild gene pool is inevitable as the traditional and wild relatives of cultivated crop species house valuable sources of resistance genes to various biotic and abiotic stresses.

Need for Valuation

Biodiversity has value that may reside in the satisfaction that people get from using directly or indirectly, now or in the future, or in the concerns that humankind has some wider responsibility towards other living things. The essential reason that values are important is that biodiversity competes with humankind for space. The conservation of biodiversity is a cost bearing exercise, looking to human needs, certain trade-offs are also required in the biodiversity management. Therefore valuation is required to decide the trade-offs and also to bring into the notice of public at large that biodiversity is valuable and it needs to be conserved and used carefully. Predominantly the value of biodiversity is implicit; the absence of apparent value combined with absent or poorly defined property rights creates a problem of over exploitation and unregulated use. Another aspect is quantification in terms of monitory and social gains, something that cannot be precisely quantified or is difficult to monitor and evaluate is easy to disregard. Absence of an economic value for biodiversity means that they fail to compete on a level playing field with forces that are driving their decline.

Demonstrating the value of biodiversity is a fundamental step in conservation (Pearce and Moran 1998). OECD identifies the valuation of biodiversity benefits as one of the pillars of the institution's strategy and work (OECD 2001). The importance of recognising biodiversity's value and thus valuation tools is enshrined in the Convention on Biological Diversity (CBD, 1992)[2] also. Importance of valuation is furthermore emphasized in the Conference of Parties (COP) Decision IV/10 that "economic valuation of biodiversity is an important tool for well-targeted and calibrated economic incentive measures". CBD is also taking into account the economic, socio-cultural and ethical valuation in development of relevant incentive measures. Valuation should be an integral part of biodiversity policies for finalizing conservation plans to make a basis for sustainable use.

Criteria for Valuation of Agro-biodiversity

Based on CBD: Quantification of Value and Priority Setting

Resources for conservation are limited, for whatever reason, so that setting priorities is important. The priority setting will probably differ if the aim is to conserve diversity rather than resources. But even if the aim is to conserve diversity, priority setting is complicated because it does not necessarily follow that resources should be allocated first to the scarcest or most threatened diversity, even though this is a widely recommended procedure (Myers *et al.*, 2000). Prioritizing action according to the degree of threat of extinction could ignore the reason why the biodiversity is severely threatened in the first place. If the cause of extinction is not very amenable to policy measures, allocating resources to conservation is likely to be wasteful anyway. This suggests an approach based on cost-effectiveness rather than scarcity.

Value Indicators for Agro-biodiversity

Components of Diversity: Richness, Evenness, Composition and Interaction

Species Richness

A systematic inventory of the number of species contained within an area. This is the commonest method for rapid impact statements about the change in diversity. Van Kooten (1998) notes that the measurement of biodiversity involves three aspects: scale, the component aspect and the viewpoint aspect.

---

2 CBD Article 11 calls on the Signatory Parties to "....adopt economically and socially sound measures that act as incentives for the conservation and sustainable use of components of biological diversity".

1. *Scale*: The scale element is made up of the alpha diversity, beta diversity and gamma diversity. Alpha diversity is species richness within a local ecosystem. Beta diversity reflects the change in alpha diversity as one move from one ecosystem to another across a landscape. Gamma diversity pertains to species richness at a regional or geographical level. This is a more global concept and a measure that is much more dependent on global shocks rather than the local ones (*e.g.* forest fires) that affect alpha and beta diversity.
2. The component element of measurement concerns the identification of what constitutes a minimum viable population for the survival of a species. This is akin to setting safe minimum standards for species.
3. The viewpoint issue refers to the existence of many viewpoints, ranging from practical through to moral and aesthetic. Perlman and Adelson (1997) discuss the assignment of values in more detail. They note that viewpoints are necessarily subjective and value-laden and that some value criteria have theoretical and legal standing irrespective of either their deliberate use or their ethical foundations.

Two refinements are suggested to the species richness indicator of diversity. One is to restrict the count to certain combinations of species only. Counting a limited set of certain species is a shortcut for representing overall diversity of other species that are not counted (Fjeldsa, 2000). Alternatively and perhaps additionally, species can be assigned importance weights using taxonomic information. Therefore following may be given emphasis;

1. *Species Evenness*: It is the distribution of populations of various species within ecosystem. Species evenness may matter far more than richness.
2. *Species Composition*: The particular species that is present. It is important as witnessed by the effects of introducing new species to ecosystems where they were previously absent.
3. *Species Interaction*: The relationships between species-species interaction (trophic interaction) can influence ecosystem function, for example by taking nitrogen up-take by plants.

Diversity-Resilience Linkage

Ecosystems come under threat from various shocks and stresses such as climate change or natural disasters and calamities. It is widely thought that systems that are more diverse have more capability to respond to such shocks, whereas those with low diversity are more likely to collapse without recovery. Diversification of crops in farming adopts exactly the same idea and farmers may diversify even though it reduces overall productivity. More diverse systems may also be more resistance to species invasions (Chapin *et al.*, 2000). The diversity-resilience linkage gives rise to the notion of an insurance value of diversity. More strictly, since risk tends to refer to contexts where probabilities of stress and shocks are known, the insurance is against uncertainty, *i.e.* a context where risks often are not known in any actuarial sense (Perrings, 1995).

Methods of Valuation of Agro-biodiversity

It is well known that, without biological diversity, human life would cease. Hence biodiversity has a life support function. Dasgupta (2000) remarks "The value of an incremental change to the natural environment is meaningful because it presumes humanity will survive the change to experience it. The reason (that) estimates of the total value (of the environment) should cause us to balk is that if environmental services were to cease, life would not exist". Economic value is linked to cost-benefit analysis (CBA), although its uses range far more widely than this. Basic rules of CBA are presented in Table 11.1.

Table 11.1: Basic Rules of CBA

Situation	*Recommendation*
One policy option relative to the status quo	Accept such policy if benefits exceed costs
Policy options are mutually exclusive and one must be chosen	Select that policy with the highest net benefits
Various policies can be chosen as part of an overall programme of change	Accept all policies with a ratio of benefits to costs greater than unity until the available budget is exhausted
	Use rate-of-return on investment wherever feasible to rank projects

Majority voting and intensity of preferences are the two forms of decision rule to deal with situations where people cannot agree unanimously on a course of action. Market place provides a very powerful indicator of preferences. Willingness to pay (WTP) and opportunity cost (OC) are the two important features of market place preferences with direct relevance to the process of economic valuation. The net sum of all the relevant WTPs defines the total economic value (TEV) of any change in well-being due to a policy or project. Figure 11.1 shows the characterization of TEV by types of value. Use values relate to actual use of the good in question, planned use or possible use. Non-use value refers to willingness to pay to maintain some good in

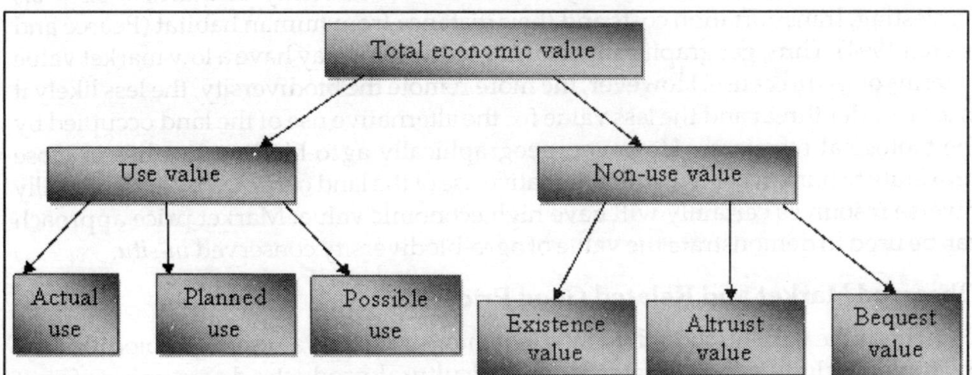

Figure 11.1: Total Economic Value
***Source*: Adapted from OECD 2002 with Slight Modification**

existence even though there is no actual, planned or possible use. Differentiating use and non-use values is important because the latter can be large relative to the former, especially when the good in question has few substitutes and is widely valued. In addition, non-use value remains controversial, so that it is important to separate it out for presentational and strategic reasons. Different valuation methods available for biodiversity valuation are presented schematically in Figure 11.2.

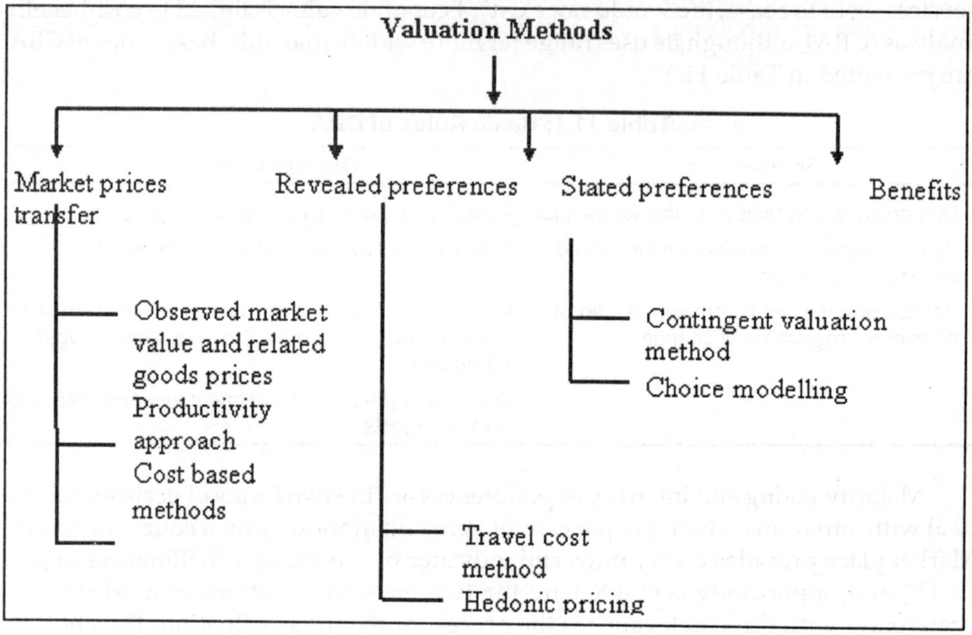

Figure 11.2: Economic Valuation Methods for Agro-biodiversity

Market Prices Approach

Market value for biodiversity depends on availability of markets *vis a vis* harvesting, transportation costs and their distance from human habitat (Pearce and Moran 1994). Thus, geographically remote biodiversity may have a low market value in terms of its direct use. However, the more remote the biodiversity, the less likely it is to be under threat and the less value for the alternative use of the land occupied by the biological resources. However, geographically agro-biodiversity lies in close proximity to human habitat, thus alternative use of the land occupied by agriculturally diverse resources certainly will have high economic value. Market price approach can be used to demonstrate the value of agro-biodiversity conserved *in-situ*.

Observed Market and Related Good Prices

This is the widely studied method to demonstrate the value of agro-biodiversity. Examples include genetic material for agricultural products, drugs, minor forest products etc. price adjustment in terms of deducting production and transportation costs from the observed market price, price corrections for price distortions, externalities and internationally traded products.

The Productivity Approach

This method values biological resources as inputs by observing the physical changes in environmental quality and estimating what differences these changes will make to the value of goods and services that are marketed. An example is a change in wetland size that leads to a change in water quality that reduces the quantity of fish caught. This lost market value can be estimated using market price information. The difference in the value of output resulting from the change being the value attributed to the amount of lost wetland. The production function is the formal representation of this relationship between the change in environment as an input and the change in the production of a marketed output. The method can be undertaken with a varying degree of rigour applied to the derivation of the physical relationship (or dose response) between the inputs (environmental assets and other man made inputs like capital and labour), and the valued output.

Cost-based Methods

This approach includes replacement, restoration, relocation and preventative expenditure costs approaches. Essentially they assess the costs of different measures that would ensure the maintenance of the services provided by the environmental asset that is being valued. These would aptly suit for demonstrating the economic value of agro-biodiversity conserved *ex-situ* like gene sanctuary, germplasm conservation habitats etc. Bockstael *et al.*, 2000 point out, the replacement cost is only a valid measure if three conditions are met: 1) that a human engineered system provides functions that are equivalent in quality and magnitude to the natural function; 2) the human-engineered solution is the least cost alternative way of performing the function; 3) that individuals in aggregate would in fact be willing to incur these costs if the natural function were no longer available.

Opportunity cost analysis is prominent in market price approaches to economic valuation. Two contexts for opportunity cost-analysis can be distinguished. In the first, some indicator of biodiversity is traded-off against the (opportunity) cost of biodiversity conservation. In the second, the biodiversity indicator and the cost-indicator are supplemented by other criteria that may be thought relevant to the conservation decision (multi-criteria analysis).

Revealed Preferences

These methods use observed behaviour to infer the environmental value. Contrary to the market price approach, the relevant prices in the markets are affected by the non-market asset. They include travel cost method, random utility model, hedonic pricing and averting behaviour models. They rely on surrogate market that provides a behavioural trial to identify the environmental value of interest.

Travel Cost Method

This method may not be useful for demonstrating the value of agro-biodiversity as it is mostly suitable for those biological resources meant for recreation that include wildlife and landscape appreciation.

Hedonic Pricing

This is based on the idea that a private good can be viewed as a bundle of characteristics, each with its own implicit price, some of which may be non-market in nature. Individuals express their preferences for a particular non-market attribute by their selection of a particular bundle of characteristics. These preferences will be reflected in the differential prices paid for the private good. The approach then applies econometric technique to data on private good characteristics and prices to derive the relationship between the attributes of the good and its market price and from there estimate implicit prices for non-market characteristics. A closely related application of this method is found in the area of plant breeding and crop improvement (Evenson 1990; Gollin and Evenson 1998). An illustrative example of the application of hedonic pricing approach is presented in Figure 11.3.

Stated Preference Methods

For many environmental goods, to demonstrate their values, a market must be constructed using questionnaires. This is the essence of stated preference methods.

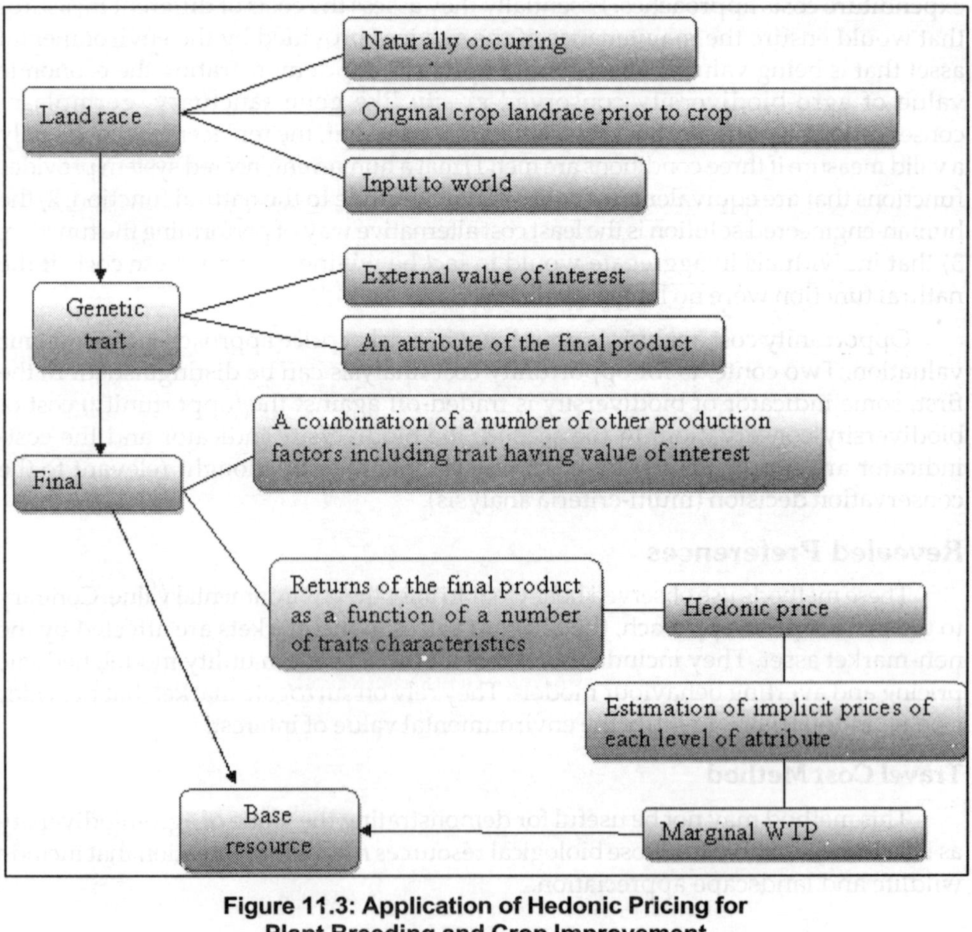

Figure 11.3: Application of Hedonic Pricing for Plant Breeding and Crop Improvement

An important feature of these methods is that they can help reveal values that are not revealed using other methods. In particular, stated preferences can uncover non-use values. They use questionnaires that are targeted at a sample of individuals and that seek to elicit, directly or indirectly, the individual's monetary valuation of a change in a given marketed good.

Contingent Valuation Method

A survey technique that attempts direct elicitation of individuals (or households) preferences for a good or service. This method has been used extensively in the valuation of biological resources including rare and endangered species, habitats and landscapes. Economic valuation of agro-biodiversity using CVM is hypothetically presented Figure 11.4.

Theoretical evidence of CVM was given by Stevens *et al.*, 2000. Suppose that individual utility associated with environmental quality, EQ, can be expressed as a function of income, *Y*, and EQ attributes such as water quality, wildlife habitat

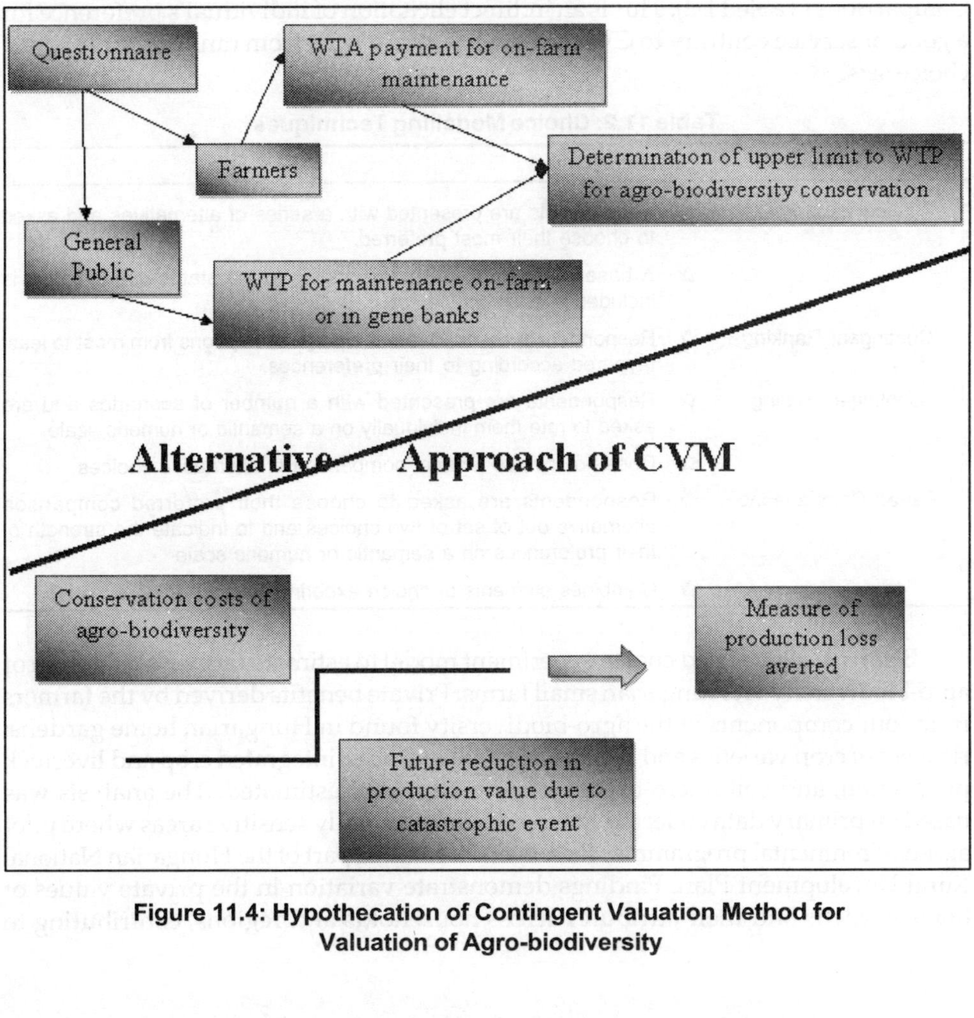

Figure 11.4: Hypothecation of Contingent Valuation Method for Valuation of Agro-biodiversity

preserved, and cost. In dichotomous choice CVM, individuals are asked to undertake activities on their own property to improve

EQ that cost them a predetermined amount, $*N*. The value of utility, observed by the researcher, when amount *N* is paid is:

$$U_1 = U(D_1\text{-}Y\text{-}N) + e_1$$

where, D_1 is a vector of EQ attributes and *e* is a random variable.

The WTP probability can then be written as:

$$P_r = G(dv)$$

where G is the probability function for the random component of utility ($e_1 e_0$).

Choice Modelling

This method encompasses a range of stated preferences techniques that take a similar approach to valuing environmental goods. The term includes Choice experiments, Contingent ranking, Contingent rating and the method of Paired Comparisons (Table 11.2). This is an indirect elicitation of individual's preference for a good or service contrary to CVM method. It infers WTP from ranking or ratings of choice sets.

Table 11.2: Choice Modelling Techniques

Technique	*Description*
Choice Experiment	☆ Respondents are presented with a series of alternatives and asked to choose their most preferred.
	☆ A baseline alternative, corresponding to the status quo situation is included in each choice set
Contingent Ranking	☆ Respondent is asked to rate the proposed options from most to least favoured according to their preferences.
Contingent Rating	☆ Respondents are presented with a number of scenarios and are asked to rate them individually on a semantic or numeric scale
	☆ Does not involve a direct comparison of alternative choices
Paired Comparisons	☆ Respondents are asked to choose their preferred comparison alternative out of set of two choices and to indicate the strength of their preferences on a semantic or numeric scale
	☆ Combines elements of choice experiments and rating exercise

Birol *et al.*, 2006 used choice experiment model to estimate farmer's valuation of agro-biodiversity on Hungarian small farms. Private benefits derived by the farmers from four components of the agro-biodiversity found in Hungarian home gardens: richness of crop varieties and fruit trees; crop landraces; integrated crop and livestock production; and soil micro-organism diversity were estimated. The analysis was based on primary data collected in three environmentally sensitive areas where pilot agri-environmental programmes have been initiated as part of the Hungarian National Rural Development Plan. Findings demonstrate variation in the private values of home gardens and their attributes across households and regions, contributing to understanding the potential role of home gardens in these agri-environmental schemes. This study has implications for sustaining agro-biodiversity in transitional economies.

Benefits Transfer

Benefits transfer involves borrowing an estimate of willingness to pay from one site (the study site) and applying it to another (the policy site). This method avoids the cost of engaging in primary studies whereby WTP is estimated with one or more of the techniques described earlier and valuable time can be saved. The essential problem with this is its reliability. The main procedure for validation involves transferring a value and then carrying out a primary study at the policy site as well. Ideally, the transferred value and the primary estimate should be similar. If this exercise is repeated until a significant sample of studies exists in which primary and transferred values are calculated for policy sites, then there would be a justification for assuming that transferred values could be used in the future without the need to validate them with primary studies. An alternative procedure is to conduct a meta-analysis on existing estimates of WTP. At its simplest, a meta-analysis might take an average of existing estimates of WTP, provided the dispersion about the average is not found to be substantial, and use that average in policy site studies. Finally, benefits transfer could be tested by estimating WTP before an actual project is implemented and then revisiting the area later when the project is complete to see if people behaved according to their stated WTP.

A wide array of methods is available for economic valuation of biodiversity. These techniques are suitable for valuation of different components of biodiversity either singly or in combination of two or more. Whichever may be the technique applicable to biodiversity valuation, the same cannot be used as such for agro-biodiversity valuation. The technique to be applied should be decided on a case to case basis encompassing all the factors which ultimately serve the cause of CBD. Privatization would provide compensation to those who safeguard agro-biodiversity, thus stimulating conservation without public investment while providing an idea of genetic resources users' willingness to pay for conservation. But, privatization if achieved through the use of intellectual property rights (IPRs) fail to reward local people for their important contributions (of knowledge and resources) to the products for which industry is awarded patent protection. On the other hand, benefit sharing arrangements would be an alternative on the grounds that they are the easiest means to create a market for genetic resources. Contracts between producers of genetic resources (*e.g.* farmers) and private users (*e.g.* biotechnology companies) are a way to avoid the monopoly-related problems associated with IPR. Model agreements for 'agro- biodiversity prospecting' now exist and should always exist as a step towards the fair and equitable share, sustainable use and conservation of agro-biodiversity.

Conclusion

- ☆ Agro-biodiversity is a human engineered ecosystem and is an important component of natural biodiversity.

- ☆ Economic value of biodiversity is implicit due to the absence of apparent value combined with absent or poorly defined property rights creating a problem of over exploitation and unregulated use.
- ☆ If degree of threat of extinction is set as a priority for conservation, can ignore the reason why the biodiversity is severely threatened in the first place. If the cause of extinction is not very amenable to policy measures, allocating resources to conservation is likely to be wasteful.
- ☆ The different components of diversity (species richness, species evenness, species composition and species interaction) and diversity- resilience linkage are the indicators for setting up criteria for valuation of biodiversity.
- ☆ Economic value is linked to cost-benefit analysis and total economic value includes both use and non-use value.
- ☆ Economic valuation techniques of agro-biodiversity are broadly based on four approaches
  - *Market prices*: Prices occur in the market for the agro-biodiversity asset
  - *Revealed preferences*: Prices are revealed in some other market
  - *Stated preferences*: WTP derived from questionnaires
  - *Benefits transfer*: values borrowed from existing studies
- ☆ Privatization provides compensation to those who safeguard agro-biodiversity and benefit sharing arrangements are a way to avoid the monopoly-related problems associated with IPR.

References

Bockstael, N. Freeman, M., Kopp, R. Portney, P. and Smith, V. (2000). On measuring economic values for nature, *Environmental Science and Technology*, **34:** 1384-1389.

Caicedo, A., Williamson, S., Hernandez, R.D., Boyko, A. and Fledel-Alon, A. (2007). Genome wide patterns of Nucleotide polymorphism in domesticated rice. *PLoS Genetics* **3:** e163.

Chapin III, F.S., Zavaleta, E.S., Eviner, V.T., Naylor, R.L., Vitousek, P.M., Reynolds, H.L., Hooper, D.U., Lavorel, S. Sala, O.E., Hobbie, S.E., Mack, M.C. and Diaz, S. (2000). Consequences of changing biodiversity. *Nature* **405:** 234-242.

Dasgupta, P. (2000). Human Well-Being and the natural environment, Department of Economics, University of Cambridge, Cambridge, mimeo.

Evenson (1990). Genetic resources: measuring economic vlue. In: Vincent J, Crawford E and Hoehn (eds) Valuing Environmental Benefits in Developing Countries, Michigan State University, East Lansing.

Evenson, R.E. and Gollin (1997). Genetic resources, international organizations, and rice varietal improvement. *Economic Development and Cultural Changes* 45 **(3):** 471-500.

Fjeldsa, J. (2000). The relevance of systematics in choosing priority areas for global conservation. *Environmental Conservation* 27 **(1):** 67-75.

Gollin, D. and Evenson, R.E. (1998). An application of hedonic pricing methods to value rice genetic resources in India. In: Evenson RE, Gollin D and Santaniello V (eds). Agricultural values of plant genetic resources. Wallingford: CAB International, pp 139-150.

Maxted, N. and Kell, S.P. (2009). Establishment of a Global Network for the *In Situ* Conservation of Crop Wild. Relatives: Status and Needs. FAO Commission on Genetic Resources for Food and Agriculture, Rome, Italy. 266 pp.

Myers, N.R., Mitteremeier, C., Mitteremeier, G. da Fonseca and Kent, J. (2000). Biodiversity hotspots for conservation priorities. *Nature* 403 24 February, 853-859.

O.E.C.D. (2001). OECD Environmental strategy for the first decade of the 21st Century, OECD, Paris.

O.E.C.D. (2002). Handbook of Biodiversity valuation: A guide to policy makers, OECD, Paris.

Pearce, D. and Moran, D. (1998). The economics of biological diversity conservation. In: Fiedler, P.L., Kareiva, P.M. (Eds.), Conservation Biology for the Coming Decade. Chapman & Hall, New York, pp. 384–409.

Perlman, D. and Adelson, G. (1997). Biodiversity: Exploring values and priorities in conservation, Oxford: Blackwell.

Perrings, C. (1995). biodiversity conservation as insurance, In: T. Swanson (ed), The economics and ecology of biodiversity decline, Cambridge: Cambridge University Press, 69-78.

Stevens, T.H., Belkner, R., Dennis, D., Kittredge, D. and Willis, C. (2000). Comparison of contingent valuation and conjoint analysis in ecosystem management. *Ecological Economic* **32:** 63-74.

Van, Kooten K. (1998). Economics of conservation biology: a critical review, *Environmental Science and Policy* **1:** 13-25.

Zhu, Q., Zheng, X., Luo, J., Gut, B. and Ge, S. (2007). Multilocus analysis of nucleotide variation of Oryza sativa and its wild relatives: severe bottleneck during the domestication of rice. *Mol Biol Evol* **24:** 875-888.

Chapter 12

Chilli (*Capsicum* spp.): A Diverse Crop with Innovative Uses

Jyoti Pandey[1]*, *K. Srivastava*[1] *and Sanjeet Kumar*[2]

[1]*Department of Genetics and Plant Breeding, Institute of Agricultural Sciences, Banaras Hindu University, Varanasi, U.P., India*

[2]*Scientist, Pepper Breeding, World Vegetable Center, Taiwan*

Chilli (*Capsicum* species), a popular crop of New World origin, is cultivated for its fruits valued for colour, flavour, spice, vegetable and nutrition that it provides to the several food items (Kumar *et al.*, 2006). Botanically, chilli plants are dicotyledonous and short-lived perennial herb of Solanaceae family and are commercially grown as an annual and as perennial in kitchen gardens. Among the five cultivated throught the world species of the genus *Capsicum*, *C. annuum* is most commonly cultivated either for pungent fruits *i.e.* chilli (synonyms: hot pepper, American pepper, chile, azi, cayenne, paprika etc.) or for non-pungent fruits sweet pepper (synonyms: capsicum, paprika, bell pepper, Shimla mirch). India, China, Korea, Hungary, Spain, Nigeria, Thailand, Turkey, Kenya, Sudan, Uganda, Japan, Ethiopia, Indonesia, Pakistan, Mexico are the major chilli and sweet peppers producers. The majority of cultivars grown in Asian, Central and Latin American countries are pungent, while in European countries cultivation of less pungent and non-pungent peppers are more common. Sweet pepper is often called bell pepper because majority of non-pungent cultivars grown worldwide produces bell shaped fruits with four lobes. Green fruits are one of the richest sources of antioxidative vitamins such as Vitamin

* Corresponding Author E-mail: pandey_jyoti123@yahoo.com, pandeyjyoti123@gmail.com

A, C and E for the first, vitamin C was purified from *Capsicum* fruits in 1928 by Hungarian biochemist Albert Szent Gyorgyi, which helped him to receive Nobel Prize of physiology and medicine in 1937. In this review, recent taxonomic status of pepper (*Capsicum* spp.) and updated innovative uses of this crop have been described.

The Genus *Capsicum*

The name genus *Capsicum* perhaps derived from the Latin word 'capsa', meaning chest or box because of the shape of fruits, which enclose seeds very neatly, as in a box (Berke and Shieh 2000). The *Capsicum* has been known since the beginning of civilization in the Western Hemisphere. It has been a part of the human diet since 7500 BC and chilli is among the oldest cultivated crops of the Americas. The genus *Capsicum* ($2n = 24$) encompasses a diverse group of plants producing pungent or non-pungent fruits. At present, it is widely accepted that the genus consists of approximately 25 wild and 5 cultivated species. This genus was domesticated at least five times by prehistoric peoples in different regions of South and Center America, resulting in five domesticated species, *viz.*, *C. annuum* L., *C. baccatum* L., *C. chinense* Jacq., *C. frutescens* L., and *C. pubescens* R. & P. (Bosland 1992). On the basis of gene flow through natural and conventional hybridization, *Capsicum* species are grouped in three species complexes (Table 12.2). Except that of *C. pubescens*, wild forms of the remaining four cultivated species are known. The *C. annuum* was domesticated in highland of Mexico and includes most of the Mexican chile (syn. chilli), most of the chilli of Asia and Africa, sweet peppers of temperate countries. However, due to the non-adaptability of *C. annuum* in lowland tropics of Latin America, its cultivation was replaced by *C. frutescens* and *C. chinense* (Pickersgill 1997). The cultivation of *C. baccatum* and *C. pubescens* are mostly restricted to Latin American countries like Peru, Bolivia, Columbia and Brajil. Although chilli was introduced in India during 1700 BC by Portuguese in Goa, however, it has rapidely spread in all parts of countries C. *annuum* is most widely cultivated in India, *C. frutescens*, *C. chinense* and *C. baccatum* are also grown in specific regions, especially in North-East region and state of Kerela.

Species Dilemma of *C. chinense* and *C. frutescens*

In 1776, Jacquin gave the name *C. sinence* and Linnaeus in 1753 gave the name *C. frutescens*. Smith and Heiser in 1957 studied on *C. sinence* and emphasized that *C sinence* differs from *C. frutescens* and gave the new name *C. chinense*, but it is not related to China. From the several studies it was observed that these two species are morphologically similar. However, another study revealed that *C. chinense* and *C. frutescens* are same species with different botanical varieties (Eshbaugh *et al.*, 1993). For the first time, Smith and Heiser (1957) studied on both the species and concluded that *C. chinense* and *C. frutescens* are morphologically different species. After crossing with each other, they found very less amount of seed set and F_1 plants had very less viable pollens and non-viable seeds. Nevertheless, many researchers (Eshbaugh 1993, McLeod *et al.*, 1983, Pickersgill 1988) considered that both are same species. Hence this has been the subject of arguments between the Capsicum taxonomists that *C. chinense* and *C. frutescens* are the same or distinct species (Smith and Heiser 1957). In the recent study, however, based on morphological data, RAPD markers and sexual compatibility it has been demonstrated that *C. chinense* and *C. frutescens* are the different species (Baral and Bosland 2004).

Identification of Cultivated Species

All five cultivated species are represented by genotypes with pungent (hot pepper) or non-pungent (sweet pepper) fruits and have a large variability for fruit size/shape and often genotypes with similar fruit morphology exist across the species. Hence assigning a given genotype to a cultivated species based on fruit size, shape and pungency is difficult. In most of the textbooks, a number of overlapping morphological markers have been assigned to a cultivated species, which further complicate the identification of species. Nonetheless, certain flower and fruit descriptors may be used to assign a genotype to a cultivated species without much doubt (Table 12.2). Recently, RAPD markers specific to *C. chinense* and *C. frutescens* have also been reported (Baral and Bosland 2004), however, these markers should be validated in a series of accessions.

Table 12.1. Three species complexes with some representative

Spp. Complex	*Species*
C. annuum complex	*C. annuum** L., *C. frutescens** L., *C. chinense** Jacq., *C. chacoense* Hunz., *C. galapagoense* Hunz.
C. baccatum complex	*C. baccatum** L., *C. praetermissum* Heiser & Smith, *C. tovarii*
C. pubescens complex	*C. pubescens** Ruiz & Pav., *C. cardenasii* Heiser & Smith, *C. eximium* Hunz.

* Cultivated species.

Table 12.2: Distinguishable Morphology of Five Cultivated Species of *Capsicum*

Species	*Distinguishable Morphological Feature/s*
C. annuum	White corolla and white filaments
C. frutescens	Yellow/greenish corolla and purple filaments
C. chinense	Annular constriction on pedicel attachment and yellow/greenish corolla
C. baccatum	Yellow or greenish yellow spots on corolla
C. pubescens	Hairy stems/leaves and black/brown seeds

Evolutionary Significance of Pungency

Unlike other members of nightshade (*Solanaceae*) family, like. tomato, eggplant and potato are characterized by the presence of phenols responsible for a protection mechanism (non edible). Chilli leaves lack phenols, and it believed that nature has provided pepper capsaicinoids (responsible for pungency unique to the *Capsicum* genus) pathway to protect plants from the enemies, especially mammals. This could be viewed as an analogue of phenol pathway present in other members of nightshade family (Kumar *et al.*, 2006). The evolutionary significance of pungency in the fruits of chilli has actually been explained and demonstrated as directed deterrence (capsaicin) hypothesis, wherein capsaicin in fruits function selectively to discourage seed predators like mammals without deterring beneficial seed dispersers like birds (Tewksbury and Nabhan 2001).

Fruits with Various Market Types

In the genus *Capsicum*, enormous morphological variability exists for flower morphology, especially corolla colour, anther colour, fruit colour, size, shape and pungency. Based on fruit size, shape and degree of pungency, a large number of horticultural types are recognized worldwide and at least 20 types are largely cultivated at large scale in one and other parts of the world. Some of these fruit types such as ancho, bell, jalapeño, pasilla, New Mexican, yellow wax have specific traits for processing, fresh use, flavour and pungency (Bosland and Votava, 2000). The breeding objectives for quality traits of hot pepper and sweet pepper could be described on the basis of five market types (Poulos 1994). In India also, a number of genotypes with specific fruit size, shape and attributes are commercially cultivated in different regions (Table 12.3).

Table 12.3: Commercial Cultivation of Pepper in India: Fruits of Various Market Types

Stage of Harvesting	*Consumption Pattern*	*Cultivar Type*	*Species*	*Preferred Fruit Type/Size*	*Degree of Pungency*
Red ripe fruits	Spice (intact fruits or powder) and hybrids	Landraces, improved populations	*C. annuum*, *C. chinense*[2] *C. frutescens*[2]	Cayenne (10-12× 2-3 cm)	Highly pungent with more colour retention
Green/ mature fruits	Vegetable	Improved populations and hybrids	*C. annuum*	Bell (6-8×4-5 cm)	Non-pungent
Green fruits	Intact fruits or sauce preparation	Landraces, improved populations and hybrids	*C. annuum* *C. chinense*[2] *C. frutescens*[2]	Cayenne (6-8×2-3 cm)	Mild to highly pungent
Red ripe fruits (Paprika)[1]	Oleoresin extraction	Landraces (*e.g.*, Tomato Chilli, variants of Bayadagi chilli)	*C. annuum*	Cayenne with very less capsaicin and high oleoresin	High oleoresin with no pungency
Red ripe fruits	Pickle preparation	Landraces	*C. annuum*	Jalapeno, but with thin pericarp	Mild to non-pungent
Red ripe fruits	Specific flavour	Landraces (*e.g.*, Dello of NEH region)	*C. baccatum*[2]	Bell shaped with distinct flavour	Regional preferences

1 In international trade, paprika refers to pepper lines suitable for non-pungent oleoresin extraction, however in Hungary and several central European countries, paprika is synonym of pepper and invariably used for both hot and sweet peppers.

2 Only landraces with variable fruit sizes and shapes are cultivated at limited scale.

Innovative Uses

Consumption of fresh green fruits of pungent fruits or non-pungent fruits may be considered as vegetable uses of the genus *Capsicum*. Similarly, consumption and use of dry fruits in various forms (*e.g.* intact fruits, grinded powder etc.) and value-added processed products (*e.g.* capsaicin extracts, oleoresin extracts, processed pickles etc.) may be considered as spice uses of chilli (Kumar *et al.,* 2006). Many other innovative ways and the uses of a large number of cultivars within the five cultivated species have grown exponentially (Bosland 1996). Thus fruits of the genus *Capsicum* have as many versatile uses as its diversity (Table 12.4).

Table 12.4: Versatile and Innovative Uses of Pepper

I. Fresh uses: Immature-green and mature-red fruits

- ☆ Green or red ripe fruits of hot pepper with variable degree of pungency are invariably added in most of the South Asian curries.
- ☆ Immature or mature fruits of sweet peppers are exclusively prepared as vegetable.
- ☆ Immature sweet pepper fruits are added in many Chinese cuisines.
- ☆ Immature mild pungent fruits are deep fried with gram flour and consumed in India.
- ☆ Fresh green non-pungent or mild pungent fruits are consumed as salad.
- ☆ In Philippines, leaves are added to soup and stew and consumed because unlike leaves of tomato and eggplant, leaves of *Capsicum* species do not contain phenols. The upper shoots of the plants are sold in bunches, just like other leafy vegetables (Bosland 1999).

II. Fresh processing: Sauce, paste, pickles, beer!

- ☆ Green or red ripe fruits with variable degree of pungency are used to prepare sauce.
- ☆ Red ripe and mild pungent fruits are stuffed with certain spices in North Indian states and prepared as pickles. Similarly, green fruits are also pickled in edible oils. Red ripe fruits are also stored in vinegar/citric acid for several years.
- ☆ In US, mild pungent fruits are used prepared as salsa and consumed with snacks.
- ☆ Red ripe fruits are used in the preparation of tomato ketchup for improving its colour.
- ☆ The Black Mountain Brewing Co. in Arizona developed a pepper beer with an idea to produce a spicy beer for a local Mexican restaurant and idea worked (Bosland 1993).

III. Dried spice: Mature whole fruits and powder

- ☆ Dry intact fruits or grounded powder are invariably added in the preparation of almost all South Asian chicken, egg and vegetable curies.

IV. Colouring and flavoring agents: Oleoresins (carotenoids) extracts or powder

- ☆ Paprika oleoresin (colour extracts from non-pungent fruits) is natural colouring agent, therefore, it is considered to be among the best substitute of synthetic colour used in food and cosmetic industries.
- ☆ Cosmetic industry uses non-pungent oleoresin to prepare its products.
- ☆ In food processing industries, especially in meat industry, concentrated oleoresin is added to the processed meat to impart attractive colour.
- ☆ In beverage industries, oleoresins are used to improve colour and flavour of its products.
- ☆ In certain region of the world (*e.g.* Japan and South Korea) oleoresins are mixed with chicken feed in order to impart attractive red colour to chicken skin and red colour to yolk.
- ☆ Oleoresins are mixed with the feed of flamingoes in zoo and koi in aquariums for improving the feather colour (Bosland 1996).

Contd...

Table 12.4–*Contd...*

V. Ethno-botanical/traditional medicine: Fruit extracts and powder (pungent fruits)

☆ Traditionally, fruits are consumed to stimulate digestion (stimulates the flow of saliva and gastric juice), raise body temperature and cures common cold.

☆ Mayas mix fruits with corn flour to produce 'chillatolli', a treatment for common cold. Mayas also use them to treat asthma, coughs, and sore throats. The Aztecs used fruit pungency to relieve toothaches (Bosland 1999). In many African countries, fruits are consumed with the belief that it improves the complexion and increases passion (Bosland and Votava 2000).

☆ Fruits are added to rose-gargles to cure pharyengitis. Fruits are also consumed as it also has carminative effects. The West Indian native, soak fruits in water, add sugar and sour orange juice and drink it during fever (http.//www.dominion.com).

☆ In Columbia, the Tukano native pore a mixture of crushed fruits and water into their noses to relive a hangover and effectiveness of a night of dancing and drinking alcoholic beverages (Bosland 1999).

☆ In Columbia and India, victims of snakebite are given pungent fruits to taste in order to sense the functioning of nervous system caused due to snake venom. In similar fashion, freshly crushed fruits or powder are used to reduce swelling and draw out poison of bee strings, spider bites and scorpion strings (Dewitt *et al.*, 1998).

VI. Modern medicine/pharmaceuticals: Capsaicinoids and carotenoids

☆ The pharmaceutical industry uses capsaicin as a counter-irritant balm for external application (Carmichael 1991).

☆ Capsacinoids (mainly capsaicin) are active ingredient in 'Heet' and 'Sloan's Liniment', massage liniments used for sore muscles (Bosland 1996).

☆ Capsaicinoids are used in the preparation of powder, tincture, plaster ointments and medicated wools.

☆ Pharmaceutical industry uses capsaicinoid extracts to prepare certain drugs (sprays), which are applied externally to stop pains of arthritis (rheumatoid arthritis, osteoarthritis), artily diseases (peripheral neuropathies) and to relive cramps (Cordell and Araujo 1993, Bosland 1996).

☆ Application of creams containing capsaicin reduces post-operative pain for mastectomy patients and its prolonged use helps in reducing the itching of dialysis patients, pains from shingles (Herpes zoster) and cluster headaches (Bosland 1996).

☆ Carotenoids found in fruits (b-carotene, acyl derivatives of capsanthin, acyl derivatives of capsorubin) have been shown to be inhibits LDL oxidation in vitro with probable lowering the "atherogenic" LDL subfraction production (Medvedeva *et al.*, 2003).

☆ Capsanthin and capsorubin (major carotenoids exclusively present in pepper fruits) can improve the cytotoxic action of anticancer chemotherapy and considered to be potential of carotenoids as possible resistance modifiers in cancer chemotherapy (Maoka *et al.*, 2001, Molnar *et al.*, 2004).

☆ Lutein, zeaxanthin, capsanthin, crocetin and phytoene have showed more potent anticarcinogenic activity than beta-carotene and useful for cancer prevention and may be applicable as the concept of 'bio-chemoprevention', which involves transformation-assisted method for cancerchemoprevention (Nishino *et al.*, 2002).

☆ The water extract of 'paradicsompaprika' (mainly containing capsanthin) has been considered as a new anticancer agent and fat soluble component of this drug has been regarded as an anti-promoter of cancer (Mori *et al.*, 2002).

☆ Capsaicin has recently been tried as an intravesical drug for overactive bladder (bladder cancer) and it has also been shown to induce apoptotic cell death in many cancerous cells (Lee *et al.*, 2004).

Contd...

Table 12.4–*Contd...*

VII. Insecticide/repellent: Capsaicinoids

- ✩ Capsaicin extracts are used as an effective repellent against mice damaging the underground cables and protect germinating seeds from squirrels (Bosland 1996).

VIII. Spiritual: Whole fruits

- ✩ In India, fruits are stringed on a thread along with a lime fruit and hung on the entrance of houses/shops with the belief that it will keep evil away (Kumar and Rai 2005).
- ✩ Red dry fruits are used in a desire to remove the bad consequences of evil eyes on younger babies in North Indian states.
- ✩ Traditionally, in New Mexico of US, mature fruits are stringed (called 'ristras') and hung on the entrance of house as a symbol of hospitality (Bosland 1992).

IX. Ornamental: Whole plants or fruits

- ✩ Certain genotypes of pepper are grown for their attractive plant shape, dense and colourful foliage and fruits. Several colours of fruit (at various maturing stage) can be found on a single plant making plant a very attractive ornamental (Bosland and Votava 2000).

X. Defense/punishment: Capsaicin extracts/ fruit powder

- ✩ In India, traditionally villagers keep fruit powder in house as a self-defense weapon against dacoits.
- ✩ Sprays containing capsaicin are emerging as a safe weapon for armless people, especially for women in metros of many countries including India.
- ✩ Now a day, capsaicin spray has replaced the maces and tear gas in the police departments of many countries to control unruly mobs and criminals.
- ✩ In Mexico, India and other Latin American countries, pepper powder is rubbed on children's thumps to prevent sucking (Dewitt *et al.*, 1998). Similary in India, fruit paste is applied on mother's nipple to get rid of prolonged breast feeding.
- ✩ Maya threw chilli powder into the eyes of young girls who stared at boy or men and they squirt fruit juice on the private parts of unchaste women (Dewitt *et al.*, 1998).

Conclusion

The impact of the discovery by Columbus of a pungent spice (chilli) in the America was beyond imagination as it was confused with black pepper of the East Indies. Nevertheless, today hot peppers dominate the world spice trade and are cultivated everywhere in the tropical and subtropical regions (Eshbaugh 1993). Similarly, sweet peppers have become indispensable vegetables in the temperate regions and are gaining vast popularity in the tropical regions as well. Furthermore, it has also emerged as an industrial crop as fruits are used as raw materials in the food, feed, cosmetic and medicine industries (Kumar *et al.*, 2006). The recent discovery of new medicinal properties of carotenoids and capsaicinoids present in pepper fruits could be visualized as huge potential of this crop of New World origin to become an even more versatile crop in world agriculture

References

Baral, J.B. and Bosland, P.W. (2004). Unraveling the species dilemma in *Capsicum frutescens* and *C. chinense* (Solanaceae): a multiple evidence approach using morphology, molecular analysis and sexual compatibility. *J. Amer. Soc. Hort. Sci.* **129:** 826-832.

Berke, T. and Shieh, S.C. (2000). Chilli peppers in Asia. *Capsicum Eggplant Newsletter* **19:** 38-41.

Bosland, P.W. (1992). Chiles: a diverse crop. *Hort. Tech* **2:** 6-10.

Bosland, P.W. (1996). Capsicums: innovative uses of an ancient crop. In: Janick J (ed.), *Progress in New Crops.* ASHS Press, Arlington, VA 479-487.

Bosland, P.W. and Votava, E.J. (2000). *Peppers: Vegetable and Spice Capsicums.* CABI Publishing, Wallingford, UK.

Bosland, P.W. (1993). Breeding for quality in *Capsicum. Capsicum Eggplant Newsletter* **12:** 25-31.

Bosland, P.W. (1999). Chiles: a gift from a fiery God. *HortScience* **34:** 809-811.

Carmichael, J.K. (1991). Treatment of herpes zoster and postherpetic neuralgia. Amer. *Family Physician* **44:** 203-210.

Cordell, G.A. and Araujo, O.E. (1993). Capsaicin: identification, nomenclature, and pharmacotherapy. *Ann. Pharmacother.* **27:** 330-336.

Dewitt, D., Stock, M.T. and Hunter, K. (1998). *The Healing Powers of Peppers.* Three Rivers Press, New York.

Eshbaug, W.H., Guttam, S.I. and McLeod, M.J. (1993). The origin and evolution of domesticated Capsicum species. *J. Ethnobiol.* **3:** 49-54.

Jacquin, N.J. (1976). Hortus Botanicus Vindobonesis 3 pl. 82.

Kumar, S. and Rai, M. (2005). Chiles in India. Chili Pepper Institute Newsletter XXII: 1-3.

Kumar, S., Banerjee, M.K. and Kalloo, G. (2000). Male sterility: mechanisms and current status on identification and characterization in vegetables. *Veg. Sci.* **27:** 1-24.

Kumar, S., Kumar, R. and Singh, J. (2006). Cayenne/American pepper (*Capsicum* species). In: Peter KV (ed), *Handbook of Herbs and Spices, Vol.* 3. Woodhead Publishing, Cambridge, UK, pp. 299-312.

Lee, J.S., Chang, J.S., Lee, J.Y. and Kim, J.A. (2004). Capsaicin-induced apoptosis and reduced release of reactive oxygen species in MBT-2 murine bladder tumor cells. *Arch. Pharm.* Res. **27:** 1147–1153.

Linnaeus, C. (1753). Species plantarum. Laurentii, Salvi, Stolkholm.

Maoka, T., Mochida, K., Kozuka, M., Ito, Y., Fujiwara, Y., Hashimoto, K., Enjo, F., Ogata, M., Nobukuni, Y., Tokuda, H. and Nishino, H. (2001). Cancer chemopreventive activity of carotenoids in the fruits of red paprika *Capsicum annuum* L. *Cancer Letter* **172:** 103-109.

McLeod, M.J., Guttam, S.I., Eshabugh, W.H. and Rayle, R.E. (1983). An electrophoretic study of the evolution in *Capsicum* (Solanaceae). *Evolution.* **37:** 562-574.

Medvedeva, N.V., Andreenkov, V.A., Morozkin, A.D., Sergeeva, E.A., Prokofev, I. and Misharin, A. (2003). Inhibition of oxidation of human blood low-density lipoproteins by carotenoids from paprika. *Biomed. Khim.* **49:** 191-200.

Molnar, J., Gyemant, N., Mucsi, I., Molnar, A., Szabo, M., Kortvelyesi, T., Varga, A., Molnar, P. and Toth, G. (2004). Modulation of multi-drug resistance and apoptosis of cancer cells by selected carotenoids. In Vivo **18:** 237-244.

Mori, T., Ohnishi, M., Komiyama, M., Tsutsui, A., Yabushita, H. and Okada, H. (2002). Growth inhibitory effect of paradicsompaprika in cancer cell lines. *Oncol Rep.* **9:** 807-810.

Nishino, H., Murakosh, M.I.I.T., Takemura, M., Kuchide, M., Kanazawa, M., Mou, X.Y., Wada, S., Masuda, M., Ohsaka, Y., Yogosawa, S., Satomi, Y. and Jinno, K. (2002). Carotenoids in cancer chemoprevention. *Cancer Metastasis Rev.* **21:** 257-264.

Pickersgill, B. (1988). The genus capsicum: A multidisciplinary approach to the taxonomy of cultivated and wild plants. *Biologisches Zentralblatt.* **107:** 381-389.

Pickersgill, B. (1997). Genetic resources and breeding of *Capsicum* spp. *Euphytica* **96:** 129-133.

Poulos, J.M. (1994). Pepper breeding (*Capsicum* spp.): achievements, challenges and possibilities. *Pl. Breed. Abst.* **64:** 143-155.

Smith, P.G. and Hesser, C.B. (1957). Breeding behavior of cultivated peppers. *Proc. Amer. Soc. Hort. Sci.* **70:** 286-290.

Tewksbury, J.J. and Nabhan, G.P. (2001). Directed deterrence by capsaicin in chillies. *Nature* **412:** 403-404.

Chapter 13

Prospects of Microbial Diversity for Cereal and Oilseed Crops during Storage

Alka Pandey[1] *and Nitin Joshi*[2]

[1]*Lecturer, Amity Institute of Food Technology, Amity University, Noida – 201 303, U.P.*

[2]*Former Research Associate, Department of Entomology & Agricultural Zoology, Institute of Agricultural Sciences, B.H.U., Varanasi – 221 005, U.P.*

What is Microbial Biodiversity?

In very simple manner 'biodiversity' can be understand by the degree of variation of life forms within a given ecosystem, biome, or an entire planet. Biodiversity is a measure of the health of ecosystems. Microbial diversity is a part of that ecosystem. The growth of microbes depends on the biotic and abiotic factors of any ecosystem. For some crops, development of natural microflora is highly desirable for the development of flavour precursor compounds *e.g.* for cocoa beans and tea leaves. Likewise development of microflora is also required during aging of cheese and some alcoholic beverages for the maturation of flavour associated with the products. However, development of natural microflora during storage of seed grains indicates sign of deterioration. Storage is a man made ecosystem where pH, moisture content, water activity (a_w), oxidation-reduction potential, available nutrient, temperature, relative humidity (RH), types and numbers of already existing microorganism in storage etc. jointly made the environment of particular species of microbes. For example, a stored grain bulk is a man-made ecological system in which deterioration is an ongoing process, resulting from interactions among physical, chemical, and biological variables. Damages by insects, fungi, and sprouting cause economic losses to grain producers, merchandisers and processors each year (Harein and Meronuck, 1995). Here, microbial diversity will be discussed for cereal and oilseed crops during storage. However, other crops like fruits and vegetable crops, pulses, coffee and cocoa beans are also important for storage point of view but for Indian mass population cereal grains and oilseeds are the major and cheapest source of nourishment and hunger satisfaction. Rice, wheat, sorghum, maize and barley are the major cereal crops whereas rapeseed, groundnuts, sunflower and soybeans are the major oilseed crops of India. They may act as nutrition/ medium or substrate for the growth and reproduction of undesirable microorganism during storage, if not stored, properly. Their undesirable growth leads to qualitative and quantitative losses like mycotoxins development, loss of germinability, discoloration etc. This chapter deals mainly with the spoilage of cereal grains and oilseeds caused by microorganisms during storage and the measures of quantification of spoilage as well as advanced methods of storage to prevent and/or minimize that losses.

Storage Losses

In tropical developing countries, a large proportion of the crop is harvested under humid and warm climatic conditions and most small farmers lack equipment for drying grains (Mendoza *et al.*, 1982). Consequently, the crop is stored while still relatively moist and warm, which results in rapid deterioration of the grains, mainly because of growth of molds. Even when the crops are sun dried after harvest, exposure to high relative humidity during open storage may result in moisture uptake by the stored grains, with resulting enhanced deterioration (Landers and Davis, 1986). Post harvest losses of food grains, caused by insect infestation and mold activity, have been conservatively estimated at 10–15 per cent (Grolleaud, 2002). Molds growing on grains present a second threat, through production of mycotoxins, the secondary metabolites produced by fungi that grow on a wide range of agricultural commodities including cereals and oilseeds (Epstein *et al.*, 1970). Mycotoxins pose a serious health risk to both humans and animals (Van Rosenburg, 1977; Vedman, 2004).

Types of Losses

Losses may be considered in terms of either quantity or quality. Quantitative loss is a physical loss of substance as shown by a reduction in weight or volume. It is the form of loss that can most readily be measured and valued whereas qualitative losses include appearance changes, nutritional degradation, loss of germination capacity, presence of insect fragments and mold contamination (Sinha and Muir, 1973). Some of these losses are difficult to detect visually (Lacey *et al.*, 1980) and is perhaps best identified through comparison with well defined standards. Nutritional loss and loss of seed are both aspects of quality losses. Microorganisms are associated with both types of losses which includes weight loss, generation of off flavour and off smell (microbiological rancidity or ketonic rancidity), loss of seed viability and monetary loss. Under humid storage conditions, however, the grains may deteriorate rapidly, resulting in qualitative and quantitative losses, and this deterioration is accelerated at higher temperatures.

Microorganisms Involved During Storage

Bacteria

Bacteria found on seeds generally require moisture content in equilibrium with 100 per cent relative humidity for active growth. Such environmental conditions are unusual in storage bins. As a result, bacteria are not involved in storage losses other than in very wet grains in the final stages of microbiological heating. However, bacterial decay of wet grain from harvest to dryer can lead to serious quality losses (Anon, 2003). The minimum a_w that can support active growth of most Gram-negative and Gram-positive bacteria are 0.97 and 0.90, respectively (Adams and Moss, 2000). The dominant bacterial flora on plant surfaces are Gram-negatives like *Erwinia, Pseudomonas* and *Xanthomonas* (Flannigan, 1987) with a smaller number of Gram-positive bacteria, such as *Lactobacillus* and *Leuconostoc* (Kaspersson *et al.*, 1988; Adams and Moss, 2000) that may become important in the production of fermented feed. However, species of *Bacillus, Enterococcus, Lactococcus, Pediococcus* and *Weisella* were also present on cereal grains at the time of harvesting. Presence of *Enterobacteriaceae* is indicative of the general hygiene status of the feed. Generally, control measures during grain storage do not focus on bacteria, as they are not regarded to be problem organism as discussed previously (Olstorpe, 2008).

Yeast

Yeast cells are best known for their contribution to society through their fermentation of bread, alcoholic beverages, and other products *i.e.* during processing of plant based products. Many studies have also been published on the spoilage of food and feed by yeasts (Middelhoven & van Balen, 1988; Fleet, 1992; Loureiro and Malfeito-Ferreira, 2003). Yeasts of different genera such as *Candida, Cryptococcus, Pichia, Rhodotorula* and *Sporobolomyces* have been isolated from grains at harvest (Flannigan, 1987). However, the significance of their presence has not been examined in cereal grains, as filamentous fungi are usually considered to be the main agents of pre and post harvest spoilage of grains (Lacey, 1989; Lacey and Magan, 1991). Therefore, the importance of yeasts associated with pre and post harvest deterioration of food grains needs further investigation.

Fungi

Fungi are the major cause of deterioration during storage of food grains. They may cause total deterioration of grain mass because they produce secondary metabolites such as mycotoxins that render the product unsafe for human and animal consumption. Fungi involved in the deterioration of cereal grains and oilseeds and other agricultural produce have been classified as field fungi and storage fungi depending on the time of their invasion and colonization of grains and whether they occur before or after the harvest (Jain, 2008).

Field Fungi

The field fungi infect their development on maturing/drying plants in the field or later when the harvest is lying in the field. They may be categorized as obligate parasites *e.g.* smuts, bunts and ergot or facultative parasites such as *Alternaria, Bipolaris, Curvularia, Cladosporium, Epicoccum, Helminthosporium, Fusarium* and *Nigrospora* etc (Balhara *et al.*, 2006). Some of the well known toxicogenic molds like *Aspergillus flavus* and *A. parasiticus* are also known to produce aflatoxins in the maize cobs and groundnut pods even in the fields. As far as the processing part for value addition of food crops is concerned, presence of aflatoxin in grains and oilseeds are major point of concern.

Storage Fungi

The common storage fungi include several species of *Aspergillus* and *Penicillium*. Other genera of storage fungi are *Absidia, Chaetomium, Mucor* and *Rhizopus*. Each has a different relative humidity requirement, and thus its development is an indicator of the moisture content of the stored grains. On the basis of temperature and water activity (a_w) relationships of storage microorganisms, they have been classified into following seven physiological groups (Table 13.1).

Ecology of Fungi in Grain

Field fungi include species of *Cladosporium, Alternaria, Epicoccum* and *Fusarium* (Magan and Lacey, 1984; Flannigan, 1987; Lacey and Magan, 1991). *Cladosporium* species are among the most abundant components of daytime summer air-borne spores. Species of this genus are widespread on the ears of cereals at harvest. *Alternaria alternata* is, after *Cladosporium* spp., probably the most common airborne fungal spore. *Alternaria* may colonise cereal crops soon after emergence and penetrate the kernel sub-epidermally. This makes it tolerant of fungicides, and *Alternaria* species can be isolated from most grains at harvest (Lacey, 1989). *Fusarium* spp. are important pathogens of cereal grains, causing various infections such as scab, ear rot or head blight. They may also produce mycotoxins, such as deoxynivalenol and various trichothecenes, in the grain, both pre and post harvest (Lacey *et al.*, 1999; Aldred and Magan, 2004). Depending on the storage conditions of cereal grains, growth of typical storage fungi may occur. These fungi are present at low levels before harvest, and are principally species of *Aspergillus* and *Penicillium*. Ruminants eating *P. roqueforti* infested feed displayed symptoms such as lack of appetite, ketosis, paralysis and spontaneous abortions (Häggblom, 1990). *Aspergillus* spp. are characteristic colonisers of stored products, different species vary considerably in their growth requirements; thus, the dominance of certain species may be indicative of previous storage conditions (Lacey, 1989). On the other hand, it is clear that the concept of field and storage flora should not be carried too far. The terminology was first used in northern temperate regions, whereas in warmer, more humid climates, the species distribution between field and storage fungi differs (Lacey and Magan, 1991). The species composition of field and storage flora may also vary with grain storage method. *Cladosporium*, typically regarded as field flora, was detected as storage fungi by Kaspersson *et al.* (1988) in some of their studies. Lacey and Magan (1991) stated that *Fusarium* spp. could occur

Table 13.1: Classification of Storage Fungi on the Basis of Temperature and Water Activity (a_w) Relationship

Sl.No.	Groups	Temperature Range (°C)	Water Activity	Example	Note
1.	Psychrotolerant	10-35	–	*P. brevicompactum, P. chrysogenum, P. cyclopium, P. expansum, P. notatum* and *P. viridicatum* and *Chrysosporium pannorum.*	Most of these fungi are common contaminants of stored products in temperate climate.
2.	Mesophilic	20–50	≤0.75	*Aspergillus restrictus group, Eurotium repens, E. amstelodamy, E. chevaleri, E. rubrum* and *Wallemia sebi*	In dry grain storages these species act as primary invaders of grains and oilseeds.
3.	Lower mesophilic	20–37	0.75–0.79	*Aspergillus versicolor*	–
4.	Slightly xerophillic	–	≥0.90	*P. corylophilum, P. glabrum, P. palitans, P. rugulossum, Fusarium, Mucor, Rhizopus, Scopulariopsis, richoderma* and *Trichothecium*	–
5.	Upper mesophilic	5-50	0.75-0.90	(*i*) Moderately xerophilic (0.75–0.79 a_w) comprising *Aspergillus candidus, A. flavus, A. ochraceus, A. terreus, A. temarii, A. niger, A. wentii* etc. Most of the species are well known for toxin production, (*ii*) Slightly xerophilic (0.80–0.89 aW) comprising some species of *Penicillium* and *Byssochlmys* and a few of *Paecilomyces.* (iii) Hydrophilic (0.90 a_w) that include species of *Absidia, Rhizopus* and *Syncephalastrum.*	Similar to mesophic fungi, micro-organisms of this group are thermo-tolerant in nature but can be distingui-shed in their water requirement.
6.	Thermophilic	Upto 60	0.80–0.89	*Humicola stellata, Malbranchea sulphurea, Rhizomucor pusillus* and *Sporotrichum thermophillum* etc. represent this group.	Mainly storage fungi
7.	Extremely thermophilic	Upto 70	≥0.90	Some actinomycetes belonging to the genus *Micropolyspora, Saccharomonospora, Streptomyces, Thermomonospora* etc.	They do not play any role in the initial colonization of the stored products but many species of these genera cause serious allergic conditions on the sensitized subjects, particularly the farm workers.

Adapted from Lacey *et al.*, 1980 and Jain, 2008.

as storage flora, when a_w is high and temperatures are low. Fungi seldom occur on grains in isolation, but usually as a mixed consortium of bacteria, yeasts and filamentous fungi (Magan *et al.*, 2003).

Effects of Storage Fungi on Seeds

Invasion of seeds by storage fungi leads to an increase in respiration, water content and temperature, resulting in damage to the embryo and eventually loss of seed viability (Neergard, 1977). Seeds with a high rate of respiration, their own and of the fungi, attain high temperature and moisture. Due to invasion of these fungal species, the stored seeds are likely to clump together. Fungi produce mycotoxins in the stored grains. The major deleterious effects of storage fungi are as follows (Anon, 2003; Balhara *et al.*, 2006; Jain, 2008 and www.fao.org, 2011):

Reduced Viability and Discolouration

Seed viability may be significantly reduced before the fungus is visible even with the aid of a microscope. Weakening or killing of the seed embryo precedes any discoloration. The seed will not germinate if discoloration is easily detected. The speed at which seed viability is lost depends on the storage fungi. *A. flavus* may kill the entire infected seed lot of rice within 3 months of storage. In contrast, seed lots infected with *A. restrictus* may not loose all germinability for as long as 8 months.

Change in Texture and Flavour

Change in texture and flavour rendering the produce unacceptable by fermentation which converts carbohydrates to acids and gases; by putrifaction which breaks proteins and microbial and hydrolytic rancidity of fats which converts fats into acids and acid derivatives as the activity of hydrolytic rancidity increased considerably. This is especially important for the processing of oilseeds (*e.g.* groundnuts, oil palm kernels) and fatty products like rice-bran etc.

Grain Heating (Thermogenesis)

The respiration of moist grain was once thought to be responsible for the heating of stored grains. Research has subsequently proven that the metabolic processes of storage fungi are responsible for this phenomenon. Ferdinand Cohn was the first who had shown that molding barley seedlings in an insulated container could heat to 60°C and that it was probably due to metabolic activity. Later H. Miehe conducted a series of experiments on hay and demonstrated the role of microorganisms in heating during storage. According to him, the level of heating caused by the given organism depends upon its own maximum growth temperature. He stated that 'if there are no thermophiles present in self-heating hay, the temperature should measure only to the point where mesophiles are no longer able to thrive *i.e.*, 45°C. If thermophilic forms are present they would take over at 40-45°C and their metabolism may carry the heating to a higher level 60-70°C. A comprehensive account of the water and temperature produced by the activity of different microorganisms in stored products is given in Table 13.2. During thermogenesis, a continuous accumulation of heat due to microbial activity may raise the temperature of stored products to approximated 70°C. Non-biological process may further raise the temperature to ignition under certain conditions. Microbial thermogenesis can be more rapid if thermophilic fungi receive optimum growth conditions under storage. Tropical climate favours early development of thermophilic and thermo tolerant fungi on stored products. Maximum attainable temperature may cause spontaneous combustion in storage bulk.

Table 13.2: Storage Microorganisms in Relation to Likely Water Activity (a_w) and Maximum Temperature Rise during Storage

Predominant Microorganisms	*Approximate a_w*	*Maximum Temp. (°C)*
Aspergillus restrictus	0.7–0.6	Ambient
Eurotium spp.	0.8–0.9	35
A. versicolor group, Scopulariopsis brevicaulis, Streptomyces griseus	0.9–0.95	40
A. candidus, Penicillium spp.	0.95	45
Absidia spp., A. nidulans, Streptomyces albus	0.98	50–55
A. fumigatus, Rhizomucor pusillus, Malbranchea sulfurea, Humicola lanuginose, Talaromyces thermophilus, Saccharomonospora viridis, Saccharopolyspora rectivirgula (Micropolyspora faeni), Thermoactinomyces spp.	1.0	65

Musty Odors, Caking and Decay

These characteristics are indicative of the advanced stages of spoilage, detectable by the unaided eye or nose. Substantial levels of fungal growth occur before it becomes readily apparent. There is generally visible growth on the grains before the musty odors are detected. Caking is a result of webbing of fine threadlike mycelium between and within the kernels. The caking may sometimes be only a few inches thick, consisting of rotten kernels and fungal mycelium, while the bulk of the grain underneath remains sound. Regardless of the depth, caking represents the final stages of decay.

Food Intoxification

Production of mycotoxins may lead to poisiong (food intoxification) in human beings and animals. Aflatoxin, produced by the common storage fungus *Aspergillus flavus* is known, for example, to cause liver collapse in certain domestic animals. As aflatoxin is one of the most well-known mycotoxin in tropical and sub-tropical areas, therefore its details are also covered separately.

Mechanism of Seed Deterioration

The details of various studies to explain the mechanism of seed deterioration are given by Balhara *et al.* (2006) as follows:

The maintenance of high quality in stored grains has always been a problem in conditions that permit the invasion of the seeds by fungi. The mechanism of seed deterioration is not fully understood. Even under ideal conditions of storage, loss in seed viability can not be checked completely. High moisture and high temperature during storage are highly detrimental factors for the survival of most seeds. The primary causes of seed deterioration can be grouped into two categories-

1. Seed tissues may deteriorate due to natural aging
2. Deterioration also may be caused by invasion of seed and damage to tissues by pests and pathogens.

Seed deterioration can begin in the field after the seeds have reached physiological maturity, particularly if the harvest is delayed by wet weather. Physiological changes, which occur during storage in association with deterioration, are: delayed germination, reduced seedling growth rate, decreased tolerance to adverse germination conditions, and loss of germinability (Abdul-Baki and Anderson, 1972). Biochemical changes associated with seed deterioration include alterations in metabolic activity and memebrane degradation, low O_2 uptake and high CO_2 output (Kharluki and Agarwal, 1984).

Membrane Disintegration

Cell membranes are constituted of a certain proportion of unsaturated lipids. In presence of $O_{2,}$ the lipids may react to form free radical intermediates and unstable peroxides (Agarwal, 1988). This results in destruction of lipid itself, the formation of insoluble protein complexes by cross linkage and damage to cell membrane (Dablani and Agarwal, 1983a).

Seeds leach out a number of water-soluble compounds when imbibed in water. Leaching increases during seed deterioration. Analysis of seed lechate for soluble sugars, amino acid and electrolytes has been used to access seed quality. Negative correlation has been reported between the leaching of sugars and electrolytes and seed viability (Dadlani and Agarwal, 1983b) and seedling vigour (Mullett and Wilkinson, 1979).

Membrane deterioration, in addition to the loss of solute control, results in dispersal of highly ordered system of membrane associated enzymes (Stewart and Bewley, 1980). This may cause, among other effects, a shift in respiratory pathway in hexose oxidation during germination.

Enzyme Activity

In oilseeds, seed coat is made up of cellulose and pectin. Hence, the ability of storage fungi to secrete cellulytic and pectinolytic enzymes is important for invasion and colonization of seed tissues. Lipolytic enzymes liberated by fungi act upon triglycerides of oil thereby releasing free fatty acids and causing hydrolytic rancidity (Lalita Kumari *et al.*, 1971). Fungal infestation causes biochemical changes in nutritional value of the oilseeds. Protein molecules are altered by microorganism's enzymatic action and new proteins of microbial origins are synthesized. In addition, depletion of some enzymes and intensification of others and/or production of new multiple molecular forms of enzymes takes place.

Aflatoxins

Aflatoxins are toxic secondary metabolites produced by certain strains of *Aspergillus flavus* and *Aspergillus parasiticus*. Toxicity studies have demonstrated their carcinogenic, hepatotoxin, teratogenic and immunosuppressive activity (Betina, 1984).

Their occurrence, structure and chemical properties of aflatoxins have been discussed previously by various authors in various ways. In a very simplest form they are described in Dictionary of Food Science and Technology (Anon, 2005) as under:

- **Aflatoxin B_1 :** A potent hepatocarcinogen. Toxic to many species, including humans, birds, fish and rodents.
- **Aflatoxin B_2 :** Moderate hepatocarcinogen compared with aflatoxin $B_{1.}$
- **Aflatoxin B_3 :** Hepatocarcinogen with a toxicity similar to that of aflatoxin $B_{1.}$
- **Aflatoxin D_1 :** Carboxylated product of aflatoxin B_1. Possesses less toxicity than aflatoxin $B_{1.}$
- **Aflatoxin G_1 :** A potent hepatocarcinogen with a toxicity similar to that of aflatoxin $B_{1.}$
- **Aflatoxin G_2 :** Dihydroxylated product of aflatoxin G_1. Possese less toxicity than aflatoxin $G_{1.}$
- **Aflatoxin M_1 :** Metabolic product of aflatoxin B_1 in animals, usually excreted in the milk of cattle and other mammalian species that have consumed aflatoxin B_1 contaminated foods or feeds. Possess less toxicity than aflatoxin B_1.
- **Aflatoxin M_2 :** Metabolic product of aflatoxin B_2 in animals. Usually excrete in the milk of cattle and other mammalian species that have consumed aflatoxin B_2 contaminated foods and feeds. Possess less toxicity than aflatoxin B_2.
- **Aflatoxin P_1 :** Demethylated and hydroxylated product of aflatoxin B_1 found in animals. Very weak toxin compared to aflatoxin B_1.
- **Aflatoxin Q_1 :** Main metabolite of aflatoxin B_1 found in humans and primates.

Symptoms of Aflatoxicosis

Mycotoxins in general and aflatoxins in particular are being studied as health hazard for humans. The ingestion of food containing aflatoxin may have serious adverse health effects in human. Diseases in animals and human beings resulting from consumption of aflatoxins are called aflatoxicosis. Aflatoxicosis causes acute liver damage, liver cirrhosis, induction of tumours, attack on central nervous system, skin disorders and hormonal effects (Pitt, 1989). Epidemiological studies have indicated a relationship between aflatoxin intake and incidence of liver cancer in several developing counteries. In India, Aflatoxicosis and liver cancer in human have also been reported by Bilgrami and Sinha (1986). Indian childhood cirrhosis (ICC) is the most common cause of death of children under the age of 5 years. The characteristic features of the disease involve low grade fever; mild abdominal distension followed by enlarged liver with characteristic leafy borders, the disease may progress to jaundice, ascites, fibrosis, cirrhosis and hepatic comma (Yadgiri *et al.*, 1970, Amla *et al.*, 1971). Children exposed to aflatoxins through mother's milk and food such as parboiled rice and unrefined peanut oil may acquire ICC (Amla *et al.*, 1970). Linsell and Peers (1977) observed the possibility of infection by hepatitis B virus prior to the incidence of primary liver cancer in aflatoxin exposed persons. The studies revealed that aflatoxin could be a major disease determinant in primary liver cancer and hepatitis B virus served as co factor in the etiology.

Acute Toxicity

The general symptoms of aflatoxicosis are edema of the lower extremities, abdominal pain and vomiting. The principal target organ for aflatoxins is the liver. After the invasion of aflatoxins into the liver, lipids infiltrate hipatocytes and leads to necrosis or liver cell death. The aflatoxin metabolites react negatively with different cell proteins and inhibited protein synthesis and carbohydrate and lipid metabolism (Balhara *et al.*, 2006).

Chronic Toxicity

Aflatoxin B_1 is mainly related to chronic toxicity. Chronic aflatoxiosis results in congested liver with haemorrhagic and necrotic zones, proliferation of hepatic parenchyma and epithelial cells of the duct and congestion of kidneys that show occasional haemorrhagic enteritis. Liver damage is apparent due to yellow colour that is characteristic of jaundice and the gall bladder will become swollen. The chronic aflatocicosis reduces growth rate and reproductive efficiency (Balhara *et al.*, 2006).

Cellular Effects

Aflatoxins have high affinity for nucleic acids and polynucleotides and inhibit the synthesis of nucleic acids. Aflatoxin B_1 is genotoxic, producing adducts in humans *in vivo*. In human cell culture, it produces DNA damage, gene mutation and chromosomal abnormality (IARC 56).

Immune Suppression

In low range, aflatoxins would be immunomodulatory (Pestka and Bondy, 1994). Aflatoxins reacts with T-cell and resulted in immune suppression and decrease the activities of vitamin K and phagocytic activity in macrophages.

Detection Methodology of Aflatoxin

In view of the evidence concerning the effects, particularly the carcinogenic effects of aflatoxin in several animal species, and in view of the association between aflatoxin exposure levels and human liver cancer, incidence observed in some parts of the world, exposure to aflatoxins should be kept as low as practically achievable. An awareness of the level of contamination of aflatoxin in natural products can only be obtained by developing good analytical methodologies for detecting aflatoxin in foods, mixed feeds and ingredients, animal tissue, blood, urine and milk. The aflatoxin detection methods can be divided into three categories:

1. Rapid presumptive tests to identify samples from agriculture products such as corn, peanut lots that may contain toxin,

2. Rapid screening procedures to determine the presence or absence of toxin, and
3. Quantitative methods to determine aflatoxin levels.

The presumptive test for aflatoxin in corn is the black light test or Bright Greenish-Yellow fluorescent test (BOY) based on the fluorescence under ultraviolet light (365 nm.) associated with *Aspergillus flavus* and *A. parasiticus*. Rapid screening tests have included mini column methods that can be done in a laboratory with minimal facilities, and thin layer chromatography (TLC). Quantitative methods to determine aflatoxin levels involves extraction, purification of extract, and measurement of the toxin by thin layer Chromatography (TLC) using visual comparisons with a standard or densitometry or high pressure liquid chromatography (HPLC). Detailed methodology for the detection of mycotoxins analysis has been described in the recent manual 'Methods of analysis of foods for mycotoxins' published by All India Food Processors' Association, New Delhi which is based on A.O.A.C. (Association of Official Analytical Chemists, Washington) procedures.

Recent Advances in Safe Storage System

Globally, there is increasing demand for high quality and safe food, free of chemical and physical contaminants and pathogens. Grain growers and users must maintain and protect their harvested grain from insect and microbial damage (Sinha, 1995). Under dry conditions, grains (paddy rice, maize, etc.) can be stored for extended periods provided that there is no insect infestation or microbial activity. With increasing consumer concerns over the potential health and environmental hazards of pesticides, as well as insect resistance, non-chemical control methods will become increasingly important in future pest management strategies. Many different temperature based technologies may have application in various niches of post harvest storage, handling, and processing, with potential for combination with other non-temperature based technologies. For example, grain from the field may be disinfected by electric fields, microwaves, or fluidized bed drying. Whole grain may be processed in a plant disinfested with a carbon di-oxide/heat sterilization treatment. Applications which provide continued control, such as ambient aeration and chilled aeration can easily be combined with quick, non-residual disinfestation procedures such as electric fields, microwaves, and various heat treatments in an integrated pest management programme. There are many different combinations, which could be employed, and more research needs to be conducted to determine which treatments are most effective and economical in various applications and combinations. Electric fields and microwaves would require greater efficiencies to cover the increased capital and maintenance costs incurred or consumer and environmental demands for chemical-free products to compete as cost effective alternatives to current control methods.

Practical aspects of application are also important in determining the feasibility of these technologies. Public perception is possibly the most important factor in determining the value of various applications. For example, microwave technology has been widely accepted by the general populace, and would most likely be easy to market as a safe alternative to chemicals. The following conclusions were drawn by

Mason and Strairt (2011) after reviewing various past and present research efforts on the suitability of temperature based technologies for the control of insects and fungi in stored crops.

Microwaves

In the frequency range of 2.45 to 10.6 GHz microwaves require higher grain temperatures than temperatures required in the range of 11 to 90 MHz to achieve complete mortality. Application of microwaves is limited due to insufficient penetration depth.

Variation in Temperature

Low temperatures can slow mold development, thereby increasing storage time. Generally, mortality in pests is dependent upon cooling rate as well as final temperature. However, molds are able to grow at sub-freezing temperatures but their activity is significantly reduced. High temperatures applied in either fluidized or sprouted beds, can quickly cause destruction of molds. Treatment time is dependent upon heating rate and final temperature. The treated commodity requires re-cooling to slow re-infestation and product quality must not be altered.

Controlled Atmosphere Storage

This technique involves oxygen deficient environments which can improve the effectiveness and efficiency of traditional and alternative pest control measures when combined with them. Burning of hydrocarbons, nitrogen flushing, use of lime, activated charcoal are some of the techniques used to create atmosphere of desirable gaseous composition (Pandey, 2009).

Irradiation

Irradiation is non-thermal method of preservation of raw and packaged foods in which product is exposed to a predefine doses of particular radiations (X-rays, β rays etc.) for some direct and in-direct effects of radiation. In direct effect, radiation directly kills the genetic material of organisms whereas, in in-direct effects, radiation produces peroxides radicals from water (H_2O) molecules which is a universal component of all biological material like food grains etc. these peroxides upon further reaction produce a well known biological poison *i.e.* hydrogen peroxide (Potter and Hotchkiss, 2010).

Sun Drying

Under tropical conditions, grains can be sun-dried to intermediate moisture levels *e.g.* maize cobs is about 18 per cent. A feasible technology to store such intermediate moisture grains, including maize, is known as self regulated modified atmosphere (Richard-Molard *et al.*, 1987). In this technology, storage of various grains at intermediate moisture levels (15–18 per cent) in sealed containers results in depletion of oxygen and enrichment in CO_2, as a result of respiration of the grains, insects and microorganisms (Navarro and Donahaye, 2005). In addition, the limited microbiological activity may result in production of volatile fatty acids (VFAs). Both the anaerobic conditions and the VFA inhibit fungal development (Moon, 1983;

Weinberg *et al.*, 1993). This technology is eco-friendly and does not involve the use of antifungal chemicals.

Biosensors

CO_2 sensors can effectively monitor spoilage prior to the time that spoilage would be detected by traditional methods. In-lab and pilot bin experiments as well as tests in large commercial storage structures have been successfully conducted. Improvements in grain quality, storage management and processing have been accomplished that increased food safety and reduced costs. Some of these include: Evaluation of ozone for reducing off-odors, control of insect pests, microbial loads, and mycotoxin content has been conducted. Ozone and hydrogen peroxide were evaluated for treating *Fusarium* head blight infected malting barley (http://oardc.osu.edu/nc213/POW_2008to2013.pdf, 2008).

Image Analysis

It is a latest technology can be developed for more accurate, non-destructive measurement of grain quality traits for evaluation of suitability for various end uses. Visible and non-visible range on electromagnetic spectrum used to detect the internal and external quality attributes of cereals, pulses and oilseeds including fungal contamination (Choudhary, 2006).

Hermetic Storage System

Scientifically, hermetic means a system that is water vapour and air proof (Potter and Hotchkiss, 2010). Since the size of microbes is greater than water vapour molecules, therefore a concept of hermetic storage system shows that grains which are not exposed to air and water ingression would not be exposed to microbes. The research on hermetic storage system for various crops is described as under.

Rice and Rice Seeds

As a result of extensive studies at IRRI (Rickman and Aquino, 2004) and later by PhilRice (Sabio *et al.*, 2006), over the last 10 years, the benefits of storing both rice and rice seeds in hermetic storage are now well understood and in widespread use, particularly in Asia (Villers *et al.*, 2006). These cocoons (storage systems) are used by the National Food Authority of the Philippines, to safely store rice paddy for up to one year. Hermetic storage applications for rice and/or rice seed are currently found in Cambodia, East Timor, Indonesia, India, Pakistan, Philippines, Sri Lanka, and Vietnam (Montemayor, 2004).

Wheat and Barley

Hermetic storage of wheat in Hermetic Bunkers with capacities ranging from 10,000 to 20,000 tonnes was first introduced in the early 1990. Hermetic storage of wheat, stored at or below its critical moisture content of 12.5 per cent, provides storage without significant degradation of quality, including maintenance of baking qualities, for up to 2 years (Navarro *et al.*, 1993). In Cyprus such Bunkers allowed quality preservation of barley for 3 years, with total losses of 0.66 per cent to 0.98 per cent, and with germination remaining above 88 per cent (Varnava and Muskos, 1997).

Maize

Cocoons are widely used in Rwanda, Ghana and the Philippines for storing both shelled and unshelled maize, in capacities ranging from 50 to 1050 tonnes. Similar quality preservation results were obtained for maize when stored in 60 kg capacity Super Grain bags. The large flexible hermetic storage units are generally used at the village level, but also as strategic reserves to prevent famine at the district level (Navarro, 2006 and Montemayor, 2004). Not only this but hermetic storage is also achieved in specially constructed plastic structures suitable for long-term storage systems, as well as intermediate storage of cereals, pulses, coffee and cocoa have been developed and applied. Flexibility, transportability, ease of erection, simplicity of operation and maintenance and durability are distinct advantages.

Conclusion

The microorganisms associated with the stored cereal grains and oilseeds are mostly undesirable due to public health safety concerns. Spoilage in grain post harvest is initiated by insufficient drying or by subsequent moisture increases due to poor storage equipment. To prevent molding and rotting in tropical and subtropical regions, grains should be dried immediately after harvest, and cooled or treated with antifungal chemicals such as propionic acid. However, humid conditions and economic constraints prevent rapid drying to safe moisture levels (14 per cent and below). The ecosystem that forms in the sealed containers with the self regulated atmosphere, and the inter-relationship between grain respiration and microbial activity has not yet been fully explored. Therefore, it is suggested to keep detailed records of each and every year of microflora of storage bins and environmental condition etc. Accurate records can help in identifying potential storage problems and in planning preventative action. This would also help for documentation for ISO certification like Quality management system (QMS) certification.

Whatever new technologies are developed, it is important to ensure public acceptance, compliance with food and environmental regulations, and the development of proper safety procedures, in addition to achieving acceptable, cost-effective control. There is a need to combine the various methods to prevent storage losses by IPM programme which are feasible, economical and eco-friendly.

References

Abdul-Baki, A.A. and Anderson, J.D. (1972). Physiochemical and biochemical deterioration of seeds. T. T. Kozlowski (ed.), Seed Biology. Academic Press, New York. pp 283-316.

Adams, M.R. and Moss, M.O. (2000). Food Microbiology. IInd ed. Cambridge, UK: The Royal Society of Chemistry.

Agarwal, P.K. (1988). Seed deterioration during storage. Proc Int. Cong PI Physiol New Delhi. pp 1271-1278.

Aldred, D. and Magan, N. (2004). Prevention strategies for trichothecenes. *Toxicology Letters*. **153:** 165-171.

Amla, I., Kamala, C.S., GopalaKrishna, G.S., Jayaraj, A.P., Sreehivasamurthy, V. and Parpia, H.A.B. (1971). Cirrhosis in children from peanut meal contaminated by aflatoxin. *Am. J. Clin. Nutrition.* **24:** 609.

Amla, I., Kumari, S.Y., Sreenivasmurthy, V., Jayaraj, A.P. and Parpia, H.A.B. (1970). Role of Aflatoxin in Indian childhood Cirrhosis. *Indian Pediatr.* 262-270.

Anonymous, (2003). Managing Rice Decay during Storage in Rice quality workshop, 2003. www. plantsciences. ucdavis.edu/uccerice/rice_quality/C-8RiceDecay2003.pdf

Anonymous, (2005). Dictionary of Food Science and Technology, Complied and edited by the International Food Information Service. Blackwell Publishing, pp 07.

Balhara, M.T., Kapoor, R. and Bhatnagar, A.K. (2006). Seed Mycroflora and Aflatoxin Contamination in Mustard during Storage. Chapter 22: Current Concepts in Botany, edited by K. G. Mukerji, Manoharachary, C. I. K. International Publishing House Pvt. Ltd. New Delhi, pp 325-329.

Betina, V. (1984). Mycotoxins, production, Separation and Purification. Developments in Food Science, 8: 89-130.

Bilgrami, K.S. and Sinha, K.K. (1986). Aflatoxin in India: 1 In. Aflatoxin in Maize. A Proceedings of the Workshop of CIMMYT, EL Batan, Mexico. pp 349-357.

Choudhary, R. (2006). Bio-imaging and sensing applications in dairy and food industries. *Beverage & Food World.* 33 **(3):** 21-24.

Dablani, M. and Agarwal, P.K. (1983a). Mechanism of soybean seed deterioration. *Plant Physoil. & Biochem.* **10:** 23-30.

Dablani, M. and Agarwal, P.K. (1983b). Factors influencing leaching of sugars and electrolytes from carrot and okra seeds. *Scientia Hort.* **19:** 39-44.

Epstein, E., Steinberg, M.P., Nelson, A.I., and Wei, L.S. (1970). Aflatoxin production as affected by environmental conditions. *Journal of Food Science.* **35:** 389–391.

Flannigan, B. (1987). The microflora of barley and malt. In: *Brewing Microbiology* (eds FGPriest and ICampbell) Elsevier Appl. Sci., London, pp 83–125.

Grolleaud, M. (2002). Post harvest losses; discovering the full story. FAO Corporate Document Repository, FAO Agro Industries and Post-harvest Management Service (AGSI). /http://www.fao.org/ DORCEP/004/AC301E/AC301E00.htmS.

Häggblom, P. (1990). Isolation of Roquefortine C from feed grain. *Applied and Environmental Microbiology.* 56, 2924-2926.

Harein, P. and Meronuck, R. (1995). Stored grain losses due to insects and molds and the importance of proper grain management, *In* V. Krischik, G. Cuperus & D. Galliart, eds. *Stored Product Management*, pp 29-31. E-912. CES. Div. Agric. Sci. Nat. Res. OSU. USDA. FGIS. ES. APHIS. 242 pp.

http://oardc.osu.edu/nc213/POW_2008TO2013.pdf, 2008. Marketing and Delivery of Quality Grains and BioProcess Coproducts as on June, 2011.

IARC (Intrernational Agency for Research on Cancer) (1993). Monograph 56. IARC Monograph series on the evaluation of carcinogenic risks to humans. Lyon, France.

Jain, P.C. (2008). Microbial degradation of grains, oilseeds, textiles, wood, corrosion of metals and bioleaching of mineral ores. nsdl.niscair.res.in/bitstream/123456789/558/1/ MicrobialDegradation.pdf.

Kaspersson, A., Lindgren, S. and Ekström, N. (1988). Microbial dynamics in barley grain stored under controlled atmosphere. *Animal Feed Science and Technology* **19:** 299-312.

Kharluki, L. and Agarwal, P.K. (1984). Evidence for participation of pentose phosphate at books.google.co.in/books/isbn=8188237647 pp 336.

Lacey, J. and Magan, N. (1991). Fungi in cereal grains: Their occurrence and water and temperature relationships. In: Chelkowski, J. (Ed.) *Cereal grain. Mycotoxins, fungi and quality in drying and storage.* Amsterdam, The Netherlands: Elsevier Science Publishers. pp. 77-118.

Lacey, J. (1989). Pre- and post-harvest ecology of fungi causing spoilage of foods and other stored products. *Journal of Applied Bacteriology*. 67, 11-25.

Lacey, J., Bateman, G.L. and Mirocha, C.J. (1999). Effects of infection time and moisture on development of ear blight and deoxynivalenol production by *Fusarium* spp. in wheat. *Annals of Applied Biology*. **134:** 277-283.

Lacey, J., Hill, S.T. and Edwards, M.A. (1980). Microorganisms in Stored Grains : Their Enumeration and Significance. *Trop. Stored Prod. Information*. **39:** 19-33.

Kumari, Lalitha, Govindswany, D.C.V. and Vidyasekaran, P. (1971). Effect of seed-borne fungi on the physico-chemical properties of groundnut oil. *Indian Phytopathology*. **24:** 283-289.

Landers, K.E. and Davis, N.D. (1986). Influence of atmospheric gases on aflatoxin production by Aspergillus flavus in peanuts. *Phytopathology*. **57:** 1086–1090.

Linsell, C.A. and Press, F.G. (1977). Aflatoxin and liver cell cancer. *Trans Ray. Soc. Trop. Med. Hyg*. **71:** 471-473.

Loureiro, V. and Malfeito-Ferreira, M. (2003). Spoilage yeasts in the wine industry. *International Journal of Food Microbiology* **86:** 23-50.

Magan, N. and Lacey, J. (1984). Effect of temperature and on pH on water relations of field and storage fungi. *Transactions of the British Mycological Society* **82:** 71-81.

Magan, N., Hope, R., Cairns, V. and Aldred, D. (2003). Post-harvest fungal ecology: Impact of fungal growth and mycotoxin accumulation in stored grain. *European Journal of Plant Pathology*. **109:** 723-730.

Mason, L.J. and Strait, C.A. (2011). Stored Product Integrated Pest Management with Extreme Temperatures available at cipm.ncsu.edu/IPMtext/chap6.pdf as on 05/08/2011.

Mendoza, E., Rigor, A.C., Mordido, J.R.C.C. and Marajas, A.A. (1982). Grain quality deterioration in on-farm level of operation. In: Teter, N.C., Frio, A.S. (Eds.), Progress in Grain Protection. Proceedings of the Fifth Annual Grains Post-harvest Workshop, 19–21 January 1982, Chiang Mai, Thailand, 107–117 pp.

Middelhoven, W.J. and Van Balen, A.H.M. (1988). Development of the yeast flora of whole-crop maize during ensiling and during subsequent aerobiosis. *Journal of the Science of Food and Agriculture* **42:** 199-207.

Montemayor, R. (2004). World Grain Magazine, November, 2004 Issue. Better Rice Store. IRRI. Manilla, Philippines. pp 43-45.

Moon, N.J. (1983). Inhibition of the growth of acid tolerant yeasts by acetate, lactate and propionate and their synergistic mixtures. *Journal of Applied Bacteriology*. **55:** 453–460.

Mullett, J.M. and Wilkinson. (1979). The relationship between amounts of electrolyte lost on leaching seeds of *Pisum sativum* and some parameters of plant growth. *Seed Science & Technology*. **7:** 393-398.

Navarro, S. (2006). Modified Atmospheres for the control of stored-product insects and mites. In: Heaps, J. W. (Ed) Insect Management for Food Storage and Processing, Second Edition, AACC International, St. Paul, MN, USA, 105-146 pp.

Navarro, S. And Donahaye, E.J. (2005). Innovative environmentally friendly technologies to maintain quality of durable agricultural products. In: Ben Yehoshua, S. (Ed.), Environmentally Friendly Technologies for Agricultural Produce Quality. CRC Press, Boca Raton, FL, USA, pp. 204–260.

Navarro, S., Varnava, A. and Donahaye, E. (1993). Preservation of grain in hermetically sealed plastic liners with particular reference to storage of barley in Cyprus, In: Navarro, S., Donahaye, E. (Eds). Proceedings International Conference on Controlled Atmosphere and Fumigation in Grain Storages, Winnipeg, Canada, June 1992, Caspit Press Ltd., Jerusalem, pp. 223-234.

Neergard, P. (1977). Seed Pathology. John Wiley. New York. pp 839.

Olstorpe, M. (2008). Feed Grain Improvement through Biopreservation and Bioprocessing, Doctoral diss., Dept. of Microbiology, SLU. Acta Universitatis agriculturae Sueciae vol. 2008: 77.

Pandey, A. (2009). Controlled Atmosphere Storage: A protection tool for storage. In XIth Indian Agricultural Scientists and Farmers Congress at Iswar Saran Degree College, Allahabad. pp. 23-24

Pestka, J.J. and Bondy, G.S. (1994). Immunotoxic Effects of mycotoxins. J. D. Miller (ed.). Mycotoxins in Grain, Compounds other than Aflatoxin Eagas Press, St. Paul Mn. pp. 339-358.

Pitt, J.T. (1989). Toxigenic Aspergillus and Penicillium Species. A Paper Presented on the International Training course on mycotoxin Prevention and Control. Dept. of Agriculture, Bangkok, Thailand.

Potter, N.N. and Hotchkiss, J.H. (2010). Food Science, Chapter XI: Irradiation, Microwave and Ohmic Processing of Foods. V^{th} edition, and Aspen publication, pp 245-255.

Richard-Molard, F., Diawara, D.B. and Cahagnier, B. (1987). Susceptibility of cereal microflora to oxygen deficiency and carbon dioxide concentration. In: Donahaye, E.J., Navarro, S. Eds., Proceedings of the Fourth International Working Conference on Stored-Products Protection, 21–26 September 1986. Tel Aviv, Israel, Maor-Wallach Press, Caspit, Jerusalem, Israel, pp. 85–92.

Sabio, G.C., Dator, J.V., Orge, R.F., Julian, D.D.T., Alvindia, D.G., Miranda, G.C. and Austria, M.C. (2006). Preservation of Mestizo 1 (PSB Rc72H) Seeds Using Hermetic and Low Temperature Storage Technologies. GrainPro Document SL2329GCS1206. 2006. 10th International Working Conference on Stored Product Protection 452 Julius-Kühn-Archiv, 425.

Sinha, R.N. (1995). The stored-grain ecosystem. In: Jayas, D.S., White, N.D., Muir, W.E. (Eds.), Stored Grain Ecosystems. Marcel Dekker, New York, pp. 1–32.

Sinha, R.N. and Muir, W.E. (1973). Grain Storage: Part of a System. The Avi Publ. Co. Inc., Westport, CT.

Stewart, R.R.C. and Bewely, J.D. (1980). Lipid peroxidation associated with accelerated ageing of soybean axes. *Plant Physiology*. **65:** 245-248.

Van Rosenburg, S.J. (1977). Role of epidemiology in the elucidation of mycotoxins health risks. In: Rodricks, J.V., Hesseltine, C.W., Mehlman, M.A. (Eds.). Mycotoxins in Human and Animal Health. Pathotox Publishers, Park Forest South, IL, USA, pp. 699–711.

Varnava, A. and Mouskos, C. (1997). 7-Year results of hermetic storage of barley under PVC liners: losses and justification for further implementation of this method of grain storage. In: Donahaye, E. J., Navarro, S., and Varnava, A. (Eds) Proceeding of International Conference Controlled Atmosphere and Fumigation in Stored Products, 21-26 April 1996, Printco Ltd., Nicosia, Cyprus, pp. 183-190.

Vedman, B. (2004). Mycotoxins in the animal production chain. In: Proceedings of the 2nd World Mycotoxin Forum, Nordwijk, the Netherlands, February 2004. European Mycotoxin Awareness Network, pp. 275–280.

Villers, P., de Bruin, T. and Navarro, S. (2006). Development and Applications of the Hermetic Storage Technolcgy, In: Lorini, I., Bacaltchuk, B., Beckel, H., Deckers, D., Sundfeld, E., P., d.S.J., Biagi, J.D., Celaro, J.C., Faroni, L.R.D.A., Bortolini, L.d.O.F., Sartori, M.R., Elias, M.C., Guedes, R.N.C., da Fonseca, R.G., Scussel, V.M.(Eds). Proceedings of the Ninth International Working Conference on Stored Product Protection, 15-18 October 2006, Campinas, Brazil, Brazilian Post-harvest Association, Campinas, Brazil, pp. 719-729.

Weinberg, Z.G., Ashbell, G., Hen, Y. and Azrieli, A. (1993). The effect of applying lactic acid bacteria at ensiling on the aerobic stability of silages. Journal of Applied Bacteriology. **75:** 512–518.

www.fao.org/docrep/x5036e/x5036E0p.htm. (2011). Fungal damage in durable foodstuffs with special reference to storage in the tropics as on June, 2011.

Yadgiri, B., Reddy, V., Tupule, P.G., Sri Kanta, S.G. and Gopalan, C. (1970). Aflatoxin and Indian childhood cirrhosis. *American Journal of Clinical Nutrition.* **23:** 94.

Chapter 14

Soil Nematode Biodiversity Aspects

*Virendra Kumar Singh**

Division of Plant Pathology, S.K. University of Agricultural Sciences and Technology, DLRSS, Dhiansar – 181 133, Jammu

Nematodes commonly called as round worms, eel worms or thread worms, are the most numerous multicellular organisms, occurring in diverse habitats, belonging to diverse trophic groups. Nematodes constitute on the largest and diverse groups of metazoans on earth and occupy an important role in the food-web. More than a dozen kinds of nematodes with distinct food habits as bacterial, fungal and algal feeders, parasites of animals and plants and predators are found in large numbers in polyspecific communities in wide variety of substrates and habitats all over the earth. Their biodiversity in terms of numbers and species composition vary greatly in relation to various abiotic and biotic factors. Of the estimated over 100,000 species of nematodes about 80 per cent are not yet known to science, Nematodes parasitize and cause diseases in humans, cattle, birds, fish and in most of the economically important plant and animal species. Now a days the term biodiversity has been a political issue because the aims of biodiversity research are to provide information for the conservation and long-term social and economic utilization of the diversity of life.

However precise meaning of the word biodiversity has evolved considerably over two decades. Originally, the term was restricted to a marriage of the study of ecological diversity and the new discipline of genetic diversity. Most biodiversity

---

* E-mail: virendra_singh16@yahoo.com

researchers found this to be far too restrictive. Later a number of authors in a variety of discipline in their zeal of finding a powerful new "buzzword" have defined the term in various ways. Harper and Hawks worth, 1994 defined biodiversity more restrictively as an expression of the variety of living things, at genetic, species and ecosystem levels. Enumeration of the number of species or genotypes in a biological community is a simple measure of biodiversity. However, the relative abundance of the different components within a community and the interaction that occur between the components are also important attributes of biodiversity. They did not include evolutionary phase in biodiversity.

Terms, in the convention on biological diversity, defines, biodiversity as "The variability among living organisms from all sources, including inter alia, terrestrial marine and other aquatic, ecosystems and the ecological complexes of which they are part; this includes diversity within species, between species and of ecosystems". This definition represents three aspects of diversity, namely ecosystems, species, and genes. Hence, Biodiversity essentially includes study of taxonomy and classification, biogeography, ecology, genetics population biology, and evolutionary biology. Nematodes constitute one of the most abundant and ubiquitous, simple looking, yet most diverse groups of animals both morphologically and ecologically. Four of every five metazoans is a nematode. Commonly known as roundworms, eelworms or threadworms, they are typically small vermiform animals, generally colorless, metanerically non-segmented, and lacking a true body cavity and appendages, They represent the evolutionary beginning of complex animal body systems. With an estimated 1 million species, only insects rival nematodes in biodiversity. Nathan Augustus Cobb, the father of the science of nematology aptly wrote in 1913, " if all the matter in the universe except the nematodes were swept away, our world would still be dimly recognizable, and if, as disembodied spirits, we could then investigate it, we should find its mountains, hills, valleys, rivers, lakes and oceans represented by a film of nematodes."

The science of nematology in the world as a whole and particularly in India, is rather young compared to the other sister plant protection disciplines. The plant parasitic nematodes were not even reported from India in modern literature until the beginning of the twentieth century. There are several references in Rig-veda and Yajurveda (about 3000 BC) to 'Krimi", their injury to plants and animals and methods of their control by several ways including herbs and solar heat etc. In Sanskrit 'Krimi' means 'worm', which most probably could be nematodes. The ancient Chinese medical literature mentioned and guineaworms, serious parasites of man. In European, the ear-cockle nematodes were recorded in 1743 and the importance of cyst and root-knot nematodes as the cause of soil sickness and unthrifty crops had been realized in the post-half of the nineteenth century. Barber (1901) reported root-knot nematodes in India but organized nematological research in Indian began only in 1966. Ever since the ancient Chinese and Indian records of nematodes as parasites of man, nematodes have been abhorred as harmful animals causing diseases and debilitation of all kinds of animals and plants. Entomopathogenic nematodes that feed on insects play an important role in the management of important insect pests of many agricultural and horticultural crops. They are environmentally safe and have a good market value.

Types of Nematodes

Nematode biodiversity includes several types of nematodes in earth and sea. They can be found in almost all environments from the polar regions to deserts and hot springs, and from high mountain peaks to the deep ocean sediments. They vary in their size and food habits. Scientists in agriculture, veterinary, fisheries, health and marine biology have paid attention to nematodes feeding on bacteria, fungi and algae, predatory nematodes and parasites of insects,mollusks,fish,birds,cattle,man and plants of all kinds. Mostly nematodes are thread like, but females of some species attain different shapes. In size nematodes vary from 0.2 mm to 8 metre in length, but their diameter is much smaller (1/10 to 1/500 of length). Soil and plant parasitic nematodes are mostly very small and microscopic, 0.2 to 5 mm long and 0.01 to 0.05 mm diameter. The various species of true plant parasites may be migratory or sedentary, ectoparasites, semiendoparasites or end parasites generally of root and other underground plant parts or of aboveground leaf, stem and flower of plants.

Identification

Traditionally only presumed plant pathogenic nematodes in soil samples have been identified to species and attempted identification of total nematode faunae. Parasitic nematodes have to be accurately identified for designing precise and effective management schedules. The beneficial nematodes also need to be accurately identified for their efficient deployment. The Phylum Nematoda is classified into two main classes, Adenophorea and Secernentea. Each Class is divided into 2-3 sub-classes and about 10 Orders (Maggenti, 1983). The orders have super families, families, genera and species.

Identification is mainly based on morphology, habitats and food habits. Stereoscopic binocular microscope with about 40-100X magnification and transmitted visible light is usually sufficient to identify most nematodes upto generic level. High resolution compound research microscope with oil immersion 100X objective and 6–15X eyepieces and transmitted visible light with appropriate filters is required to identify most nematodes, especially the ones found in soil, upto species level. In cases of closely resembling species the scanning electron microscopy is helpful. The biochemical differentiation of very closely related species, such as by banding patterns of proteins and enzymes of nematodes have been found useful in differentiation and correct identification. Various molecular markers like random amplified polymorphic DNA (RAPDs), PCR-RFLP of internal transcribed spacers of mitochondrial or ribosomal DNA, amplified fragment length polymorphism (AFLP) micro satellites etc. have been used for rapid identification and also for resolving taxonomic ambiguities and determining the phylogenetic relationship of various groups of nematodes (Vrain *et al.*, 1992, Hugall *et al.*, 1999, Gaur *et al.*, 1996, Sabir *et al.*, 2001). Analysis of DNA has also been used for understanding the phylogenetic relationships amongst the various taxa of nematodes. A new classification scheme has been proposed by De lye and Baxter (2002) emphasizing monophyletic taxa based on phylogenetic analyses. This system is transition state and is yet to become popular amongst nematologists.

	A-Enoplia	Enoplida	
		Isolaimida	
1-Adenophorea	Mononchida	Predators	
		Dorylaimida	Omnivores
		Stichosomida	
	B-Chromadoria	Chromadrorida	
		Monhysterida	Predators
		Araeolaimida	Predators
		Desmodorida	
		Demoscolecida	
		Rhabditida	Bacterial Feeders
		Strongylida	Animal Parasites
	A- Rhabditia	Ascaridida	Animal Parasites
		Drilonematida	
		Camallanida	
2-Secernentea	**B**-Spiruria	Spirurida	
		Filaroidea	Animal parasites
	C-Diplogasteria	Diplogasterida	Bacterial feeders
		Tylenchida	Fangal, algal and plant feeders
		Aphelenchida	Fungal and plant feeders

Figure 14.1: Classification of Nematodes (after Maggenti, 1983)

Diversity of Nematodes and Nematode Faunae

In the animal Kingdom, nematodes are placed between the platy helminthes and annelida. Much of the Nematode biodiversity is marine (50 per cent), while 25 per cent are free living, 15 per cent are parasites of vertebrates and invertebrates, and only 10 per cent are phytoparasites. Nematodes are generally identified on the basis of their morphology, the various groups also have distinct have distinct feeding behaviors, consuming particular kinds of food, determined by the kind of feeding apparatus, oesophagus and other features. Since the abundance of different food types is related to the ecological conditions in the microcosms, the relative abundance of different functional groups or taxa of nematodes provides bioindicators to assess the soil, water and other habitats. The most recent review of nematodes feeding groups (Yeasts *et al.*, 1993) accepted the following groups:

1. Plant feeding
2. Fungal (hyphal) feeding
3. Bacterial feeding
4. Substrate ingestors
5. Predators on protozoa, nematodes, rotifers, encheitreids etc.

6. Algal feeding
7. Dispersal stages parasites of vertebrate and invertebrate animals.
8. Omnivores.

Habitats

Nematodes are basically aquatic animals that adjust naturally to a variety of terrestrial habits provided a thin film of water is present. They are found in abundance in soil right from the sea shores to agricultural lands, grasslands, plantations, hills and mountains. They can thrive well in various moisture conditions from frozen tundra and Antarctica to the hottest deserts or osmotic conditions sea water to fresh water. Their food ranges from bacteria and diatoms to the most evolved of plants and animals. They also subsist quite well in the aqueous organic solutions that exist inside plants and animals. The species composition, population densities and proportionate dominance vary greatly in relation to the food and other abiotic and biotic factors. An analysis of the nematode community at a given place gives a fair idea of many of the habitat characteristics.

The fresh waters have been inhabited by an infinite number and variety of described and undescribed nematode species. Habitat adaptations to extreme conditions are exemplified by the occurrence of fresh water species in desert waters, and in hot springs. For example, *Aphelenchoides parietinus* was found in hot springs at Yellowstone park at a temperature of 61 °C. The fresh water nematodes impact their environment in several important ways. In the fresh water food chain nematodes serve as a source of nutrition for numerous invertebrates, small vertebrates, and a variety of fungi. The fresh water nematodes are a vital part of the fresh water environment serving as a link in the food chain, as consumers, and as a guiding factor in the quality and purity of drinking water in North America. Agro-ecosystems characteristically undergo periodic disturbances. The practices like tillage, fertilizers, pesticides, crop cycles and monocultures etc. result in decreasing nematode biodiversity. This also results in decrease or elimination of antagonists. In a complex soil ecosystem a nutrient flush is rapidly used by consumers including bacteria, protozoa, nematodes etc. and converted into biomass. Nematodes help in releasing nutrients from bacterial biomass. Different taxa of nematodes within the same or different functional group acquire dominance in a typical succession as the soil physio-chemical conditions change gradually. In a sustainable system a high diversity of soil biota is requited to prevent leaching by uptake of nutrients into biomass and to reduce the population growth of plant parasites and other pathogens.

Role of Nematodes in Nutrient Mineralization

Laboratory experiments and field studies have since demonstrated that those nematodes that feed on bacteria and fungi play important roles in influencing the turnover of the soil microbial biomass and thus in the availability of plant nutrients (Bard gett *et al.*, 1999). It has been estimated that approximately 40 per cent of nutrient mineralization in certain ecosystems is due to nematodes and other soil fauna as they feed on microbial populations (De Ruiter *et al.*, 1993)nematodes play a major role in decomposition and martinet cycling in soil food webs. Recent findings have shown the importance of non-plant parasitic nematodes, which constitute more than 75 per cent of the nemic biodiversity even in agricultural lands. The microbivorous nematodes are now considered important in organic matter decomposition and regulation of soil health (Gaur *et al.*, 2003) or as indicators of soil pollution. Nematodes on nutrient mineralization result directly from the excretion of ingested nutrients not used in tissue production. The effects result directly from modification of the microbial community, inoculation of new substrates with microorganisms, accelerated turnover of microbial cells and leakage into the rhizosphere of nutrients from feeding sites (Bardgett and Griffiths, 1997; Griffiths and Bardgett, 1997; Yeats *et al.*, 1999). Ingham *et al.* (1985) demonstrated that soil microcosms containing Pelodera or Acrobeloides had higher bacterial densities than similar microcosms without nematodes. Plants growing in such microcosms grew faster and initially took up more nitrogen because of increased nitrogen mineralization by bacteria. The fungal feedings nematodes also help in nutrient mineralization in soil. Studies at IARI, New Delhi have shown that the rate of carbon substrate utilization and release of ammonical and nitrate nitrogen were significantly enhanced in presence of *Cephalobus persegnis, Mesodiplogaster,Cranganorensis* and *Panagrolaimus* spp. in the present of wheat straw and green gram stover (Guar *et al.*, 2003). Indian nematologists have generated information on nematode biodiversity in agricultural lands in wide range of agro climates, which would also be relevant to the other countries in Asia.

Nematode Biodiversity as Soil Health Indicator

In the study of the impact of pasture contamination by the heavy metals Cu, Cr and As, on nematode communities. Yeates *et al.* (1994) found that diversity (described using the Shannon-Weaver index) declined with increasing levels of metal contamination. They also described a shift in dominance from plant-feeding nematodes in control uncontaminated soil to bacterial-feeding nematodes in the highly contaminated soil and a general increase in the proportion of predatory nematodes as the soils became increasingly contaminated. Weiss and Larink (1991) found a similar increase in the abundance of predatory nematodes in soil following the addition of sewage sludge and heavy metals to soil.

Nematodes as Model Systems for Nutritional Aspects

Nematodes have been found to be ideal models for studying the role of nutrition on physiology and biochemical events related to development, reproduction, growth and ageing. Inability to synthesize sterols makes nematodes excellent models to study sterol metabolism in eukaryotic organisms (Bolla, 1987). Free-living nematodes can be used as models to determine the metabolic relationship and differences in sterol metabolism between parasitic nematodes and their hosts. This could lead to development of biological control mechanisms, which interfere with the parasite's sterol metabolism without interfering with that of the host.

Nematodes as Indicators of Toxic Environmental Conditions

Free living nematodes have several characteristics with make them outstanding indicator organisms for determination of the presence of toxic contaminants in aquatic, marine, and terrestrial environments. Nematodes have a permeable cuticle, which allows them to respond with a range of reactions to pollutants and correspond with the restorative capacity of soil ecosystems (Saly and Ragala, 1984, Wasilewska, 1974, 1989, Neher, 2001). Some nematodes have resistant cryptobiotic stages or cysts that allow them to survive inactively during adverse environmental conditions. Nematodes have stock proteins that are highly conserved (Hashmi *et al.*, 1997). Expression of these proteins is enhanced when exposed to stresses such as heat, metal ions, or organic toxins (Kammenga *et al.*, 1998). Probably these proteins could serve as biomarkers for ecotoxicological assessment of soils (Kammenga *et al.*, 2000). Since the 1970s, nematodes have been used as environmental biomonitors for aquatic systems. *Panagrellus redivivus* has been used to detect toxin concentrations in environments that affect molting and organisms size through stimulation, inhibition, or lethality, and provides a rapid bioassay that costs less than 10 per cent of a salmonella bioassay. Increase in air-borne lead pollution caused a decrease in the number of species and decrease in larger species like dorylaims, feeding on mosses in the Po river in Italy (Zullini & Peretti, 1986. Polluted irrigation water may also have similar effects. Increase in lead and copper pollution due to exhausts from a metallurgical plant decreased the nematode abundance, diversity and maturity indices were decreased in the affected forest agro-ecosystems (Popvici and Korthals, 1995). Acid rains were reported to have modified the abundance of bacterial feeders vis-à-vis root fungal feeders. Pollutants in irrigation water also affected the nematode biodiversity (Zullini, 1976). The various agrochemicals including pesticides, fertilizers and amendments as well as organic and other wastes like sewage sludge etc. carry unintended heavy metals and other pollutants. Contaminants like arsenic, cadmium, copper, chromium, nickel, lead, zinc, silver etc. often increase and adversely affect soil biodiversity including typical changes in the community structure of nematodes (Wyss and Larink, 1991 and Yeates and Bongers, 1999).

Role of Nematodes in Global Climate Change

Global Warming is likely to influence the distribution of nematodes. A number of simulated and real field experiments have been carried out, mostly in Europe and Japan whose findings clearly show the profound influence climate change can affect nematode communities and decomposition pathways in soil at different altitudes. The Himalayas and coastal Indian provide excellent opportunity to study and nematode fauna in relation to ecological succession and climate change. New problems may increase in newer areas depending upon species host preferences and changing food sources. Boag & Neilson (1996) derived a correlation between temperature and distribution of *Longidorus caespiticola*. The model forecasted extension of the distribution in UK if global warming occurs, suggesting the need for enforcement of phytosanitary regulations to stop the introduction of some nematodes and their associated diseases. There is scope for studying nematode biodiversity in relation to climate change in the Indio-China region given the geographical and climatic diversity.

Nematode Biodiversity Losses

Several cultural practices and adverse climatic conditions can adversely influence the nematode biodiversity. Various crop practices like crop rotation, application of plant protection chemicals, inter cultivation practices are recommended for reducing the harmful nematodes, but the same could be harmful to the beneficial nematodes also, thus resulting in the loss of the biodiversity of the beneficial species. The release of industrial effluents into marine waters, oil spillages, adverse climatic conditions like sunamis, cyclones, tornadoes, earthquakes, and extreme temp conditions can also result in loss or dispersal of biodiversity. Indiscriminate usage of pesticides and fertilizers can result in ground water pollution due to leaching and adversely affect the nematode biodiversity. These can particularly affect the entomopathogenic and free-living nematodes.

Nematode Biodiversity Conservation Approaches

Avoiding excessive damage to biodiversity while applying agricultural and other practices affecting the habitat of nematodes is important. In-situ and ex-situ conservation methods will also be required to conserve selected species. Safer management of nematode pests and other soil-borne pests and pathogens is essential for maintaining production. A variety of methods have been evolved to treat the soil and crops to kill them or reduce their numbers and invasion. Many of these practices have broad spectrum and besides reducing the target species these also tend to eliminate or suppress the non-target fauna and flora. The predatory, saprozoic microbivorous and fungivorous nematodes are also killed when toxic pesticides or other lethal treatments are applied. Nematicides, could be harmful to nematode biodiversity. However, at least in the developing countries including India these are not available or are too costly. Most of the halogenated hydrocarbon and other earlier nematicides have been banned due to concerns about environmental and health hazards. Use of nematicidal pesticides is very limited. Some organo-phosphates and carbamates are applied for the control of insect pests. Rarely, these are also used against nematodes. Small doses of pesticides, such as carbofuran, carbosulfan, triazophos etc. are advised for nursery-beds for transplanted crops and seed treatment of direct seeded crops. Ecologically safer methods of nematode management are advocated in India (Gaur, 1995). These include modified tillage practices, organic manuring, nematode suppressive cropping systems, biological control, physical cleaning and sanitation, hot water treatments, soil solarizations, modified water and nutrient management and minimized and targeted pesticide use. Since these practices individually may not give desirable level of control, attempts are made to develop judiciously planned combinations of more than one practices, *i.e.*, integrated nematode management packages suited to specific crop, pest and locations. Such packages help in keeping the pest population densities below the economic injury threshold.

Conclusion

Soil nematode biodiversity is high but typically only six functional groups are recognized. Biodiversity will give before hand information about authentically identified nematode fauna population of both noxious and useful in a given area. It has predictive value for estimating of crop loss in given area and time, for specific crop by nematodes. The information will be helpful to take decision for most suitable method in managing nematode in specific crop with least damage to environment. Assessment of biodiversity of soil and plant parasitic nematode has the potential to provide useful insight into the health and functioning of soil. It has public appeal and is compatible with the ideals of biodiversity conservation and ecologically sustainable development. In many jurisdictions such assessment will be difficult due to inadequate systematic knowledge of the nematode fauna in all studies appropriate resources will be necessary to permit adequate identification.

References

Barber, C.A. (1901). A tea eelworm disease in south India, Deptt. of Land Records and Agriculture India, Agricultural branch Vol. II Bulletin No. 45, Madras.

Bardgett, D.R. and Griffiths, B.S. (1997). Ecology and Biology of soil protozoa, nematodes and micro arthropods. In: (eds. J.D. Van Elas, J.T. Trevors, and E.M.H. Wellington) Modern Soil Microbiology. Marcel Decker, New York, pp. 129-163.

Bolla, R.J. (1987). Nematodes as model systems for nutritional studies. In: (eds. Veech, J.A. and Dickson, D.W.) Vistas on Nematology.pp 424-432 SON, USA.

Gaur, H.S. (1995).Some ecological considerations in Integrated nematode management. In: Nematode Pest Management–An appraisal of Ecofriendly Approaches, G. Swarup, D.R. Dasgupta & J.S. Gill (Eds) Nematol. Soc. India, New Delhi. pp 26-28.

Gaur, H.S., Anju, Kamra, Sheela, M.S. and Kaul, R.K. (2003). Status of Nematode Management in IPM, In Recent Advances in Integrated Pest Management (eds. A. Singh, T.P. Trivedi, H.R. Sardana, O.P. Sharma & N. Sabir). (Proceedings and Recommendations of the NATP Interactive Workshop on Integrated Pest Management, Feb 26-28, 2003) pp.139-149.

Gaur, H.S., Mende, Nicola von and Perry, R.N. (1996). Differentiation of two groups of species of the genus *Meloidogyne* by polymerase chain reaction and restriction fragment length polymorphism of ribosomal DNA.Afro-Asian Journal of Nematology, **6:** 50-54.

Griffiths, B.S. and Badgelt, R.D. (1997). Interactions between microbe-feeding invertebrates and soil micro-organisms. In: (eds. J.D. Van Elas, J.T. Trevors, and E.M.H. Wellington) Modern Soil Microbiology. Marcel Dekker, New York pp. 165-182.

Hashmi, G.S., Hashmi, S., Grewal, P.S. and Gaugler, R. (1997). Polymorphism in heat shock protein gene (hsp 70) in entomopathogenic nematodes (Rhabditida). Journal of Thermal Biology **22:** 143-149.

Hugall, A., Stanton, J. and Mortiz, C. (1999). Reticulate evolution and the origins of ribosomal internal transcribed spacer diversity in apomictic Meloidogyne Molecular Biology and Evolution **16:** 157-164.

Ingham, R.E., Trofymow, J.A., Ingham, E.R. and Coleman, D.C. (1985). Interactions of bacteria, fungi and their nematode grazers effect on nutrient cycling and plant growth Ecological Monographs **55:** 119-140.

Kammenga, J.E., Arts, M.S.J. and Oude-Breuil, W.J.M. (1998). HSP60 as a potential biomarker for toxic stress in the nematode Plectus acuminatus. Archives of Environmental Contamination and Toxicology **34:** 253-258.

Kammenga, J.E. Dallinger, R., Donker, M.H. Kohler, H.R. Sinonnen, V. Triebskorn, R. and Weeks, J.M. (2000). Biomarkers in terrestrial invertebrates for ecotoxicological soil risk assessment. Review of Environmental Contamination and Toxicology **54:**253-258.

Maggenti, A.R. (1983). Nematode higher classification as influenced by species and family concepts. In: Concepts, in Nematode systematic. pp. 25-40. In: A.R. Stone, H.M. platt and L.F. Khalil). The systematic association special Vol. **22:** Academic Press. London.

Neher, D.A. (2001). Nematode communities as ecological indicators of agroecosytem. In: (eds.S.R. Gliessman) Agro-ecosystem Sustainability. Developing Practical Strategies pp. 105-120. Boca Raton. FL: CRC Lewis Press.

Popovici, J. and Korthals, G. (1995). Soil nematodes used in the detection of habitat disturbance due to industrial pollution. Stud. Univ. Babes Bolyai Biol (1993) **30:** 37-41.

Sabir, N.,Gaur, H.S. and Naved, S.H. (2001) Molecular methods in nematode phulogenetics In (eds. M.S. Jairajpuri and P.F. Rahman) Nematode Taxonomy. pp.69-91. Maulana Azad National Urdu University, Hyderabad.

Saly, A. and Ragala, P. (1984). Free-living Namatode-bioindicators of the effects of chemization on the soil fauna. Sborrik Uvtiz Oechrana Rostlin **20:** 15-21.

Vrain, T.C., Wakachuk, D.A., Levesque, A.C. and Hamilton, R.I. (1992). Intraspecific rDNA restriction fragment length polymorphism in the *Xiphinema americanum* group. Fundamental and Applied Nematology **15:** 563-573.

Wasilewista, L. (1974).The structure and function of soil nematode communities in natural ecosystems and agrocenoses. Polish Ecological Studies **5:** 97-145.

Wasilewska, L. (1989). Impact of human activities nematodes. In: (eds. C. Clarbolm, and B. L. Dordrecht) Ecology of Arable Land, pp. 123-132. The Netherlands Kluwer Academic.

Yeates, G.W. and Bongers, T. (1999). Nematode biodiversity in agroecosystems. Agriculture, Ecosystems and Environment. **74:** 113-135.

Yeates, G.W., Bongers, T., de Goeda, R.G.M., Freckman, D.W. and Georgieva, S.S. (1993). Feeding habits in nematode families and genera-an outline for soil ecologists. Journal Nematology **25:** 315-331.

Yeates., G.W., Saggar, S., Hedley, C.B. and Mercer,C.F. (1999). Increase in 14C-Carbon translocation to the soil microbial biomass when five plant parasitic nematodes intact roots of white clover. Nematology **1:** 295-300.

Zullini, A. (1976). Nematodes as indicators of river pollution. Nematologia Mediterranean, **4:** 13-22.

Zullini, A. and Peretti, E. (1986). Lead pollution and moss inhabiting nematodes of an industrial area Water, Air and Soil Pollution **27:** 403-410.

Chapter-15

Catenaria anguillulae Sorokin as a Biological Control Agent of Nematodes

S.S. Vaish

Department of Mycology and Plant Pathology, Institute of Agricultural Sciences, Banaras Hindu University, Varanasi – 221 005, U.P.

Use of agrochemicals for increasing food production has become the characteristic feature of the modern agriculture worldwide. However, excessive use of agrochemicals and fertilizers cause not only various types of pollution problems by entering into food chain but also exert deleterious impact on natural biological equilibrium by eroding flora and fauna which are essential component of an ecosystem. The changes in natural biological equilibrium result in new pathogen problems (Duddington, 1957; Chen *et al.*, 2003; Hu and Cao, 2008; Pandey, 2011). Several such examples of nematode diseases *viz.*, golden nematode of potato (*Globodera rostochinensis*), cereal cyst nematode (*Heterodera avenae*), root knot disease of rice, wheat and barley (*Meloidogyne graminicola*) threatening our agriculture are experienced by farmers and agricultural scientists worldwide. Many diseases known to be less important have become more important with the adoption of new agricultural practices with intensive use of agrochemicals over a long period. The modern agriculture has neglected use of organic manures which maintain not only soil biodiversity but also supplement variety of nutrients including major and minor elements required for good plant

* E-mail: shyam_saran@rediffmail.com

growth. In recent years, therefore greater emphasis is being given on biological control of plant pathogens including plant parasitic nematodes duly integrated with other disease management approaches. Among the nematophagous fungi, *Catenaria anguillulae* is the first zoosporic nematophagous fungus that has received good attention of a large number of workers after it was first reported by Sorokin in 1876 that caused a disease in epidemic form in a nematode population eventually killing most of them.

The pathogenic nature of *C. anguillulae* to nematodes was later confirmed by several workers (Couch, 1945; Sterling and Platzer, 1978; Jaffee and Shaffer, 1987; Singh and Gupta, 1986). However, a few workers considered it to be a weak parasite (Boosalis and Mankau, 1965; Sayre and Keeley, 1969). Boosalis and Mankau (1965) reported for the first time that Dorylaimids are more susceptible to *C. anguillulae* than Tylenchids. Similarly, some saprophytic nematodes are reported to be highly susceptible to *Catenaria* infection (Sayre and Keeley, 1969; Singh *et al.*, 1996). It is established that in general, motile stages of Tylenchids are tolerant to *Catenaria* infection, however, it is highly pathogenic to young females of *Heterodera cajani*, *H. sorghi* and *Meloidogyne javanica in-vitro* and also to the different stages of *H. sorghi* and *M. graminicola* in severely infested fields under natural condition (Singh and Gupta, 1986; Singh *et al.*, 1996; Singh *et al.*, 2007). While working with a virulent isolate of *C. anguillulae*, Singh *et al.* (1996) reported that the fungus grew in the living adults of *Xiphinema basiri* and *Seinura* sp. as well as males of *M. javanica* and *H. cajani*, causing paralysis of the nematode body following death. This fungus revealed restricted growth endobiotically in nematodes belonging to the Tylenchida without causing death for few to several days indicating balanced parasitism. A little exposure of these nematodes to moisture stress resulted in their death and development of sporangia following addition of few drops of water, which clearly indicated the true virulent nature of the fungus (Singh *et al.*, 1996). Needham (1743) reported *Anguina tritici* (Steinbuch) Filipjev from England as the first plant parasitic nematode causing ear-cockle disease of wheat. This nematode is found to be severely infected and killed by *C. anguillulae*. The motile second stage juveniles (J_2s) of *A. tritici* are readily colonized predominantly at the excretory pore by zoospores of *C. anguillulae* which infects live nematodes and produces sporangia within the nematode body resulting into their death (Singh *et al.*, 2008). Since *A. tritici* belongs to the Tylenchida, and the nematodes belonging to this group are known to be very tolerant to *Catenaria* infection, the observations on parasitism of motile J_2s of *A. tritici* contradict the generalization made by Boosalis and Mankau in 1965.

Characteristics of *Catenaria anguillulae*

Catenaria anguillulae belongs to the family Catenariaceae under Blastocladiales of Chytridiomycota (Couch, 1945). The growing thallus consists of branched and unbranched, aseptate or septate hyphae (5-13 um wide) forming sporangia in chains connected with usually two cell isthmuses (Figure 15.1). Because of catenate sporangia, the fungus was designated as *Catenaria* by Sorokin (1876). Sporangia are spherical, elliptical, oval, triangular, pyriform or sub-pyriform in shape. The size of sporangia depends on substrate, nematode, culture media and growing condition.

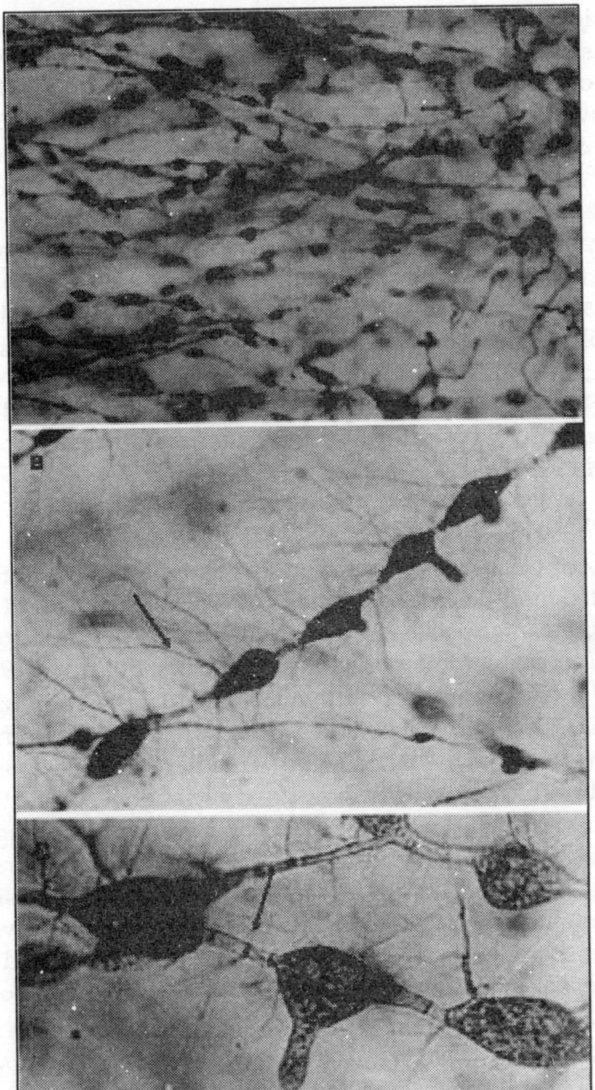

Figure 15.1: Showing Characteristic Features of *Catenaria anguillulae* in a Culture Medium
A = Sporangia in chains; B = rhizoids; C = two celled isthmuses.
Mag. 80X (A), 200X (B), 500X (C)

However, the sporangial size commonly varies between 15-70 × 12-36 um in nematodes. Isthmuses are hyaline two celled 6-34 um long. *C. anguillulae* is also characterized by the presence of rhizoids on developing sporangia and hyphae growing in culture media and within nematode body. This fungus produces pyriform or pear shaped zoospores (measuring 3.2-5.6 × 5.6-8.0um in size) having single whiplash posterior flagellum (up to 26 um). These zoospores colonize mostly at the natural openings of nematodes namely mouth, excretory pore, vulva or anus. After colonization, zoospores germinate by putting forth a germ tube that grows into a septate hypha within nematode body endobiotically (Figure 15.2). After septa formation, each cell develops into a sporangium which on maturation produces a large number of zoospores. The zoospores are liberated through a discharge tube. Just before release of zoospores, the shining tip of a discharge tube gelatinizes facilitating zoospore liberation usually singly. Occasionally, a small vesicle is also formed in which few zoospores accumulate momentarily and escape into water film showing very fast movement. The fungus is a facultative parasite having ability to grow as a saprophyte in soil on plant debris as well as on nematodes and other small animals as a parasite (Vaish and Singh, 2000; Singh *et al.*, 2002; Gupta *et al.*, 2005; Gleason *et al.*, 2010). The fungus completes its life cycle in nematode body *i.e.*, from zoospore colonization at the natural openings to liberation of zoospore within 24 hours (Singh *et al.*, 1993).

Variability of *C. anguillulae*

Morphology of *C. anguillulae* particularly of sporangia is highly variable. In view of the reported morphological variations, Couch (1945) while studying taxonomy and life cycle of *Catenaria* species concluded that the earlier descriptions under the name *C. anguillulae* represented more than one species. Sparrow (1960) and Karling (1977) gave similar comments on the fungus owing to its morphological variations. Size of sporangia, discharge-tubes and isthmuses of the different isolates of *C. anguillulae* grown on culture media and oilcake media (0.5 per cent) as well as growth of the fungus in different nematodes show great variation (Couch, 1945; Vaish, 1999; Gupta, 2002; Gupta *et al.*, 2004, Gupta *et al.*, 2005). In general, sporangia of larger size usually long, flask shaped with very long discharge–tube are formed on oilcake media particularly on the oilcake aggregates/pellicles. In general, a single discharge-tube from a sporangium is a characteristic feature of *Catenaria*, however 2-3 discharge tubes from a larger sporangium is quite common on some oil-cake media. The variation in sporangial size and shape is also related with different species of nematodes as well as their size.

Morphology of sporangia of *C. anguillulae* also vary greatly in response to salt concentration and temperature variations (Gupta, 2002; Gupta and Singh, 2002; Gupta *et al.*, 2004). Isolates loving high temperature producing tubular/filamentous sporangia at moderate temperature produce almost spherical sporangia at 40°C. Similarly, the isolates producing filamentous sporangia produce well developed elliptical/ spherical sporangia in response to 0.5 per cent salt concentration. The role of temperature has been critically examined and reported for isolates of the fungus, exhibited growth at a range of 13 to 44°C (Gupta *et al.*, 2004). This fungus is also able to grow at a wide range pH ranging from 5-9 with maximum growth at pH 7. The variation in temperature requirement clearly specifies reasons as why population of *C. anguillulae* is found during all months of a year.

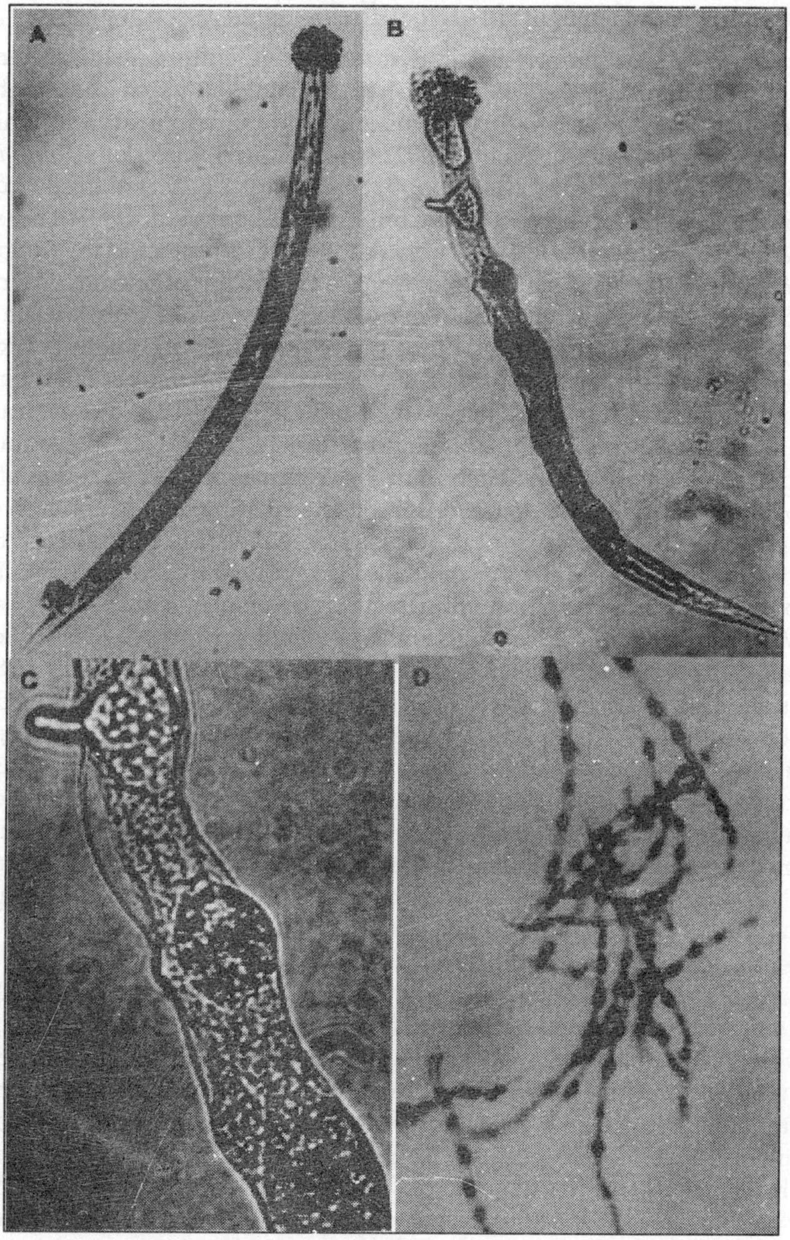

Figure 15.2: Showing Zoospore Colonization and Endobiotic Growth of *Catenaria anguillulae*

A = Zoospore colonization at mouth and anal region of second stage juvenile of *Heterodera cajani*, B = Endobiotic growth of *C. anguillulae* depicting chain of sporangia in a second stage juvenile of *Heterodera cajani*, C = Larger view of sporangia within body of *Heterodera cajani*, D= endobiotic growth of *C. anguillulae* within second stage juveniles of *Anguina tritici*. Mag. 160X (A), 260X (B), 800X(C), 85X (D)

Distribution of *Catenaria anguillulae*

C. anguillulae is a widely distributed facultative endoparasite of nematodes (Barron, 1977; Persmark *et al.*, 1995; Vaish and Mir, 2002; Vaish and Singh, 2002). Persmark *et al.* (1995) reported this fungus to be wide spread in agricultural soils of Central America. Vaish and Singh (2002) studied distribution of *C. anguillulae* in Indian soils in detail. The examination of 490 soil samples collected from 39 different locations revealed the occurrence of the fungus in 451 soil samples. All the samples collected from the alluvial soil and tarai soil showed the presence of the fungus while three soil samples out of 25 soil samples from red soil, 9 out of 20 from desert soil, 11 out of 25 from saline soil, 5 out of 15 from peat soil, one out of 5 from laterite soil and 2 out of 10 soil samples from mixed black and red soil did not contain *C. anguillulae*. From the intensive survey on the distribution in soils of 12 states of India *viz.*, Delhi, Haryana, Jammu and Kashmir, Jharkhand, Madhya Pradesh, Maharashtra, Orissa, Punjab, Rajasthan, Tamil Nadu, Uttaranchal and Uttar Pradesh, it was found that the fungus is widely distributed in Indian soils. The recovery of *C. anguillulae* to the extent of 92 per cent from the soil samples collected during different month of a year also clearly revealed that the fungus remains in active stage through out the year. This indicates that the fungus is the integral component of soil biodiversity and it might play definite role in maintaining the population of nematode below a certain level. Wide distribution of this fungus is also evident from growth over a wide range of pH and temperatures (Stirling and Platzer, 1978; VoB and Wyss, 1990; Stephen, 1992; Gupta *et al.*, 2004); infection of wide range of nematodes with varying degree of virulence (Sayre and Keeley, 1969; Singh *et al.*, 1996, Singh *et al.*, 1998); natural parasitism of nematodes (Barooti *et al.*, 1985; Gupta and Singh 2002; Singh *et al.*, 2007); colonization of dead roots (Karling, 1934; Stephen, 1992) and its facultative endoparasitic nature. These attributes clearly reflect that *C. anguillulae* is an integral component of the soil bio-diversity, having a definite role in maintaining the population of nematodes going beyond a certain level as evidenced by a steep decline in population of *Heterodera sorghi* in a sorghum sick plot after a severe out break of this fungal disease (Singh and Gupta, 1986; Singh *et al.*, 2007). Under natural conditions, as high as 60 per cent infection of *Hemicriconemoides* by *C. anguillulae* was recorded in samples collected from Bearmann's funnel in April 1984 from soils around the citrus roots. We have evidence that *C. anguillulae* sometimes appear in epidemic form on soil nematodes. During 2004, three field soils were extracted for nematodes 2 days after the summer rain in the month of May. The nematode population extracted from one of the fields incorporated with cow dung manure, showed 100 per cent motile healthy nematodes after 48 h of soil processing. Soil suspensions observed after 1 day showed that 8-10 per cent of the nematodes were parasitized by *C. anguillulae* while after 2 days 25 per cent of the nematodes were parasitized. After 4 days of incubation, more than 60 per cent of the population was parasitized by *C. anguillulae*. In other fields, the parasitism of nematodes was between 5 per cent and 10 per cent after 4 days. This clearly indicates that *C. anguillulae* regulates the population of nematodes in nature (Singh *et al.*, 2007).

Isolation, Purification, Quantification and Rapid Virulence Testing of *C. anguillulae*

Techniques for selective isolation, semi-quantification and rapid virulence testing of C. anguillulae were described by Singh *et al.* (1998). Since some workers have been unable to isolate *C. anguillulae* from soil (*e.g.*, Couch, 1945) despite its wide distribution of this fungus, selective isolation technique was developed using second stage juveniles (J_2s) of *A. tritici* for easy and quick isolation of virulent isolates in almost pure population from a large number of soil samples for studies on biological control of nematodes. Different steps of selective isolations are described below:

1. Collect 500 g soil samples from the top 15 cm of the soil profile from different locations in separate polyethylene bags.
2. Homogenize soil and transfer 50g soil into a 250 ml conical flask containing 100 ml sterile water.
3. Shake the conical flask gently to get soil suspension.
4. Thereafter, filter the soil suspensions through muslin cloth to separate the larger soil particles.
5. Subsequently, filter again the soil suspensions overnight through tissue paper supported on a wire mesh over 9 cm Petri-dishes to get suspensions free from course particulate matter.
6. Finally make the collected suspension of each soil sample up to 100 ml.
7. From each processed soil sample's suspension, transfer one ml soil suspension into a 50 mm Petri-dish/cavity block and dilute by adding 2 ml sterilized distilled water.
8. Prepare nematode suspension of J_2s of *A. tritici* by teasing a single water soaked gall after its surface sterilization with 0.1 per cent $HgCl_2$ followed by five washing with sterile distilled water in a sterilized 50 mm cavity block containing 1 ml of sterilized distilled water.
9. Finally add two drops of the thoroughly washed nematode suspension (approx. 200 individuals) as bait for *C. anguillulae* into each Petri-dish/ cavity block.
10. One ml sterilized distilled water containing 2 ppm streptomycin sulphate should also be added to each Petri-dish/cavity block to reduce the bacterial contamination.
11. Incubate the Petri-dishes/cavity blocks at room temperature (25-31°C) for 48 h.
12. On observation, *C. anguillulae* can be seen in some of the J_2s of *A. tritici* with developing or fully developed sporangia in chains connected by isthmuses.
13. Transfer a single infected juvenile of *A. tritici* into each of several cavity blocks containing sterilized water to get an almost pure population of an isolate for further studies.

Couch (1945), while studying the taxonomy and life of *Catenaria* spp., concluded that the earlier descriptions under the name of *C. anguillulae* Sorokin represented more than one species. Sparrow (1960) and Karling (1977) gave similar comments on this fungus owing to its morphological variation. Since these workers probably did not isolate the fungus from zoospores of a single sporangium, morphological variation may be due to mixed infection in the nematode. Therefore, a technique was developed to isolate the fungus from zoospores of a single sporangium in artificial medium based on non-synchronous release of zoospores from sporangium of *C. anguillulae* (Singh, 1989). The different steps of this technique are given below:

1. Collect the second stage larvae of *Heterodera cajani* and *Meloidogyne javanica* (any other small nematode).
2. Wash the collected larvae 4-5 times in cool sterile water and kill in sterilized hot water then decant water and replace with one ml sterile water.
3. Transfer one *C. anguillulae* infected larva to the cavity block and incubate for 18 h at 27°C to allow the development of the fungus.
4. Transfer a single *Catenaria* infected larva of *Heterodera cajani* or *Meloidogyne javanica* on to the cavity slide containing 2-3 drops of sterile distilled water.
5. Take out the infected larva after the release of zoospores from a single sporangium and replace by a single heat killed larva.
6. Incubate the cavity slide for 18 h at 27-30°C to allow the development of the fungus.
7. Larvae thus infected can be used for multiplication of the fungus for the subsequent uses.

The present technique is better than earlier methods, as the larvae of *Heterodera* spp. or *Meloidogyne* spp. are relatively smaller. Usually 7-10 sporangia are formed compared with the larger nematodes (>50), and thus the collection of zoospores from a single sporangium becomes easier. Also, it is difficult to isolate a single hypha from cultures of *C. anguillulae* on the medium preferred by Couch (1945) as the mycelial growth is often very compact.

Semi-quantification method was developed for assessing the population of the fungus using J_2s of *A. tritici* as bait in relation to various organic additives or agrochemicals used in soil for raising crops. The total numbers of infected juveniles are counted after 48 h of baiting and percentage infection calculated as follows:

$$\text{Percentage infection} = \frac{\text{Total number of infected juveniles}}{\text{Total number of inoculated juveniles}} \times 100$$

Rapid virulence technique was developed to screen a large number of isolates relatively quickly which are specific to certain species of nematodes in which there is no need to culture *C. anguillulae* using infected J_2s of *H. cajani, M. incognita* and *A. tritici* as standard inocula in order to avoid any possibility of loss of virulence of the pathogen *in vitro* pathogenicity test. The different steps are given below:

1. Prepare nematode suspension of J_2s of *A. tritici* by teasing a single water soaked gall after its surface sterilization with 0.1 per cent $HgCl_2$ followed by five washing with sterile distilled water in a sterilized 50 mm cavity block containing 1 ml of sterilized distilled water.
2. Collect the second stage larvae of *Heterodera cajani* and *Meloidogyne incognita* by crushing cysts or from egg masses hatched after incubation for 5-6 days at 27±2°C.
3. The test nematode species can be isolated from roots or soil round roots of different plants as described by Southey (1970).
4. Collect the test nematode species separately in cavity blocks and surface sterilize with 0.2 per cent sodium hypochlorite and wash three times with sterilized water.
5. Inoculate the cavity blocks containing 50 nematodes of a nematode species with 10 juveniles (*H. cajani* and *M. incognita*) and six juveniles of *A. tritici* with well developed *C. anguillulae* thalli before emergence of zoospores.
6. Take four cavity blocks as replicate for each inoculum type of a nematode species.
7. Observe the inoculated nematodes daily for 7 days and record the mortality.

However, isolation, purification and pathogenicity testing by using different levels of zoospores (Sayre and Keeley, 1969),infested poppy seeds (Jaffee and Shaffer, 1987) or agar disc containing the fungus (VoB and Wyss, 1990) as inocula involves a lot of time and inadequate as standard inocula to get precise results. It may be seen that zoospore's numbers even if taken may not give a precise inoculum because the motility of the zoospores is itself highly variable. For rapid virulence testing, there is no need to culture the fungus. The entire process of virulence testing can be completed in 12-13 days if data collection is complete within 8 days.

Maintenance of Isolates of *C. anguillulae*

Isolates of *C. anguillulae* can be maintained by transferring 20 J_2s of *A. tritici* infected with *Catenaria* having fully developed sporangia prior to release of zoospores into each of several 5ml screw cap bottles having about 2000 motile J_2s of *A. tritici* in 2.5 ml of sterilized distilled water at 5±1°C (Vaish and Singh, 2011). Culture media

Nutritional Requirements of *C. anguillulae*

Early studies on the nutrition of *C. anguillulae* have clearly shown that this fungus does not utilize inorganic sources of Nitrogen. This fungus requires organic source of nitrogen be it of plant or animal origin for its good growth. Nolan (1970) studied in detail about the nutritional requirement and suggested a medium. Studies by Stephen (1992) indicated good growth of the fungus on several oil-cake media. Vaish (1999) prepared and tested several media based on chemical composition of oil-cakes and brans and brought out a synthetic medium for the growth of *C. anguillulae*. Of the 24 media, initially prepared for the growth of *C. anguillulae* having various compositions, based on chemical constituents of oil-cakes and brans, only

Medium No.18 initiated the growth of the fungus. The initiation was so little that it could not be measured after 21 days. This medium was further modified into three major media by varying concentrations of minerals, of which, Medium No.18C recorded 25mm growth of the fungus in 20 days. When Medium No. 18C was modified with various combinations of amino acids, a combination of only three amino acids namely arginine, L-methionine and leucine supported best growth of *C. anguillulae*, which was 26.4mm on 15th day of inoculation. At the outset of the experiment the amount of dextrose was 12g/litre. Out of the total 8 carbon sources tested equivalent to 10g dextrose, dextrose supported best growth of the fungus (29.5 mm on 14th day of inoculation). Further on dextrose supplemented medium, the growth of the fungus was thick, compact and exhibiting some aerial turf. Studies on different level of dextrose exhibited that a concentration of 5g/litre supported best growth of *C. anguillulae* (33.8 mm on 15th day of inoculation) followed by 10g/litre (Vaish, 1999). These studies clearly showed that *C. anguillulae* utilizes nitrogen in organic form as arginine, L-methionine and leucine supported good growth of the fungus. Among vitamins, choline was found as an essential vitamin along with the other vitamins. The studies on nutritional requirement provide a good base for the selection of substrates for its mass culture.

Screening of Substrates for Mass-Culturing of *C. anguillulae*

Different oil-cakes agar media *viz.*, linseed oil-cake agar, mustard oilcake agar, neem oil-cake agar @ 0.5 per cent when compared with beef extract agar, Emerson agar and Yeast peptone soluble starch (YPSS) agar for their effect on growth of *C. anguillulae* revealed the maximum growth of the fungus on linseed oil-cake agar medium. The fungus showed typical characters of the fungus and clear visibility of morphological details (Stephen, 1992; Vaish, 1999, Gupta *et al.*, 2005).

Sixteen media were tested using the substrates namely bran of wheat, rice, pea, pegionpea, lentil and chickpea; meals of maize, soybean and chickpea; straws of mustard, linseed, wheat and rice; sawdust, farm yard manure and mollases at the concentration of 6 per cent followed by blending brans and straws in a warring blender to get them in powder form and respective media were made by adding requisite amount of the substrates in desired volume of distilled water along with Agar-agar @ 1.5 per cent. The experiment proved wheat bran to be the best substrate supporting excellent growth of the fungus followed by F.Y.M. showing good growth of *C. anguillulae* (Vaish, 1999; Vaish and Singh, 2000). Wheat bran with sand in the ratio of 12:88 along with 0.5 per cent linseed oil- cake saturated with distilled water was found to be the best for mass culture of *C. anguillulae*.

Good growth of the fungus on wheat bran may possibly be due to higher number of amino acids, minerals and vitamins available in the bran. However, other substrates did not support good growth owing to lack of some essential nutrients, presence of some toxic principles and absence of enzymes necessary for break down of complex materials into utilizable form.*e.g.*, lake of leucine, methionine and high conc. of Mg (may be lethal)in rice bran attributed to reduced growth of the fungus.

Bio-control Potential of *C. anguillulae* Against Plant Parasitic Nematodes

From several experiments it is proved that *C. anguillulae* is a virulent pathogen of several nematodes (Sayre and Keeley, 1969; Esser and Ridings, 1973; Stirling and Platzer, 1978; Prasad and Dayal, 1986; VoB and Wyss, 1990; Singh *et al.*, 1996, Vaish, *et al.*, 1997, Singh *et al.*, 2008, Gupta *et al.*, 2004). However the degree of virulence of *C. anguillulae* varies with different species of nematodes. Mortality of *Xiphinema basiri* indicated that all the isolates were pathogenic, however, the percentage mortality varied with the isolates (Singh *et al.*, 1998; Vaish, 1999). The isolate No.1 was most virulent followed by isolate No. 5, 4, 3 and 2. It is experienced that among the plant parasitic nematodes *Xiphinema* is one which shows higher degree of susceptibility to *C. anguillulae*. Therefore, it will be worthwhile that in the first phage of rapid virulence testing, *Xiphinema* may be used for testing virulence of *C. anguillulae* and the highly virulent isolates may be used against other genera of nematodes.

Application of mass-cultured *C. anguillulae* for management of *H. cajani* in pot at the rate of 1, 1.5 and 2g per pot has been found to reduce the number of cysts and 2nd stage juveniles of *H. cajani* significantly (Vaish *et al.*, 1997). It has been noted that isolates differ in their effect in controlling the nematode. Therefore, it is ideal to select a virulent isolate before mass culture and its subsequent use for the biological control of the nematode pathogen. Performance of mass-cultured *C. anguillulae* as biological control agent against plant parasitic nematodes associated with croton roots resulted in reduction in the population of native plant parasitic and saprophytic nematodes (Vaish, 1999). However, reduction in the population of *Xiphinema* was most conspicuous (70 per cent).

Attribute of *Catenaria anguillulae* as a Biological Control Agent

C. anguillulae possesses several traits related to its bio-control potential against plant parasitic nematodes which include its capacity of fast multiplication as the fungus completes its life cycle, *i.e.*, zoospore to zoospore in nematode body, only within 24 hours following colonization by the zoospores at the natural openings, say mouth, excretory pore anal region and vulva and subsequently produces abundant zoospores for parasitizing the nematodes, wide distribution, facultative nature, selective stimulation in its population in response to soil amendments with oilcakes (Vaish, 1999; Singh *et al.*, 2002; Singh *et al.*, 2007), good saprophytic ability and virulence to plant parasitic nematodes. Its parasitic nature established on the fact that it colonizes, infects and develops in living nematodes. Moreover, it is best suited to Indian conditions and other tropical and sub-tropical countries as it grows luxuriantly at temperatures ranging between 20-35°C (Optimum25-30 °C) which prevail in most part of the country for longer duration in a year.

In nature *C. anguillulae* has been found to play a great role in regulating the population of nematodes, which is evident from the field observations on *H. sorghi*, *Hemicriconemoides* and *Meloidogyne graminicola* (Singh and Gupta, 1986; Singh *et al.*, 1996; Singh *et al.*, 2007)

Future Prospects

1. Detail studies on physico-chemical properties of soil such as level of organic matter, soil type, soil pH etc. in relation to population dynamics of *C. anguillulae* need attention to understand its ecology.
2. The cause of susceptibility of nematodes seems to be due to secretion and accumulation of excretory substances. The quality and quantity of such chemicals warrant examination of its chemical nature and also the amount that seems to be highly chemo-tactic attracting zoospores in water. In spite of the motility of nematodes and higher dilution of secreted chemicals in water, the zoospores are able to sense and track the natural openings. Thus, it is important to understand the role of chemicals present in secretion and ability to attract the zoospores of *C. anguillulae.*
3. Interrelationship between native population of *C. anguillulae* and its parasitism in phytonematodes under field conditions preferably in the rhizosphere of various crop plants should be elaborately worked out with respect to different seasons and types of soils etc., for establishing the role of endemic population of this fungus in regulating the population of plant parasitic nematodes.
4. Selective stimulation of *C. anguillulae* and parasitized nematode population following soil amendment with oil-cakes, particularly neem oil-cake commonly used in Indian agriculture will provide a sound base for inclusion of this practice as an important component of integrated nematode management.
5. Efforts for further screening of substrates in order to find a substrate better than wheat bran are needed to be continued. Screening of more substrates for selective stimulation of *C. anguillulae* in agricultural fields should be further extended. The substrates should be compatible with the prevailing agricultural practices.
6. Isolates of *C. anguillulae* showing higher degree of virulence have been found to be fast growing. This should be confirmed using a large number of Isolates and if confirmed so, the fast growing isolates may be selected for effective bio-control purpose.
7. Further refinement of the developed synthetic medium for the growth of the fungus should be undertaken.
8. Further improvement of mass culture technique will be desirable to get more vigorous growth of the fungus in mass culture medium.
9. A great deal of work is required in connection to performance test of more virulent isolates of *C. anguillulae* against locally important nematode pest of agricultural crops and trees with and without organic substrate(s).
10. Work on the aspect of formulations, methods of application, time of application and packaging system with respect to mass cultured *C. anguillulae* will be required prior to its delivery to farmers.

11. Cost/benefit ratio should also be worked out.
12. Last but not the least, demonstrations on performance of mass culture of *C. anguillulae* in farmers' fields will also be of great importance.

Conclusion

Kind and quantum of work on various aspects *C. anguillulae* carried out by several researchers so far clearly reflects a big hope in this fungus to utilize it for the management of plant parasitic nematodes in integration with other disease management approaches. Now, techniques are available on isolation, purification, quantification, maintenance, virulence test and mass culture to carry out work on this fungus against nematodes. *C. anguillulae* is widely distributed in soils and has a good potential as a biological control agent against nematodes, however, still there is a need to carry out work for the collection of meaningful data on the various issues mentioned earlier.

References

Barooti, S.B., Daneshpazuk, B. and Torabi, M. (1985). Two predaceous and parasitic fungi to nematodes in Iran. *Iranian Journal of Plant Pathology* 21: 13-16.

Barron, G.L. (1977). The nematode–destroying fungi, Topics in Mycobiology-1, Canadian Biological Publications Ltd., Guelph, Canada.

Boosalis, M.G. and Mankau, R. (1965). Parasitism and predation of soil microorganisms. In: K.F. Barker & W.C. Snyder (eds), Ecology of soil-borne plant pathogens. California University Press. Berkeley, USA, pp. 374-391.

Chen, L., Li, Q. and Liang, W. (2003). Effect of Agrochemicals on nematode community structure in a soybean field. *Bulletin of Environmental Contamination and Toxicology.* **71:** 755-760.

Couch, J.N. (1945). Observations on the genus *Catenaria. Mycologia* 37: 163-193.

Prasad, Geeta. and Dayal, R. (1986). Studies on nematophagous fungi VIII: two parasites on Amphidorylaimus sp. *Proceedings National Academy of Science India* **56 (B,I):** 45-47.

Duddington, C.L. (1957). The friendly fungi: A new approach to the eelworm problem. Faber and Faber, 24 Russell London. pp.188.

Esser, R.P. and Ridings, W.H. (1973). Pathogenicity of selected nematodes by Catenaria anguillulae. *Proceedings of soil and crop science society of Florida* **33:** 60-64.

Gleason, F.H., Marano, A.V., Johnson, P. and Martin, W.W. (2010). Blastocladian parasites of invertebrates, *Fungal Biology Reviews* **24:** 56-67

Gupta, R.C. (2002). Studies on variability and biocontrol potential of *Catenaria anguillulae*. Ph.D. Thesis, Institute of Agricultural Sciences, Banaras Hindu University, Varanasi, India

Gupta, R.C. and Singh, K.P. (2002). Effect of sodium chloride on biology of *Catenaria anguillulae*. *Mycobiology* **30:** 219-224.

Gupta, R.C., Vaish, S.S., Singh, R., Singh, N.K. and Singh, K.P. (2004). Effect of temperature, pH and sub- culturing on virulence of *Catenaria anguillulae* Sorokin. *Current Nematology* **15:** 101-108.

Gupta, R.C., Vaish, S.S., Singh, R.K., Singh, N.K. and Singh, K.P. (2005). Oilcakes as media for growing *Catenaria anguillulae* Sorokin- a facultative endoparasite of nematodes. *World Journal of Microbiology and Biotechnology* **21:** 1181-1185.

Hu, C. and Cao, Z.P. (2008). Nematode community structure under compost and chemical fertilizer management practice, in the north china plain. *Experimental Agriculture* **44:** 485-496.

Jaffee, B.A. and Shaffer, R.L. (1987). Parasitism of *Xiphinema americanum* and *X. rivesi* by *Catenaria anguillulae* and other zoosporic fungi in soil solution, Baermann funnels or soil. *Nematologica* **33:** 220-231.

Karling, J.S. (1934). A saprophytic species of *Catenaria* isolated from the roots of *Panicum variegatum*. *Mycologia* **26:** 528-543.

Karling, J.S. (1977). Iconogrphia Chytridiomycetearum. Lubrecht & Cramer. Monticello, N.Y. 414 pp.

Needham, T. (1743). A letter concerning certain chalky tubulous concretions called malm; with some microscopical observations on the farina of the red lily, and of worms discovered in smutty corn. *Philosophical Transactions of the Royal Society of London* **42:** 634-641.

Nolan, R.A. (1970). Sulfur source and vitamin requirements of the aquatic phycomycete, *Catenaria anguillulae*. *Mycologia* **62:** 568-577.

Pandey, S.K. (2011). Studies on root knot disease of barley. M. Sc. Ag. Thesis, Institute of Agricultural Sciences, Banaras Hindu University, Varanasi,1-99.

Persmark, L., Mondoza, N.M. and Jansson, H.B. (1995). Nematophagous fungi from agricultural soil of Central America. (Nematropica) **25: 117-124.**

Sayre, R.M. and Keeley, L.S. (1969). Factors influencing *Catenaria anguillulae* infections in a free living and a plant parasitic nematode. *Nematologica* **15:** 492-502.

Singh, K.P. and Gupta, P. (1986). Observations on *Catenaria anguillulae* parasitic on *Heterodera sorghi*. *Advances in Biological Research* **4:** 240-250.

Singh, K.P. (1989). Artificial culture of *Catenaria anguillulae* from monosporangial zoospores. *Mycological Research* **92:** 107.

Singh, K.P., Stephen, R.A. and Makesh Kumar, T. (1993). Development of Catenaria in *Heterodera cajani*. *Mycological Research* **97:** 957-960.

Singh, K.P., Stephen, R.A and Vaish, S.S. (1996). Pathogenicity and development of *Catenaria anguillulae* on some nematodes. *Mycological Research* 100: 1204-1206.

Singh, K.P., Bandyopadhyay, P., Stephen, R.A., Vaish, S.S. and Makesh Kumar, T. (1998). Techniques for selective isolation, semiquantification and rapid virulence testing of *Catenaria anguillulae. Mycological Research* **102:** 658-660.

Singh, K.P., Bandyopadhyay, P., Stephen, R.A., Vaish, S.S., Makesh Kumar, T. and Gupta, R.C. (2002). Growth and population dynamics of *Catenaria anguillulae* in relation to oil cakes. *Indian Phytopathology* **55:** 286-289.

Singh, K.P., Jaiswal, R.K. and Kumar, Niranjan (2007). *Catenaria anguillulae* Sorokin: a natural biocontrol agent of *Meloidogyne graminicola* causing root knot disease of rice (*Oryza sativa* L.) *World Journal of Microbiology & Biotechnology* **23:** 291-294.

Singh, K.P., Kumar, Niranjan., Singh, K.D. and Vaish, S.S. (2008). *Catenaria anguillulae* the most virulent of *Anguina tritici*. In: Zhang W., Bai F., Zhong J.J., Yang S. (Eds.) Abstracts of the 13th International Biotechnology Symposium and Exhibition. Dalian, China

Sorokin, N. (1876). Note Sur les vegetaux parasites des *Anguillulae. Annels des Sciences Naturalles, Botanique* **64:** 62-71.

Southey, N.W. (1970). Laboratory Methods for work with Plant and Soil Nematodes. Ministry of Agriculture, Fisheries and Food, Technical Bulletin No.2. HMSO: London.

Sparrow, F.K.J.R. (1960). Aquatic Phycomycetes. Ann Arbor: The University of Michigan Press.

Stephen, R.A. (1992). Studies on *Catenaria anguillulae* Sorokin, Ph.D. Thesis, Institute of Agricultural Sciences, Banaras Hindu University, Varanasi,1-98.

Stirling, A.M. and Platzer, *E.G.* (1978). *Catenaria anguillulae* in the mermithid nematode *Romanomermis culcivorex. Journal of Invertebrate Pathology* **32:** 348-354.

Vaish, S.S. (1999). Studies on *Catenaria anguillulae* Sorokin as biocontrol agent against some plant parasitic nematodes. Ph. D. Thesis, Banaras Hindu University, Varanasi, India.

Vaish, S.S., Gupta, R.C. and Singh, K.P. (1997). Pathogenicity and performance test of *Catenaria anguillulae* on *Heterodera avenae. Current Nematology* **18:** 1-6.

Vaish, S.S. and Mir, A.A. (2002). *Catenaria anguillulae* Sorokine a facultative endoparasite of nematodes from soils of cold region of Leh (Ladakh).Indian Journal of Plant Pathology **20:** 87-89.

Vaish, S.S. and Singh, K.P. (2000). Growth of *Catenaria anguillulae* on different substrates for its mass culture. *Indian Phytopathology* 53: 269-272.

Vaish, S.S. and Singh, K.P. (2002). Distribution of *Catenaria anguillulae,* a facultative endoparasite of nematodes in soils from different locations of India. *World Journal of Microbiology and Biotechnology* **18:** 65-67.

Vaish, S.S. and Singh, K.P. (2011). Techniques for single zoospore isolation and maintenance of virulence of *Catenaria anguillulae* Sorokine: a ubiquitous fungal endoparasite of nematodes. In: Zhong, J.J., X u, Jian-He., Hua, Q., Bao, J., Wang, Jing-Ye., Wang, Y., Ou, Hong-Yu., Yu, Hui-Lei., Li, Chun-Xiu., Du, Zhi- Qiang., Tan, Gao- Yi., Sun, Jie., Ni, Yan.,(Eds.) Book of Abstracts. Asian Congress on Biotechnology (11-15 May, 2011). pp. 82.

VoB, B. and Wyss, U. (1990). Variation between strains of the nematophagous endoparasitic fungus *Catenaria anguillulae* Sorokin. Factors affecting parasitism *In–Vitro. Journal of plant diseases and Protection* **97**: 416-430.

Chapter 16

Pesticides from Plants and Organic Origin: IPM Tools of Future

Jai Prakash Rai

Assistant Professor in Plant Protection, Department of Mycology and Plant Pathology (KVK), Institute of Agricultural Sciences, Banaras Hindu University, Varanasi – 221 005, U.P.

Scientific developments have done a great service to the development of mankind in almost every sphere of the material life. Nobody can deny the great contributions of science to the improvement and advancement of the society. Agriculture and crop cultivation cannot remain an exception in which scientific developments have brought revolution redefining it in the new context. Chemical pesticides have played a major role in securing food supplies the world over. Development of the chemical called as dichloro-diphenyl-trichloroethane brought a revolution in the area of pest control. Development of DDT and its effectivity in pest management, led us to the era of use of chemicals in pest management. The joy of such an invention left the world overwhelmed and all the scruples were absolutely ruled out as is reflected in the following statement:

"The most discussed of the new insecticides is dichloro-diphenyl-trichloroethane, shortened to DDT but also called Guesarol. This compound has remarkable power to

* E-mail: drjaibhu@gmail.com

kill insects, particularly body lice-the 'cooties' of World War I. The prevalence of typhus, carried by body lice, in the Mediterranean theater of this war has emphasized its value. DDT's effectiveness in war may well be overshadowed by its value in peace. Painstaking investigations have shown it to be signally effective against many of the most destructive insects that feed upon crops." (Scientific American, July 1944).

To be acceptable, pesticides must not have strong toxicity toward non-target organisms, especially humans. Yet, to be efficient, they must be highly toxic toward their intended targets. The mechanism of this type of selectivity is often the targeting of a molecular target site that is found only in the pest or, if in other organisms, is particularly vulnerable in the pest; *e.g.*, an enzyme form that is significantly different from that of other organisms (Duke *et al.*, 2010).

Chemicals are, beyond doubt, the cheapest and miraculously effective means of pest management and hence are still the most popular means to manage pests but since last some decades, the voices against this so treasured a measure of pest management are becoming louder with data showing harm done to the ecology and environment by these chemicals to varying extents because excessive use of these chemicals has led to increased environmental pollution, harmful effects on human health and resistance among pests to the chemicals used. This has driven the researchers to start searching for less harmful yet effective alternatives.

With increasing pressure on several components of environment and ecology that has emanated from the unscrupulous use of inorganic plant protection chemicals in agriculture that is so much talked of in the modern times with examples being ubiquitous and so much conspicuous, we are now compelled to seek for the environmentally and toxicologically safer, more selective and efficacious pesticides. There is, therefore, the need to screen for safe and effective biodegradable pesticides with non-toxic effects on non-target organisms. In the last two decades, considerable efforts have been directed towards screening of plants, in order to develop new botanical insecticides from the vast store of chemical substances in them as alternatives to the existing synthetics, which are associated with phytotoxicity, vertebrate toxicity, pest resistance and resurgence, widespread environmental hazards and high costs (Asogwa *et al.*, 2010).

Increased dependence of agriculture on chemical alternatives of plant nutrition as well as plant protection has already brought us on a stage where we cannot even think of withdrawing these alternatives all of a sudden. Increased depletion/ exhaustion of nutrients from soil and increased pesticide resistance have worsened the problem to a greater extent. This is all a chaotic situation where drawing a simple inference shall be even greater an error.

We are now advocating production of organic food possibly to correct the measures that were once highly recommended. We detest use of chemicals in plant protection today more than we loved them once. All of this can still be corrected, if planned efforts are done in order to achieve the goal over a suitable span of time.

No matter what we do, pests shall be there as long as crops exist. In other words, the struggle between mankind and the so called 'agricultural pests' shall remain continued till there is clash of interests of both the parties. Both shall claim the crop as their birthright. Man is mighty, of course, but pests are not going to lift their claim so easily. It is a long-long war between man and crop pests/pathogens and the existence of either of the parties depends on the outcome of this war. However, it seems to be an endless process and for this, long term approaches are required to keep an upper hand in the long run.

One of such long term approaches is search for environmentally and toxicologically safer compounds, as stated earlier, for their use against crop pests. Such compounds are present in the nature itself, only there is a need to improve their efficacy and afterwards, validate them for their use in plant protection.

With all the honour to the people seeking exactitude and with a word of apology, I shall take the liberty to loosely call all the pest inhibiting principles from plants and organic origin as 'pesticides' henceforth, only to reduce confusion between several terms that describe specific modes of action in pest inhibition.

Medicinal and pesticidal principles present in several plants and pesticides from organic origin have always been an answer to the pest problems of resource poor farmers and villagers. In this regard, Neem is perhaps the most utilized plant for its pest inhibiting properties. Not only neem but tens of thousands of secondary products of plants have been identified with pesticidal properties to one or more pests and there are estimates that hundreds of thousands of these compounds exist in nature. There is growing evidence that most of these compounds are involved in the interaction of plants with other species-primarily the defense of the plant from plant pests. Thus, these secondary compounds represent a large reservoir of biochemicals with pesticidal activity to varying extents. This resource is largely untapped for use as pesticides. Natural pesticides are active principles derived from plants for the management of human and animal pest organisms or it can be said to be biologically active ingredients, principally derived from plants, for the management of human and animal pest organisms (Ivbijaro, 1990). With the growing global demand for environmentally sound pest management strategies; there is a need to develop alternative pesticides with minimal or non-ecological hazards. Botanical pesticides are easily biodegradable (Devlin and Zettel, 1999) and their use in crop protection is a practically sustainable alternative. It maintains biological diversity of predators (Grange and Ahmed, 1988) and reduces environmental contamination and human health hazards. The use of plant extracts to control destructive insect pests or disease vectors is not new.We shall try to get an overview of pesticidal nature of certain plants and pesticides of organic origin that have been utilized or have a good prospect of being utilized for pest control serving as a substitute to the synthetic compounds in the present chapter.

Compounds Derived from Plants with Pesticidal Potential

Herbicidal Compounds

In nature, intraspecific struggle among living organisms is not uncommon. This has also been tapped by mankind, although to a negligible extent. In the world of plants, an almost similar phenomenon is termed as 'allelopathy'. We have several examples of plants with properties inhibitory to the normal development of certain other plants. This is also true that almost all the plants produce secondary compounds that are phytotoxic to varying levels. However, the concentration and amount of these compounds is negotiable to the natural adjustments between biota of the given region and perhaps this adjustment is the reason which is operating behind the fact that these particular compounds provide the producer species a competitive advantage over other species (which are less tolerant to the compound) in only a relatively few cases. Out of these few seemingly prospective examples, still few of the allelochemicals have been actively pursued as herbicides. Even then, the natural compounds of these allelochemicals further need to be modified in order to suit the purpose. Although no allelochemicals are used as herbicides but one major class of herbicides, the triketones were developed from a triketone allelochemical, leptospermone. Leptospermone, a constituent of the bottlebrush plant, is an inhibitor of hydroxyphenylpyruvate dioxygenase (HPPD), an enzyme required for synthesis of tocopherols and plastoquinone in plants (Duke *et al.*, 2010). A derivative of the terpenoid allelochemical 1,8-cineole, with the common name of cinmethylin, is being commercially developed. Other very weakly phytotoxic compounds from plants such as benzoic acids can be made much more active in their herbicidal properties by halogen substitutions. Several benzoic acid derivatives such as dicamba (3,6-dichloro-2-methoxybenzoic acid) are widely used as herbicides.

During our attempts to search for potentially phytotoxic compounds from plants a few highly phytotoxic plant-produced compounds have been discovered. However, none have been developed as herbicides so far. The sesquiterpenoid lactone, artemisinin from Artemisia annua L., was found to inhibit plant growth as did the commercial herbicide cinmethylin. Other compounds, such as 2,4-dihydroxy-1,4-benzoxazin-3-one are as active as plant growth inhibitors similar to many herbicides. Many photodynamic compounds, such as hypericin, are produced by plants which are strongly phytotoxic, provided they can be introduced into the plant cell. These compounds have comparatively less chance to be developed as pesticides because, in the presence of light, they become toxic to all kind of living organisms. However, any plant can be made to generate phytotoxic levels of photodynamic porphyrin compounds followed by treating the plant with both d-aminolevulinic acid, a natural porphyrin precursor, and 2,2'-dipyridyl, a synthetic compound. This comparatively safe combination of compounds is being developed as the 'laser' herbicide. Several classes of commercial herbicides have recently been shown to act by causing accumulation of phytotoxic levels of protoporphyrin IX, a photodynamic chlorophyll and heme precursor in target plant species. Thus, a natural product, and not the synthetic herbicides is the acutely toxic compound in such cases. Application of protoporphyrin IX alone to plant issues, however, is not effective, apparently because it does not reach the proper cellular compartments in sufficient quantity, which is, in fact, the site of action of these compounds.

The major problem with plant-produced phytotoxins as potential herbicides is that in the native state, they are generally only weakly active compared to their commercial counterparts. Most of the known allelochemicals need to be applied at the rates of more than 10 kg/ha to achieve significant levels of weed control whereas, most recently marketed herbicides would achieve the same level of control at levels three orders of magnitude smaller. This is not unexpected, because production of highly phytotoxic compounds would lead to strong autotoxicity unless the producing plant develops metabolic or physical mechanisms to cope up with its own phytotoxins. Some of the more potent allelochemicals are toxic to the producing species themselves and this autotoxicity has been implicated in vegetation shifts. Microbial conversion of relatively non-phytotoxic compounds in the soil to highly phytotoxic derivatives has also been well documented.

As far as development of potential herbicides from phytotoxic compounds produced by plants is concerned, it is a fact that plants have been much more successfully exploited as sources of pesticides for pests other than weeds. This can be attributed to several factors. The selection pressure caused by pathogens and herbivorous pests has probably been more acute and intense than that caused by plant competitors. A plant species can effectively compete with a plant foe in many ways other than by poisoning them and having to manage the problem of possible autotoxicity since they share most of the physiological and biochemical pathways and sites of action of phytotoxic compounds acting as pesticides. On the other side, pathogens and herbivorous pest, being not closely related with plants in physiological and biochemical means, have many potential physiological and biochemical sites of action for pesticides that the plant producing phytotoxic compounds does not share. Biosynthesis of a compound to affect any of these unrelated and unshared sites reduces the chance of autotoxicity. Thus, the chemical option is generally a more attractive option in responding to a herbivorous pest or pathogen that can rapidly devour or invade the plant than it is in responding to a plant competitor.

There are many more examples of use of chemical warfare among plants to win the competition. The attempts, however, need to be intensified with simultaneous improvement and refinement of the technology used for this purpose in order to get desired results in quest for a suitable herbicidal principle from plant origin.

Insecticidal Compounds

Plant products and their analogues are an important source of agrochemicals used for the control of insect pests (Cardellina, 1988). As indicated earlier, success in development of insecticides from plant products or those of organic origin is comparatively more than that of herbicides. It is evident throughout the history that plant products and natural substances have successfully been exploited for their properties as antifeedants, insect repellents and in some cases, insecticides. Pyrethroids can be named as the most successful group of insecticides ever developed from a plant product. Several Chrysanthemum spp. were known for their insecticidal properties since long back in Asian Continent. Even today, powdered dried flowers of these plants are sold as insecticides. Further studies on elucidation of the chemical structures of the six terpenoid esters (pyrethrins) responsible for the insecticidal activity of these plants, paved the way for successful development of many synthetic analogues which were patented and marketed in due course of time. Synthetic pyrethroids have better photostability and in general are more active than their natural counterparts in plants.

Many terpenoids have also been demonstrated to possess insecticidal or other insect-inhibiting properties. The most talked of examples of this kind include azadirachtin and other terpenoids of the limonoid group from the families Meliaceae and Rutaceae which have been proved as potent growth inhibitors of several insect species. The insecticidal effect of neem has been proved on several insect groups, including Lepidoptera Diptera, Coleoptera, Homoptera and Hemiptera species (Sadre *et al.*, 1983).

Commercial use of Nicotine and nornicotine, the components of several members of the genus Nicotiana, as insecticides has successfully been done. The chief commercial source of these compounds is N. rustica. Some other natural analogues of nicotine have been demonstrated to possess considerable insecticidal properties and one of such analogues, anabasine or neonicotine, has been developed as an insecticide from the shrub, *Anabasis aphylla*, in the Soviet Union. Synthetic variations of nicotine, such as 5'-methylnornicotine have been shown to be effective insecticides. An alkaloid from the tropical shrub, *Ryania speciosa* known as Ryanodine, has been used as a commercial insecticide against the pest, European corn borer. Similarly, an alkaloid from *Physostigma venenosum* called Physostigmine, was the compound upon which carbamate insecticides were designed. Among some other alkaloids, furoquinoline alkaloid such as dictamine and beta-carboline alkaloid such as harmaline, can also be cited which are potent photosensitizing compounds and are highly toxic to insect larvae in sun light. However, the relative high cost toxicity to mammals, and limited efficacy have restricted the use of natural alkaloid as insecticides and also, this is perhaps the major hurdle in bringing them into application. Further endeavours to remove this barrier with increased efficacy may open a great new world of natural insecticides.

Another group of commercial insecticides from plants which were popular during 1930s include preparations from the roots of genera Derris, Lonchocarpus and Tephrosia which were known to contain rotenone as the insecticidal principle. Rotenone is a flavanoid derivative which affects mitochondrial respiration inhibiting it strongly. It acts as a contact or stomach poison to insects. It is highly toxic to fish and therefore, its use in areas with possibilities of waterways contamination should be avoided. Simultaneously, it is moderately toxic to humans and most animals but non-toxic to honeybees. Still it is not safe for all the beneficial insects and may be toxic for some. Formulations of rotenone lose effectiveness within approximately one week after application. Although, presence of certain phenolic compounds in plant tissues has been shown to confer resistance in the host plants against insects as well as plant pathogens and also some of them have been demonstrated as strong insect growth inhibitors and antifeedants, none other of them has been proved to be commercially viable option as an insecticide.

Accumulation of toxic levels of photodynamic porphyrin compounds in plants caused by delta-aminolevulinic acid (ALA), in combination with 2,2'-dipyridyl, has motivated the researchers to assess impact of the combination of these compounds on insect species. Larvae of several insect species were fed these compounds and thereafter, such larvae were exposed to light. It was seen that larvae were killed rapidly which strengthened the activity of these compounds upon insects as insecticides. Protoporphyrin IX, the same compound caused to accumulate in plants by certain photobleaching herbicides as stated elsewhere in this chapter, is the same porphyrin which is responsible for the toxicity of these compounds to insects. Other photodynamic compounds from plants such as polyacetylenes have also been demonstrated to be acutely toxic to insects. However, their general toxicity would probably disqualify them from their commercial value and therefore, use as insecticides.

The goal of managing damage caused by insect pests by compounds of plant or organic origin can also be achieved by means other than rapid death of the pest concerned. It may include plant products with compounds acting as insect repellents, those able to alter feeding behavior of the pest, behavior during mating and oviposition and its growth and development ecdysis (moulting). Most of the insect repellent compounds are volatile terpenoids such as terpenen-4-ol. Other terpenoids may act as attractants with some cases of the same terpenoid acting as repellent to a certain species of an insect (usually pest) with simultaneous action as attractant to some more beneficial species of insect. Geranium oil (geraniol) can be cited as a good example of this kind of a compound which repels houseflies and simultaneously attracts honeybees. Several compounds belonging to many different chemical classes have been reported to act as insect antifeedants. A sesquiterpenoid from Polygonum hydropiper known as Polygodial has been proved as strong inhibitor of aphis feeding. Likewise, several steroids from plant origin which are close analogues of the insect moulting hormone called ecdysterone, are able to prevent moulting in the pest species. The mechanisms of inhibition of moulting by certain other chemically unrelated terpenoids are unknown. Although not exactly insecticidal but plant terpenoids acting as locomotor excitants, biting or piercing suppressants, ovipositioning deterrents or mating behaviour disruptants have been described. Effective sterilization of insects has been observed by more than a dozen terpenoid juvenile hormone mimics from plant origin. Therefore, plants contain a multitude of compounds which have potential for commercial development as an important tool of insect pest management.

Fungicidal Compounds

Plants do not have an immune system like animal body to combat pathogenic attacks from microorganisms. Instead, they have to rely completely on chemical protection with secondary compounds to defend themselves from pathogens. Phytoalexins are the compounds that are formed when host plant cells come into contact with pathogenic fungus and inhibit the establishment and growth of the latter. Characterization of many of such secondary compounds has been done and it is being proved that certain compounds which although may not be directly toxic to the pathogen, yet they confer resistance to the host plant by strengthening the defense mechanism of the plant and thus, playing an active role in prevention of the disease in plants. There are evidences that certain synthetic molecules used in plant protection act by inducing the production of chemicals involved in active defense mechanism of the host plant against a given pathogen.

Several compounds derived from plants have been demonstrated to be strong elicitors of phytoalexins. Certain oligosaccharide components of cell walls from stressed or dying higher plant cells which act as elicitors are examples of such type of compounds. Further investigations on phytoalexins elicitors derived from plants could possibly enable us to use them as fungicides. Involvement of several isoflavanoid compounds such as glyceollin in soybean, phaseolin in garden bean and pisatin in pea, in protection of these crops from pathogens has been shown. Several other compounds have also been identified which are either confirmed or suspected to act as phytoalexins. Some of them have even been able to provide protection against pathogenic fungi under field conditions proving their merit. Better protection of bean seedlings from rust was observed with foliar application of the phenolic lactone juglone, a product of several walnut species, as compared to some commercial fungicides. Wyerone, an acetylenic acid derivative produced by legumes as a phytoalexin is known to have a wide fungicidal spectrum against plant pathogens and has been effectively tested against fungal infection of crop plants. In spite of a great array of many antifungal and antibacterial compounds, plant products have not been exploited to much extent in the development of antimicrobial pesticides. This field of investigation is quite promising and seems to have a great potential in development of newer and relatively safer compounds for protection of crop plants from diseases.

Other Compounds

As far as other group of pests is concerned, nematodes, snails, slugs and rodents are the next major pests of crops in queue. These groups of pests are often overlooked and exact estimates of losses from these are lacking in most of the cases.

Although, substantial progress has been achieved regarding knowledge about parasitic nematodes of crop plants, yet still we depend mostly on the regular cultural measures for keeping them below economic injury level. There are a few chemicals which are used in managing the alarmingly damaging cases. These chemicals are potentially harmful to the ecology and environment, thus posing another kind of threat. Knowledge, identification and characterization of plant products with antinemic properties is rather a less explored area which needs further investigations in order to develop effective tools of nematode management.

There are several plant species, which are known to be highly resistant to nematodes. Some important ones include Neem (Azadirachta indica), marigold (Tagetes spp.), Castor (Ricinus communis) and chrysanthemums (Chrysanthemum spp.) etc. The active principles of all of these plant products responsible for their antinemic activity have not so far been identified and characterized and therefore, no commercial formulation of these plant derived products could be made available for practical use. In case of several members of family Asteraceae (family Compositae), alpha-terthienyl, a photodynamic compound is attributed for the antinemic activity of the roots.

Saponins derived from plants are, in general, toxic to snails. Resistance of certain legumes to snails and slugs has been attributed to cyanogenic glicosides. However, here also, the case is similar as with plant derived nematicides and no commercial product is available for use in management of pests in these categories.

Crude Preparations of Plants as Pesticides

Crude preparations of plants have been used for managing pests to a certain extent, although as a weak alternative to their chemically characterized counterparts. It is an age old practice and was the principal means to combat pests in pre-chemical period in agriculture. While retreating back to the nonchemical organic agricultural production, we have great expectations from crude preparations of plants, (even if the active pesticidal principle could not be characterized and reproduced in refined form, as has been done with pyrethroids and several others mentioned previously) for pest management.

The crude preparations can be essential oils, crushed coarse materials including stem, stembark, roots, rootbark, leaves, flowers, fruits, seeds, seed kernels and so on. The formulations can either be in the form of powder (dust), crude oil extracts, ethanol extracts, aqueous extracts or commercial formulation. This chapter shall be incomplete if we omit this part of plant products which are being popularized as pesticides of the future.

Essential oils of many plants have been shown to possess pesticide properties against a number of pests. Essential oils from natural plant products are easy to extract, biodegradable and do not persist in soil and water.

Eucalyptus is particularly useful as it possesses a wide range of desirable properties for pest management and is regarded as non-toxic to humans. Its essential oil can act directly as a natural insect repellent and a study by Batish *et al.* (2008) lists numerous pieces of research that demonstrate this property. For example, previous research has found that eucalyptus essential oil can protect plants against rice weevils, moths and mushroom flies. The study also lists examples of research which have found that eucalyptus essential oil is toxic to microbes including bacteria and fungi. Eucalyptus essential oil could therefore have a role to play in the protection of crops against mould, mildew and wood rot fungi. In addition, when applied in a vapour form, eucalyptus essential oil has potential to manage weeds, especially as its toxicity appears to be species-specific. Further, essential oils or any crude preparation of plant products is a complex mixture of components (as against chemical pesticides which are usually a single product) and therefore, it is unlikely that pests will become resistant to them.

According to the Cornell Cooperative Extension's Pest Management Guidelines, oil extracted from geranium plant (oil of geranium/geraniol) is considered a minimum-risk pesticide. It may be used as an active ingredient in reducing the presence of insects and other pests on plants in the garden. The oil of geranium has a sweet fragrance and therefore, is also used in several perfumes and household deodorizing sprays. This property of the oil makes it a pheromone apart from insecticide, insect repellent, dog and cat repellent. In pure form, it is toxic and therefore, must be handled with due care.

Based on essential oils of plants an organic pesticide with the commercial name of Bioganic® Lawn and Garden Spray Multi-Insect Killer has been developed in USA. This pesticide is composed entirely of plant and tree oils (4 per cent each of thyme, clove, sesame, with the remaining 88 per cent consisting of unspecified combinations of water, soybean oil, wintergreen oil and lecithin). Similarly, combination of canola oil with pyrethrins has been commercialized with the name of PyolaTM which is recommended against flea beetle, cucumber beetles, Mexican bean beetles, squash bugs, aphids, mites and Colorado potato beetle. Attempts for development of similar products are being intensified in other regions of the world.

Neem is one of the most utilized plants for its insecticidal and antimicrobial properties. Various crude formulations of neem that act as pesticides have been very lucidly described by Asogwa *et al.* (2010). Apart from neem, several other plants have been shown to contain pesticidal principles. Somda *et al.* (2007), confirmed the potential of aqueous extract of lemongrass (Cymbopogon citratus) as sorghum seed disinfectant against the two important seed borne fungal pathogens, *viz.* Colletotrichum graminicola the pathogens of sorghum anthracnose and Phoma sorghina which is involved in the head mould complex with Fusarium moniliforme.

Future Prospects

Plants are a rich source of biochemicals with almost any property one can imagine. Pesticides are no exceptions and hunt for such principles in nature shall, no doubt, involve a great investment of time, effort and money. These pesticidal principles from nature may be used directly depending on their suitability and efficacy or, they may provide us at least a lead to prepare their synthetic analogues or templates that can be used optimally. Today, the talk of the period is production of organic food and we shall require a great deal of safer pesticides or natural pesticides which do not pose any threat to the ecology and environment. It is going to be a big issue in the coming times of toxicologically more aware society. Synthetic pesticides, although, not totally thrown out but they are going to have very limited application and use. Attempts for search and development of effective natural pesticides shall be expedited with the advancement in the fields of chemical and biological technology. More accurate results with increased reliability shall enhance the chances of discovery of newer pesticidal principles in the times to come.

References

Asogwa, E.U., Ndubuaku, T.C.N., Ugwu, J.A. and Awe, O.O. (2010). Prospects of botanical pesticides from neem *Azadirachta indica* for routine protection of cocoa farms against the brown cocoa mired-*Sahlbergella singularis* in Nigeria. *Journal of Medicinal Plants Research* **4 (1):** 001-006.

Batish, D.R., Singh, H.P, and Kohli, R.K. ***et al.*** (2008). Eucalyptus essential oil as a natural pesticide. *Forest Ecology and Management.* **256:** 2166-2174.

Cardellina, J.H. (1988). Biologically active natural products in the search of for new agrochemicals. In: Gulter, H. G. (ed), Biologically active natural products: potential use in agriculture. American Chemical Society, Washington, publication pp. 305-311.

Devlin, J.F. and Zettel, T. (1999). Ecoagriculture: Initiatives in Eastern and Southern Africa. Weaver Press. Harare.

Duke, S.O., Cantrell, C.L., Meepagala, K.M., Wedge, D.E., Tabanca, N. and Schrader, K.K. (2010). Natural Toxins for use in Pest Management. *Toxins* **2:** 1943-1962.

Grange, N. and Ahmed, S. (1988). Handbook of plants with pest control properties. John Wiles & Sons. New York.

Ivbijaro, M.F. (1990). Natural Pesticides: Role and Production Potential in Nigeria. National workshop on the pesticide Industry in Nigeria University of Ibadan, Sept. 24–27, p. 24.

Sadre, N.L., Deshpande, V.Y., Mendulkar, K.N. and Nandal, D.H. (1983). "Male *Azadirachta indica* in different species" Proc 2nd Int. Neem Conf. Rauischholzhausen. p. 482.

Somda, I., Leth, V. and Sérémé, P. (2007). Evaluation of lemongrass, eucalyptus and neem aqueous extracts for controlling seed borne fungi of sorghum grown in Burkina Faso. *World Journal of Agricultural Sciences* **3(2):** 218-223.

Chapter 17

Mushrooms: Nutritional and Medicinal Properties

R.C. Ram[1] and Dayaram[2]***

[1]Department of Mycology and Plant Pathology, Institute of Agricultural Sciences, Banaras Hindu University, Varanasi – 221 005, U.P.

[2]Department of Microbiology, Faculty of Basic Sciences & Humanities, R.A.U. Pusa, Samastipur, Bihar

Mushrooms are considered to be healthy food because of their relatively high and qualitatively good protein content and because of their good vitamins, minerals and low fat content. Mushrooms have been recommended by FAO as food that contributes to the protein nutrition of developing countries which depends largely cereals. Mushrooms are important constituents of minor forest produce, that grow on the most abundant biomolecule of this biosphere, that is, cellulose. Presently mushrooms are regarded as a macro-fungus with a distinctive fruiting body which can be either epigeous or hypogenous and large enough to be seen with the naked eyes and to be picked by hand (Chang and Miles, 1992). Only fruiting body of the mushroom can be seen whereas the rest of the mushroom remains underground as mycelium. Geologically, mushrooms existed on the earth even before man appeared on it, as evidenced from the fossil records of the lower cretaceous period. Thus anthropologically speaking, there is every possibility that man used the mushrooms as food when he was still a food gatherer and hunter on the chronology of cultural evolution. Mushrooms offer tremendous applications as they can be used as food

E-mail: *rcrbhumpp@yahoo.com; **dayaram@sify.com

and medicines besides their key ecological roles. They represent as one of the worlds greatest untapped resources of nutrition and palatable food of the future.

Mushrooms have been found effective against cancer, cholesterol reduction, stress, insomnia, asthma, allergies and diabetes (Bahl, 1983). Due to high amount of proteins, they can be used to bridge the protein malnutrition gap. Mushrooms as functional foods are used as nutrient supplements to enhance immunity in the form of tablets. Due to low starch content and low cholesterol, they suit diabetic and heart patients. One third of the iron in the mushrooms is in available form. Their polysaccharide content is used as anticancer drug. Even, they have been used to combat HIV effectively (Nanba, 1993; King, 1993).

Table 17.1: Proximate Composition (per cent fresh weight) of the Cultivated Mushrooms

Mushrooms/ Vegetable	*Moisture*	*Protein*	*Fat*	*Carbo-hydrate*	*Fiber*	*Ash*	*Clorie*
Agaricus bisporus	90.1	2.9	0.3	5.0	0.9	0.8	36
Pleurotus sajor-caju	90.2	2.5	0.2	5.2	1.3	0.6	35
Volvariella volvacea	90.1	2.1	1.0	4.7	1.1	1.0	36
Cabbage	91.9	1.8	0.1	4.6	1.0	0.6	27
Cauliflower	90.8	2.6	0.4	4.0	1.2	1.0	30
Potato	74.7	1.6	0.1	22.6	0.4	0.6	97

Source: Rai and Sohi, 1988.

Mushrooms as a Source of Food

The mushrooms are consumed basically for their texture and flavor, and Mushrooms have from nutrition point of view, a distinct place in human diet which otherwise consist of items either of plant or animal" origin. Mushrooms are perhaps the only fungi deliberately and knowing consumed by human being and they complement and supplement the human diet with various ingredients not encountered in or in deficient in food items of plant and animal origin. Chang and Buswell (1996) have coined the term mushroom nutraceuticals and pharmaceuticals. Nutraceuticals are the functional food enriched / modified and consumed as normal diet to provide health giving benefits.

Man has been hunting for the wild mushrooms since antiquity (Cooke, 1977). Thousands of years ago, fructifications of higher fungi have been used as a source of food (Mattila *et al.*, 2001) due to their chemical composition which is attractive from the nutrition point of view. During the early days of civilization, mushrooms were consumed mainly for their palatability and unique flavors (Rai, 1994, 1997). Present use of mushrooms is totally different from traditional because, lot of research has been done on the chemical composition of mushrooms, which revealed that mushrooms can be used as a diet to combat diseases. The early history regarding the use of mushrooms in different countries has been reviewed by number of workers (Buller, 1915; Rolfe and Rolfe, 1925; Singer, 1961; Atkinson, 1961; Bano *et al.*, 1964;

Jandaik and Kapoor, 1975; Bano and Rajarathnam, 1982; Abou *et al.*, 1987; Houghton, 1995). Rolfe and Rolfe (1925) mentioned that the mushrooms like *Agaricus campestris, Morchella esculenta, Helvella crispa, Hydnum coralloides, Hypoxylon vernicosum* and *Polyporus mylittae* were used much earlier in India. The oriental use of mushrooms is older than the European (Lambert, 1938). Lintzel (1941, 1943) recommended that 100 to 200 g of mushrooms (dry weight) is required to maintain an optimal nutritional balance in a man weighing 70 kg. Bano *et al.* (1963) determined the nutritive value of *Pleurotus flabellatus* as 0.974 per cent ash, 1.084 per cent crude fibre, 0.105 per cent fat, 90.95 per cent moisture, 0.14 per cent non-protein nitrogen and 2.75 per cent protein. Bano (1976) suggested that food value of mushrooms lies between meat and vegetables. Crisan and Sands (1978) observed that mushrooms in general contain 90 per cent water and 10 per cent dry matter. More so, the protein content varies between 27 and 48 per cent. Carbohydrates are less than 60 per cent and lipids are between 2 to 8 per cent. Orgundana and Fagade (1981) indicated that an average mushroom is about 16.5 per cent dry matter out of which 7.4 per cent is crude fiber, 14.6 per cent is crude protein and 4.48 per cent is fat and oil. Of several thousand mushroom species known worldwide, only around 2000 are considered edible, of which about 20 are cultivated commercially with only 4 to 5 under industrial production (Chang, 1990). There is also a significant difference in the nutrient contents of pileus verses stalks (Zakia *et al.*, 1993).

Proteins

The value of protein is determined by the kinds of amino acids that form protein. Mushrooms contain all the essential amino acids as well as the most commonly occurring non-essential amino acids and amides. Mushrooms are rich in lysine and tryptophan, the two essential amino acids that are deficient in cereals. The most nutritious mushrooms are almost equal in nutritional value to meats and milk. Protein is the main body building constituent of our food. Protein content of mushrooms depends on the composition of the substratum, size of pileus, harvest time and species of mushrooms (Bano and Rajarathnam, 1982).

Protein is an important constituent of dry matter of mushrooms (Aletor, 1995 Alofe *et al.*, 1995) Florczak and Lasota, 1995; Zrodlowski, 1995; Chang and Buswell, 1996). Mau and Flegg (1975) reported the digestibility of mushroom protein to be as high as 71 to 90 per cent. Protein content of the mushrooms has also been reported to vary from flush to flush (Crisan and Sands, 1978). Purkayastha and Chandra (1976) found 14 to 27 per cent crude protein on dry weight basis in *A. bisporus, Lentinus subnudus, Calocybe indica* and *Volvariella volvacea*. Purkayastha and Chandra (1976) found 14 to 27 per cent crude protein on dry weight basis in *A. bisporus, Lentinus subnudus, Calocybe indica* and *Volvariella volvacea*. Haddad and Hayes (1978) indicated that protein in *A. bisporus* mycelium ranged from 32 to 42 per cent on the dry weight basis. Abou *et al.* (1987) found 46.5 per cent protein on dry weight basis in *A. bisporus*. Samajipati (1978) found 30.16, 28.16, 34.7 and 29.16 per cent protein in dried mycelium of *A. campestris, Agaricus arvensis, M. esculenta* and *Morchella deliciosa* respectively. On dry matter basis, the protein content of mushrooms varies between 19/100 and 39/100 g (Weaver *et al.*, 1977; Breene, 1990). In terms of the amount of crude protein, most other foods including milk (Chang, 1980). On a dry weight basis, mushrooms normally contain 19 to 35 per cent proteins as compared to 7.3 per cent in rice, 12.7 per cent in wheat, 38.1 per cent in soybean and 9.4 per cent in corn (Crisan and Sands, 1978; Li and Chang, 1982; Bano and Rajarathnam, 1988). Verma *et al.* (1987) reported that mushrooms are very useful for vegetarian because they contain some essential amino acids which are found in animal proteins. The digestibility of *Pleurotus* mushrooms proteins is as that of plants (90 per cent) whereas that of meat is 99 per cent (Bano and Rajarathnam, 1988). Rai and Saxena (1989a) observed decrease in the protein content of mushroom on storage. The protein conversion efficiency of edible mushrooms per unit of land and per unit time is far more superior compared to animal sources of protein (Bano and Rajarathnam, 1988).

Table 17.2: Essential Amino Acids (per cent crude protein) in Edible Mushrooms

Amino Acid	*Agaricus bisporus*	*Pleurotus sajor-caju*	*Volvariella volvacea*
Leucine	7.5	7.0	4.5
Isoleucine	4.5	4.4	3.4
Valine	2.5	5.3	5.4
Tryptophan	2.0	1.2	1.5
Lysine	9.1	5.7	7.1
Threonine	5.5	5.0	3.5
Phenyl alanine	4.	5.0	2.6
Methionine	0.9	1.8	1.1
Histidine	2.7	2.2	3.8

Source: Bano and Rajarathnam, 1982; Li and Chang, 1982.

Mushrooms in general have higher protein content than most other vegetables (Bano and Rajarathnam, 1988) and most of the wild plants (Kallman, 1991). Sharma *et al.* (1988) reported 14.71 to 17.37 per cent and 15.20 to 18.87 per cent protein in the fruiting bodies of *Lactarious deliciosus* and *Lactarious sanguiffus* respectively. Mushrooms contain all the essential amino acids required by an adult (Hayes and Haddad, 1976). Gupta and Sing (1991) reported 41.4 per cent essential amino acids in *Podaxis pistillaris*. Friedman (1996) reported that the total nitrogen content of dry mushrooms is contributed by protein amino acids and also revealed that crude protein is 79 per cent compared with 100 per cent for an ideal protein.

Carbohydrates

The carbohydrate content of mushrooms represents the bulk of fruiting bodies accounting for 50 to 65 per cent on dry weight basis. The mannitol, also called as mushroom sugar constitutes about 80 per cent of the total free sugars, hence it is dominant (Tseng and Mau, 1999; Wannet *et al.*, 2000). Mc-Connell and Esselen (1947) reported that a fresh mushroom contains 0.9 per cent mannitol, 0.28 per cent reducing sugar, 0.59 per cent glycogen and 0.91 per cent hemicellose. Water soluble polysaccharides of mushrooms are antitumor (Yoshioka *et al.*, 1975). Carbohydrates of *Agaricus bisporus* were reported by Crisan and Sands (1978). Raffinose, sucrose, glucose, fructose and xylose are dominant in it. (Singh and Singh, 2002). Florezak *et al.* (2004) reported that *Coprinus atramentarius* (Bull.: Fr.) Fr. contain 24 per cent of carbohydrate and free sugars amounts to about 11 per cent. on dry weight basis.

Fats

In mushrooms, the fat content is very low as compared to carbohydrates and proteins and it is important for patients of obesity. The fats present in mushroom fruiting bodies are dominated by unsaturated fatty acids. Singer (1961) determined the fat content of some mushrooms as 2.04 per cent in *Suillus granulatus*, 3.66 per cent n *Suillus luteus* and 2.32 per cent in *A. campestris.* Hugaes (1962) observed that mushrooms are rich in linolenic acid, which is an essential fatty acid. Total fat content in *A. bisporus* was reported to be 1.66 to 2.2/100 g on dry weight basis (Maggioni *et al.*, 1968). Ogundana and Fagade (1981) indicated that mushrooms have 4.481 per cent fats on dry weight basis. Kanwar *et al.* (1990) has reported a fat content of 11.52 per cent in the *Amanita ceasarea* fruiting bodies on dry weight basis. In 100 g fresh matter of *A. bisporus* (Lange) Sing and *Pleurotus ostreatus* (Jacq: Fr.) Kumm, the content of fatty compounds were found to be 0.3 and 0.4 g respectively (Manzi *et al.*, 2001), but on dry weight basis, it is 2 and 1.8 g respectively (Shah *et al.*, 1997). Pedneault *et al.* (2006) reported that fat fraction in mushrooms is mainly composed of unsaturated fatty acids.

Vitamins

Mushrooms are good source of vitamins (Table-3) such as vitamin 'B_1' (Thiamine), vitamin 'B_2' (Riboflavin), niacin, biotin and vitamin 'C' (Ascorbic acid). Mushrooms are one of the best sources of vitamins especially Vitamin B (Breene, 1990; Mattila *et al.*, 1994; Zrodlowski, 1995; Chang and Buswell, 1996; Mattila *et al.*, 2000). Vitamin content of edible mushrooms has been reported by Esselen and Fellers (1946), and Litchfield (1964). Manning (1985) gave a comprehensive data of vitamin content of mushrooms and some vegetables. According to Mattila *et al.* (1994), wild mushrooms contains much higher amounts of vitamin D2 than dark cultivated *A. bisporus.* Mushrooms also contain vitamin C in small amounts (Sapers *et al.*, 1999; Mattila *et al.*, 2001) which are poor in vitamins A, D, and E (Anderson and Fellers, 1942).

Table 17.3: Vitamin Content in Different Species of Mushroom

	Mushrooms	*Content mg/100g D. wt.*			
		Thiamine	*Riboflavin*	*Niacin*	*Ascorbic Acid**
(*i*)	*Agaricus bisporus*	1.1	5.0	55.7	81.9
(*ii*)	*Lentinus edodes*	7.8	4.9	54.9	0.0
(*iii*)	*Volvariella volvacea*	0.32-0.35	1.63-2.97	64.8	20.2
(*iv*)	*Pleurotus* spp.	1.16-4.8	4.7*	46.1	0.0

* Adapted from Eli V. Crisan and Anne Sands (Chang and Hayes, 1978).

Source: Chadha and Sharma, 1995.

Minerals

Mushrooms are also good source of minerals such as potassium, phosphorus and sodium and contain low but available form of Iron, Potassium and Sodium ratio is very high which is desirable for patients of hypertension. Minerals and heavy metals content of *Pleurotus* spp. are given in Table 17.4.

Table 17.4: Minerals Content of *Pleurotus* Mushrooms

Species	K	P	Mg	Na	Ca	Fe	Cd	Zn	Cu	Pb	Hg
	(mg/100g. D wt.)							(ppm)			
P. Sajor-caju	3260	760	221	60	20	12.4	0.3	29	12.2	3.2	0
P. Sajor-caju	2240	565	156	256	40	2.8	-	2.6	0.5	-	-
P. eous	4570	1410	242	78	23	9.0	0.4	82.7	17.8	1.5	0
P. flabellatus	3760	1550	292	75	24	12.4	0.5	56.6	21.9	1.5	0
P. florida	4660	1850	192	62	24	18.4	0.5	11.5	15.8	1.5	0

Source: Rai 1995.

The fruiting bodies of mushrooms are characterized by a high level of well assimilated mineral elements. Major mineral constituents in mushrooms are K, P, Na, Ca, Mg and elements like Cu, Zn, Fe, Mo, Cd form minor constituents (Bano and Rajarathanum, 1982; Bano *et al.*, 1981; Chang, 1982). K, P, Na and Mg constitute about 56 to 70 per cent of the total ash content of the mushrooms (Li and Chang, 1982) while potassium alone forms 45 per cent of the total ash. Abou-Heilah *et al.* (1987) found that content of potassium and sodium in *A. bisporous* was 300 and 28.2 ppm. respectively. *A. bisporus* ash analysis showed high amount of K, P, Cu and Fe (Anderson and Fellers, 1942). Kaul (1978) reported that *M. esculenta* contains Ca (0.5776 mg), P (3.313 mg), Fe (1.213 mg) and K (3.831 mg). Varo *et al.* (1980) reported that *A. bisporus* contains Ca (0.04 g), Mg (0.16), P (0.75 g), Fe (7.8 g), Cu (9.4 mg), Mn (0.833 mg) and Zn (8.6 mg) per kilogram fresh weight. Mushrooms have been found to accumulate heavy metals like cadmium, lead, arsenic, copper, nickel, silver, chromium and mercury (Schmitt and Sticher, 1991; Mejstrick and Lepsova, 1993; Wondratschek and Roder, 1993; Kalac and Svoboda, 2000; Svoboda *et al.*, 2001; Issilogglu *et al.*, 2001; Malinowska, 2004). The mineral proportions vary according to the species, age and the diameter of the fruiting body. The mineral content of wild edible mushrooms has been found higher than cultivated edible mushrooms (Aletor, 1995; Mattilla *et al.*, 2001; Rudawska and Leski, 2005).

Medicinal Importance

Numerous reports have shown that mushroom and mushroom products have significant medicinal properties such as immuno-modulation, anti-cancer, anti-oxidant, blood pressure lowering, cholesterol lowering, liner protectine, antifibrotic, anti-inflatiory, anti-diabetic and anti-microbial activities. *Ganoderma lucidum* is the most popular medicinal mushroom in China and has been used for a wide range of health benefits, from preventive measures and maintenance of health to regulation and treatment of chronic as well as acute ailments (Chang 1995). One of the most interesting mushroom derived products is the polysaccharides which exhibit promising immuno-modulatory and anticancer effects. Numerous anti-tumour polysaccharides have been discovered from mushroom. These antitumour agents of polysaccharides nature *i.e. Lentinus, schizophyllan* and polysaccharides proteins complexes (PSK & PSP) were isolated from *Lentinus edodes, Schizophyllum commene* and *Coriolus versicolar*, respectively. Many mushrooms polysaccharides have shown antiviral activities including human viruses. Cholesterol is negligible, fat and calories are found less amount in the Mushroom. So it is comfortable for patients of hypertension and blood pressure and for diabeties patient, due to lack of sugar in mushroom. Some genera of mushrooms having chormous medicinal quality. As a medicine mushroom *Ganoderma lucidum* have been traditionally used in Japan and China. Cosmetic product and tonic beverages have been produced in China from *Ganoderma lucidum.* Antitumer effect have been reported by the extract of various edible fungi including *Lentinus edodes, Flammulina valutipes, Pleurotus ostreatus, Agaricus bisporus, Pholiota nemak, Aricholoma matsutak and Auricularia auricula.*

In China, *Ganoderma lucidum* considered as a symbol of happy augury, good fortune, good health, longevity and even immortality. It can reduce blood pressure blood cholesterol and blood sugar level and inhibited the platelets aggregation. *Ganoderma lucidum* (Curt. Fr) P. Karst, known as Reishi or Mannentake in the Japan and Ling Zhi in China is the most important commercial medicinal mushroom. It has been reported to passes a plethora of pharmacological properties (Rai, 1997; Wasser and Weis 1999; Willard, 1990). The mushroom was highly regarded in the Chinese Japanese system of the medicine expectedly its artificially cultivation started in these countries and then spread to Korea, Malaysia and Thailand. It has been estimated that now some American firms have also started its production and marketing. Current world production of this mushroom is around. 8000 tonnes, half of which comes from China. World trade in this mushroom is around 2 billion US $ and as per one estimate trade in India around 120 crore annually. (Business world, 11 Nov. 2002). Malaysian companies are bringing their produce processing packaging and marketing in India through multilevel marketing system.

The compounds extracted from *Agaricus bisporus, Lentinus edodes. Comprinus comotus and Oudemensiella mucida* have been reported to have antifungal and antibacterial properties. Various pharmaceutical compounds have been isolated from several mushrooms. (Buswell and Change 1993) some mushroom genera are given in (Table 17.5) which having medicinal properties.

The mushrooms are consumed basically for their texture and flavour, and there is a little awareness about their nutritional and medicinal attributes. Mushrooms have also been investigated by the nutritionists after the Second World War when food shortage stimulated the research on nutritional value of the food items. Medical mycology is as old as traditional uses of mushrooms. They have been used in medicine since the Neolithic and Paleolithic eras (Samorini, 2001). First century Greek physician Dioscorides, included the lurch polypore, (*Fomitopsis officinalis* (Villars: Fr.) Bond and Singer, *Polyporaceae*; syn. *Laricifomes officinalis* (Villars: Fr.) in his *De Materia medica* known then as *Agaricum* and latter as the Quinine conk. It was used for the

Table 17.5: Various Pharmaceutical Components Isolated from Several Mushroom are Listed in Table

Pharmacodynamic	*Component*	*Species*
Antibacterial effect	Hirsuitic acid	*Many species*
Antibiotic	E-B,Methoxyacrylate	*O. radicata*
Antiviral effect	Polysoceharid, Protein	*L. edodes* and *Polyporaceal*
Cardoic tonic	Volvatoxin, Flammutoxin	*Volvariella, Flammulina*
Decrease cholesterol	Eritadenine	*Collybia velutipes*
Decrease blood pressure	Triterpene	*G. lucidum*
Decrease level of blood	Peptide, glycogen Ganoderan *Glucon*	*G. lucidum*
Antitfrombus	5' AMP, 5'GMP	*Psalliota hartensis*
Inhibition of PHA	r-GHP	*P. hartensis, L. edodes*
Anti tumer	B-glucan RNA complex	*Many* spp. *Hysizygus marmoreus*
Increase secretion of bile	Armillarisia A	*Armillaria tabescens*
Analgestic and Sedetive effect	Marasmic acid	*Marasmices androsaceus*

Source: Chadha and Sharma 1995.

treatment of "consumption", a disease now known as tuberculosis. Although mushrooms as medicine have been used in China since 100 A.D. (Gunde, 1999), but it was only in 1960 that scientists investigated the basic active principles of mushrooms which are health promoting. Mushrooms have been used in health care for treating simple and age old common diseases like skin diseases to present day complex and pandemic disease like AIDS. They are reputed to possess anti-allergic, anticholesterol, anti-tumor and anti-cancer properties (Jiskani, 2001). The first successful research discovered the antitumor effects of the hot water extracts from several mushrooms (Ikekawa *et al.*, 1969). The main components proved to be polysaccharides especially–D- glucans. Chihara *et al.* (1969) isolated from the shiitake fruiting bodies, an antitumor polysaccharide, which was named *lentinan*. Bahl (1983) reported that mushrooms cure epilepsy, wounds, skin diseases, heart ailments, rheumatoid arthritis, cholera besides intermittent fevers, diaphoretic, diarrhea, dysentery, cold, anesthesia, liver disease, gall bladder diseases and used as vermicides. Most of the mushroom drugs are now available in tablet form in China (Yang *et al.*, 1993). In underdeveloped countries where protein malnutrition has taken epidemic proportions, Food and Agricultural Organization (F. A. O.) has recommended mushroom foods to solve the problem of malnutrition (Sohi, 1988). Mannentake (*Ganoderma lucidum*) are known to lower blood pressure and serum cholesterol concentration of hypertensive rats (Kabir *et al.*, 1988). *Lentinus tigrinus* and *G. lucidium* are proved anticholesterolmic (Ren *et al.*, 1989). *Lentinus edodus* has been used to enhance vigour, sexuality, energy and as an anti aging agent. Lentinan sulphate obtained from *Lentinus* species inhibits HIV (Gareth, 1990). Jong *et al.* (1991) reported that mushrooms cause regression of the disease state. Mushroom medicines are without side effects (Sagakami *et al.*, 1991).

Puffballs have been used in urinary infections (Buswell and Chang, 1993). Maitake extract has been shown to kill HIV and enhance the activity of T-helper cells (Nanba, 1993; King, 1993). *Ganoderma* nutriceuticals have exhibited promising antiviral effects like, anti-hepatitis B (Kino *et al.*, 1989), anti-HIV (Kim *et al.*, 1993; Liu and Chang, 1995). Dreyfuss and Chapela (1994) reported hundreds of secondary metabolites of fungal origin possessing biological activity. Mushrooms act as biological response modifiers by promoting the positive factors and eliminating the negative factors from the human body and thus regarded as the fourth principal form of the conventional cancer treatment (Yang *et al.*, 1993). *G. lucidium* (Fr.) Karst is believed to act as an anti-inflammatory agent (Stavinoha *et al.*, 1991); acts as antidiabatic (Teow, 1997). It is also used by Indian tribals for treating joint pain (Harsh *et al.*, 1993). Hobbs (1995) reported various medicinal uses of mushrooms like reishi, cordyceps, enoki, maitake, lion's mane and splitgill for cancer treatment; shiitake, blazei, reishi, enoki, cordyceps, maitake, mesima and oyster were found effective against cholesterol reduction. Reishi, cordyceps, shiitake and maitake is used for reducing stress. Lion's mane has been used for memory improvement; reishi for inducing sleep, cordyceps for physical endurance and sexual performance, reishi, cordyceps, chaga and lion's mane for asthma and allergy treatment. Shiitake, cordyceps, chaga, shiitake and turkey tail as liver protectants; reshi, maitake, turkey tail and shitake for treating diabetes. It is also believed to be a good health elevator (Mizuno, 1996). *Auricularia* species were used since times for treating hemorrhoids and various stomach ailments (Chang and Buswell, 1996). *Pleurotus tuber-regium* mushroom have been used for curing headache, high blood pressure, smallpox, asthma, colds and stomach ailments (Oso, 1997; Fasidi and Olorumaiye, 1994). It has been reported that *P. ostreatus* lowers the serum cholesterol concentration in Wani *et al.*, 2601 rats (Bobek *et al.*, 1996). PSK, an anticancer drug from the mushroom, *Coriolus versicolor* accounted for 25.5 per cent of the country's total sales in Japan in 1987 as anticancer drug (Chang and Buswell, 1996). Puffballs (*Clavatia, Lycoperdon*) have been used for healing wounds (Delena, 1999). Pharmaceutical substances with potent and unique health enhancing properties have been isolated from mushrooms (Wasser and Weis, 1999). Fresh mushrooms are known to contain both soluble and insoluble fibres; the soluble fibre is mainly beta-glucans polysaccharides and chitosans which are components of the cell walls (Sadler, 2003). Soluble fibre present in mushrooms prevents and manages cardiovascular diseases (Chandalia *et al.*, 2000). Wasser (2005) reported that mushroom health supplements are marketed in the form of powders, capsules or tablets made of dried fruiting bodies, extracts of mycelium with substrate, biomass or extract from liquid fermentation. *P. sajor-caju* has been found to be inductive for growth of probiotic bacteria (Oyetayo *et al.*, 2005). *Cordyceps sinensis* also treated as half caterpillar and half mushroom has been known and used for many centuries in traditional Chinese medicine. *Cordyceps* has been used to induce restful sleep, acts as anticancer, antiaging, and antiasthama agents besides proved effective for memory improvement and as sexual rejuvenator (Sharma, 2008).

Antioxidant Activity

As in India, mushrooms have traditionally been used in China and Japan also for the medicinal and tonic properties and in recent years there has been an upsurge interest in this aspect. Anti-tumor effects have been reported by the extracts of various edible fungi including *Lentinus edodes, Flammulina velutipes, Pleurotus ostreatus, Agaricus bisporus, Pholiota nameko, Tricholoma matsutake* and *Auricularia auricula*. Recently, maitake (*Grifola frondosa*) and shiitake (*Lentinus edodes*) mushrooms have been reported to be inhibitory to AIDS virus in USA and Japan (Namba, 1993). *Coriolus versicolor* is a medicinal mushroom widely prescribed for the prophylaxis and treatment of cancer and infection in China (Chu *et al.*, 2002).

Antioxidants are chemical compounds that protect cells from the damage caused by unstable molecules known as free radicals. Free radicals are powerful oxidants and those chemical entities that contain unpaired electrons. They are capable of randomly damaging all components of the body, *viz.* lipids, proteins, DNA, sugars and are involved in mutations and cancers (Przybytniak *et al.*, 1999). The nascent oxygen is trapped by enzymes like superoxide dismutase, catalase and glutathione peroxidase. Over production of free radicals creates oxidative stress. The antioxidants are an important defense of the body against free radicals and mushrooms which are rich sources of these antioxidants (Mau *et al.*, 2004; Puttaraju *et al.*, 2006; Ferreira *et al.*, 2007; Oyetayo *et al.*, 2007). Waxy cap mushroom extracts (*Hygrocybe coccinea)* are inhibitory to sarcoma (Ohtsuka *et al.*, 1997). Immunoceticals isolated from more than 30 mushroom species have shown anticancer action in animals (Wasser and Weis, 1999). Schizophyllan from *Schizophyllum commune* is effective against head and neck cancer (Kimura *et al.*, 1994; Borchers *et al.*, 1999). Antioxidant property of compounds is correlated with their phenolic compounds (Velioglu *et al.*, 1998). Kim and kim (1999) reported that mushroom extracts possess DNA protecting properties. *G. lucidum* extracts can trap number of free radicals 2602 J. Med. Plant. Res. (Jones and Janardhanan, 2000). Mau *et al.* (2001) found antioxidant properties of several ear mushrooms. Many species of mushrooms have been found to be highly potent immune enhancers, potentiating animal and human immunity against cancer (Wasser and Weis, 1999; Borchers *et al.*, 1999; Kidd, 2000; Feng *et al.*, 2001). Tyrosinase from *A. bisporus* is antioxidant (Shi *et al.*, 2002). Lakshmi *et al.* (2005) determined antioxidant activity of *P. sajor caju*. Russell and Paterson (2006) observed that triterpenoides are the main chemical compounds in *G. lucidium*. Camptothecin is responsible for antioxidant properties in *G. lucidum* (Zhou *et al.*, 2007).

Conclusion

In developing countries malnutrition is one of the major problems because most of the population remains under the economic line. Mushrooms are considered to be healthy food because of their relatively high and qualitatively good protein content and because of their good vitamins, minerals and low fat content. Mushrooms have been recommended by FAO as food that contributes to the protein nutrition of developing countries which depend largely on cereals. The mushrooms are consumed basically for their texture and flavors, and there is a little awareness about their nutritional and medicinal attributes. Mushrooms have also been investigated by the nutritionists after the Second World War when food shortage stimulated the research on nutritional value of the food items. Diverse benefits of mushrooms towards humans by the words of the father of medicine that is, Hippocrates "Let food be your medicine and medicine be your food". This saying aptly suits mushrooms, as they have tremendous medicinal food, drug and mineral values; hence they are valuable asset for the welfare of humans.

References

Abou-Heilah, A.N., Kasionalsim, M.Y. and Khaliel, A.S. (1987). Chemical composition of the fruiting bodies of *Agaricus bisporus. Int. J. Expt. Bot.*, **47:** 64-68.

Aletor, V.A. (1995). Compositional studies on edible tropical species of mushrooms. *Food Chem.*, **54:** 265-268.

Aletor, V.A., and Aladetimi, O.O. (1995). Compositional studies on edible tropical species of mushrooms. *Food Chem.*, **54:** 265-268.

Alofe, F.V., Odeyemi, O. and Oke, O.L. (1995). Three edible mushrooms from Nigeria: Their proximate and mineral composition. *Plant Foods for Hum. Nutr.*, **49:** 63-73.

Anderson, E.E. and Fellers, C.R. (1942). The food value of mushrooms *(A Campestris). Proc. Am. Soc. Hort. Sci.*, **41:** 301.

Atkinson, C.F. (1961). Studies of American Fungi Mushrooms Edible, Poisonous, Hafner publishing Co, New York, pp. 322.

Bahl, N. (1983). Medicinal value of edible fungi. In: Proceeding of the International Conference on Science and Cultivation Technology of Edible Fungi. *Indian Mushroom Science* II, pp. 203-209.

Bano, Z. (1976). Nutritive value of Indian mushrooms and medicinal practices. *Eco. Bot.*, **31:** 367-371.

Bano, Z., and Rajarathanam, S. (1982). *Pleurotus* mushrooms as a nutritious food. In: Tropical mushrooms–Biological Nature and cultivation methods, (Chang ST, Quimio, TH, eds.) The Chinese University press, Hongkong, pp. 363-382.

Bano, Z. and Rajarathanum, S. (1988). Pleurotus mushroom part II. Chemical composition nutritional value, post harvest physiology, preservation and role as human food crit. *Rev. Food Sci. Nutr.*, **27:** 87-158.

Bano, Z., Ahmed, R. and Srivastava, H.C. (1964). Amino acids of edible mushrooms, *Lepiota* sp. and *Termitomyces* sp. *Indian j. Chem.*, **2:** 380-381.

Bano, Z., Bhagya, S. and Srinivasan, K.S. (1981). Essential amino acid composition and proximate analysis of Mushroom, *Pleurotus florida. Mushrooms News Lett. Trop.*, **1:** 6-10.

Bano, Z., Srinivasan, K.S. and Srivastava, H.C. (1963). Amino acid composition of the protein from a mushroom (*Pleurotus flabellatus*). *Appl. Microbiol.*, **11:** 184-187.

Borchers, A.T., Stern, J.S. and Hackman R.M. (1999). Mushrooms, tumors, and immunity. *Proc. Soc. Exp. Biol. Med.*, **221:** 281-293.

Breene, W.M. (1990). Nutritional and medicinal value of specialty mushrooms. *J. Food Protect.*, **53:** 883-894.

Buller, A.H.R. (1915). The fungus lore of the Greeks and Romans.-Trans. Br. Mycol. Soc., **5:** 21-26.

Buswell, J.A. and Chang, S.T. (1993). Edible mushrooms attributes and applications. In: *Genetics and breeding of edible mushrooms* (Chang, S.T.J. Buswell, J.A and Miles PG (Eds). Gordon and Breach, Philadelphia, pp. 297-394.

Chadha, K.L. and Sharma, S.R. (1995) Mushroom Research in India. History, Infrastructure and Achievements. In : *Advances in Horticulture*. K.L. Chadha and S.R. Sharma (Eds), Malhotra Publish House, New Delhi. **13** : 1-33

Chandalia, M., Garg, A., Lutjohann, D., von Bergmann K., Grundy, S.M. and Brinkley, L.J. (2000). Beneficial effects of high dietary fiber intake in patients with type 2 diabetes mellitus. *N. Eng. J. Med.*, **342**:1392-1398.

Chang, S.T. (1980). Mushroom as human food, *Bio Science* **30:** 339-401.

Chang, S.T. (1982). Prospects for mushroom protein in developing countries. In: *Tropical Mushroom*–Biological Nature and Cultivation Methods (Chang ST, Quimio TH. eds.), Chinese University Press, Hong Kong, pp. 463-473.

Chang, S.T. (1990). Future trends in cultivation of alternative mushrooms. Mush. J. **215:** 422-423.

Chang, ST. (1995). Ganoderma the leader in production and technology of mushroom nutraceulticals. Recent Advances in *Ganoderma lucidum* Research. Seoul. Korea; *The pharmaceutical society of Korea*, 43-52.

Chang, S.T. and Buswell, J.A. (1996). Mushroom Nutriceuticals. *World J. Microbiol. Biotechnol.*, **12:** 473-476.

Chang, S.T. and Miles, P.G. (1992). Mushroom biology–A new decipline. Mycologist, **6:** 64-65.

Chu, K.K., Ho, S.S. and Chow, A.H. (2002). *Coriolus versicolor*: A medicinal mushroom with promising immunotherapeutic values. *J. Clin. Pharmacol.* **42:** 976-984.

Cooke, R.C. (1977). Fungi, Man, and his Environm*ent*, Largman, London, New York, pp. 144.

Crisan, E.W. Sands (1978). A Nutritional value. In: Chang ST and Hayes WA (eds.). The biology and cultivation of edible mushrooms. Academic press, New York, pp. 172-189.

Delena, T. (1999). Edible and useful plants of Texas and South west–A practical guide university of Texas press, pp. 542.

Esselen, W.B. and Fellers, C.R. (1946). Mushrooms for food and flavor. *Bull. Mass. Agric. Exp. Sta.*, p 434.

Fasidi, I.O. and Kadiri, M. (1990a). Changes in nutrient contents of two Nigerian mushrooms. *Termitomyces robusts* (Beeli) Heim and *Lentinus subnudus* (Berk), during sporophore development. Die Nahrung, **34:** 141-420.

Ferreira, I.C.F.R., Baptista, P., Vilas-Boas, M. and Barros, L. (2007). Free-radical scavenging capacity and reducing power of wild edible mushrooms from northeast Portugal: individual cap and stipe activity. *Food Chem.*, **100:** 1511-1516.

Florezak, J., Karmnska, A. and Wedzisz, A. (2004). Comparision of the chemical contents of the selected wild growing mushrooms. *Bromatol. Chem. Toksykol.*, **37:** 365-371.

Florezak, J. and Lasota, W. (1995). Cadmium uptake and binding by artificially cultivated cultivated (*Pleurotus ostreatus*). *Bromatol. Chem. Toksykol.*, **28:** 17-23.

Friedman, M. (1996). Nutritional value of proteins from different food sources. A review *J. Agric. Food Chem.*, **44:** 6-29.

Gareth, J.E.B. (1990). Edible Mushrooms in Singapore and other South East Asian countries. *The Mycologist*, **4:** 119-124.

Gunde-Cimmerman, N. (1999). Medicinal value of the genus *Pleurotus* (fr). P Karst (*Agaricales* s.l. *Basidiomycetes*). *Int. J. Med. Mush.*, **1:** 69-80.

Gupta, S. and Sing, S.P. (1991). Nutritive value of mushroom *Podaxis pistillaris. Indian J. Mycol. Plant Pathol.*, **21:** 275-276.

Hadded, N.A. and Hayes, W.A. (1978). Nutritional factors and the composition of the *Agaricus bisporus* mycelium. *Mushroom Science*, **10:** 715-722.

Harsh, N.S.K., Rai, B.K. and Tiwari, D.P. (1993). Use of *Ganoderma lucidum* in folk medicine. *J. Trop. Biodivers.*, **1:** 324-326.

Hayes, W.A. and Haddad, N. (1976). The food value of the cultivated mushrooms and its importance in industry. *Mushroom J.*, **40:** 104- 110.

Hobbs, C. (1995). Medicinal mushrooms. an exploration of tradition, healing and culture, Botanica Press, 10226, Empire Grade, Santa Cruz, CA, 95060.

Houghton, W. (1995). Notices of fungi in the Greek and Latin author–Ann. Mag. Nat. His., **15:** 22-29.

Hugaes, D.H. (1962). Preliminary Characterization of the lipid constituents of the cultivated mushroom Agaricus campestris. *Mush. Sci.*, **5:** 540- 546.

Ikekawa, T., Uehara, N., Maeda, Y., Nakanishi, M. and Fukuoka, F. (1969). Antitumor activity of aqueous extracts of edible mushrooms. *Cancer Res.*, **29:** 734-175.

Issilogglu, M., Yilmaz, F., and Merdivan, M. (2001). Concentrations of trace elements in wild edible mushrooms. *Food Chem.*, **73:** 163-175.

Jandaik, C.L. and Kapoor, J.N. (1975). Cultural studies on some edible fungi. *Indian J. Mushrooms*, **1:** 22-26.

Jiskani, M.M. (2001). Energy potential of mushrooms. Dawn Econ. Bus. Rev., p. 4.

Jones, S. and Janardhanan, K.K. (2000). Antioxidant and antitumor activity of *Ganoderma lucidum* (curt ex Fr.). P. Karst-Reshi (Aphyllophoromycetieae) from south India. *Int. J. Med. Mushrooms*, **2:** 195-200.

Jong, S.C. and Birmingham, J.M. (1991). Medicinal benefits of the mushroom Ganoderma. *Adv. Appl. Microbiol.*, **37:** 101-134.

Kabir, Y., Kimura, S. and Tamura, T. (1988). Dietary effect of *Ganoderma lucidum* mushroom on blood pressure and lipid levels in spontaneously hypertensive rats (SHR). *J. Nutr. Sci. Vitaminol.*, **34:** 433-438.

Kalac, P. and Svoboda, L. (2000). A review of trace element concentrations in edible mushrooms. *Food Chem.*, **69:** 273-281.

Kallman, S. (1991). Nutritive value of Swedish wild plants. Svensk Bot. Tidskr., **85:** 397-406.

Kanwar, N., Sharma, B.M. and Sing B.M. (1990). Nutritive value of *Amanita caesarea* (Scop. ex. Fr.) Quel. *Indian J. Mycol. Plant Pathol.*, **20:** 249-250.

Kaul, T.N. (1978). Nutritive value of some edible *Morchellaceae. Ind. J. Mushroom*, **4:** 26-34.

Kim, B.K., Kim, H.W. and Choi, E.C. (1993). Anti-HIV activity of *Ganoderma lucidum. J. Biol. Chem.*, **264:** 472-478.

Kim, K.C. and Kim, I.G. (1999). *Ganoderma lucidum* extract protects DNA from strand breakage caused by hydroxyl radical and UV irradiation. *Int. J. Mol. Med.*, **4:** 273–277.

Kimura, Y., Tojima, H. And Fukase, S. (1994). Clinical evaluation of sizofilan as assistant immunotherapy in treatment of head and neck cancer. *Acta Otolaryngol.*, **511:** 192-195.

King, T.A. (1993). Mushrooms, the ultimate health food but little research in U. S to prove it. Mushroom News, **41:** 29-46.

Kino, K.Y., Yamaoka, K., Watanabe, J., Kotk, S.K. and Tsunoo, H. (1989). Isolation and characterization of a new immunomodulatory protein Zhi-8 (LZ-8) from *Ganoderma lucidum. J. Biol. Chem.*, **264:** 472- 478.

Lakshmi, B., Tilak, J.C., Adhikari, S., Devasagayan, T.P.A. and Janardhanan, K.K. (2005). Evaluation of antioxidant activity of selected Indian Wani *et al.*, 2603 mushrooms" *Inter J. Pharm. Biol.*, **42:** 179-185.

Lambert, E.B. (1938). Principles and problems of mushroom culture. *Bot. Rev.*, **4:** 397-426.

Li, G.S.F. and Chang, S.T. (1982). Nutritive value of *Volvariella volvacea*, In: Tropical mushrooms–Biological nature and cultivation methods (Chang ST, Quimio TH (eds)) Chinese university press Hong Kong, pp. 199-219.

Lintzel, W. (1941). The nutritional value of edible mushroom proteins. *Biochem. Acta.*, **308:** 413-419.

Litchfield, J.H. (1964). Nutrient content of morel mushroom mycelium: B vitamin composition. *J. Food Sci.*, **29:** 690-691.

Liu, F.O. and Chang, S.T. (1995). Antitumor components of culture filtrates from *Tricholoma sp. World J. Microbiol. Biotechnol.*, **11:** 486-490.

Maggioni, A., Passera, C., Renosto, F. and Benetti, E. (1968). Composition of cultivated mushrooms (*Agaricus bisporous*) during the growing cycle as affected by the nitrogen source in compositing. *J. Agr. Chem.*, **16:** 517-519.

Malinowska, E., Szefer, P. and Faradays, J. (2004). Metals bioaccumulation by bay Bolete, *Xerocomos badius* from selected sites. *Poland Food Chem.*, **84:** 405-416.

Manning, K. (1985). Food value and chemical composition. Flegg PB Spencer DM, Wood DA (Eds). The biology and technology of the cultivated Mushroom. John Willey and sons, New York, pp. 221-230.

Manzi, P.A., Agguzzi, A. and Pizzoferrato, L. (2001). Nutritional mushrooms widely consumed in Italy. *Food Chem.,* **73:** 321-325.

Mattila, P., Konko, K., Eurola, M., Pihlawa, J.M., Astola, J., Vahteristo, Lietaniemi V., Kumpulainen, J., Valtonen, M. and Piironen, V. (2001). Contents of vitamins, mineral elements, and some phenolic compounds in cultivated mushrooms. *J. Agric. Food Chem.,* **49:** 2343-2348.

Mattila, P.H., Piironen, V.I., Uusi- R. and Koivistoinen, P.E. (1994). Vit. D contents in edible mush. *J. Agr. Food Chem.,* **42:** 2449-2453.

Mattila, P.K., Konko, M., Eurola, J., Pihlava, J., Astola, L., Vahteristo, V., Hietaniemi, J., Kumpulainen, N., Valtonen, V. and Piironen, V. (2000). Contents of vitamins, mineral elements and some phenolic compounds in the cultivated mushrooms. *J. Agric. Food Chem.,* **49:** 2343-2348.

Mau, C.N., Huang, S.J. and Chen, C.C. (2004). Antioxidant properties of methanolic extracts from *Grifola frondosa, Morchella esculenta* and *Termitomyces albuminosus* mycelia. *Food Chem.,* **87:** 111-118.

Mc Connel, J.E. and Esselen, W.B. (1947). Carbohydrate in cultivated mushrooms. *Food Res.,* **12:** 118-121.

Mejstric, V. and Lepsova, A. (1993). Applicability of fungi to the monitoring of environmental pollution by heavy metals. In: B. Markert (ed) Plants as biomonitors. Germany: VCH weinheim, pp. 365-378

Mizuno, T. (1996). Oriental medicinal tradition of *Ganoderma lucidum* (Reishi) in India. In: *Ganoderma lucidum* (Mizuno,T and Kim,B.K eds.). Li Yang Pharm. Co. Ltd., Seoul, Korea, pp. 101-106.

Nanba, H. (1993). Maitake mushroom the king mushroom. *Mushroom News,* **41:** 22-25.

Ohtsuka, S., Ueno, S., Yoshikumi, C., Hirose, F., Ohmura, Y., Wada, T., Fujii, T. and Takahashi, E. (1997). Polysaccharides having an anticarcinogenic effect and a method of producing them from species of *Basidiomycetes.* UK Paten No. 1331513.

Orgundana, S.K. and Fagade, O. (1981). The nutritive value of some Nigerian edible mushrooms. In: Mushroom Science XI, Proceedings of the Eleventh International Scientific Congress on the Cultivation of Edible Fungi, Australia, pp. 123-131.

Oyetayo, F.L. (2007). Potential antioxidant properties of Nigerian edible 2604 *J. Med. Plant. Res. mushrooms.* Agro food Industry, Hi-tech., **18:** 44 -45.

Oyetayo, V.O. and Oyetayo, F.L. (2005). Preliminary investigation of health promoting potentials of *Lactobacillus fermentum* OVL and *Plerotus sajor caju* administered to rats. *Pakistan J. Nutr.,* **4:** 73-77.

Przybytniak, G. and Ambroz, H. (1999). Free radicals, their identification and determination. In: Marciniec B and Zak S. (eds.) Analytical methods in studies of pollutants and hazardous. Bydgoszcz: BTN, p. 17.

Purkayastha, R.P. and Chandra, A. (1976). Amino acid composition of protein of some edible mushroom growth in synthetic medium. *J. Food Sci. Technol.*, **3:** 13-17.

Puttaraju, N.G., Venkateshaiah, S.U., Dharmesh, S.M., Urs, S.M. and Somasundaram, R. (2006). Antioxidant Activity of Indigenous Edible Mushrooms. *J. Agric. Food Chem.*, **54:** 9764-9772.

Rai, R.D. (1994). Nutritional and medicinal values of mushrooms. In: *Advances in Horticulture.* (Chadha KL, Sharma SR eds.), Malhotra publishing house, New Delhi, pp. 537-551.

Rai, R.D. (1995). Nutritional and Medicinal Values of Mushrooms. In : *Advances in Horticulture* (S.R. Sharma and K.L. Chadha Feds) Malhotra Publishing House, New Delhi. **13** : 537-557.

Rai, R.D. (1997). Medicinal mushrooms. In: Advances in Mushroom Biology and Production (Rai RD, Dhar BL, Verma RN ed.) Mushroom society of India. NRCM, Solan, H.P., pp. 355-368.

Rai, R.D. and Saxena, S. (1989a). Biochemical changes during the post harvest storage of button mushroom (*Agaricus bisporus). Curr. Sci.*, **58:** 508-10.

Ren, L., Visitev, A.V., Grekhov, A.N., Tertov, V.V. and Tutelyan, V.A. (1989). Antiatherosclerotic properties of macrofungi. *Voprosy Pictaniya*, **1:** 16- 19.

Rolfe, R.T. and Rolfe, F.W. (1925). The Romance of the fungus world. Chapman and Hall Ltd. London, pp. 309.

Rudawska, M. and Leski, T. (2005) Macro and micro elemental contents in fruiting bodies of wild mushrooms from the Netecka forest in west–central Poland. *Food Chem.*, **92:** 499-502

Russell, R. and Paterson, M. (2006). *Ganoderma*–A therapeutic fungal factory *Phytochemistry J. Phytochem.*, **67:** 1985-2001.

Sadler, M. (2003). Nutritional properties of edible fungi. *Br. Nutr. Found. Nutr. Bull.* **28:** 305-308.

Sagakami, H., Aohi, T., Simpson, A and Tanuma, S. (1991). Induction of immunopotentiation activity by a proteinbound polysaccharide, PSK. *Anticancer Res.*, **11:** 993-1000.

Samajipati, N. (1978). Nutritive value of Indian edible mushrooms. *Mushroom Sci.*, **10:** 695-703.

Samorini, G. (2001). Fungi Hallucinogeni. Studi etnomicologici. Telesterion. Dozza Ed., Bologna, Italy, Shen (Guo J, Cheng HY, Wei X. eds.), pp. 250.

Sapers, G.M., Miller, R.L., Choi, S.W. and Cooke, P.H. (1999). Structure and composition of mushrooms as affected by hydrogen peroxide wash. *J. Food Sci.*, **64:** 889-892.

Schmitt, H.W. and Sticher, H. (1991). Heavy metal compounds in soil.In Merian,E (ed.). Metals and their compounds in the environment.Weinheim: VCH Verlagsgessellschaft, pp. 311-326.

Shah, H., Iqtidar, A. and Khalil, Jabeen, S. (1997). Nutritonal composition and protein quality of Pleurotus mushroom. *Sarhad. J. Agric.*, **13:** 621-626.

Sharma, R.P., Kaisth, K.R. and Lakhanpal, T.N. (1988). Protein and Mineral content of two edible *Lactarius species. Ind. J. Mushrooms*, **14:** 44-47.

Sharma, T.K. (2008). Vegetable caterpillar, Science Reporter. 5th May ISBN 0036-8512. National institute of science communication and information resources (NISCAIR), CSIR, pp. 33-35.

Shi, Y.L., James, A.E., Benzie, I.F.F. and Buswell, J.A. (2002). Mushroom derivedpreparation in the prevention of H2O2–induced oxidative damage to cellular DNA. *Teratoegensis* Carcinogenesis *Mutagenesis*, **22:** 103-111.

Singer, R. (1961). Mushrooms and Truffles, Leonard Hill Books Ltd., p. 272.

Singh, N.B. and Singh, P. (2002). Biochemical Composition of *Agaricus bisporus. J. Indian Bot. Soc.*, **81:** 235-237.

Sohi, H.S. (1988). Mushroom culture in India, Recent research findings. *Indian Phytopath*, **41:** 313-326.

Svoboda, L., Zimmermannova, K. and Kallac, P. (2001). Concentrations of Mercury, Cadmium, Lead, and Copper in the fruiting bodies of the edible mushrooms in an emission area of a copper smelter and a mercury smelter. *Sci. Total Environ.*, **246:** 61-67.

Tanaka, M., Kuei, C.W., Nagashima, Y. and Taguchi, T. (1998). Application of antioxidative maillrad reaction products from histidine and glucose to sardine products. *Nippon Suisan Gakkaishil*, **54:** 1409-1414.

Tseng, Y.H., and Mau, J.L. (1999). Contents of sugars free amino acids and free 5-nucleotides in mushroom, *Agaricus bisporus*, during the post harvest storage. *J. Sci. Food Agric.*, **79:** 1519-1523.

Varo, P., Lahelman, O., Nuurtamo, M., Saari, E. and Koivistoinen, P. (1980). Mineral element composition of Finish Food. VII Postal, Vegetables, fruits, berries, nuts and mushrooms. *Acta Agric. Scandinavica Supplement*, **22:** 107-113.

Velioglu, Y.S., Mazza, G., Gao, L. and Oomah, B.D. (1998). Antioxidant activity and total phenolics in selected fruits, vegetables, and grain products. *J. Agric. Food Chem.*, **46:** 4113-4117.

Verma, R.N., Singh, G.B. and Bilgrami, K.S. (1987). Fleshy fungal flora of N. E. H. India- I. Manipur and Meghalaya. Indian Mush. Sci., **2:** 414-421.

Wannet, W.J.B., Hermans, J.H.M., Vander Drift, C. and Op den Camp, H.J.M. (2000). HPCL detection of soluble carbohydrates involved in mannitol and trehalose metabolism in the edible mushroom, *Agaricus bisporus. J. Agaric. Food Chem.*, **48:** 287-291.

Wasserm, S.P. (2005). Reishi or Lingzhi (*Ganoderma lucidum*). Encyclopedia of Dietary Supplements, Marcel Dekker, Germany, pp. 603-622.

Wasser, S.P., and Weis, A.L. (1999). Medicinal properties of substances occurring in higher Basidiomycetes mushrooms: current perspectives (Review). *Int. J. Med. Mushrooms*, **1:** 1-62.

Weaver, K.C., Kroger, M. and Kneebone, L.R. (1977). Comparative protein studies on nine strains of *Agaricus bisporus* (Lange) *J. Food Sci.*, **42:** 364-366.

Willard, T. (1990). Reishi Mushroom : herb of spiritual potency and Medical wonder, Isaquah.

Wondratschek, I. and Roder, U. (1993). Monitoring of heavy metals in soils by higher fungi.In B. Market (Ed). *Plants as biomonitors*, pp. 365-378.

Yang, Q.Y., HU, Y.J., LI, X.Y., Yang, S.X., LIU, J.X., Liu, T.F., XU, G.M. and Liao, L.M. (1993). A New Biological Response Modifier–PSP. (In: CHANG ST, BUSWELL JA, CHIU SW (eds.) Mushroom Biology and Mushroom Products. The Chinese University Press, Hong Kong, pp. 247-259.

Yoshioka, Y, Ikekawa, T. Nida, M. and Fukuoka, F. (1975). Studies on antitumor activity of some fractions from *basidiomycetes* I. An antitumor acidic polysaccharide fraction of *Pleurotus ostreatus* (Fr.) *Quel. Chem. Pharm. Bull.*, **20:** 1175-1180.

Zakia, B. and Rajarathnam, S. (1994). Mushrooms-Human nutrition and health. In: Microbes for better Living. MICON 94, *35th AMI Cong.*, 9- 12 Nov., pp. 395- 399.

Zakia, S.A., El-Kattan, M.H., Hussein, W.A. and Khaled, A.M. (1993). Chemical composition and processing potential of oyster mushroom, *Pleurotus ostreatus*. *Egypt J. Agric. Res.*, **71:** 621-631.

Zhou, Z. Lin, J. Yin, Y. Zhao, J. Sun, X. and Tang, K. (2007). Ganodermataceae: Natural products and their related pharmacological functions. *Amer. J. Chin. Med.*, **35:** 559-574.

Zrodlowski, Z. (1995). The influence of washing and peeling of mushrooms *Agaricus bisporus* on the level of heavy metal contaminations. *Pol. J. Food Nutr. Sci.*, **4:** 23-33.

Chapter 18
Diseases of Mushroom

Vinit Pratap Singh and Gopal Singh

Department of Plant Pathology,
Sardar Vallabhbhai Patel University of Agriculture, and Technology,
Meerut – 250 110, U.P.

Like all other crops, mushrooms are also affected adversely by a large number of biotic and abiotic agents/factors. Among the biotic agents, fungi, bacteria, viruses, nematodes, insects and mites cause damage to mushrooms directly or indirectly. Diseases may appear in substrate, compost and casing soil. At any phase of growth an undesirable growth or development of certain moulds can occur and can adversely affect the final mushroom yield. Some common and important diseases of mushroom have been given as follows:

Fungal Diseases

Dry Bubble

Dry bubble is also known as *Verticillium* disease, brown spot and La mole. This is the most common and serious fungal disease of mushroom crop. If it is left uncontrolled, disease can totally destroy a crop in 2-3 weeks (Fletcher *et al.*, 1986). It was first reported in India during 1973. The disease delays the pin-head formation, reduces the number of sporophores and the yield of white button mushroom. It is more common when the cropping period is extended beyond 60 days. Artificial inoculation with the pathogen at the time of spawning and at different loads of inoculums had delayed pinhead formation by 5 days and reduced the number and weight of fruit bodies by 2.26-47.2 per cent and 2.19-38.01 per cent, respectively (Sharma and Vijay, 1993).

Symptoms

The symptom varies with the time of infection. Whitish mycelial growth is initially noticed on the casing soil which has a tendency to turn greyish yellow. If infection takes place in an early stage, typical onion shaped mushrooms are produced. Sometimes they appear as small-undifferentiated masses of tissue upto 2cm in diameter. When affected at later stage, crooked and deformed mushrooms with distorted stipes and with tilted cap can be seen. When a part of the cap is affected harelip symptom is noticed. Affected mushrooms are greyish in colour. If the infection occurs at later stage, grey mouldy fuzz can be seen on the mushrooms. Sometimes little pustules or lumps appear on the cap. On fully developed sporophores, it produces localized light brown depressed spots. Adjacent spots coalesce and form irregular brown blotches. Diseased caps shrink in blotched area, turn leathery, dry and show cracks.

Fungus

Verticillium fungicola var. fungicola: The fungus produces numerous hyaline, single–celled, thin–walled, oblong to cylindrical conidia and measure 3.5 to 15.9 x 1.5 to 5.0 um in size. Conidia are produced on lateral or terminal and vertically branched conidiophores. Condiophores are 220 to 800 x 1.5 to 5.0 um in size. Conidia accumulate in clusters surrounded by sticky mucilage.

Mode of Spread

The major source of contamination is debris and dust on the floor of the mushroom house. It is carried to the mushroom farm by infested casing soil. Conidia are produced in sticky clusters and sticks easily to any contact. It can spread to other growing rooms by spores in air, by mites, phorid and sciarid flies, equipments, hands and clothing of workers. Conidia are spread by splashing and running water. Excess water running off the beds carries the conidia to lower beds or to the floor of the mushroom house. When the floor dries, air movement over the floor surface carries the spores. The fungus is soil-borne and conidia survive in moist soil for one year. It also perpetuates as resting mycelium in the infected sporophore and spent compost. High humidity (90-95 per cent), lack of proper air circulation, delayed harvesting and temperature above 17°C favour the development and spread of disease.

Management

Use of sterilized casing soil, pasteurized compost and proper disposal of spent compost helps in reducing the incidence of the disease. Affected patches may be sprayed with 2 per cent commercial formalin. Spraying with mancozeb (0.25 per cent) or zineb (0.25 per cent) at 10 days interval controls the disease. A proper environmental condition like relative humidity to 80 to 85 per cent and temperature up to 14°C helps to reduce the disease incidence.

Wet Bubble

Wet bubble is also called as *Mycogone* disease, La mole, white mould and bubble disease. Wet bubble was first reported from Jammu and Kashmir India in 1978. It is a serious disease of white button mushroom when it develops early in the crop.

Symptoms

The symptoms vary with the time of infection. Smith (1924) recognized two main symptom types, infected sporophores and sclerodermoid masses, which he considered to be the result of infection by *M. perniciosa* at different stages in the development of the sporophores. Thus, when infection took place before the differentiation of stipe and pileus the sclerodermoid form resulted, whereas, infection after differentiation resulted in the production of thickened stipe with deformation of the gills. Clear brown drops caused by putrefying bacteria exude from the bubble at very high humidity or moisture. If the conditions are dry the destroyed mass becomes dry. When the beds are infected the disease appears in spots. White mycelial patches occur on the surface of casing following infection of a developing mushroom below the casing surface. When the infection occurs at a later stage of mushroom development, brown streaks are formed on the stalk and gills. The affected gill shows white mycelial growth.

Fungus

The disease, wet bubble, is caused by Mycogone perniciosa and the perfect stage is Hypomyces perniciosa. Mycelium of the pathogen is white, compact, felt- like. Hyphae branched interwoven, septate, hyaline, 3.5m broad. Conidiophores short, slender, branched, hyaline measuring 200 x 3-5m and having sub-verticillate to verticillate brances which bear thin walled, one-celled conidia measuring 5-10 x 4-5m. Large two-celled chlamydospores present; upper cell warty, thick walled, globose, bright coloured measuring 15-30 x 10-20m, lower cell hyaline, smooth and measure 5-10 x 4-5m.

Mode of Spread

The infection can be air-borne, water borne or may be mechanically carried by mites and flies. Chlamydospores have been reported to survive for a long time (upto 3 years) in casing soil and may serve as the primary source of inoculum. The fungal conidia produced on the infected mushroom spread by air, water splashes, flies, mites and by pickers. The optimum temperature for the mycelial growth is 25°C and the fungus infects few wild fleshy fungi also.

Management

Adoption of strict hygienic measures reduce wet bubble incidence. The casing soil should be sterilized properly before use. Spraying with zineb 0.3 per cent or mancozeb 0.3 per cent at weekly interval controls the disease. Spray of benomyl or chlorothalonil or thiabendazole 0.2 per cent is also recommended. Immediately after the spray the mushroom house should be kept closed for 8 to 10 hours. Later, ventilation should be provided.

Green Moulds

One of the most common and destructive diseases in mushroom cultivation is the green mould which is mainly caused by different species *of Trichoderma, Penicillium* and *Aspergillus*. Among these moulds, *Trichoderma* spp. induce significant quantitative and qualitative losses in the yield of *Agaricus bisporus, Pleurotus* spp., *Auricularia,*

Calocybe indica and *Lentinula edodes*. Kligman (1950) was the first to report the presence of *Trichoderma* in mushroom compost.

Symptoms

Trichoderma spp. are associated with green mould symptoms in compost, on casing soil, in spawn bottles and on grains after spawning. *Trichoderma viride* attacks the spawning tray and reduces the spawn run. It appears as green patches on the spawned and cased trays. When it appears on the casing soil it reduces the pinhead formation of the mushroom. *T. viride* causes reddish brown discolorations of the stipe and sunken lesions on the pileus. *T. koningii* grows as a cottony weft of grey mycelium over the casing surface. It also produces purple brown spots with a dry cracked surface. Infected caps turn brown. *T.hamatum* grows during spawn run. On young pin heads enlarged spots occur leading to cracking of cap. Later stipe is also infected. Under high humid conditions it causes brown spots on the caps.

Fungi

Different *Trichoderma* spp. are responsible for green moulds in composts and casing soil. They are *Trichoderma hamatum, T harzianum, T.koningii* and *T.viride. T.hamatum*. Mycelium is hyaline, septate and branched. Conidiophores branches at right angles. Phialides are pin-shaped and arise singly or in whorls. Phialospores are dark green and measure 2.5 to 7.5 x 2.5 to 3.0mm. Phialospores of *T. harzianum* are pale green and measure 4.0 to 6.5 x 3.5mm. Phialospores of *T. viride* are single celled, green, thin walled and measure 2.5 to 5.0 x 2.4mm. It produces chlamydospores in old cultures.

Mode of Spread

Green mould generally appears in compost rich in carbohydrates and deficient in nitrogen. If the compost is tampled too hard in the beds, or the filling weight is too high, this can make the peak heating difficult. This is certainly the case with compost which has a short texture and which might also have too high moisture content, resulting in improper pasteurization and conditioning of compost. Frequent use of formalin also tends to promote the development of green moulds (Sharma *et al.*, 1999). Different sources of primary inoculum of *Trichoderma* spp. could be dust particles, contaminated clothings, animal vectors especially the mite, *Pygmephorus mesembrinae*, mice and sciarid flies, air-borne infection, infected spawn, surface spawning, contamination of compost by handling and machinery and equipments at the mushroom farm. High relative humidity and low pH of the casing soil favour green moulds.

Management

Proper pasteurization and conditioning of compost check the green moulds. Supplements should be sterilized properly before their use. Dead mushrooms and cut stalks should be removed from mushroom house promptly and destroyed. Proper hygienic conditions should be maintained during mushroom growing. Spraying with zineb 0.2 per cent or carbendazim 0.1 per cent or thiabendazole 0.2 per cent or treatment with calcium hypochlorite 15 per cent on the used soil control the green moulds. *T. lignorum* was killed by exposing soil to 70°C for one hour. *T.lignorum* and *T. koningii* were eliminated by holding the compost at 60°C for 12 hours and then at 40°C for five days.

Cobweb Disease

It is also known as soft decay, *Dactylium* disease and mildew disease. It causes soft rot or decay of fruiting body. It was first reported on white button mushroom by Seth (1977) from Himachal Pradesh. This disease causes great damage to mushroom houses where humidity is high. The disease caused 51.7 per cent loss in yield (Sharma, 1992).

Symptoms

Small, circular, white patches of mycelium of the fungus appear on the surface of casing. Later the fungus grows as a fluffy white mould on the mushroom. The diseased mushroom turns brown and rots. The mycelium of the pathogenic fungus turns pink or red.

Fungus

Cladobotryum dendroides (*Dactylium dendroides;* Perfect stage *Hypomyces rosellus*). Mycelium is hyaline, branched and septate. Conidiophores are erect, simple or branched in many whorls. Conidia are single, elongate, 2 to 3 septate, slightly constricted at the base and measure 20 to 30 x 10 to 12.5 u in size.

Mode of Spread

The pathogen is soil-borne and is introduced into the crop by soil contamination or by farm workers. In the farm spores spread through air. High humidity and temperature favours cobweb disease.

Management

Sterilization of the casing mixture at 50°C for 4 hrs or disinfection of casing soil by benomyl (150g/100m² casing area) controls the disease. The relative humidity and temperature during picking should not exceed 90 per cent and 65°F. Dusting between flushes with zineb or mancozeb at 100g/100m² or spraying with formalin 0.2 to 0.3 per cent prevents the fungal attack.

Competitor/Indicator/Weed Moulds

False Truffle

It is also known as calve's brain disease. This is the most dreaded competitor in mushroom beds. In India, Sohi *et al.* (1965) observed false truffle causing serious losses to mushroom crops when the compost temperature in the trays reached beyond 22-24C. False truffle was reported from Uttar Pradesh, Himachal Pradesh, Punjab and Haryana in India and it is a limiting factor in cultivation of *A. bitroquis*. The disease is of common occurance during February or early March in *A. bisporus* in the plains of the Northern India and during summer months in *A. bisporus* and *A. bitorquis* in hilly regions of the country.

Symptoms

The disease appears as small weft of white to cream mycelium on the surface of the compost or under the casing soil. Later it becomes thick and develop into white, solid, wrinkled, round to irregular mass resembling brain or peeled walnut-like structure called ascocarp. Ascocarp appear in masses and raise the casing soil gently. Ascocaps are spherical to irregular, white to cream initially and turn brown at maturity. They finally disintegrate into a powdery mass emitting a chlorine-like smell. The fungus does not allow the mushroom mycelium to grow and compost turns dull brown. The spawn in affected patches turns soggy and disappears.

Fungus

Diehliomyces microsporus Diel. and *Lambert (Pseudobalsamia microspora)*. Ascocarps are fleshy. They contain many asci. Asci are oval or sub-spherical in shape with short or long stalks. Each ascus measures 19 to 27 x 10.5 to 15mm in size and contains three to eight ascospores. Ascospores are spherical, sulphur coloured with one distinct oil drop. Ascospore is 6.5mm in diameter. Chlamydospores may be noticed in the hyphal web of ascocarp.

Mode of Spread

Ascospores from casing soil and in wooden trays of previous crops are the sources of infection. Ascopores can survive for a periods of 5 years in soil and spent compost and mycelium for 6 months (Sharma, 1998) and thus serve as the major source of primary inoculum. Spread of the ascospores occurs in drainage water and on air-borne debris. Ascospores germinate at 30°C and its germination is stimulated by presence of actively growing mycelium of the mushroom. Optimum growth of the fungus has been recorded at 26-28°C.High compost temperature greatly favours the truffle development.

Management

Strict hygienic measures should be followed. Compost should be prepared only on concrete floors. Compost temperature during spawn run should not exceed 21 to 24°C. During cropping temperature should be kept below 18°C. Casing soil which contains ascospores should not be used. Young truffles can be picked and buried or burnt before fruit bodies turn brown. Drying of wood works and trays help eradication of the fungus. Initial infection can be checked by treating the affected patches with 2 per cent formalin solution (Sohi, 1988).

Olive Green Mould

The first evidence of the occurrence of *C. olivaceum* in India was provided by Gupta *et al.* (1975) at the mushroom farm at Chail, Kasauli and Taradevi. Another species, *C. globosum*, was later reported from mushroom farms in HP, Delhi and Mussorie (Thapa *et al.*, 1979). It causes yield loss ranging from 10 to 50 per cent in *A. bisporus*.

Symptoms

The fungus appears as greyish white mycelium in the compost or a fine aerial growth on the compost surface 10 days after spawning. Soon olive green to brown and pin head sized pethecia are formed. Spawn growth is delayed and reduced. Compost which does not support spawn will support the growth of *Chaetomium spp.*

Fungi

Chaetomium globosum and C. olivaceum. C. olivaceum: Perithecia are superficial, opaque, globose, thin, membranous with an apical tuft of dark bristles or setae. Asci are clavate and evanescent. Ascopores are dark brown, broadly ovate, umbonate at both ends. They measure 9.0 to 12.5 x 7.0 to 9.5mm in size. *C. globosum:* Perithecia are scattered or gregarious and broadly ovate or ellipsoid. Often they are pointed at the base and thickly clothed with slender flexuous hairs. Asci are oblong to clavate and evanescent. Ascospore dark, broadly ovoid, faintly apiculate at both ends and measure 8-9.5x6-8mm.

Mode of Spread

Compost and casing soil are the major source of infection. Ascospores are spread by air flows, clothes and other materials used in mushroom farm. These fungi are able to survive at higher levels of ammonia. The growth of *C. olivaceum* is favoured by alkalinity of the compost.

Management

Olive green mould can be prevented by good composting in Phase-I and control of environment during phase II. It is essential to control the peak heat conditions to avoid anaerobiosis. There should be sufficient time for peak-heating and sufficient supply of fresh air during pasteurization. Higher temperatures (above 60°C) for longer time should be avoided. Spraying with zineb 0.2 per cent controls the spread of the disease. (Sohi, 1986)

Brown Plaster Mould

Brown plaster mould caused by *Papulospora byssina* was first reported from India. It causes 90 to 92 per cent loss in *Agaricus bisporus.*

Symptoms

It is first noticed as whitish mycelial growth on the exposed surface of compost and casing soil in trays as well as on sides in bags due to moisture condensation. In course of time the colour changes to light brown to cinnamon brown and to rust colour at the end. In severe cases, no mushroom mycelium grows on places where plaster mould occurs.

Fungus

Papulospora byssina. Mycelium is septate and brown. It produces clusters of brown, many celled, spherical bulbils measuring 60 to 130 x 30 to 190mm.

Mode of Spread

Primary infection comes through air-borne bulbils or containers, compost and casing soil and workers. Wet compost and improper pasteurization of compost, higher temperature during spawn run and cropping favour the development of the fungus.

The fungus is commonly seen when gypsum is added to compost lesser than the required quantity.

Management

Composting should be made with addition of recommended level of gypsum. Peak heating should be of sufficient duration and at proper temperature. Compost should not be too wet before or after peak heating. Spraying on the affected patches with 2 to 4 per cent formalin reduces the spread. Spraying with carbendazim 0.1 per cent, benomyl 0.1 per cent, thiophanate methyl 0.1 per cent and carboxin 0.1 per cent are recommended for its control.

Yellow Moulds

All these fungi produce yellow mycelial growth in the compost. Yellow moulds on white button mushroom have been reported from Jammu and Kashmir (Kaul *et al.*, 1978), Punjab (Garcha *et al.*, 1987), Haryana and Himachal Pradesh in India.

Symptoms

These fungi form a yellowish brown corky mycelial layer at the interphase of compost and casing. The yellow moulds may develop in a layer below the casing (mat disease), form circular colonies in the compost (confetti) or they may be distributed throughout the compost (vert-de-gris). It becomes visible when it develops its stroma. They grow as competitors. They reduce the food supply to the mushroom or kill its mycelium by toxic metabolites.

Fungi

Chrysosporium luteum (Ces.) Gram, *C. sulphureum* (Cost & Matr.) Corm and *Myceliophthora lutea* Cost. *M. lutea:* Mycelium is white at first and turns yellow to dark, septate and branched. It produces following there kinds of spores:

1. Conidia which are smooth, ovoid and terminal.
2. Chlamydospores, which are smooth and thick-walled, and
3. Chlamydospores, which are spiny and thick-walled.

Mode of Spread

The primary infection is through air, chicken manure, spent compost and improperly sterilized wooden trays. The secondary spread is through flies, mites, water splashes, picking and tools used in mushroom growing. Yellow moulds prefer the same conditions as the mushroom fungus but are favoured for wet and improperly pasteurized compost. Severity of yellow moulds increased in the compost with 70 per cent moisture and 19 to 20°C temperature.

Management

Good farm hygiene, proper pasteurization of compost and casing layer reduces their incidence. Uses of air filters check the fungus. Spraying with benomyl 0.2 percent or copper oxychloride 0.2 per cent are also recommended. Spraying with calcium hypochlorite solution 1.5 per cent eradicates the moulds (Sohi, 1986).

Sepedonium Yellow Mould

Incidence of yellow mould disease caused by the fungus, *Sepedonium* was reported in India (Thapa *et al.*, 1991) and its incidence ranged from 5 to 20 per cent. Sometimes it caused total failure of the white button mushroom crop.

Symptoms

The mould occurs in the compost as white growth. Later the colour changes to yellow or tan. Generally it occurs at the basal layers of compost or at the bottom of the cropping bags. The fungus causes distortion of mushrooms probably due to production of volatile toxins. These toxins also inhibit mycelial growth of the mushroom.

Fungi

Sepedonium chrysosporium and *S. maheshwarianum (Hypomyces chrysospermum).* Mycelium is septate, hyaline and 3 to 5 cm in diameter. Conidiophores are erect and bear lateral simple or botryose cluster of branches, which are septate. At the tip of these branches conidia are borne singly.

Mode of Spread

Primary infection is from spent compost or improperly sterilized wooden trays. Spores spread through air. The fungus survives through thick-walled chlamydospores. Higher nitrogen content in the form of chicken manure favour the fungus. More wetness leads to its development in the lower layer of compost.

Management

Proper pasteurization of compost and provision of air-filters during spawning and spawn running (to prevent the entry of spores) reduces the incidence. Incorporation of carbendazim 0.5 per cent in the compost effectively controls it.

Ink Caps or Ink Weeds

Ink caps (*Coprinus* spp.) appear generally during spawn run in North India on white button mushroom. When the peak heating takes place at too low temperature these weedy mushrooms are commonly seen.

Symptoms

Ink caps appear in the compost during spawn run or in newly cased beds. The stipe of these weedy mushrooms is slender with bell shaped caps. Caps are cream coloured at first but later turn to bluish black. Sometimes it occurs in cluster. They decay and form a black shiny mass.

Fungi

Coprinus atramentarium, C commatus, C.fimetarious and *C.lagopus* occur in white button mushroom beds. Caps are 1.5 to 4.0 cm wide and campanulate. Surface of the cap is white when the cap expands its margin splits. Gills are 6 to 10 cm long, free, first white but soon turn black on liquefying. Stipe is 5 to 10 cm long, 2 to 3mm in diameter, hollow white, shiny, fragile and tapers upwards with a small bulb at the base. Basidiospores are black, elliptic and measure 8 to 12 x 3 to 5um in size.

Mode of Spread

The primary infection is through improperly pasteurized compost or casing soil. Normally ink caps are seen when the compost has excess nitrogen in the ammoniacal form, and when insufficient quantities of gypsum are added to the compost. They may also occur when the compost is wet and poor in feature.

Management

Proper pasteurization of compost and casing soil, avoiding excessive watering and rouging out the weedy mushrooms from the beds immediately after the emergence are the important measures of control. Before filling the trays the compost should be freed from ammonia. If the fruiting bodies of the weedy fungus are formed in large numbers in spawned trays then the compost should be repasteurized at 60°C for two hours and then recased.

White Plaster Mould

Occurrence of white plaster mould has been reported from different parts of India and it causes about 35 per cent loss in white button mushroom yield. It is not noticed in the beds infected with brown plaster mould or yellow moulds.

Symptoms

The fungus appears as white dense patches on the compost or casing soil. The patches may be small to more than 50 cm in diameter. After a week, the white growth changes to light pink. Due to its attack spawn run is affected.

Fungus

Scopulariopsis fimicola (Cost. & Matr.) *(Oospora fimicola).* Mycelium is septate. Conidiophores are short and branched. Annellospores are ovate, globose, round or show truncation stuff to avellaneious in mass. They occur in chains or in cluster and measure 4.8 to 9.0 x 4.8um in size.

Mode of Spread

Improper fermentation of manures and chilling of the beds cause occurrence of this fungus. The fungus is favoured by over-composted compost, which retains ammonia smell and has a pH of 8.2. Excessive moisture and inadequate ventilation favour of fungus.

Management

Proper composting and addition of optimum quantities of gypsum and water reduce the incidence. Removal of the fungal plaster and spraying with benomyl 0.1 per cent or spot application of formalin four per cent solution reduce the incidence.

Cinnamon Mould

It is also known and Cinnamon brown mould or brown mould and it has been reported to occur in Punjab, Himachal Pradesh and J&K in India.

Symptoms

The fungus appears as large circular patches of white aerial mycelium on the compost or casing. The colour changes to light yellow or to light golden brown. When the spores mature colour changes to cinnamon brown. The fungus produces many cup-like fleshy fruit bodies on the beds. The fungus inhibits the mycelium of the mushroom, depletes the nutrients in the compost and disfigure the fruiting bodies resulting in delay cropping and reduction in yield.

Fungi

Chromelosporium fulva (Syn. *Ostracoderma fulva* and *C. ollare*. *Peziza ostracoderma* is the perfect stage. Apothecia are small, discoid, dark brown, gelatinous and measure 1 to 2 cm in dia. Stem is 5 to 9 mm long. Asci are cylindrical and measure 80 to 160 x 8 to 12 mm. Each ascus contains eight ascospores in a single row. They are hyaline, ellipsoid and measure 8 to 12 x 4 to 8mm in size. Paraphyses are hyaline.

Mode of Spread

Primary infection is through casing soil and damp wood. The ascospores spread through air. The disease is favoured by over-pasteurized compost, high moisture content and by presence of excess ammonia in the compost.

Management

Casing soil should be sterilized and moisture content should be maintained properly. Newly cased beds can be protected by spraying with zineb. Spraying the trays and surrounding areas with sodium pentachlorophenol (3 to 5 lb/450 litres of water) also give protection against cinnamon mould.

Lipstick Mould

It is also called as ***red lipstick mould***. In India it has been reported from Punjab (Garcha *et al.*, 1987) and Himachal Pradesh (Sohi, 1988). In white button mushroom it usually occurs in crops previously attacked by virus disease.

Symptoms

The mould appears as fine, cottony white mycelial growth in the cracks or crevices of casing soil or in the sides of compost trays. As the spores of the fungus matures the colour changes to cherry red and finally to dull orange or buff. The fungus inhibits mycelial growth of the mushroom.

Fungus

Sporendonema purpurascens (Ben.) Mason & Hughes. Mycelium is white, septate and gets segmented into chains of one-celled, short, red, cylindrical spores with truncate ends.

Mode of Spread

Spent compost and casing soil are important sources of primary infection. The chicken manure is suspected to carry this fungus. The fungus is spread by pickers and water splashes.

Management

Cropping bed temperature exceeding 70°F should be avoided. It can be controlled effectively by heating with steam at 144°F for 90 minutes or by dusting with calcium hypochlorite or by using benomyl or zineb. Drenching the casing with zineb (1 lb/ 450 litres of water at 4.5 litres/100 sq. ft of bed) is also effective against lipstic mould.

Lilliputia Mould

In India incidence of *Lilliputia refula* (Perfect stage–*Gliocladium prolificum* Bainer) of 1 to 40 per cent has been reported from Delhi and Himachal Pradesh. It restricts the spawn spread in the compost prepared for white button mushroom and reduces the yield.

Mode of Spread

Chicken manure and horse manure are the sources of infection. Mycelium is viable for three months at 10°C whereas cleistothecia are viable for nine months.

Management

Spraying with zineb 0.02 per cent controls the mould.

Pink Mould

Pink mould has been reported from Himachal Pradesh (Seth and Munjal, 1981) and Jammu and Kashmir in India. During white button mushroom cultivation it appears first as white growth on the casing soil and later it turns pink.

Fungus

Cephalothecum roseum. Mycelium is branched and septate. Conidiophores are erect, branched and slightly swollen at the tip. Conidia are hyaline to pink, single, pear-shaped, two-celled and measure 11 to 18 x 7.5 to 9.5 um. The apical cell of the conidium is larger.

Mode of Spread

The conidia are spread through air.

Management

Pink mould is controlled by spraying twice at 9 to 10 days interval captan or thiram 0.04 per cent on the casing soil.

Bacterial Diseases

Bacterial Blotch

It is otherwise known as bacterial spot. It has been reported on white button mushroom from many countries including India. The disease is common in mushroom farms where there is poor ventilation system. Farm yard manure is used as one of the components in casing mixture often leads to bacterial blotch incidence.

Symptoms

The pathogen produces 2 to 3 mm deep pale yellow spot on the mushroom. Later the colour changes to golden yellow or chocolate brown. Its occurrence is noticed from early buttons stage onwards. It is also seen even on stored mushrooms. When pin heads are attacked they turn completely brown. Under favorable moisture conditions spots enlarge and cover the entire mushroom cap and stipe. Affected mushroom become sticky. Severely diseased mushrooms are distorted. Splitting at the blotched area is also noticed.

Bacterium

Pseudomonas fluorescens biotype G. (Syn. *P. tolaasii, Phytomonas tolaasii* and *Bacterium tolaasii*).

Mode of Spread

Casing and air-brone dust are the primary sources of infection. Bacteria on the mushroom cap will reproduce easily when moisture of free water persists for more than three hours after watering. High relative humidity and low temperature favour the infection. The bacterium survives between crops on the mushroom spores, surfaces in debris, peat, chalk and on tools used in mushroom production. Secondary spread is through hands of pickers, tools, ladder, implements, debris, sciarid flies and mites.

Management

Casing materials before and after mixing should be stored in areas free from the pathogen. Diseased mushrooms should be removed and destroyed. Preventive measures should be taken to check spread through picker's hand and watering. Adequate hygienic measures reducing the relative humidity (85 per cent) in the room by minimizing the number and volume of water sprays, improved circulation system (fresh air into cropping room) followed by spraying with chlorinated water (100-150 ppm, 0.5 litres/m^2) at three or four days interval to control disease. Spraying with streptomycin 200 ppm or oxytetracycline 300 ppm is effective in reducing the disease.

Bacterial Rot

The bacterium, *Pseudomonas alcaligens* is the incitant of bacterial rot in *Pleurotus sajor-caju*. The symptoms include water-soaked spots and yellowish brown discolouration of young sporophores and rotting of matured sporophores. Rotting starts from the centre of the sporophore towards periphery. The gills on the lower surface turn yellow. The caps become crinckled and rolls upward and inward.

Dipping in streptocycline solution (more than 100 ppm) or formalin (25 ppm) controlled this bacterium.

Brown Spot

Pseudomonas stutzeri is reported as competitive bacterium on paddy straw substrate used for the cultivation of *Pleurotus sajor-caju*. It induced brown spots in the substrate and caused 27 to 61 per cent yield reduction.

Dipping in streptocycline solution (more than 100 ppm) or formalin (25 ppm) controlled this bacterium.

Yellow Blotch

Yellow blotch on *P. sajor-caju* is reported to be caused by *Pseudomonas agarici* and it caused complete crop failure in Solan, Himachal Pradesh in India. The disease appears as blotches of varying sizes on pileus. The blotches are depressed, yellow and hazel-brown or orange in colour. If the disease is noticed during primordial stage the entire crop will be lost. The infected fruit bodies rot, become shiny and emit bad odour at higher temperature and humid conditions.

Spraying with oxytetracycline 400 ppm, streptocycline 400 ppm or sodium hypochlorite 400 ppm effectively controls the bacterium.

Virus Diseases

In 1948, a serious infectious disease of button mushroom was observed in the USA on a farm in Pennsylvania run by the La France brothers, and thus it was known as *La France* disease. In India, virus-like diseases were reported in white button mushroom and oyster mushroom. In 1967 in Netherlands, 4.5 per cent or about 790,000kg of mushroom was lost due to virus disease.

Symptoms

Mycelium disappears after the normal spread. Pinheads development is slow and they are also small. Pinheads appear late and below the surface of the casing layer. Sporophores have off-white caps and mature early. Caps may be small and flat. Stipes are slightly bent and elongated or watery in nature. Diseased mushrooms are loosely attached to substrate. The gills become hard. Sometimes diseased mushroom give out musty smell.

Viruses

Spherical virus particles of 24 to 26 nm diameter have been reported in *Pleurotus ostreatus, P. sapidus, P. columbinus* and *P. florida*. A polyhedral virus measuring 34 nm in diameter has been reported on *Volvariella volvacea*. Rod and spherical viruses have been reported on *Lentinus edodes*.

Mode of Spread

The viruses spread through mycelium, spores and germ tubes of mushrooms and through vectors. Phorid flies (*Megaselia halterata*) and mites help in introduction of virus particles to trays free from virus infection.

Management

Agaricus bitorquis has been reported to be immune to all viruses affecting *A. bisporus*.

References

Bahl, N. and Chowdhary, P.N. (1980). Podospora faurelii, a new competitor in the mushroom cultivation (Volvariella volvacea) Curr. Sci. **50:** 37

Bhatt, N. and Singh, R.P. (2000). Incidence and lossess in yield by fungal pathogens encountered from the beds of A. bisporus. Indian. J. Mush. **18:** 46-49

Doshi, A., Sharma, S.S. and Trivedi, A. (1991). Problems of competitor moulds and insect-pests and their control in the beds of *Calocybe indica* P&C. *Adv. Mush. Sci.*p57

Garcha, HS. (1978). Diseases of mushroom and their control. *Indian Mush. Sci.***1:** 185-191.

Gularia, D.S. (1976). A note on the occurrence of brown blotch of cultivated mushrooms in India. *Indian J. Mush* 2**(1):** 25.

Guleria, D.S., Thapa, C.D. and Jandaik, C.L. (1987). Occurrence of diseases and competitors during cultivation of *A. bitorquis* and their control. Natl. Symp. Adv. Mycol. PU Chandigarh pp55-56

Jandaik, C.L., Sharma, V.P. and Raina, R. (1993). Yellow blotch of *Pleurotus sajor-caju* (fr) Singer- a bacterial disease new to *India. Mush. Res.* **2:** 45-48

Nair, N.G. (1976). Diagnosis of musheroom virus diseases. Austr. Mush. Growers Assoc. J. 2 **(5):** 22- 24

Seth, P.K. and Dar, G.M. (1989). Studies on *Cladobotryum dendroides* causing cobweb disease of *A. bisporus* and its control. Mush. Sci. 12 **(2):** 711- 723

Sharma, S.R. 1994. Survey for diseases in cultivated mushrooms. Ann. Rep. NRCM, pp23

Sharma, S.R. and Kumar, S. (2000). Viral diseases of mushrooms. In: Diseases of Horticultural Crops Vegetables, Ornamentals and Mushrooms (Verma, LR and Sharma RC eds). Pp 166-178.Sceintific Publishers Jodhpur.

Sohi, H.S. (1986). Diseases and competitor moulds associated with mushroom culture and their control. Extension Bull. No. 2- 12p

Tewari, R.P. and Singh, S.J. (1984). Mushroom virus disease in India. *Mush. J.* **142:** 354-355

Tewari, R.P. and Singh, S.J. (1985). Studies on virus diseases of white button mushroom in India. *Indian J. Virol.* **1:** 35-41

Upadhyaya, R.C., Sohi, H.S. and Vijay, B. (1987). *Cladobotryum apiculatum*- a new mycoparasite of *Pleurotus* beds. *Indian Phytopath* **40:** 294

Vijay, B., Gupta, Y. and Sharma, S.R. (1993). Sepedonium maheshwarianum- a new competitor of *A. bisporus. Indian J. Mycol. Pl. Path.* **23:** 121.

Chapter 19

Phylogenetic Inference: Genes or Proteins?

Poonam Bhargava, Shivani Chandra and M. Krishna Mohan*

Birla Institute of Scientific Research, Statue Circle, Jaipur, Rajasthan

Systematics is the field of science that deals with the diversity of life and the relationship between life's components. Systematists practice systematic methods of reconstructing phylogenies in order to understand the pattern and process of evolution. A phylogeny is a pattern of common ancestry reflecting this evolutionary process. Phylogenetics, deals with identifying and understanding the evolutionary relationships among the different kinds of life on earth, both living (extant) and dead (extinct). Evolutionary theory states that similarity among individuals or species is attributable to common descent, or inheritance from a common ancestor. Thus, the relationships established by phylogenetics describe a species evolutionary history and, hence, its phylogeny, the historical relationships among lineages or organisms or their parts, such as their genes.

Classic phylogenetics dealt mainly with physical or morphological features–size, color, number of legs, etc. Modern phylogeny uses information extracted from genetic material–mainly DNA and protein sequences. The characters used are usually the DNA or protein sites (a site means a single position in the sequence). The relationships between species are then deduced from well conserved blocks in the alignment of several sequences, one from each examined species.

* Corresponding Author E-mail: pbhargava16@gmail.com

Phylogenetic Inference

Phylogenetic Inference is the estimation of the hierarchy of common descent. It is believed that all species on earth come from a common ancestor. The data collected from a pool of species, is used to reconstruct the history of speciation events that lead to their emergence. Phylogenies are useful for organizing knowledge of biological diversity, for structuring classifications, and for providing insight into events that occurred during evolution.

Since species originate over evolutionary time scales, by the splitting or branching process inherent in reproductive isolation their origination can not be observed directly. Instead, relative origination timing of species is inferred by comparing features that they possess. Data consisting of DNA and protein sequences is used to establish a relationship among distantly related individuals, as a result of their shared common ancestry

Cladograms and Phylogenetic Trees

Phylogenetic patterns generated from branching processes may be represented in at least two different ways: cladograms and phylogenetic trees. *Cladograms* are branching diagrams that illustrate patterns of phylogenetic relationships. The pattern of branching itself is the focus of a cladogram; the relative lengths of branches in cladograms have no special significance. Time is included in cladograms only in a relative sense, in the inter-nested structure of the cladogram itself.

A phylogenetic tree, also known as a phylogeny, is a diagram that depicts the lines of evolutionary descent of different species, organisms, or genes from a common ancestor. Furthermore, because these trees show descent from a common ancestor, and because much of the strongest evidence for evolution comes in the form of common ancestry, one must understand phylogenies in order to fully appreciate the overwhelming evidence supporting the theory of evolution. Because species originate over evolutionary time scales, by the splitting or branching process inherent in reproductive isolation, their origination cannot be observed directly. Instead, inference is drawn in the relative timing of origination of species by comparing features that they possess. Special methods of phylogenetic inference, have been developed over the last several decades, to compare features present in a given group of species to allow us to figure out how they are related to one another. These methods are based on the same simple principle that we use unconsciously to guess at human family relationships. Closely related individuals tend to share a larger number of similar features than do distantly related individuals, as a result of their shared common ancestry.

Phylogenetic Tree Basics

In a phylogenetic tree, each node with descendants represents the most recent common ancestor of the descendants, and edge lengths correspond to time estimates. Each node in a phylogenetic tree is called a taxonomic unit. Internal nodes are generally referred to as Hypothetical Taxonomic Units (HTUs) as they cannot be directly observed.

There are two kinds of phylogenetic trees; rooted and unrooted. A *rooted* phylogenetic tree is a directed tree with a unique node corresponding to the (usually imputed) most recent common ancestor of all the entities at the leaves of the tree. The most common method for rooting trees is the use of an uncontroversial outgroup– close enough to allow inference from sequence or trait data, but far enough to be a clear outgroup. The root represent the ultimate ancestor of the group of sequences (includes hierarchy). In the case of unrooted trees, branching relationships between taxa are specified by the way they are connected to each other, but the position of the common ancestor is not. Therefore, *Unrooted* tree lacks an "outgroup" or sequence from an accepted common ancestor. While *unrooted* phylogenetic trees can be generated from rooted ones by omitting the root from a rooted tree, a root cannot be inferred on an unrooted tree without either an outgroup or additional assumptions (for instance, about relative rates of divergence).

Methods of Constructing Trees

There are three main methods of constructing phylogenetic trees. Phenetic methods based on *distances* and cladistic methods based on *characters.*

- ✰ Distance-based methods such as neighbour-joining,
- ✰ Parsimony-based methods such as maximum parsimony, and
- ✰ Character-based methods such as maximum likelihood or Bayesian inference.

Distance Based Methods

The method measures the pair-wise distance/dissimilarity between two genes, the actual size of which depends on different definitions, and constructs the tree totally from the resultant distance matrix. The goal of distance based methods is to identify a tree that positions the neighbors correctly and that also has branch lengths which reproduce the original data as closely as possible. Finding the closest neighbors among a group of sequences by the distance method is often the first step in producing a multiple sequence alignment. This method employs the number of changes between each pair in a group of sequences to produce a phylogenetic tree of the group. The sequence pairs that have the smallest number of sequence changes between them are termed "neighbors." On a tree, these sequences share a node or common ancestor position and are each joined to that node by a branch.

Parsimony Based Methods

The phylogenetic analysis using maximum parsimony produces trees with no evolutionary root. This method predicts the evolutionary tree that minimizes the number of steps required to generate the observed variation in the sequences. For this reason, the method is also sometimes referred to as the minimum evolution method. A multiple sequence alignment is required to predict which sequence positions are likely to correspond. These positions will appear in vertical columns in the multiple sequence alignment. For each aligned position, phylogenetic trees that require the smallest number of evolutionary per cent changes to produce the observed sequence changes are identified. This analysis is continued for every position in the sequence alignment. Finally, those trees that produce the smallest number of changes overall for all sequence positions are identified. This method is used for sequences that are quite similar and for small numbers of sequences, for which, it is best suited. The algorithm followed is not particularly complicated, but it is guaranteed to find the best tree, because all possible trees relating a group of sequences are examined. For this reason, the method is quite time-consuming and is not useful for data that include a large number of sequences or sequences with a large amount of variation.

Character Based Methods

Character-based methods use the individual substitutions among the sequences to determine the most likely ancestral relationships. Character-state methods keep track of the amino acid or nucleotide at a give site in a sequence. They start with the known sequences and attempt to reconstruct the history of changes that had to take place from a common ancestor. Each branch on a tree of this kind has a length equal to the number of substitutions (or mutations) required to get from one node to the next. The program looks at large numbers of possible trees and chooses those that have the shortest total number of steps. The character-state methods tend to be more computers intensive. These include parsimony methods such as PAUP (Phylogenetic Analysis Using Parsimony), PROTPARS and DNAPARS in PHYLIP etc.

Nomenclature of Phylogenetic Trees

- **Node:** represents a taxonomic unit. This can be either an existing species or an ancestor.
- **Branch:** defines the relationship between the taxa in terms of descent and ancestry.
- **Topology**: the branching patterns of the tree.
- **Branch length**: represents the number of changes that have occurred in the branch.
- **Root**: the common ancestor of all taxa.
- **Distance scale**: scale that represents the number of differences between organisms or sequences.
- **Clade**: a group of two or more taxa or DNA sequences that includes both their common ancestor and all of their descendents.
- **Operational Taxonomic Unit (OTU)**: taxonomic level of sampling selected by the user to be used in a study, such as individuals, populations, species, genera, or bacterial strains.

rRNA Based Phylogenetic Analysis

Recently, genotypic classification based on nucleotide sequence comparison of ribosomal RiboNucleicAcid (rRNA) genes has become available as an additional taxonomic tool (Olsen and Woese, 1993). Sequencing ribosomal RNA (rRNA) genes is currently the method of choice for phylogenetic reconstruction, nucleic acid based detection and quantification of microbial diversity. It is assumed that ribosomal RNA must have been present since very early in the development of life forms, because it is essential for protein synthesis. Also, 16S rRNA, along with the 23S rRNA, has properties which predestine it as a universal phylogenetic marker. Every living organism, prokaryotes as well as eukaryotes, contains it either as a 16S or an 18S molecule (23S or 28S, respectively), and it has always the same function. Therefore, it is assumed that the rRNA genes contain a large number of highly conserved sequence patterns. There are a number of sequence differences which did not impair on the functioning of the ribosome, however, and which were maintained over evolutionary times. These can be used to distinguish phylogenetically different organisms. There are regions on the 16S rRNA which are quite conserved and others which are variable. Comparing the differences in the base sequence of 16S rRNA genes is, therefore, an excellent means to study evolutionary changes and phyolgenetic relatedness of organisms.

Comparative analyses of rRNA sequences, initiated in the 1970s, suggest that the living world is divided into three domains: Eucarya, Archaea (formerly archaebacteria), and Bacteria (formerly eubacteria). Ribosomal DNA (rDNA) sequences have been aligned and compared in a number of living organisms, and this approach has provided a wealth of information about phylogenetic relationships. Studies of rDNA sequences have been used to infer phylogenetic history across a very broad spectrum, from studies among the basal lineages of life to relationships among closely related species and populations. The reasons for the systematic versatility of rDNA include the numerous rates of evolution among different regions of rDNA (both among and within genes), the presence of many copies of most rDNA sequences per genome, and the pattern of concerted evolution that occurs among repeated copies. These features facilitate the analysis of rDNA by direct RNA sequencing, DNA sequencing (either by cloning or amplification), and restriction enzyme methodologies. Constraints imposed by secondary structure of rRNA and concerted evolution need to be considered in phylogenetic analyses, but these constraint do not appear to impede seriously the usefulness of rDNA. An analysis of aligned sequences of the four nuclear and two mitochondrial rRNA genes identified regions of these genes that are likely to be useful to address phylogenetic problems over a wide range of levels of divergence.

Molecular phylogeny increasingly supports the understanding of organismal relationships and provides the basis for the classification of microorganisms according to their natural affiliations. Comparative sequence analysis of ribosomal RNAs or the corresponding genes currently is the most widely used approach for the reconstruction of microbial phylogeny. The highly and less conserved primary and higher order structure elements of rRNAs document the history of microbial evolution and are informative for definite phylogenetic levels. An optimal alignment of the primary structures and a careful data selection are prerequisites for reliable phylogenetic conclusions. rRNA based phylogenetic trees can be reconstructed and the significance of their topologies evaluated by applying distance, maximum parsimony and maximum likelihood methods of phylogeny inference in comparison, and by fortuitous or directed resampling of the data set. Phylogenetic trees based on almost equivalent data sets of bacterial 23S and 16S rRNAs are in good agreement and their overall topologies are supported by alternative phylogenetic markers such as elongation factors and ATPase subunits. Besides their phylogenetic information content, the differently conserved primary structure regions of rRNAs provide target sites for specific hybridization probes which have been proven to be powerful tools for the identification of microbes on the basis of their phylogenetic relationships.

The Drawbacks and the Need of an Alternative Marker

As it is very clear from the above discussion that the use of molecular sequences for determining phylogeny largely relies on the assumption that changes in gene sequences occur in a time dependent manner (Felsenstein, 1982, Fitch and Margoliash, 1967, Kimura, 1983, Swofford and Olsen, 1990). This gives rise to the concept of evolutionary clock (136). This is the reason wh;y some sequences which change slowly are very well suited for the assessment of evolutionary history which obviously spans a very very long period. In the November 1999 issue of *ASM News*, p. 752-757, Ludwig and Schleifer state that the 16S rRNA-based phylogenies provide the best available method for understanding the evolutionary relationships among *Bacteria*. Since 16S rRNA was the first molecule examined for studying prokaryotic phylogenies, and has played a seminal role in evolutionary studies, the trees based on 16S rRNA have become the widely accepted "gold standards" for such purposes. However there are several associated problems with the use of gene sequences for phylogeny.

The first problem is the degeneracy of the code. Say for example two amino acids (Met and Trp) are encoded by atleast two codons differing only in there third position. This change in the third position does not change the protein sequence and does not caste any effect on the function carried out by these proteins. So this position is found to be different in very closely related species (Felsenstein, 1988). However in distantly related ones this position could have changed so many times that its importance is mostly nullified. Thus inclusion of this third base in the phylogenetic analysis generally leads to the noise in the signal.

Secondly the G+C content of species which is generally considered to be same for a given species appear to be another problem when using gene sequences for the phylogeny. The G+C content of different species is known to differ greatly (this is often true for two species within the same genus as well), and it is generally homogenized over the entire genome. In the protein-coding sequences, these differences in the G+C contents are accommodated by selective changes (*i.e.*, codon preferences) in the third codon positions. The species which are rich in G+C show a strong preference for codons that have G or C in the third position (often >90 per cent), whereas species with low G+C content predominantly utilize the codons with A or T in these positions. Thus, two unrelated species with similar G+C contents (*e.g.*, either very high or very low) may have very similar bases in the third codon positions. If phylogenetic analysis is carried out based on nucleic acid sequences, these species may show a strong affinity for each other but for the wrong reason (Steel *et al.*, 1993, Hashimoto *et al.*, 1994). Thus, the third codon positions, rather than being informative, can introduce major bias into the analyses (Gupta, 1998).

Thus while rRNA phylogenes have played a major role in understanding the evolutionary relationships among organisms, the answers to some of the central issues in prokaryotic phylogenies will likely emerge from consideration of other molecular sequences and by alternative approaches.

Protein Markers

After critical survey of completed genome sequences it has been suggested that there are only a limited number of genes that can serve as alternative markers for phylogenetic analysis. These are those sequences which share sufficient sequence similarity to be recognized as orthologous or paralogous.

Tatusov *et al.* (1997) have shown that majority of the universally conserved cluster of orthologous groups (COGs) belong to the information storage and processing proteins which appear to hold promise for future phylogenetic analysis. 110 COGs are present in all genomes with eight additional genes present in all prokaryotes. Additionally there are 126 COGs which are found in the remaining five microbial genomes excluding mycoplasmas. More than half of the information processing COGs contain ribosomal proteins which are small and therefore not sufficiently informative for the inference of global phylogeny. Thus one has to consider three major points before considering any protein as a marker (i) they should be of sufficient size and complexity to provide information, (ii) they should share sufficient sequence to be considered as orthologous or paralogous and (iii) a sufficient database on there sequence information should exist so that phylogenetic evaluation can be made. These constraints leave us with only a very few genes to consider. For example actin, a- and b-tubulin, glyceraldehyde-3-phosphate dehydrogenase, elongation factor, recA and hsp60 are some of the markers that sufficiently fulfill the requirements of the phylogentic inference while for others the ubiquity requirement cannot be assessed because of the limited number of sequences.

Elongation Factor

Elongation factor specially -1a has assumed a key importance when alternative markers are concerned. Since it has a central role in protein biosynthesis (Yager and von Hippel 1987). it is one of the widely studied molecule with a large database. Additionally it also interacts with some cytoskeletal proteins, especially actin (Durso and Cyr 1994; Shiina *et al.*, 1994). Additionally elongation factors assume importance as alternative marker because they are functionally different from rRNAs. A general assumption is that all the elongation factors are paralogous molecules resulting from ealy gene duplications. Some relevant phylogenetic questions have been successfully addressed using this marker, including the phylogenies of some groups of metazoans (Kobayashi, Wada, and Satoh 1996; McHugh 1997), the sisterhood of the fungal and animal clades (Baldauf and Palmer 1993), and the monophyly of the slime molds and their inclusion within the eukaryotic tree (Baldauf and Doolittle 1997).

Roger *et al.*, in there study of eleongation factor state that except for Ciliophora and Alveolata, all major eukaryotic groups are recovered as monophyletic by this molecule. In addition, the Archaebacteria appear to consistently root the eukaryotic EF-1a subtree on the diplomonad lineage. This was earlier attributed to poor sampling and taxonomic methods used. However, a better sampling of protists, especially the Ciliophora, did not improve the resolution of the EF-1a tree (Moreira *et al.*, 1999), suggesting that sparse taxonomic sampling is not a suf?cient explanation for the instability of the EF-1a tree. The instability of deeper nodes within the eukaryotic EF-1a subtree could also stem from the relatively few alignable positions available for analysis. Like taxonomic sampling, the number of positions is positively correlated with accuracy of phylogenetic reconstruction.

The EF-1a data set in the study of Roger *et al.* (1999) contained 435 alignable positions, which is relatively few compared with ;1,000 positions of SSU rRNA that are alignable between Archaebacteria and eukaryotes and much fewer than the several thousand positions likely necessary for highly accurate reconstructions of taxonomically well sampled data sets (Philippe, Chenuil, and Adoutte 1994; Hillis 1996). A study of the performance of methods in reconstructing the correct phylogeny using simulated data sets of a comparable length and divergence to EF-1a could be useful to test whether the instability of deeper nodes in the EF-1a tree is due to an inherent limitation on the amount of phylogenetic information this molecule can contain. Finally, it is possible that the lack of resolution in the eukaryotic EF-1a tree may be due to extremely short times between the events of cladogenesis that produced the major eukaryotic groups; in this case, the internal branching order of the eukaryotic tree would be intrinsically dif?cult to recover by molecular phylogenetic methods. This idea, known as the "big-bang" hypothesis of eukaryotic diversi?cation, holds that most of the observed resolution in global molecular phylogenies of eukaryotes is an artifact resulting from faster rate sequences being attracted to extremely distant outgroup sequences (Philippe and Adoutte 1998). In the case of EF-1a, this would require that the relatively robust deeply-branching position of the diplomonad lineage is artifactual, a possibility discussed above. The incongruencies between the branching orders of major eukaryotic groups in trees of different molecules, coupled with the often large variation in the relative rates of evolution of different groups in these trees, can be considered circumstantial evidence for this hypothesis (Philippe and Adoutte 1998). Once better models of the substitution process are developed and simulation studies are applied to these data sets, the "big-bang" hypothesis may be tested.

RNA Polymerases

These are essential components of the transcription process in all organisms and the genes of the largest subunit are highly conserved and ubiquitous. RNAPs contain 2300 to 2400 amino acids and these can be clearly aligned for phylogenetic purposes. Infact no paralogous genes are known for RNAPs. In general for bacteria the intradomain topology of the trees derived from both RNAp large subunits support the 16S rRNA based trees in almost all details. Except that the position of the root of the domain. The RNAp based phylogenetic tree places the mycoplasms close to bacteria as against the thermophiles.

Many prokaryotes have multiple ribosomal RNA operons. Generally, sequence differences between small subunit (SSU) rRNA genes are minor (0,1 per cent) and cause little concern for phylogenetic inference or environmental diversity studies. For Halobacteriales, an order of extremely halophilic, aerobic Archaea, within-genome SSU rRNA sequence divergence can exceed 5 per cent, rendering phylogenetic assignment problematic. The RNA polymerase B9 subunit gene (rpoB9) is a singlecopy conserved gene that may be an appropriate alternative phylogenetic marker for Halobacteriales. We sequenced a fragment of the rpoB9 gene from 21 species, encompassing 15 genera of Halobacteriales. To examine the utility of rpoB9 as a phylogenetic marker in Halobacteriales, Walsh *et al.* (2004) investigated three properties of rpoB9 trees: the variation in resolution between trees inferred from the rpoB9 DNA and RpoB9 protein alignment, the degree of mutational saturation between taxa, and congruence with the SSU rRNA tree. The rpoB9 DNA and protein trees were for the most part congruent and consistently recovered two well-supported monophyletic groups, the clade I and clade II haloarchaea, within a collection of less well resolved Halobacteriales lineages. A comparison of observed versus inferred numbers of substitution revealed mutational saturation in the rpoB9 DNA data set, particularly between more distant species. Thus, the RpoB9 protein sequence may be more reliable than the rpoB9 DNA sequence for inferring Halobacteriales phylogeny.

Proton translocating ATPase

The catalytic subunit of the proton translocation ATPase is one another protein marker. This has the advantage of having a reasonable dataset. It also has the advantage of having nothing in common with the transcription or translation except for its own synthesis. Eubacteria, mitochondria and chloroplasts contain a proton-translocating ATP-synthase (ATPase) complex (Futai and Kanazawa,1983). The complex is composed of two portions, designated F1 and Fo. Fo is intrinsic to the membrane and forms a proton channel. F1 is an extrinsic membrane protein complex composed of five subunits of which one, the /?-subunit, contains the catalytic site of the enzyme. The primary structure of the /3-subunit seems to be highly conserved throughout evolution (Walker *et al.*, 1984; Tybulewiez *et al.*, 1984; Falk *et al.*, 1985; Zilberstein et al., 1986; Kagawa *et al.*, 1986; Curtis, 1987; Cozens and Walker, 1987; Amann *et al.*, 1988). Since it is also widely distributed among eubacteria and shows functional constancy, it should be an ideal macromolecule for deducing the phylogenetic relationship of bacteria. In general this also supports the rRNA based tree. However local discrepancies are found.

Conclusion

The above discussion clearly reveals the use of molecular sequences be it protein or gene for the phylogenetic analysis. However this very debatable science still looks for a stable and undebatable marker. All sequences analyed till date have some advantages and some disadvantages. We are looking forward to a time when sufficient numbers of molecules have been studied over a sufficient phylogenetic breadth to identify those genes that best represent organismal history, and when the corresponding history has been inferred. From that point, the emphasis will naturally shift to the exceptions. Studies will delve into questions like: What were the genetic origins of chlorophyll-based photosynthesis? Has the ability been laterally transferred? What are the origins of the genes involved in aerobiosis? Has it been reinvented, transferred, or simply inherited by the diverse species that possess it today? For these and many other questions, the answers lie in comparing the organismal tree with the individual gene trees. Thus, the conclusions drawn today from the rRNAs and other universal molecules will provide an essential reference for understanding the rest of the genome.

Acknowledgments

Poonam Bhargava is thankful to DST for financial support in the form of a project. We are also thankful to the Birla Institute of Scientific Research, Jaipur for facilities.

References

Amann, R., Sostak, P., Ludwig, W. and **Schleifer, K.H.** (1988). Cloning and sequencing of genes encoding the beta subunits of the ATP-synthases of *Enterobacter aerogenes* and *Flawbacterium ferrugineum*. FEMS Microbiol. Lett. **50**: 101-106

Baldauf, S. and Palmer, J. (1993). Animals and fungi are each other's closest relatives: congruent evidence from multiple proteins. Proc Natl Acad Sci USA **90**:11558–62.

Brown, J.R. and **Doolittle, W.F.** (1997). Archaea and the prokaryote-to-eukaryote transition.Microbiol Mol Biol Rev **61**:456–502.

Cozensa, L. and **Walker, J.E.** (1987). The organization and sequence of the genes for ATP-synthase subunits in the cyanobacterium *Synechococcus* 6301. J. Mol. Biol. **194**: 359-383.

Curtis S.E. (1987). Genes encoding Beta and Epsilon subunits of the proton-translocating ATPase from *Anabaena* sp. strain PCC 7120. J. Bacteriol. **169**: 80-86.

Durso, N.A. and **Cyr, R.J.** (1994). A calmodulin-sensitive interaction between microtubules and a higher plant homolog of elongation factor-1a. Plant Cell **6**: 893–905

Falk, G., Hampe, A. and **Walker, J.E.** (1985). Nucleotide sequence of the *Rhodospirillum rubrum* atp operon. Biochem. J. **228**: 391-407.

Felsenstein, J. (1982). Numerical methods for inferring evolutionary trees. Q Rev Biol. **57**:379–404.

Felsenstein, J. (1988). Phylogenies from molecular sequences: inference and reliability. Ann Rev Genet. **22**:521–565.

Fitch, W.M. and **Margoliash, E.** (1967) Construction of phylogentic trees: a method based on mutational distances as estimated from cytochrome c sequences is of general applicability. Science.**155**: 279–284.

Futai, M. and **Kanazawa, H.** (1983). Structure and function of proton-translocating adenosine triphosphatase (FoF I): biochemical and molecular biological approach. Microbiol. Rev. **47**: 285-312

Gupta, R.S. (1998) Protein Phylogenies and Signature Sequences: A Reappraisal of Evolutionary Relationships among Archaebacteria, Eubacteria, and Eukaryotes. Microbiol. Mol. Biol. Rev. **62**: 1435-1491.

Hashimoto, T., Nakamura, Y., Nakamura, F., Shirakura, T., Adachi, J., Goto, N., Okamoto, K. and **Hasegawa, M.** (1994). Protein phylogeny gives a robust estimation for early divergences of eukaryotes: phylogenetic place of a mitochondria-lacking protozoan, *Giardia lamblia*. Mol Biol Evol. **11**:65–71.

Hillis, D.M. (1996). Inferring complex phylogenies. Nature **383**:130–131.

Kawaga, Y., Ishizuka, M., Saishu, T. and **Nakao, S.** (1986). Stable structure of thermophilic proton ATPase beta-subunit. J. Biochem. **100**: 923-934

Kimura, M. (1983). The neutral theory of molecular evolution. Cambridge, England: Cambridge University Press.

Kobayashi, M., Wada, H. and Satoh, N. (1996). Early evolution of the Metazoa and phylogenetic status of diploblasts as inferred from amino acid sequence of elongation factor-1 alpha. Mol Phylogenet Evol. **5**:414-422.

McHugh, D. (1997). Molecular evidence that echiurans and pogonophorans are derived annelids. Proc. Natn. Acad. Sci. U.S.A. **94**:8006-8009.

Moreira, D., Le Guyader, H. and Philippe, H. (1999). Unusually high evolutionary rate of the elongation factor 1 alpha genes from the Ciliophora and its impact on the phylogeny of eukaryotes. Mol. Biol. Evol. **16**:234–245.

Olsen, G.J. and Woese,C.R. (1993). Ribosomal RNA: a key to phylogeny. Faseb J. **7**: 113-123.

Philippe, H. and Adoutte, A. (1996). How reliable is our current view of eukaryotic phylogeny? In: Protistological actualities (Eds. Brugerolle G. and J.-P. Mignot, J.-P.) Clermont-Ferrand. p. 17–33

Philippe, H., Chenuil, A. and Adoutte, A. (1994). Can the Cambrian explosion be inferred through molecular phylogeny? Dev. Suppl. 15–25.

Roger, A.J., Sandblom, O., Doolittle, W.F. and Philippe, H. (1999) An evaluation of elongation factor 1a as a phylogenetic marker for eukaryotes. Mol. Biol. Evol. **16**:218–233.

Shiina, N., Gotoh, Y., Kubomura, N., Iwamatsu, A. and Nishida, E. (1994). Microtubule severing by elongation factor 1a. Science **266**: 282–285.

Steel, M. A., Lockhart, P. J. and Penny, D. (1993) Confidence in evolutionary trees from biological sequence data. Nature **364**:440–442.

Swofford, D. L. and Olsen, G.L. (1990) Phylogeny reconstruction. In: Molecular systematic (Eds. Hillis, D. and Moritz, C.). Sunderland, Mass: Sinauer Associates, Inc., p. 411–501.

Tatusov, R.L., Koonin, E.V. and Lipman, D.J. (1997) A genomic perspective on protein families. Science. **24**:631-637.

Tybulewicz, V.L.J., Falk, G. and **Walker, J.E.** (1984). *Rhodopseudomonas blastica atp* operon : nucleotide sequence and transcription. *Journal of Molecular Biology* **178**: 185-214.

Walker, J.E., Saraste, M. and Gay, N.J. (1984). The *unc* operon : nucleotide sequence, regulation and structure of ATP-synthase. Biochimica et biophysica acta **768:** 164-200.

Walsh, D.A., Bapteste, E., Kamekura, M. and Doolittle, W.F. (2004). Evolution of the RNA Polymerase B9 Subunit Gene (rpoB9) in Halobacteriales: a complementary molecular marker to the SSU rRNA Gene. Mol. Biol. Evol. **21:** 2340–2351.

Yager, T.D. and von Hippel, P.H. (1987). Transcript elongation and termination in *Escherichia coli.* In: The Molecular and Cell Biology of *E. coli* and *S. typhimurium* (Ed. Neidhardt, E,). American Society of Microbiology, Washington, D.C. p. 1241-1275.

Zilberstein, D., Shirvan, M.H., Barile, M.F. and Rottem, S. (1986). The p-subunit of the FIFOATPase is conserved in mycoplasmas. J. Biochem. **261:** 7109-7111.

Chapter 20

Viral Diseases and its Mixed Infection in Mungbean and Urdbean: Major Biotic Constrains in Production of Food Pulses in India

Kajal K. Biswas, Avijit Tarafdar and Koushik Biswas*

Plant Virology Unit, Division of Plant Pathology, Indian Agricultural Research Institute, New Delhi – 110 012

India has the largest pulse producing area of about 24 million ha with production of grains only 14.8 million tones per annum. The productivity of pulses staggers around 600 kg/ha well below the world average productivity of 846 kg/ha over last 15-20 years. The low productivity of the pulse in India is caused by the effect of several biotic constraints, of which diseases caused by viral pathogens are most important. Further, the variation in production of pulses in India is common as pulses are cultivated mainly under rainfed conditions. The over increasing population of the country and low productivity of pulses has resulted depletion of per capita availability to about 40 gm in recent year. Thus, it is necessitated to import pulses from neighboring countries. Therefore, it is urgent need to overcome theconstrains of

* Corresponding Author E-mail: kkbiswas@mailcity.com, drkkbiswas@yahoo.co.in

pulse production, so that, our country is able to produce about 27 million tones pulses by 2015 to meet the demand of our citizens.

Mungbean and urdbean are mainly cultivated in many tropical and sub-tropical continents like Pakistan, Bangladesh, Thailand, Philippines, China, Vietnam, Indonesia and Burma. The major pulses grown in this country are chick pea (*Cicer arietinum*), pigeon pea (*Cajanus cajan*), mungbean (*Vigna radiata*), urdbean (*V. mungo*), lentil (*Lens culinaris*), peas (*Pisum sativum*), French bean (*Phaseolus vulgaris*) and cowpea (*P. unguiculata*). More than 100 viruses have been reported to causes diseases in the pulse crops. In present chapter viral disease in two important pulse, mungbean and urdbean will be discussed.

Viral Diseases in Mungbean and Urdbean

The productivity of pulses in pulse growing countries is severely affected by several emerging viral diseases; many of them are of great economic importance to cause serious economic losses (Table 20.1). However, extent of crop losses caused by viruses depends on several factors. Although there are several viral diseases infecting mungbean and urdbean reported (Table 20.2), among them thrips (*Thrip tabaci*) transmitted *Groundnut bud necrosis virus* (GBNV), whitefly (*Bemisia tabaci*) transmitted *Mungbean yellow mosaic India virus* (MYMIV), and uncharacterized leaf crinkle virus complex (LCD) are most important with great economic losses in Indian condition (Biswas and Varma, 2000; 2001; Varma and Malathi, 2003).

Table 20.1: Economic Losses Caused by Viruses in Mungbean and Urdbean in India

Crop	*Virus*	*Yield loss (per cent)*
Mungbean	MYMV	100
	ULCD	94
	GBNV	91
Urdbean	MYMV	100
	ULCD	100

Table 20.2: Plant Viruses Infecting Mungbean and Urdbean, Experimentally and in Field Condition, in India

Mungbean	*Urdbean*
Alfalfa mosaic virus (AMV)#	Cucumber mosaic virus (Cucumo) CMV#
Bean common mosaic virus (poty) BCMV#	Horsegram yellow mosaic virus
Bean yellow mosaic virus (poty) BYMV#	Limabean mosaic cucumo virus
Blackeye yellow mosaic virus (poty) (BlCMV)#	Mungbean leaf curl virus
Blackgram mottle virus (Caromo) BMoV#	Mungbean mosaic virus
Blackgram mild mottle virus (BgMMV)#	Mung and urd mosaic virus 1(MUMV-1)#
Cowpea aphid-borne mosaic virus (poty) CABMV#	Mung and urd mosaic virus -2(MUMV-2)#
Cowpea (vein) banding mosaic (Cucumo) virus CPBMV#	Mungbean yellow mosaic virus (gemini) MYMV
Cowpea (chavali) mosaic virus (CPCMV)#	Pea leaf roll luteovirus (PelRV)

Contd...

Table 20.2–*Contd...*

Mungbean	*Urdbean*
Cowpea chlorotic mottle virus (Bromo) CCMV	Groundnut bud necrosis tospovirus
Cowpea chlorotic spot tobamo virus (CPCSV) #	Groundnut bud necrosis illarvirus
Cowpea golden mosaic virus (Gemini) CPGMV	Peanut mottle poty virus (PeMoV)#
Cowpea mild mottle virus (Carla) CPMMV	Southern bean mosaic poty virus (SBMV) #
Cowpea mosaic virus (Como)CPMV#	Tobacco ring spot nepo virus (TRSV) #
Cowpea mottle virus (Carmo)CPMoV	Tomato streak ilar virus (TSV)
Cowpea necrosis virus (Cucumo) CPNV	Tomato spotted wilt tospo virus (TSWV)
Cowpea severe mosaic virus (Como) CPSMV	Urdbean leaf crinkle disease ULCD) #
Cowpea top necrosis virus CPTNV	Voandezia distortion virus(VDMV) poty
Cowpea yellow fleck virus CPYFV	Voandezia necrotic mosaic virus(VNMV) tymo

#: Seed transmitted.

Bud Necrosis

Tospoviruses are fast emerging as serious pathogens affecting the cultivation of field and horticulture crops. In India, so far four distinct *tospoviruses* namely, GBNV and PYSV from groundnut (Reddy *et al.*, 1992; Styanarayana *et al.*, 1996), WBNV from watermelon (Jain *et al.*, 1998) and IYSV from onion (Ravi *et al.*, 2005) have been identified. Necrosis disease caused by GBNV belonging to the genus tospovirus is emerging as a serious pathogen not only in legumes but also in other crops in India (Rao *et al.*, 2003; Bhat *et al.*, 2002; Sarita and Jain, 2007; Biswas *et al.*, 2009). Natural occurrence of *Tospovirus* was also observed on several vegetable and leguminous crops such as tomato (Rao *et al.*, 1985); Sastry 1982; Sabitha *et al.*, 1984; Paul Khurana *et al.*, 1990), urd and mungbean (Amin *et al.*, 1985), peas (Rao *et al.*, 1985), cowpea, chilli and brinjal (Prasada Rao *et al.*, 1987) potato (Khurana *et al.*, 1989) and soybean (Rao *et al.*, 1985) and the virus was designated as TSWV. Symptoms induced by Tospovirus in tomato plants are highly variable. The severely infected plants produced brown necrotic streaks on the petioles, stems and on terminal bud (Sabitha *et al.*, 1984).

Symptoms on mungbean and urdbean included necrosis of all plant parts including leaves, stems, petioles and growing point of plant bud and pods (Bhat *et al.*, 2001 and Biswas *et al.*, 2009). Since the fist report of GBNV in legumes, the incidence and severity of GBNV in mungbean are reported to be increasing year to year causing necrosis of all plant parts including stem, petioles, leaves, vein, buds and also pods resulting in death of plant (Saritha, 2007). Recently at experimental Farm in IARI, New Delhi, the incidence of GBNV upto 52 per cent and yield loss upto 92 per cent have been recorded in mungbean (Biswas, unpublished data). Occurrence of multiple or mixed infection with MYMV and GBNV is recently emerging as serious constraints causing tremendous crop losses in legumes like mungbean, urdbean (Biswas *et al.*, unpublished data).

Complete genome sequences of Topspoviruses revealed that the virus is partitioned into three single stranded RNA segments, L (large) RNA, M (medium) RNA and S (small) RNA (Figure 20.1). The L RNA is completely negative sense 8.9 kb nucleotide length containing single ORF in viral complementarily sense encoding the 331.5 kDa L protein as putative RNA Dependent RNA Polymerase (RdRP). The M RNA is approximately 4.8 kb and contains 2 ORFs separated by an intergenic region. The S RNA has approximately 3 kb and contains two ORFs separated by a large intergenic region. ORF at 5′ end designated as NSs codes for non structural viral protein. The NSs protein is believed to play a role of suppressor molecule in Post Transcriptional Gene Silencing (PTGS). The ORF near to 3′end designated as nucleocapsid (N) protein gene codes for the N protein which encapsidates the viral RNAs within the viral envelope.

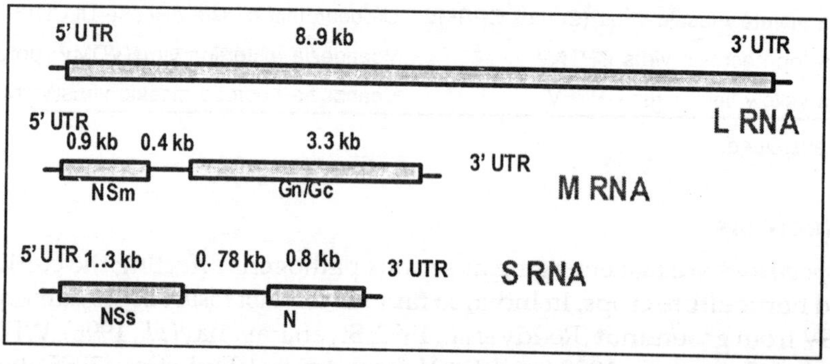

Figure 20.1: L-RNA, M RNA and S RNA of Genome of Mungbean GBNV Isolate

Yellow mosaic: In India, the yellow mosaic disease affecting variety of legume hosts is caused by two begomovirus species, *Mungbean yellow mosaic India virus* (MYMIV) and *Mungbean yellow mosaic viruses* (MYMV) belonging to the genus *Begomovirus* of the family *Geminiviridae*. Nucleotide sequence comparison revealed that there are four species causing YMD in India. Sequence analysis clearly showed that there are four species of yellow mosaic viruses, *Mungbean yellow mosaic India virus* occurring in North, Central and Eastern India, *Mungbean yellow mosaic virus* occurring in South & Western India, *Horsegram yellow mosaic virus* in South India, *Dolichos yellow mosaic* virus in all over India (Varma and Malathi, 2003). A satellite DNA β was found associated with MYMIV contributing to severity in symptoms; it was the first report of DNA β with a bipartite begomovirus.

In India, YMD was first reported in Lima bean (*Phaseolus lunatus*) in western India in 1940s. Later in 1950, YMD was seen in dolichos (*Lablab purpureus*) in Pune, Nariani (1960) observed YMD in mungbean (*Vigna radiata*) in the experimental fields at Indian Agricultural Research Institute and was subsequently observed throughout India in almost all the legume crops, such as blackgram (*V. mungo*), soybean (*Glycine max;*) pigeonpea (*Cajanus cajan*), horsegram (*Macrotyloma uniforum*) Frenchbean (*Phaseolus vulgaris*) and cowpea (*Vigna unguiculata*) (Wiliams *et al*., 1968; Nene, 1972).

It is difficult to estimate the yield loss due to YMD as it varies depending on the place, cultivar and time of infection (Nene, 1972). The loss in yield is more than 60 per cent when infection occurs within twenty days after sowing. Varma *et al.* (1992) predicted that the yield loss due to YMD could be as high as $ 300 million in an epidemic year taking blackgram, mungbean and soybean together. Now it has become a major constrain in pulse production in India causing estimated annual yield losses of around US $ 300 million in urdbean, mungbean and soybean together (Varma and Malathi, 2003).

The MYMV and MYMIV have genome organization similar to the Old World bipartite but percent nucleotide identity in DNA A and DNA B between the two species share only ~80 per cent identity in DNA A justifying their designation as distinct species. Both of the genome consists of two circular ssDNA components, designated as DNA A and B which are encapsidated separately. The ~2745 nt DNA A component codes for the coat protein and the protein for viral DNA replication and transcription. The~2616 nt DNA B encodes proteins for movement and nuclear localization. DNA A replicates autonomously and is dependent on DNA B for movement function. Replication of DNA B component is dependent on DNA A encoded proteins and both DNA A and DNA B are essential for viral pathogenecity.

DNA A has two open reading frames (ORFs) in the viral sense, (ORF AV1-coat protein; AV2-pre coat protein) and five in the complementary sense, ORF AC1-replication associated protein, Rep; ORF AC2-transcription activator protein, TrAP; ORF AC3-replication enhancer protein, REn, AC4 and AC5. ORF AV2 is present only in Old World Begomoviruses. The role of ORF AV2, AC4 and AC5 is yet to be deciphered. In DNA B, there is one ORF in viral sense strand, ORF BV1-nuclear shuttle protein (NSP) and one in complementary sense, ORF BC1-movement protein (Figure 20.2). Infectivity of cloned DNAs of MYMIV in urdbean, mungbean, cowpea, pigeonpea and soybean has been established by "agroinoculation" on leguminous hosts (Biswas and Varma, 2001). In this strategy more than one copy of DNA A and

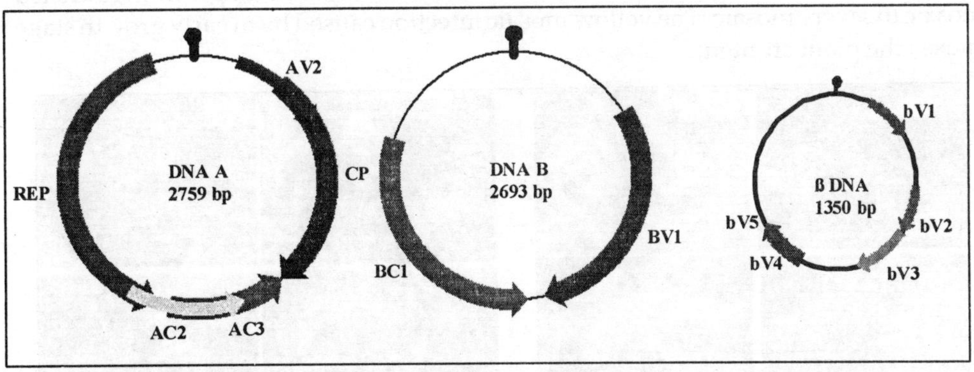

Figure 20.2: Genome Organization of Bipartite DNA A and DNA B Components of MYMIV

Arrows indicate various open reading frames (ORFs); ORFs in sense strand are AV1, AV2 in DNA A and BV1 in DNA B. ORFs in complementary sense strand are AC1, AC2, AC3, AC4 and AC5 in DNA A and BC1 in DNA B. The common region (CR) harboring the stem and loop structure is also shown.

DNA B components are cloned in a binary vector, which is mobilized to *Agrobacterium tumefaciens*.

Leaf Crinkle

Leaf crinkle disease (LCD) in pulses has been reported from India in the beginning of 19th century in Pusa Bihar. LCD in urdbean is first reported by Willams *et al*. (1968) from Delhi. This disease has been wide spread in all the urdbean growing region of India and estimated crop losses of 62-100 per cent is reported depending on the growth stage, cropping seasons and cultivars (Makkouk *et al*., 2003; Sharma *et al*., 2007). The causative agent of this disease is considered to be a virus but morphology, type of genome and mode of transmission is not yet confirmed. There are reports that ULCV is transmitted through seed, sap and varieties of insect vectors, indicating the disease may be caused by mixed infection by two or more viruses. Association of poty, tospo and tymo viruses is suspected. LCV is transmitted through seed to an extent of 18.39 per cent (Nene, 1972; Sharma *et al*., 2007).

Mixed Infection

Occurrence of multiple/mixed viral infections in mungbean and urdbean in all the pulse growing regions of India are common in *kharif* as well as in *pre-kharif* seasons. The mixed infection with MYMIV, GBNV and LCD in these crops has been characterized based on biological and molecular based diagnostic techniques in recent year (Biswas *et al*., 2009). Varied degrees of diseases caused by single as well as mixed infections have been estimated both the kharif and *pre-kharif* seasons; that varied in crops, cultivars, seasons and years. Infected crops appear various kinds of disease symptoms including yellow mosaic on leaves and pods; necrosis of buds, stem, petioles and vein; and leaf crinkle caused by individual as well as mixed infections (Figures 20.3a and b). Top necrosis caused by GBNV in early growth stage led to complete death of the plant. The severity of yellow mosaic ranges from restricted mosaic to severe mosaic. The yellow mosaic infection caused by in early growth stage causes the plant stunted.

Figure 20.3: Symptoms on Urdbean (cv.T 9) by GBNV, LCD, MYMIV+GBNV and MYMIV+LCD (a); on Mungbean (cv. Pant M-4) by LCD, GBNV+LCD, MYMV+GBNV, MYMV+LCD (b)

Recently, studies of viral disease in *Kharif* seasons in from the year of 2006 to 2009 shows that over all MYMIV incidence ranges from 8.0 to 76.0 per cent. All the mungbean cultivars are studied are susceptible to LCD with disease ranging from 4.0 to 61.0 per cent during these period of four years. Incidence of LCD increases gradually from the initial year of 2006 (5-15 per cent) to 2009 (20-32 per cent). In the case of GBNV incidence, there is a wide range of disease incidence from 1.0 to 52.0 per cent

The mixed infections by MYMIV, GBNV and LCD in *Kharif* season studied. Maximum incidence up to 11.50 per cent against YMD+GBNV, 8.2 per cent against YMD+ULCD and 17.1 per cent against YMD+ULCD respectively was recorded. The incidence of mixed infection caused by MYMIV+LCD, MYMIV+GBNV, LCD+GBNV and MYMIV+LCD+GBNV are of 1.0-26.1 per cent, 1.5-21.1 per cent, 1.63-37.5 per cent and 1.1 to 15.8 per cent respectively.

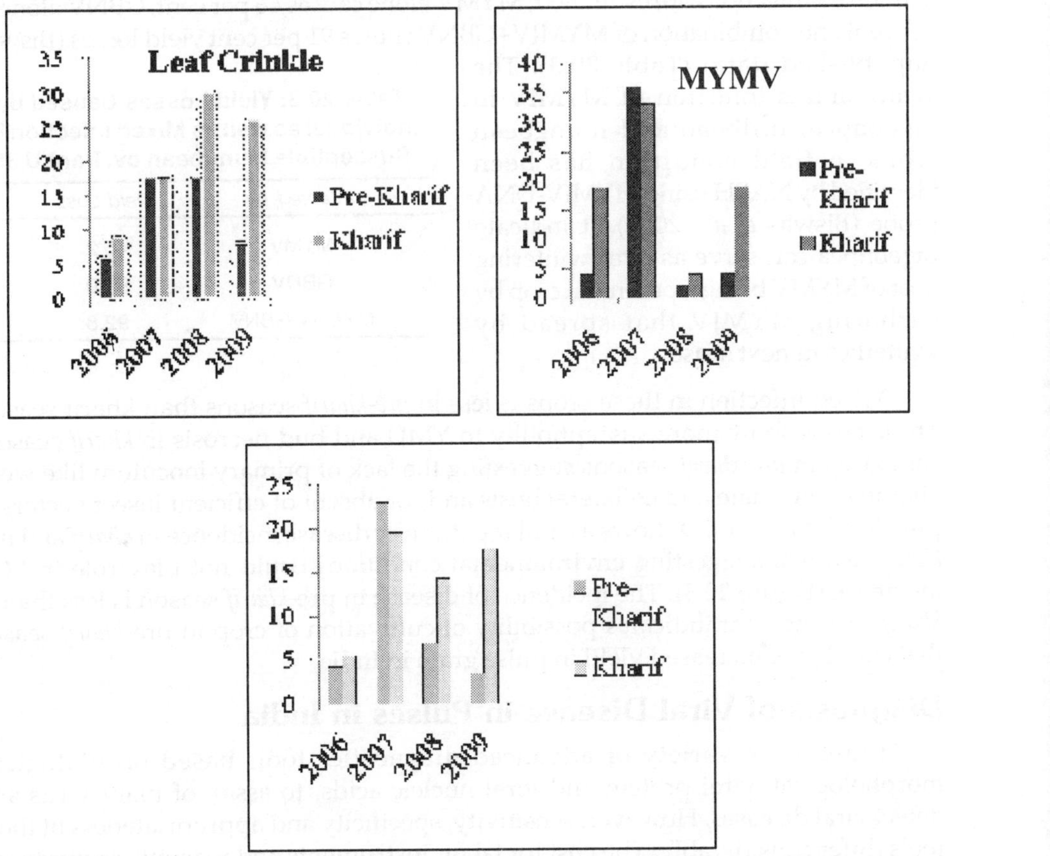

Figure 20.4: Year Wise Increase of LCD, GBNV and MYMIV in Mungbean (Unpublished data)

Viral diseases caused by MYMIV, GBNV and ULCD in urdbean in*kharif* seasons are very common: MYMIV causes various degrees of disease incidence from 1.0 to 65.5 per cent. All the urdbean crops are infected by LCD with the incidence ranging from 11.7 to 66.1 per cent. Incidence of GBNV disease can vary from cultivar to cultivar and a range from 3.0 to 8.0 per cent incidence are observed. Mixed infections with MYMIV, GBNV and ULCD are also observed; incidence of 1.0 to 21.5 by MYMIV+ULCD and 1.5- 6.5 per cent by MYMIV+ GBNV. Mixed infection by other viruses' combination is not observed. In pre-*kharif* season, occurrence of MYMIV and LCD are common in all the urdbean crops with incidence of 3.0 to 26.6 per cent by MYMIV and 4.1 to 61.6 per cent by LCD. No or very less incidence of GBNV infection may appears, indicating that occurrence of insect vectors are absent or less and environmental factors may have role for GBNV disease in pre *kharif* season. Mixed infection only with YMD+LCD is common and mixed infection with other viruses is nil or very less.

Yield losses in naturally infected mungbean crops caused by individual single and mixed infections are estimated; MYMV alone causes 46 per cent, GBNV alone 91 per cent and combination of MYMIV+GBNV shows 91 per cent yield losses (Biswas, unpublished data) (Table 20.3). The symptomless infection of MYMIV in pigeonpea, urdbean and mungbean crops in field condition has been identified by NASH using MYMIV-DNA probe (Biswas *et al.*, 2008). It indicate pigeonpea can serve as over wintering host of MYMV being a perennial crop by harboring MYMIV that spread by whiteflies in next season.

Table 20.3: Yield Losses Caused by Individual as well as Mixed Infection in Susceptible Mungbean cv. Pant U 35

Infected by	*Yield Loss*
MYMV	46.2
GBNV	91.0
MYMV+ GBNV	92.8

Mixed infection in these crops is less in *pre-kharif* seasons than kharif season. These crops show more susceptibility to YMD and bud necrosis in *kharif* seasons compared in *pre-kharif* seasons suggesting the lack of primary inoculum like weed and other alternate and collateral hosts and/or absent of efficient insect vectors in pre-*kharif* season. LCD shows a similar pattern of disease incidence in *kharif* and *pre-kharif* seasons, suggesting environmental conditions could not play role in LCD incidence (Figure 20.3). The incidence of disease in pre-*kharif* season is less than in *kharif* seasons, that indicates possibility of cultivation of crop in pre-*kharif* season that could give increased yield in pulse grain in India.

Diagnosis of Viral Disease in Pulses in India

There are a variety of advanced diagnostics tools based on biological, morphological, viral protein and viral nucleic acids, to assay of plant virus and detect viral diseases. However, sensitivity, specificity and appropriateness of these tools differ considerably. The cost for labor, instruments and reagents required is a significant factor in selection of specific diagnostic tool detection of specific viral disease. The tools exploited successfully to detect the viral disease in pulse crop in India are listed in the Figure 20.5. The diereses caused by MYMV and GBNV is detected by symptomatology, mechanical and vector transmission, electron microscopy, ELISA and PCR/RT-PCR. Success of mechanical inoculation of LCD in mungbean and urdbean is fist time reported from India (Biswas *et al.*, 2009).

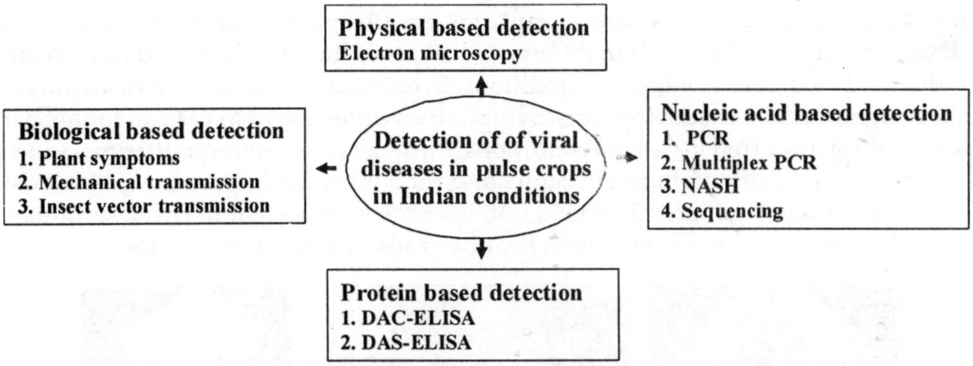

Figure 20.5: Commonly Used Diagnostic Tools for Detection of Viral Disease in Pulses of India

Characteristic symptoms caused by viruses are the initial step in diagnosis of viral disease; which are one of the oldest viral diagnostic methods. Viral infected mungbean and urdbean shows a range of symptoms variations and disease severity depending on virus strains, locations; such as yellow mosaic either on the whole leaf or on a portion of leaves appeared when plants are infected by MYMIV; whereas, leaf distortion like curling and crinkle by LCD disease; and growth distortions like stunting, necrosis of bud and stems or whole plant by GBNV (Biswas *et al.*, 2008; Biswas *et al.*, 2009). Observation of characteristic insect vector transmission is important techniques for detection of pulse viruses.

As most of economically important viral diseases are insect transmitted in non-persistent, semi-persistent or persistent manner, thus study of vector transmission is important diagnostic techniques. Transmission of disease by insect vectors thrips detects to GBNV and whitefly to MYMIV in pulses (Figures 20.6c and d). Incase of MYMIV, if adult whiteflies are given acquisition access period (AAP) for 24 h and then 8-10 viruliferous whiteflies/healthy 3 days old seedling are allowed to inoculation access period (IAP) of for 24 h, typical yellow mosaic symptoms are induced in healthy mungbean or urdbean seedlings 9-15 days (Figure 20.6b) (Biswas *et al.*, 2009; Biswas *et al.*, 2011 unpublished).

There is host specificity of most the viruses and specific viruses are restricted to a particular host or cultivars, and have very wide host ranges. Some viruses have a specific host range which may greatly assist in their identification. Mechanical inoculation using crude extract from infected hosts plants is one of the alternative methods for detection of some pulse viruses. Mechanically sap inoculation of GBNV using 0.05 M Potassium phosphate buffer, pH 7.0 supplemented with 0.05M sodium sulphite can successfully induce typical vein necrosis symptoms in Cow pea (Figure 20.6a) (cv. Pusa Komal) 15- 20 days (Biswas *et al.*, 2009).

Although, characteristic symptoms induced by viruses are the initial step of disease diagnosis, it not sufficient in accuracy of specific disease diagnosis. Similar symptoms may result from infection by more than one virus or by different kind viruses in the same host. In some cases, different strains of the same virus may individually induce different symptoms on the particular host. Symptoms development could be varied in different cultivars of the same host, and also greatly influenced by the environmental conditions. Even, many plants carry viruses but not show significant disease symptoms. Symptomless infection of MYMV in mungbean and urdbean occurred in natural conditions, those serve as reservoir of primary viral inocula for acquisition of insect vectors and then transmission to susceptible host. Thus, detection of symptomless infection in crops is important to check secondary spread of virus from infected crop to healthy crops through insect vectors.

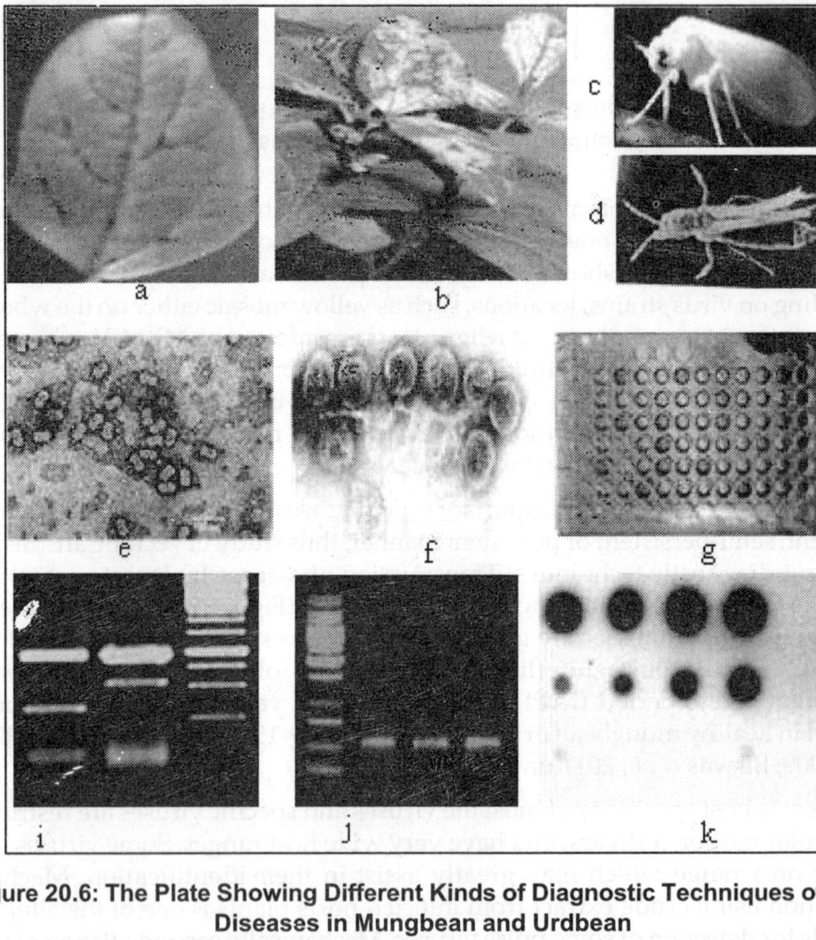

Figure 20.6: The Plate Showing Different Kinds of Diagnostic Techniques of Viral Diseases in Mungbean and Urdbean

a: mechanical sap inoculation of GBNV that causes veinal necrosis in Cowpea (cv. Pusa Komal), b: Insect vector, white flies (Bemisia tabaci) transmission of MYMIV that indicates yellow mosaic in urdbean (cv. KU 300), c & d: Viruliferous whiteflies and Thrips those are the important vectors of MYMIV and GBNV,e & f: Virus particles of MYMIV and GBNV in electron microscopy, g: Detection of GBNV by DAC-ELISA, i: Multiplex PCR for detection of GBNV and MYMIV, j: Conventional PCR for detection of MYMIV and k: detection of MYMIV through nucleic acid spot hybridization (NASH) using specific radio-labeled DNA probe

The use of electron microscope (EM) for detecting viruses in crude sap detects viruses based on topography and morphology, composition, crystallographic information etc. Use of the EM for the detection of plant viruses that occur in relatively high concentration and have elongated or rod-shaped particles is a quick, easy, reliable procedure through the 'leaf dip' technique. Recently, EM is used as regular bases to identify the pulse viruses like MYMIV and GBNV (Figures 20.6e and f). ELISA represents a very sensitive serological method to detect viruses in extracts from infected plants. This method usually used for rapid detection in relatively large numbers samples infected by viruses. In India, Direct antibody coated indirect ELISA (DAC-ELISA) and double antibody sandwiched indirect ELISA (DAS-ELISA) is preferably used for detection of pulse viruses like GBNV (Figure 20.6g) (Bhat *et al.*, 2002, Biswas *et al.*, 2009).

Nucleic acid based detection systems are greatly improved following the development of the polymerase chain reaction (PCR). This assay is relatively easy to perform but it is still not amenable to handling large numbers of samples. PCR and reverse transcriptase-PCR (RT- PCR) detection approach can detect viral infection like DNA virus MYMIV or RNA virus GBNV accurately by targeting their genomic material and used very efficiently in Indian condition (Figure 20.6j) (Biswas *et al.*, 2009; Bhat *et al.*, 2008). Multiplex PCR allows detecting more than one virus simultaneously, which is sensitive and very useful technique to identify the mixed and multiple infections in pulse crops (Figure 20.6i). Nucleic acid spot hybridization (NASH) is now a day's is very trendy accurate diagnosis technique is being used for detecting of legumes in India. Identification of different variants by host reaction and NASH is reported in Indian MYMV (Figure 20.6k) (Biswas and Varma, 2000). Synthesis of specific primers targeting viral group or virus strain specific is nowadays used for comparison of virus groups and strain through cloning of viral nucleotide and then sequencing. Sequencing of virus genome and comparative study of the sequence and phylogenetic analysis may give the actual detection of viruses and information strain demarcation (Figure 20.7).

Management of Viral Diseases

Fungicides and antibiotics are widely practiced means to manage diseases caused by fungi and bacteria, reducing invasion of them. Unfortunately, there is no such direct control method or non availability of any viricide to manage/cure/reduce diseases caused by viruses. Further, management of viral diseases is difficult due to the complex disease cycle, involving interaction of the virus, its vector and host with the environment. However these diseases can be managed by integrating the approaches like avoiding the source(s) of the infection, control of the vectors, suitable modification of the cultural practices and use of virus resistant varieties. Most of the useful procedures are designed to prevent and reduce sources of the infection within or outside of the crop with the interference of virus ecology so as to stop of delay the onset of virus incidence and to decrease the rate of progress of disease epidemic so that yield losses may be minimized. Therefore, efforts should be made to develop permanent solution to control the viral diseases by giving emphasis on conventional breeding resistance to plant virus, and to insect-vectors and genetic engineering using transgenes, which are derived from pathogen or foreign sources.

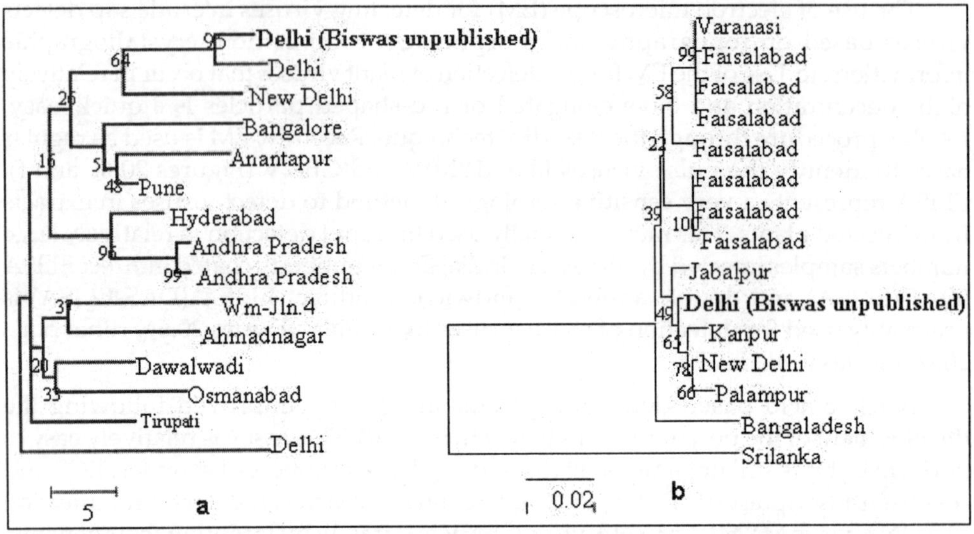

Figure 20.7: Phylogenetic Analyses of GBNV (a) and MYMIV with Other Related Viruses Based on Sequencing of Nucleotide of Respective Viruses; Prove GBNV as Tospo and MYMIV is Begomo Virus Isolates

Management through avoidance of sources of primary infection: In the case of pulses, the main sources of infection are the seeds, weed hosts, adjoining crops and volunteers. Management of weed hosts, volunteer plants and overlapping crops-particularly the ratoon crops, which play an important role in the perpetuation of viruses, is very important. Removal of weeds and volunteer plants in and around the fields reduce significantly the incidence of viruses like whitefly transmitted MYMIV and thrips transmitted GBNV, which are known to perpetuate in a large number of common weeds. Ratoon crops of pigeonpea are a major source of infection of MYMV and other viruses. Modification of date of showing, early and late can give better yield of crops particularly against MYMV as severity of diseases and yield of crops vary for the time of infection (Table 20.4).

Table 20.4: Yield Loss at Time of Infection of MYMV

Time of Infection	*Per cent Yield Losses Over Control*
At 2 weeks	83.6
At 4 weeks	52.7
At 6 weeks	34.5
At pod forming stage	73.6
Average	46.2
BN	91.0
YM+BN	92.8

Management of Insect Vector

Chemical management of insect vectors of viruses is an essential component of integrated disease management. However, application of chemical should be used judiciously and timely, that be based on the efficient monitoring of the vector population and understanding of the ecology of the viruses and insect vectors. Insecticidal treatment has been shown to be effective in the management of all groups of viruses affecting pulse crops. Biological control of vectors can also be effective. It deserves greater attention of eco-friendly management of the vector.

Management through Cultural and Agronomic Practices

Suitable modifications of agronomic practices can help in reducing the incidence of viral diseases. Introduction of practice of growing mungbean in the dry spring summer period offers good yields as the crops remained free from MYMIV. Integration of weed management in and around the field can achieve a similar effect even in the main cropping season. Various types of mulches are shown to reduce the incidence of viral diseases. Mulches will also be useful in minimizing the incidence of aphid and whitefly transmitted viruses in pulses. Mulching also improves plant growth by conserving the soil moisture, raising soil temperature and reducing competition with weeds. Intercropping and barrier cropping of pulses with graminacious crops are also shown to protect pulses from viral infections, particularly by aphid and whitefly transmitted viruses.

Management through Use of Host Plant Resistance

Conventional

Use of resistant varieties is the most practical approach for managing the viral diseases. Genetic resistance may operate through resistance to vector, transmission by vectors, resistance to multiplication and cell to cell movement of the viruses, or immunity to virus infection. Virus resistant varieties of different pulse crops are available, but due to evolution of new strains of the viruses-particularly MYMIV, host resistance frequently break down. There is a need of thorough understanding of the mechanism of resistance in different host-virus combinations so that varieties with durable resistance are evolved. The available resistant varieties need to be judiciously used as apart of integrated viral disease management strategies to extend the life span of these varieties and minimize the losses caused by viruses.

Identification of location specific virus strain and study of resistance in crops against all types of strains are needed. Crop cultivars showing resistant against to maximum number of virus strains occurring in different geographical zone, should be considered to be true resistant (Table 20.5). It has been indicated that severity of MYMV in different location of trial is different. The cultivars of mungbean and urdbean shows resistant in one location but exhibits susceptible to another location, indicating the occurrence of different MYMIV strains in various locations (Table 20.6). Therefore, crop cultivars showing resistant against disease in all the locations are considered true resistant (Table 20.7). For instances, three mungbean cvs Samrat, Meha and MUM 2, and two urdbean cvs DPU 88-31 and PDU 1 are true resistant/moderately resistant to MYMV as they show resistance in almost all the location of trials.

Non Conventional

Attempts are being made to develop transgenic mungbean, pigeonpea and urdbean for resistance to menacing viruses like MYMV and other pulses. Virus resistant transgenic lines of targeted pulse crops are expected to be available soon. These will be useful in building stable resistance to pulse viruses. Due to lack of

Table 20.5: Disease Reaction Caused by Different Variants of MYMIV

Crop	*Variety*	*MYMIV Strain*				
		Pp1	*Bg*	*MgD*	*MoL*	*Mg*
Mungbean	P 9271	MR	R	MR	I	MS
	PDM 91-242	S	I	R	MR	MS
	ML 337	R	Symptomless	R	I	I
	RMG 268	S	S	MS	S	I
Urdbean	KU 91	S	R	S	S	MS
	Pant U 19	MR	I	R	MS	MR
	UG 83-13	R	MR	R	0+	MS
	UL 257	R	Symptomless	R	I	Symptomless

HR: Highly Resistant; R: Resistant; MR: Moderately resistant; S: Susceptible; MS: Moderately susceptible; I: Immune.

Table 20.6: Resistance Against MYMV in Mungbean and Urdbean Crops at Different Location in India

Variety	*Bangalore*	*Ludhiana*	*Coimbatore*	*Pantnagar*	*Dhaulakuan*	*Kanpur*
Mungbean						
Samrat	R	MR	MR	MR	R	R
Meha	R	MR	R	MR	R	R
MUM 2	R	MR	MR	MR	R	R
T44	R	S	MR	MR	MR	MS
RMG 62	R	HS	MR	MR	R	MS
Pratap	MS	S	MR	MS	MR	MS
Vamban 1	R	S	R	MS	R	MS
LGG 407	R	S	R	MR	MR	MS
Co 6	MS	S	MR	MS	R	MR
Kopergaon	S	S	MR	MS	R	MR
SKN 155	S	S	MR	MR	R	MS
Urdbean						
TPU 4	-	MR	MR	HS	HS	HS
Uttara	-	R	R	MS	R	R
DPU 88-31	-	R	MR	HS	R	R
PDU 1		MR	MS	MS	MR	MS
Barabanki L	-	MR	MR	S	MR	MR
AKU 9904	-	HS	MR	S	HS	S

Source: Annual report of Indan Pulse Research Institue, Kanpur.

R: Resistant; MR: Moderately Susceptible; S: Susceptible; MS: Moderately Susceptible; HS: Highly Susceptible.

Table 20.7: Viral Disease Resistant Varieties in Pulses Developed Recently Against MYMIV

Crop	Resistant Variety
Mungbean	MGG443, MUM 2, Pant M 1, 2, 3, 4, PDM 11, 54, Narendra mung 1, ML 5, 131, 267, 337, 459, 513, 610, P 9271, 9272
Urdbean	DPU 1, 102, 84-14, 88-31, IPU 94-1, Pant U 19, 30, UG 218, 389, Narendra Urd, UL 310

resistance to MYMIV and GBNV in pulses, there is high a disease pressure persisting and it results in serious losses in grain production. Nowadays genetic engineering approach to modify crop plants, using viral genes like coat protein gene, movement protein gene, anti-sense RNA, RNAi approach, targeting the conserved sequences of coat protein and non coding intergenic region of viruses has proved an effective alternate method to develop virus resistant and/or tolerant cultivars in crops.

Conclusion

Viral diseases are a major problem in improving the productivity of pulse crops. The losses caused by the viral diseases need to be minimized through integrated management using all the available technology in the present conditions coupled with proper diagnosis of disease using of available diagnostic tools for diseases. Breeding of pulse varieties for durable resistance to viruses should be emphasized and the available resistant varieties should be used as a component of integrated virus disease management to minimize the chances of breakdown of their resistances. Understanding the mechanism of virus and vectors relationships needs greater attention to control the vectors. Greater attention is also required for developing transgenic lines of different pulses for resistance to viruses. In the long run these will be valuable sources of resistance to viruses. There is an urgent need to strengthen research on pulse viruses, with a focus on molecular characterization and understanding the ecology of the viruses and their vectors. As new viruses with mixed infection in pulse crops are emerging care should be taken care. The emphasis should be given on molecular markers for differentiation of vector biotypes, molecular mechanism of vector transmission, host specificity and symptom expression of viruses and molecular markers for resistance genes.

References

Amin, P.W. (1985). Apparent resistance of groundnut cultivar Robut-33-1 to bud necrosis disease. *Plant Disease* **65:** 663 665.

Bhat, A.I., Jain, R.K. and Ramiah, M. (2002). Detection of *Tobacco streak virus* from sunflower and other crops by reverse transcription polymerase chain reaction. *Indian Phytopath.* **55:** 216-218.

Bhat, A.I., Jain, R.K., Varma, A., Naresh Chandra and Lal, S.K. (2001). *Tospovirus* (es)infecting grain legumes in Delhi their identification by serology and nucleic acid hybridization. *Indian Phytopath.* **54:** 112-116.

Biswas, K.K. and Varma, A. (2000). Identification of variants of mungbean yellow mosaic geminivirus by host reaction and nucleic acid spot hybridization. Indian Phytopath. 53: 134-141.

Biswas, K.K. and Varma, A. (2001). Evaluation of resistance in blackgram (*Phaseolus mungo*) to variants of mungbean yellow mosaic geminivirus. *Indian J. Agril. Sci.* **71:** 215-218.

Biswas, K.K., Malathi, V.G. and Varma, A. (2008). Diagnosis of Symptomless *Yellow mosaic begomovirus* Infection in Pigeonpea by using Cloned *Mungbean yellow mosaic India virus* as Probe. *J. Plant Biochemistry and Biotechnogy*, 17: 9-14

Biswas, K.K., Tarafdar, A., Kumar, A., Dikshit, H.K. and Malathi, V.G. (2009). Multiple infection in urdbean (Vigna mungo) in natural condition by begomovirus, tospovirus and urdbean leaf crinkle virus complex. Indian Phytopathology 62 (1): 75-82

Jain, R.K., Pappu, S.S., Pappu, H.R., Culbreath, A.K. and Todd, J.W. (1998). Molecular diagnosis of tomato spotted wilt *Tospovirus* infection of peanut and other field and greenhouse crops. *Plant Disease* **82:** 900-904

Khurana, S.M.P., Phadtare, S.G., Garg, I.D., Singh M.N. and Bhardwaj, V.P. (1989). Potato stem necrosis epidemic due to tomato spotted wilt virus in India (ABS) Pro IV Int. PI. Virus Epidemiol. Workshop, Montpelier, France, Sept. 3-8 pp 301.

Makkouk, K.M., Kumari, S.G., Huges, J.dA., Muniyappa, V. and Kulkarni, N.K. (2003). Other lugemes. **In:** Virus and virus like disease of major crops in developing countries, (Eds Leobension., D. and Thollapilly. G), Klieur Academic Publisher, pp447-476.

Naraini, T.K. (1960). Yellow mosaic of mung (*Phaseolus aureus* L.). *Indian Phytopath.* **13:** 24-29.

Nene, Y.L. (1972). A survey of viral disease of pulse crops in Uttar Pradesh. G.B Plant University of Agriculture and Technology. Pantnagar, U.P. pp: 191.

Paulkhurana, S.M., Garg, I.D., Behl, M.K. and Singh M.N. (1990). Potential reservoirs of *Tomato spotted wilt virus* in the North West Himalayas *Nat. Acad. Sci. Letters* **13:** 297-299.

Prasada, Rao, R.D.V.J., Chakraborthy, S.K., Rajeswari, R. and Reddy, A.S. (1987). Occurrence of tomato spotted wilt virus on cowpea, chilli and brinjal. *Indian J. Plant Prot.* **15:** 117-119.

Rao, R.D.V.J.P., Babu, B.S., Sreekant, M. and Kumar, V.M. (2003). ELISA and infectivity assay based survey for the detection of peanut bud necrosis virus in mungbean and urdbean in Andhra Pradesh. *Indian J. Pl. Prot.* **31:** 26-28.

Rao, R.D.V.J.P., Rajeswari, R. Rao, M.V.B. Ragunathan, V. and Joshi, N.C. (1985). Spotted wilt of pea in India. *Indian Horticulture* **38:** 90-93.

Ravi, K.S., Kitkani, A.S. and Winter, S. 2005. Iris yellow spot virus in onion record from India.

Reddy., D.V.R., Ratna, A.S., Sudarshana, M.R., Paul, F., and Kumar, I.K. (1992). Serological relationship and purification of bud necrosis virus, a *Tospovirus* occurring in peanut (*Arachis hypogea* L.) in India. *Annals of Applied Biology* **120:** 279-286.

Sabitha, S., Ramakrishnan, G. and Jeyarajan, R. (1984). Studies with a strain of *Tomato spotted wilt virus. Madras Agric.J.* **71:** 418-419.

Sarita, R.K. and Jain, R.K. (2007). Nucleotide sequence of S and M RNA segments of a *Groundnut bud necrosis virus* isolate from *Vigna radiata* in India. *Arch. Virol.* **152:** 1195-1200

Sastry, K.S. (1982). Occurance of spotted wilt in tomato in Karnataka. *Curr. Res.* 9: 94.

Satyanaryana, T., Mitchell, S.E., Reddy, D.V.R., Kresovich, S., Jarret, R., Naidu, R.A., Gowda, S. and Demski, J.W. (1996). The complete nucleotide sequence and genome organization of the M RNA segment of peanut bud necrosis *Tospovirus* and comparison with other Tospovirus *Journal of General Virology*. **77:** 2347-2352.

Sharma, R.B., Prasad, S.M. and Kudada, N. (2007). Leaf crinkle virus disease in urdbean (Vigna mungo Linn.). Journal of Research: Birsa Agricultural University. 19 (1): 73-79.

Varma, A. and Biswas, K.K. (2009). Viral diseases of Pulse: Emerging Concerns. Souvenir on International Conference on Grain Legumes: Quality Improvement Value addition and Trade, IIPR, Kanpur, 63-69.

Varma, A. and Malathi, V.G. (2003).Emerging geminivirus problems: a serious threat to crop production. *Ann. Appl. Biol.* **142**: 145-164.

Varma, A., Dhar, A.K. and Mandal, B. (1992). MYMV transmission and control in India, In *Mungbean Yellow Mosaic Virus Disease, I Edited* by S.K. Green and D. Kim, AVRDC, Taipei, pp 8-20.

Williams, F.J., Grewal, J.S. and Amin, K.S. (1968). Serious disease of pulse crops on India in 1966. *Pl. Dis. Reptr*. **52:** 300-304.

Chapter 21

Transmission of Plant Viruses

Manisha Srivastava and U.P. Gupta*

Department of Botany,
Harish Chandra P.G. College, Varanasi – 221 001, U.P.

There are several methods by which plant viruses can be transmitted from diseased to healthy hosts where there is no involvement with another organism. Plant viruses being obligate parasites in order to survive, must be spread from one susceptible plant to another and frequently need to be introduced into living cells. They are not disseminated by wind or water. Generally, they do not cause infection unless they come in contact with the contents of a wounded living cell. The ways of virus transmission is important for following reasons:

1. A disease is recognized as a virus disease only if it can be transmitted to healthy individuals by some means.
2. How the virus spreads in the field is essential for the development of satisfactory control measures.
3. The interactions between viruses and their invertebrate and fungal vectors are of considerable biological interest.
4. Mechanical transmission is very important for the effective laboratory study of viruses.

The spread or transmission of viruses occurs in the following ways:

1. Mechanical transmission
2. Graft transmission

* Corresponding Author E-mail: sri.manisha.25@gmail.com

3. Vegetative transmission
4. Seed and Pollen transmission
5. Dodder transmission
6. Insect transmission
7. Nematode and fungus transmission

Mechanical Transmission

The application of virus-bearing fluids to the tissues of susceptible plants. The standard method of inoculating plants with viruses was to place a few drops of crude, extracted sap from the virus–infected plant upon the leaves of the plant to infected and to scratch through the drop with a needle mounted in a handle. Virus enters through wound (Trichome would suffice to enable it to enter the cell and therefore the more points of entry for virus the better) in the cell.

Gentle rubbing (Samuel, 1931) by a glass spatula with a ground glass face, dipped into the inoculum, leaf to be inoculated being supported on a filter paper held in the hand or tip of forefinger or piece of cheese cloth can also be used. During rubbing of plants, hair and epidermal cells containing virus may be broken and virus may be liberated into similarly damaged cells on healthy plants, thereby bringing about transmission.

It was found that an abrasive, dusted lightly over the leaves before inoculation or added to the inoculum, greatly increased the points of entry of the virus (Rawlins and Tompkins, 1936). Fine grade of carborundum powder or diatomaceous earth such as celite are suitable for this purpose. It is important, however, to dust the leaves lightly and to use the minimum of pressure while rubbing the leaves.

Inoculated leaves were washed with water (tap or squash bottle) immediately after inoculation, number of local lesions or points of entry of the virus was greater than unwashed leaves. One-minute rinse may cause as much as 75 per cent decrease in infection with T.M.V. (Dale, 1950). According to Yarwood (1955) short washing increased in infection but reduced it if prolonged for more than 20 seconds. Of the important plant viruses, potato virus X, tobacco mosaic virus, and cucumber mosaic virus are transmitted through sap in the field and may cause severe losses.

Mechanical transfer of virus between plants can be brought about by man, eg. PVX can survive on clothing, tools and mechanical devices used in agriculture and horticulture and these can therefore act as sources of virus when used otherwise in virus-free crops. Clothing contaminated with TMV may remain infective for several months. Viruses may also be transferred from infected to healthy plants by cutting knives when pruning. Viruses such as TMV and tomato mosaic may spread when picking out seedlings or harvesting fruits and flowers by hand. Sterilized tools and clean hands should be used for taking cuttings.

Fortunately not all viruses are mechanically transmissible. Transmission of viruses in the field by natural mechanical damage to the plant tissues is relatively rare and probably of very minor economic importance.

Graft Transmission

All viruses which are systemic in their hosts can be transmitted by grafting between susceptible and compatible plants. There are many methods of making grafts but only a few, which are suitable for use with herbaceous plants, are given here.

Detached Scion Grafting

Small shoot is trimmed through a node to a wedge and inserted in the cut stem of the stock. The scion should have most of its leaves removed and it should be inserted in the cleft of the stock so that the apical portions of the cut surfaces of the scion are just visible. The graft is then bound with bast or rubber tape and sealed with a drop of bicycle tyre solution. Plant placed in a moist for few days when graft union is formed the rubber tape will be removed.

Approach Grafting

Plants to be joined are brought together but each retains parts above and below the point of contact. Application of this method is spliced approach graft. The stock and scion are each sliced to expose the cambium; the cut surfaces are then brought together and tied. A 'tongue' is cut downwards on the stock and upwards on the scion; the tongues are then fitted together and wrapped with self–sealing crepe rubber.

Cleft Inarching

An upward cut in the stock forms the cleft and the scion is cut to a wedge to fit into the cleft. The whole must then be firmly tied.

Bottle Graft

leafy scion is approach- grafted to the stock, but the base of the scion is kept alive by immersion in a bottle of water until union is established, the base of the scion is then cut off close to the stock.

Vegetative Transmission

Virus diseases are of such paramount importance in the potato crop, in raspberry and strawberry culture in the bulb industry and in many other crops, which are produced from vegetative parts. Virus is present in nearly all organs and tissues of the plants, therefore, the parts from mother plants used for vegetative propagation eg. Cuttings, tubers, corms, bulbs or rhizomes will always contain virus and so it will be transmitted to the progeny. Apical meristems of systemically infected plants are, however, virus–free and by tissue culture virus-free plants may be grown from tips of virus-infected plants.

Seed and Pollen Transmission

Seed transmission provides a very effective means of introducing virus into a crop at an early stage, giving randomized foci of primary infection throughout the planting. Thus, when some other method of transmission can operate to spread the virus within the growing crop, seed transmission may be of considerable economic importance. Viruses may persist in seed for long periods so that commercial distribution of a seed–borne virus over long distances may occur. Table 21.1 lists the approximate frequency with which seed transmission has been found among some viruses of various groups and among viroids.

Table 21.1: Relative Importance of Seed Transmission for Viruses of Various Virus Groups

Virus Group	*No. of Members*		*Type of Potential Injury*[a]						*Seed Transmission (per cent)*[b]
	In Group	*Seed Borne*	*A*	*B*	*C*	*D*	*E*	*F*	
Alfamovirus	1	1	+	+	+				1-23
Bromovirus	6	1	+	+	+				+
Capillovirus	4	1							1-60
Carlavirus	60	2							2-90
Carmovirus	18	2							10-40
Caulimoviridae	34	1							Up to 100
Closterovirus	28	1							+
Comovirus	15	6	+	+	+				1-90
Cryptovirus	31	31							100
Cucumovirus	3	3	+	+	+				<1-1
Dianthovirus	5	0							
Enamovirus	1	1							1-2
Fabavirus	4	0							
Geminivirus	102	1							+
Hordeivirus	4	1	+	+	+		+		+
Ilarvirus	17	8				+			1-90
Luteovirus	7	0							
Marafivirus	3	0							
Nepovirus	40	17	+	+	+				3-100
Plant reovirus	14	0							
Potexvirus	36	4							1-6
Potyvirus	179	16	+	+	+	+	+	+	<1-80
Rhabdovirus	15	1							+
Sobemovirus	14	4							1-80
Tenuivirus	11	0							
Tobamovirus	17	7	+	+	+				1-20[d]
Tobravirus	3	3	+	+	+				1-35
Tospovirus	13	1							Up to 95
Tombusvirus	13	1							+
Tymovirus	23	3							+
Viroids	15	5			+	+			+

Data from Stace-Smith and Hamilton (1988), with permission, and from AAB Descriptions of Plant viruses, all members of groups tested for seed transmissibility.

a: A, survival of inoculum; B, dispersal of inoculum; C, primary inoculum source; D, contamination of germplasm lines; E, contamination of virus-free planting material; F, direct crop losses due to plants arising from infected seed. b: A plus sign (+) indicates that no percentage value was given. c: BSV, PVCV and TVCV are apparently seed transmitted in their respective hosts, but probably by activation of integrated viral sequences. d: Seed transmission of TMV probably due to contamination.

Two general types of seed transmission can be distinguished. With TMV in tomato, seed transmission is largely due to contamination of the seedling by mechanical means. This type of transmission may occur with other tobamo viruses. The external virus can be readily inactivated by certain treatments eliminating all, or almost all, seed–born infection. A small but variable proportion of the seed may be infected in the endosperm where virus may persist for many years. No TMV has been detected in the embryo of tomato (Broadbent, 1965c, Lartey *et al.*, 1997) or of *Arabidopsis* (Filho and Sherwood, 2000). The seed transmission of MNSV is assisted by the fungal vector *Olpidium bornovanus* (Campbell *et al.*, 1996). In the second and more common type of seed transmission the virus is found within the tissues of the embryo. The develop in embryo can become infected either before fertilization by infection of the gametes (indirect embryo invasion or gametic transmission) or by direct invasion after fertilization (Johansen *et al.*, 1994, Maule and Wang, 1996). Many viruses use both processes in the production of infected seed. Small seeds have been found to transmit PSbMV at a much higher rate than larger seeds (Khetarpal *et al.*, 1988). *Pea seed-borne mosaic virus* (PSbMV) is a species in the genus *Potyvirus*, family *Potyviridae*. In some cultivars of *Pisum sativum* (pea), PSbMV is seed transmitted exclusively by direct invasion of the immature embryo from the maternal tissues. The efficiency is variable dependent upon the genetic composition of both the host and the virus and environmental factors (Roberts *et al.*, 2003). More than 100 viruses are transmitted by seed to a smaller or greater extent. As a rule, only a small portion (1-30 per cent) of seeds derived from virus- infected plants of only some hosts of the virus transmit the virus. The frequency of transmission varies with the host-virus combination and with the stage of growth of the mother plant when it becomes infected with the virus. In most seed–transmitted viruses, the virus seems to come primarily from the ovule of infected plants, but several cases are known in which the virus in the seed seems to be just as often derived from the pollen that fertilized the flower. In some host-virus combinations the virus is carried in the integument of the seed and infects seedlings as they are wounded on germination. Barley stripe mosaic virus, Peanut clump, Apple stem grooving virus, Apple chlorotic leafspot, and Southern bean mosaic.

Some viruses are transmitted from plant to plant via pollen (Table 21.2). AMV is more efficiently transmitted through pollen than the ovules (Frosheiser, 1974). Seed and pollen transmission of alfalfa mosaic (AMV) viruses was investigated in annual medic species (*Medicago* spp.). For seed transmission studies with AMV, graft inoculation was used to establish early infection and maximize possible transmission rates to seedlings via seed (Pathipanowat, *et al.*, 1995). In contrast while there was 5 per cent transmission of LMV through the ovule in lettuce, there was less than 0.5 per cent infected seed produce by pollen transmission (Ryder, 1964, Hunter and Bowyer, 2008). Cryptoviruses are unusual in that they are transmitted with high efficiency through pollen and seed, but not by mechanical transmission, grafting or invertebrate vectors (Lisa *et al.*, 1986). It is suggested that the ability of viruses to infect pollen is related to their ability to invade meristematic regions (Maule and Wang, 1996, Hunter and Bowyer, 2008). In some cases, the mother plant itself may become infected. For example, Gilmer (1965) recorded an experimental tree–to–tree transmission of Sour cherry yellows virus by pollen. The pollen may be carried by humans, wind or honey bees (Converse and Lister, 1969). Natural transmission in the field may be via infected pollen and no other means, as was found for RBDV in raspberry by Murant *et al.* (1974). Virus transmitted by pollen may result in reduce fruit set, may infect the seed and the seedling that will grow from it and in some cases, can spread through the fertilized flower and down into the mother plant, which thus becomes infected with the virus. Such plant-to-plant transmission of virus through pollen is known to occur, for example, in sour cherry infected with *Prunus necrotic ring spot virus, Raspberry bushy dwarf virus, Tobacco streak virus.*

Table 21.2: Viruses and Viroids that Appear to be Spread to Other Plants by Pollen

	Viruses or Viroid	*Plant*
Virus group		
Ilarvirus	BlShV	Blueberry
	PDV	Stone fruit
	PNRSV	Stone fruit
	TSV	Various
Nepovirus	AYRSV	Artichoke
	BLMoV	Blueberry
	CLRV	Walnut, betula
Sobemovirus	SoMV	Chenopodium
Idaeovirus	RBDV	Raspberry
Viroid	ASSVd	Apple
	ASBVd	Avocado
	CSVd	Chrysanthemum
		Tomato
	CEVd	Citrus, tomato
	CCCVd	Coconut
	CbVd-1	Coleus
	HSVd	Grapevine
		Cucumber,
		Grapevine, hop, tomato[a]
	PSTVd	Potato, tomato

From Mink (1993).

a: Depends on strain of viroid.

Lettuce mosaic plants infected just before flowering produce fewer infected plants than those infected when young and plants infected after flowering produce none. The lettuce mosaic virus is also passed through the seed of groundsel, *Senecio vulgaris* L., symptoms are less but virus may be transmitted by aphids to lettuce (Broadbent, 1958). Treating the fruit-pulp with one quarter of its volume of concentrated HCl for 30 minutes best method of cleaning seeds and of eliminating any virus on the seed coat. The only treatment than of ten eliminated TMV carried internally was to heat dry seeds in an oven at 70°C, treatment for 3 days was usually enough to free seeds completely but much longer treatment failed to eliminate virus from the endosperm.

Dodder Transmission

Several plant viruses can be transmitted from one plant to another through the bridge formed between two plants by the twining stems of the parasitic plant dodder (*Cuscuta sp.*). A large number of viruses have been transmitted experimentally this way, frequently between plants belonging to families widely separated taxonomically. Dodder plants are vines and are parasitic members of the family Convolvulaceae. The virus is usually transmitted passively through the phloem of the dodder plant from the infected plant to the healthy one. Viruses such as CMV and tobacco rattle replicate in the dodder and are more efficiently transferred than TMV, which does not multiply in the dodder and is merely carried out in the dodder which acts as a passive pipeline connecting two plants. Dodder, on the other hand, can be used to transmit a virus between distantly related plants (*e.g.* Desjardins *et al.*1969). Dodder used in transmission studies may sometimes harbor an unsuspected virus. Thus, Bennett (1944) found that symptomless *Cuscuta californica* was frequently infected with a virus he called Dodder latent mosaic virus, which caused serious disease in several unrelated plant species. In addition to direct vascular connections, dodder appears to have cytoplasmic continuity with its host through plasmodesmata. Vaughn (2003) has reported that searching hyphae cells of the dodder haustorium have plasmodesmata that span walls of adjacent dodder hyphae and host cortex cells. One of the main experimental uses of dodder transmission has been to transfer viruses from hosts where they are difficult to study to useful experimental plants. Dodder is probably an insignificant factor in the transmission of economically important viruses in the field, and has rarely been used in experimental work in recent times.

Insect Transmission

The most important virus vectors are aphids, leafhoppers, whiteflies, mealybugs, thrips and beetles. These and the other groups of Homoptera, as well as true bugs, have piercing and sucking mouthparts. Beetles and grasshoppers have chewing mouthparts. Of these, the beetles are quite effective vectors of certain viruses. Insects with sucking mouthparts carry plant viruses on their stylets- **stylet-borne viruses** and can acquire and inoculate the virus after short feeding periods of a few seconds to a few minutes. Stylet-borne viruses persist in the vector for only a few to several hours. Therefore, they are also known as **nonpersistent viruses**. With some other viruses, the insect vectors must feed on an infected plant from several minutes or hours to a few days before they accumulate enough viruses for transmission. These insects can then transmit the virus after fairly long feeding periods of several minutes to several hours. Such viruses persist in the vector for a few days and are called **semipersistent viruses.** With still other viruses, the insect vectors accumulate the virus internally and, after passage of the virus through the insect tissues, introduce the virus into plants again through their mouthparts; these viruses are known as **circulative or persistent viruses**. Some circulative viruses may multiply in their respective vectors and are then called propagative viruses. Viruses transmitted by insects with chewing mouthparts (Beetles) may also be circulative or may be carried on the mouthparts.

Plant-virus-vector interactions, like plant-pathogen-insect interactions in general, show a high degree of variability (Stout *et al.*, 2006). Reviews on case studies demonstrate that virus infection of host plants may have positive, neutral, or negative effects on the vector (Belliure *et al.*, 2005; Stout *et al.*, 2006; Jiu *et al.*, 2007). Effort to investigate the factors responsible for the variability has been increasing, but so far no clear patterns have emerged. For example, Castle and Berger (1993) investigated the performance of the aphid *Myzus persicae* on potato plants infected with three types of viruses and found that the positive effects acquired by the aphid through feeding on virus-infected plants decreased with a decrease in the dependence of the virus on the aphid for transmission. However, Hodge and Powell (2008) did a similar investigation with the pea aphid *Acyrthosiphon pisum*, tic beans, and three types of viruses, and found that the performance of the aphid on virus-infected plants was not related to the dependence of the virus on the aphid for transmission. It is interesting to note that in the study by Castle and Berger (1993) the aphid acquired some benefits from the infection of the host plant even when the virus was not aphid-transmitted.

Aphids are the most important insect vectors of plant viruses and transmit the great majority (about 290) of all stylet-borne viruses. As a rule, several aphid species can transmit the same stylet-borne virus, and the same aphid species can transmit several viruses. In many cases, however, the vector-virus relationship is quite specific. In aphids transmitting stylet-borne viruses, the virus seems to be borne on the tips of the stylets, it is lost easily through the scouring that occurs during the probing of host cells and it does not persist through the molt or egg. Stylet-borne viruses are said to be transmitted in a nonpersistent manner. Alfamovirus, Caulimovirus (by *M. persicae*), Cucumovirus, Fabavirus, Macluravirus and Potyvirus. These genera include viruses with helical and isometric particles, and with DNA and RNA mono-, bi and tri-partite genomes. In the few cases of aphid transmission of circulative viruses, aphids cannot transmit the virus immediately but must wait several hours after the acquisition feeding however, once they start to transmit the virus, they continue to do so for many days after the removal of the insects from the virus source (persistent transmission). In transmission experiments, plants are involved in three was: 1) for breeding the aphids, 2) for providing virus-infected material, and 3) for providing healthy plants to test the ability of aphids to infect. It is common practice to rear virus–free aphids on a plant species 'immune' to the virus under study. However, when aphids are placed on the virus-infected plant, the change of species may influence their feeding behavior. The species and even the variety of plant used as a source of virus or as a test plant may affect the efficiency of transmission. Unstable viruses occurring in low concentration may be readily transmitted by aphid feeding, but some stable viruses such as TMV, TYMV and SBMV are not. Most experimental work on this problem has been done with TMV, with the following results: 1) aphids can not transmit TMV with via their stylets (Harris and Bradley, 1973, Froissart *et al.*, 2002); 2) under laboratory conditions they can transmit by making small wounds when they claw the surface of the leaf (Harris and Bradley, 1973); 3) they can ingest TMV from infected plants and through membranes, and release virus again in an infectious state; 4) aphid saliva does not inhibit TMV infection; and 5) when purified TMV is mixed with poly (L)- ornithine and potassium chloride, aphids can acquire TMV through a membrane and transmit it to plants via their stylets (*e.g.* Pirone, 1977, Froissart *et al.*, 2002). Perhaps the poly (L)- ornithine in some way makes the cell penetrated by the stylets susceptible to infection or facilitates the retention of the virus in the stylets. The basic concepts of virus- vector interactions were introduced by Watson and Roberts (1939). The current, most widely accepted, terminology is given in (Table 21.3), which differentiates both between and within externally and internally borne interactions. This differentiation is based on the region (s) of the vector in which the interaction (s) occurs and also takes into account the virus gene product (s) involved in the interaction. (Table 21.3) summarizes the main properties of the different kinds of relationships. Essentially, there are three stages in the transmission cycle:

1. The acquisition phase in which the vector feeds on the infected plant and acquires sufficient virus for it to be able to transmit it.
2. The latent period in which the vector has acquired sufficient virus but is not able to transmit it. For externally borne viruses there is little or no latent period.
3. The retention (transmission) period is the length of time during which the vector can transmit the virus to a healthy host.

Table 21.3: Relationships Between Plant Viruses and their Vector

Virus transmission Group					*Transmission Characteristics*				
Site in Vector	*T of t*	*Vpi with Vetor*	*A t (Max dose)*	*R t (Half life)*	*T p*	*V in Vector h.*	*L p*	*Vmv*	*TT*
Eb	Npt, sb	CHF	Sec. to min.	Min.	No	No	No	No	No
	Npt, fb (semi-persistent)	CHF	Min. to H.	H.	No	No	No	No	No
Ib	P, c		H. to days	Days to weeks	Yes	Yes	Hours to days	No	No
	P,p		H. to days	Weeks to months	Yes	Yes	Weeks	Yes	Often

T of t: Type of transmission; Vpi with vetor: Virus product interacting with vector; At: Acquisition time; Rt: Retention time; Tp: Transtadial passage; V in vector h.: Virus in vector hemolymph; LP: Latent period; Vmv: Virus multiples in vector; TT: Transovarial transmission; Eb: Externally borne; Ib: Internally borne; Npt,sb: Non-persistently transmitted, stylet-borne; Npt,fb: Non-persistently transmitted, foregut-borne (semi-persistent); P, c: Persistent, circulative; P,p: Persistent, propagative; CHF: Capsid Helper factor; Sec. to min.: Seconds to minutes; Min. to H.: Minutes to hours; H. to days: Hours to days; Min.: Minutes; H.: Hours.

Some further definitions are needed. Inoculativity is the ability of an aphid or other insect to deliver infectious virus into a healthy plant. The acquisition feed is the feeding process by which the insect acquires virus from an infected plant. The inoculative (transmission) feed is the feed during which virus is delivered in to a healthy plant. There are two phases to the interaction involved in non-persistent virus transmission: retention of the virus at a specific site and release of the virus (Gray and Banerjee, 1999). All non- persistently transmitted viruses have a simple structure of nucleic acid encapsidated in simple icosahedral or rod-shaped particles by one or more coat protein species. Thus, it is the capsid protein that is available for any interactions. Two forms of interaction have been identified in the retention phase, one in which there appears to be direct interaction between the virus capsid and the site of retention in the aphid and the other in which a non- structural virus-encoded protein is involved. This non-structural protein is termed a helper component, helper factor or aphid transmission factor (Pirone and Blanc, 1996). As well as these virus gene products that directly control virus transmission, other viral genes can affect transmission indirectly. The efficiency of transmission can be influenced by virus concentration, which is controlled by virus replication and turnover (Atreya *et al.*, 1992, Aranda *et al.*, 1996; Escaler *et al.*, 2000a). Similarly, acquisiton of a virus can be affected by the localization of the virus in the host in relation to the site (s) of interaction of the vector with the host. Viral genes, such as those involved with cell-to-cell movement, can influence virus distribution in the host.

There are about 60 subfamilies in the leafhopper family (Cicadellidae) and two of these, the Agalliinae and the Deltocephalinae, contain species that are virus vectors. The Agalliinae have herbaceous dicotyledonous hosts, while most Deltocephalinae feed on monocotyledons. There are about 15000 described species of leafhopper in about 2000 genera. Of these, only 49 species from 21 genera have been reported as being virus vectors (Table 21.4) (Nault and Ammar, 1989). There are about 20 families of plant hoppers (Fulgoroidea) but only the Delphacidae have definite virus vector species. Members of this family feed on monocotyledons, primarily members of the Poaceae. Thus, all the viruses known to be transmitted by members of this family have hosts in the *Poaceae.* These cause important diseases of cereal crops, including rice, wheat and maize. Hopper transmission of plant viruses has been reviewed by Nault and Ammar (1989) and by Nault (1997). Biological association between viruses and leafhoppers is more intimate than aphids. Leafhoppers are also vectors of diseases attributed to mycoplasmas (Phytoplasmas). Three types of associationships are seen between viruses and leafhoppers. However, in leafhoppers, propagative type of relationship is most developed. The relationship seen between green leafhopper vector *Nephotettix impicticeps* and rice tungro virus (RTSV and RTBV) and maize chlorotic dwarf virus (MCDV) with *Graminella nigrifrons* is of semi-persistent or circulative type. Virus is transmitted immediately after acquisition feeding (without any incubation period) there is no delay in infective power. Virus is lost by the vector after molting. The relationship seen between the leafhopper vector, *Circulifer tennelus* and sugar beet curly top virus appears to be of circulative type. Virus is not lost by molting but the concentration gradually decreases in the vector, hence there appears to be no multiplication. Acquisition times range from a few seconds to an hour. Longer feeding times give higher transmission rates and longer persistence in the vector. Propagative viruses have a reproductive cycle in their insect vectors. This has been concluded from transovarial transfer of viruses through insect eggs from parent to the offspring.

Rice dwarf virus is carried through the eggs of female parents, *Nephotettix cincticeps*, when female parent carried the virus. Similarly Wound tumour virus and Potato yellow dwarf virus were carried through female parent, *Agallia constricta* and Oat blue dwarf virus through *Macrosteles fascifrons* and Maize mosaic through *Peregrinus maidis* (a plant hopper).

Table 21.4: Distribution of Plant Virus Vector Among Selected *Homoptera* and *Coleoptera* Families

Order, Suborder, Family	*Common Name of Insect Group*	*Approx. No. Species Described*	*No. Vector Species*	*No. Viruses Transmitted*
Homoptera				
Auchenorrhyncha				
Cicadidae	Cicada	3200	0	0
Membracidae	Treehopper	4500	1	1
Cercopidae	Spittlebug	3600	0	0
Cicadellidae	Leafhopper	15000	49	31
Fulgoroidea	Planthopper	19000	28	24
Sternorrhyncha				
Psyllidae	Psyllid	2000	0	0
Aleyroididae	Whitefly	1200	3	43
Aphididae	Aphid	4000	192	275
Pseudococcidae	Mealybug	6000	19	10
Coleoptera				
Chrysomelidae	Leaf beetle	20000	48	30
Coccinellidae	Ladybird beetle	3500	2	7
Cucurlionidae	Weevil	36000	10	4
Meloidae	Blister beetle	2100	1	1

Nault and Ammar, 1989.

Virus can multiply in hoppers feeding on an immune host. Eggs may overwinter and provide a source of virus for spring crops in the absence of diseased plants. Thus, persistence of virus in the insect and transovarial transmission and the factors such as age of vector, time after infection, temperature, genetic variation in the leafhopper, change in properties of the virus affects transmission, may be of considerable economic importance. The biological tests and immunological tests have also shown increase in virus concentration and viral antigens in infected leafhoppers. Soluble antigen and virus particles increased rapidly.

Three genera of viruses are transmitted by whiteflies, the begomoviruses of the *Geminiviridae* family and the criniviruses and some closteroviruses of the *Closteroviridae*. Most of these viruses are found in the tropics and subtropics where they can be of substantial importance. White fly-borne diseases of legumes are of considerable importance. Most studied vector is *Bemisia tabaci*. The whitefly *Bemisia tabaci* (Gennadius) (Hemiptera: Aleyrodidae) is an insect species complex including many morphologically indistinguishable but genetically diverse groups (Boykin *et al.*, 2007), some of which have been named as "biotypes" (Perring, 2001). In many regions of the world, epidemics of plant diseases caused by begomoviruses transmitted exclusively by *B. tabaci* have been occurring soon after the invasion of the B and Q biotypes of the whitefly (Varma and Malathi, 2003; Seal *et al.*, 2006; Hogenhout *et al.*, 2008). The whiteflies which are present on the underside of leaves become viruliferous after a short feeding on the diseased plants. The virus–vector relationship of whiteflies is mostly of circulative type. Latent period may be a few hours to a day. The vectors may retain the virus from a few days to twenty-five days. Whitefly transmitted viruses produces yellow mosaic, leaf curl and rugose mosaic type of symptoms. Examples of whitefly transmitted virus diseases are soybean and mung or urdbean yellow mosaic, tobacco leaf curl, sunnhemp leaf curl, cotton leaf curl, yellow vein mosaic of okra, Abutilon yellow mosaic, Ageratum yellow mosaic and Sonchus yellow mosaic etc. Certain complexities have been described for some begomoviruses (Gray and Banerjee, 1999). TYLCV appears to persist in its whitefly vector for longer than expected from its infectivity, and is reported to be transovarially transmitted. SCLV particles have been associated with cytopathological abnormalities in some vector tissues and this virus can have detrimental effects on the vector biology and reproduction. Sexual transmission of TYLCV- is from male to female and from female to male insects has been suggested (Ghanim and Czosnek, 2000). Both the monopartite whitfly–transmitted closteroviruses and the bipartite criniviruses are transmitted in a foregut-borne, semipersistent manner. LIYV is retained in the vector for a maximum of 3 days whereas LCV and CYSDV persist for 4 to 9 days respectively. As with the aphid-transmitted closteroviruses, BYV and CTV, LIYV encodes two capsid proteins: the major protein (CP) and a minor protein (CPm). Also as with the aphid-transmitted closteroviruses, CPm is found at one end of the particles (Tian *et al.*, 1999). Purified LIYV virions could be transmitted by *B. tabaci* after in vitro acquisition and transmission was neutralized by antiserum to CPm but not by antiserum to CP. Thus, CPm is involved in the transmission of LIYV (Tian *et al.*1999).

Mealybugs, *Pseudococcus njalensis* has been reported to be most important virus vector of cacao swollen shoot virus affecting cacao tree (*Theobroma cacao*). Mealybugs feed on the phloem. The vectors are less mobile and move from plant to plant by crawling. The virus-vector relationship is of semi-persistent type. Acquisition feeding time is few hours and the virus is retained for a few days (3-4 days). Crawling nymphs are more effective vectors than adults.

Of the 5000 or so species of thrips, only 10 species, all in the family Thripidae, are vectors of plant viruses (Table 221.5). Most of these vector species are extremely polyphagous and able to reproduce on a broad range of host plants. *Thrips tabaci* is cosmopolitan, feeding on at least 140 species from over 40 families of plants. It reproduces mainly parthenogenetically. The larvae are rather inactive but the adults are winged and very active. *Thrips tabaci* feeds by sucking the contents of the sub epidermal cells of the host plant. Adults live upto about 20 days. Several generations can develop in a year. Viruses from four plant virus families or groups are transmitted by thrips (Table 21.5). The Ilarviruses, Sobemoviruses and Carmoviruses are pollen transmitted, the thrips carrying the pollen and inoculating it by mechanical damage during feeding. Tomato spotted wilt virus is reported to be transmitted by thrips (Sherwood *et al.*, 2009), *Frankliniella fusca* and *F. occidentalis*. They feed by sucking contents of sub-epidermal cells of the host. The vector transmits the virus in circulative manner and may retain infectivity for life. There are distinct levels of species and biotype specificity (Wijkamp *et al.*, 1995). *Frankliniella occidentalis* was the most efficient vector for four tospoviruses, TSWV, INSB, TCSV and GRSV. The dark form of *F. schultzei* transmitted TSWV, TCSV and GRSV, whereas the light form of this species transmitted TSWV and TCSV poorly. Only one of four populations of *T. tabaci* from different geographical regions transmitted only TSWV of the four viruses tested, and that with low efficiency. If TSWV is cultured by successive transfers only in plants, the isolated loses the ability to be transmitted by thrips.

Table 21.5: Transmission of Viruses by Thrips

Thrip Species	*Virus*	*Virus Family (Group)*	*Virus–Vector Relationship*
Frankiniella occidentalis	GRSV[a]	*Tospovirus*	PP
	INSV	*Tospovirus*	PP
	PDV	*Ilarvirus*	Pollen
	PFBV	*Carmovirus*	Pollen
	PNRSV	*Ilarvirus*	Pollen
	TCSV	*Tospovirus*	PP
	TSWV	*Tospovirus*	PP
F. fusca	TSWV	*Tospovirus*	PP
F. intensa	GRSV	*Tospovirus*	PP
	TCSV	*Tospovirus*	PP
	TSWV	*Tospovirus*	PP
F. schulzei	GRSV	*Tospovirus*	PP
	TCSV	*Tospovirus*	PP
	TSWV	*Tospovirus*	PP
Microcephalothrips abdominalis	PNRSV	*Ilarvirus*	Pollen
	TSV	*Ilarvirus*	Pollen
Thrips australis	PNRSV	*Ilarvirus*	Pollen
T. imaginis	PNRSV	*Ilarvirus*	Pollen
T. palmi	GBNV	*Tospovirus*	PP
	WSMV	*Tospovirus*	PP
T. setosus	TSWV	*Tospovirus*	PP
T. tabaci	PNRSV	*Ilarvirus*	Pollen
	SoMV	*Sobemovirus*	Pollen
	TSV	*Ilarvirus*	Pollen
	TSWV	*Tospovirus*	PP

Adapted from Ullman *et al.*, 1997. PP: Persistent propagative.

Important viruses like, cowpea mosaic, turnip yellow mosaic, southern bean mosaic, squash mosaic are transmitted by beetles (*Diabotrica spp.*). About seventy-four beetle species have been reported to be vectors. Beetles have biting type of mouthparts and during mastication they inoculate a susceptible healthy plant. The virus-vector relationship is of circulative type but there is no evidence of multiplication of virus in the vector. Increased acquisition feeding by the vector, increase transmission rate. Vectors remain viruliferous for a few days.

Fungus and Nematode Transmission

Root-infecting fungal-like organisms, the plasmodiophoromycetes *Polymyxa* and *Spongospora* transmit rod shaped or filamentous viruses (Table 21.6) the *in vivo virus* vector relationship occurs between the rod shaped viruses. Bymovirus, Furovirus and Varicosavirus genera and *O. brassicae* and the three plasmodiopheral species. The model for this relationship is based on observations on *O. brassica* and LBVV, *P. graminis* and SBWMV and *P. betae* and BNYVV (Campbell, 1996). The virus is with in the zoospores when they are released from the vegetative sporangia or resting spores and infects the new host when these zoospores establish their own infection of the root. The zoospore of *Olpidium* transmit viruses in the family Tombusviridae Rochon, (2009). The chytridiomycete *Olpidium* transmit viruses with isometric particles. This shows *in-vitro virus* vector relationship. Virions from the soil water adsorb on to the surface of the zoospore membrane and are thought to enter the zoospore cytoplasm when the flagellum is 'reeled in'. It is unknown how the virus passes from the zoospore cytoplasm to the host cytoplasm, but it is thought that this occurs early in fungal infection of the root. Reciprocal exchange of the coat proteins of TBSV and CNV showed that the coat protein is involved in the uptake of the virus by the zoospore (McLean *et al.*, 1994). Some of these viruses apparently are borne internally in, whereas others are carried externally on the resting spores and the zoospore of the fungi. The zoospores form thalli in the host cytoplasm. In the early stages of infection the cytoplasm of thalli is separated from the host cytoplasm by a membrane, but later the thalli form a cell wall. The entire thallus is converted into vegetative sporangia or resting spores. On infection of new host plants, the fungi introduce the virus and cause symptoms characteristic of the virus they transmit. Soil borne wheat mosaic virus, Peanut clump virus, Potato mop top, Beet necrotic yellow vein virus, Barley yellow mosaic virus, Tobacco necrosis virus and Lettuce big vein virus.

Brown and Weischer (1998) divided the nematode transmission of virus into seven discrete but inter related processes ingestion, acquisition, adsorption, retention, release, transfer and establishment (Visser, 2000). Ingestion is the intake of virus particles from the infected plant and although it does not require a specific interaction between nematode and virus it needs a specific interaction between the nematode and plant. In the acquisition phase the ingested viral particles are retained in an intact state and specific features on the surface of the particle are recognized by receptor sites in the nematode feeding apparatus leading to adsorption. Once adsorbed infections particles can be retained in the nematode for months or even years, but not after molting. Release of the viral particles is thought to occur by a change in pH caused by saliva flow when the nematode commences feeding on a new plant. In the

Table 21.6: Viruses and Virus-like Agents for which Fungal Vectors have been Proven or Suggested

Virus genus or group	*Virus*	*Fungal Vector*[a]					
		Obr	*Obo*	*Pgr*	*Pbe*	*Sss*	*Ssn*
Polyhedral virions, *in vitro* acquisition							
Tombusvirus	CNV		+U				
Carmovirus	MNSV		+U				
	CLSV		+U				
	CSBV		+U				
	SqNV		+ U				
Necrovirus	TNV	+ U					
	ChNV	+					
	LNV	+					
Dianthovirus	RCNMV		+				
Satellite virus	STNV	+ U					
Polyhedral virions, acquisition unknown	WYSV						+
Virion not characterized, acquisition unknown	WCLA			-			+
Virion rod-shaped, in vivo acquisition	SBWMV			+			
	OGSV			+			
	RSNV			+			
Pecluvirus	PCV			+			
	IPCV			+			
Benyvirus	BNYVV				+ U		
	BSBV				+		
Pomovirus	PMTV					+	
Bymovirus	BaMMV			+ U			
	BaYMV			+			
	OMV			+			
	RNMV			+			
	WSSMV			+ U			
Varicosavirus	LBVV	+ U					
	TSV	+					
	FLNV	+					
Other rod–shaped, not characterized, *in vivo* acquisition							
	LRNA	+ U					
	PYVA	+ U					

From Campbell (1996), Annual Reviews.

a: Vectors: Obr, *Olpidium brasssicae,* Obo, *O. bornavanus;* Pgr, *Polymyxa graminis*; Pbe, *P. betae;* Sss, *Spongospora subterranean* f.sp. *subterranean;* Ssn, *S. subterranean* f. sp. *nasturtii.*

+, Specific fungus associated with virus transmission; + U, unifungal or equivalent culture of fungus demonstrated to transmit virus, not necessarily the association of vector with virus; WCLA, watercress chlorotic leafspot agent; LRNA, lettuce ring necrosis agent; PYVA, Pepper yellow vein agent.

transfer and establishment phases the viral particles are placed in the plant cell, and start replicating and causing infection. Nematodes of the genera *Longidorus, Paralongidorus,* and *Xiphinema* transmit several polyhedral–shaped viruses known as nepoviruses, such as grape fanleaf, tobacco ring spot, and other viruses, whereas nematodes of the genera *Trichodorus* and *Paratrichodorus* transmit at least two rod–shaped tobraviruses, tobacco rattle and pea early browning. Nematode vectors transmit viruses by feeding on roots of infected plants and then moving on to roots of healthy plants. Juveniles as well as adult nematodes can acquire and transmit viruses; however, the virus is not carried through the juvenile molts or through the eggs, and after molting, the juveniles or the resulting adults must feed on a virus source before they can transmit again. Specificity of transmission does not appear to involve the ability to ingest active virus since both transmitted and non-transmitted viruses have been detected with in individuals of the same nematode species (Harrison *et al.*, 1974). Sites of retention of virus particles with in nematodes have been identified by electron microscopy of thin sections (Brown *et al.*, 1995). *Arabis mosaic virus* (ArMV) and *Grapevine fanleaf virus* (GFLV), two closely related members of the genus *Nepovirus*, family *Secoviridae*, are responsible for fanleaf degeneration disease of grapevines. ArMV is specifically transmitted by *Xiphinema diversicaudatum* and GFLV is exclusively vectored by *X. index* (Andret-Link *et al.*, 2005). Both viruses are responsible for fanleaf degeneration, one of the most severe viral diseases of grapevines, which causes serious economic losses by substantially reducing yield and affecting fruit quality (Andret-Link *et al.*, 2004a). Nepovirus particles are associated with the inner surface of the odontostyle of various *Longidorus* species and with the cuticular lining of the odontophore and esophagus of *Xiphenema* species. Tobravirus particles have been observed to the cuticular lining of the esophageal lumen.

References

Andret-Link, P. and Fuchs, M. (2005). Transmission specificity of plant viruses by vectors. *Journal of Plant Pathology* **87:** 153-165.

Andret-Link, P., Laporte, C., Valat, L., Ritzenthaler, C., Demangeat, G., Vigne, E., Laval, V., Pfeiffer, P., Stussi-Garaud, C., and Fuchs, M. (2004a). *Grapevine fanleaf virus*: Still a major threat to the grapevine industry. *Journal of Plant Pathology* **86:** 183-195.

Aranda, M.A., Escaler, M., Wang, D., and Maule, A.J. (1996). Induction of HSP70 and polyubiquitin expression associated with plant virus replication. *Proc. Natl. Acad. Sci.* USA **93:** 15289–15293.

Atreya, C.D., Atreya, P.L., Thornbury, D.W. and Pirone, T.P. (1992). Site directed mutations in the potyvirus HC-Pro gene affect helper component activity, virus accumulation and symptom expression in infected tobacco plants. *Virology* **191:** 106-111.

Bennett, C.W. (1944). Latent virus of dodder and its effect on sugar beet plants. *Phytopathology* **34:** 77-91.

Belliure, B., Janssen, A., Maris, P.C., Peters, D. and Sabelis, M.W. (2005). Herbivore arthropods benefit from vectoring plant viruses. *Ecol Lett.* 8 **(1):**70–79.

Boykin, L.M., Shatters, R.G., Rosell, R.C., McKenzie, C.L., Bagnall, R.A., De Barro, P.J. and Frohlich, D.R. (2007). Global relationships of *Bemisia tabaci* (Hemiptera: Aleyrodidae) revealed using Bayesian analysis of mitochondrial COI DNA sequences. *Mol Phylogenet Evol* 44 **(3):** 1306–1319.

Broadbent, L. (1958). *Sci. Hort.* **13:** 74-9.

Broadbent, L. (1965c). The epidemiology of tomato mosaic. XI. Seed transmission of TMV. *Ann. Appl. Biol.* **56:** 177-205.

Brown, D.J.F. and Weischer, B. (1998). Specificity, exclusivity and complementarity in the transmission of plant viruses by plant parasitic nematodes: an annotated terminology. *Fund. Appl. Nematol.* **21:** 1-11.

Brown, D.J.F., Roberston, W.M. and Trudgill, D.L. (1995). Transmission of viruses by plant nematodes. *Annu. Rev. Phytopathol.* **33:** 223-249.

Campbell, R.N. (1996). Fungal transmission of plant viruses. *Annu. Rev. Phytopathology* **34:** 87-108.

Campbell, R.N., Wipf-Scheibel, C. and Lecoq, H. (1996). Vector-assisted seed transmission of melon necrotic spot virus in melon. *Phytopathology* **86:** 1294-1298.

Castle, S.J. and Berger, P.H. (1993). Rates of growth and increase of *Myzus persicae* on virus-infected potatoes according to type of virus-vector relationship. *Entomol Exp Appl.* 69 **(1):** 51–60.

Converse, R.H. and Lister, R.M. (1969). The occurrence and some properties of black raspberry latent virus. *Phtopathology* **59:** 325-333.

Dale, J.L. (1956). *Diss. Abstr.* **16:** 1566.

Desjardins, P.R., Drake, R.J. and French, J.V. (1969). Transmission of citurs ringspot virus to citrus and non-citrus hosts by dodder (*Campestris subinclusa*). *Plant Dis. Rep.* **53:** 947-948.

Escaler, M., Aranda, M.A., Thomas, C.L., and Maule, A.J. (2000a). Pea embryonic tissues show common responses to the replication of a wide range of viruses. *Virology* **267:** 318–325.

Filho, F.M. de A. and Sherwood, J.L. (2000). Evaluation of seed transmission of turnip yellow mosaic virus and tobacco mosaic virus in *Arabidopsis thaliana. Phytopathology* **90:** 1233-1238.

Frosheiser, F.I. (1974). Alfalfa mosaic virus transmission to seed through alfalfa gametes and longevity in alfalfa seed. *Phytopathology* **64:** 102-105.

Froissart, R. Michalakis, Y. and Blanc, S. (2002). Helper component transcomplementation in the vector transmission of plant viruses. Phytopathology 92 **(6):** 576–579.

Ghanim, M. and Czosnek, H. (2000). Tomato yellow leafcurl geminivirus (TYLCV-Is) is transmitted among whiteflies (*Bemisia tabaci*) in a sex related manner. *J. Virol.* **74:** 4738-4745.

Gilmer, R.J. (1965). Additional evidence of tree-to-tree transmission of sour cherry yellows virus by pollen. *Phytopathology* **55:** 482-483.

Gray, S.M. and Banerjee, N. (1999). Mechanisms of arthropod transmission of plant and animal viruses. *Microbiol. Mol. Biol. Rev.* **63:** 128-148.

Harris, K.F. and Bradley, R.H.E. (1973). Tobacco mosaic virus: can aphids inoculate it into plants with their mouthparts? *Phytopathology* **63:** 1343-1345.

Harrison, B.D., Robertson, W.M. and Taylor, C.E. (1974). Specificity of retention and transmission of viruses by nematodes. *J. Nematol.* **6:** 155-164.

Hodge, S. and Powell, G. (2008). Do plant viruses facilitate their aphid vectors by inducing symptoms that alter behaviour and performance? *Environ Entomol.* 37 **(6):** 1573–1581.

Hogenhout, S.A., Ammar, E.D., Whitfield, A.E. and Redinbaugh, M.G. (2008). Insect vector interactions with persistently transmitted viruses. *Ann Rev Phytopathol* 46 **(1):** 327–359.

Hunter, D.G. and Bowyer, J.W. (2008). Cytopathology of Developing Anthers and Pollen Mother Cells from Lettuce Plants Infected by Lettuce Mosaic Potyvirus. *Journal of Phytopathology* **145:** 521–524.

Jiu, M., Zhou, X.P., Tong, L., Xu, J., Yang, X., Wan, F.H. and Liu, S.S. (2007). Vector-virus mutualism accelerates population increase of an invasive whitefly. PLoS ONE. 2 **(1):** e182.

Johansen, E., Edwards, M.C. and Hampton, R.O. (1994). Seed transmission of viruses: current perspectives. *Annu. Rev. Phytopathol.* **32:** 363-386.

Khetarpal, R.K., Bossennec, J.-M., Burghofer, A., Cousin, A. and Maury, Y. (1988). Effect of pea seed-borne mosaic virus on yield of field pea. *Agronomie* **8:** 811-815.

Lartey, R., Ghoshroy, S., Ho, J. and Citovsky, V. (1997). Movement and subcellular localization of a tobamovirus in *Arabidopsis*. *Plant J.* **12:** 537-545.

Lisa, V., Luisoni, E. and Milne, R.G. (1986). Carnation cryptic virus. AAB Descriptions Plant Viruses No. 315.

Maule, A.J. and Wang, D. (1996). Seed transmission of plant viruses: a lesson in biological complexity. *Trends Microbiol.* **4:** 153-158.

McLean, M.A., Campbell, R.N., Hamilton, R.I. and Rochon, D.M. (1994). Involvement of cucumber necrosis virus coat protein in the specificity of fungal transmission by Olpidium bornavirus., *Virology* **204:** 840-842.

Murant, A.F., Chambers, J. and Jones, A.T. (1974). Spread of raspberry bushy dwarf virus by pollination, its association with crumbly fruit and problems of control. *Ann. Appl. Biol.* **77:** 271-281.

Nault, L.R. (1997). Arthropod transmission of plant viruses: a new synthesis. *Ann. Entomol. Soc. Am.* **90:** 521-541.

Nault, L.R. and Ammar, E.D. (1989). Leafhopper and plant hopper transmission of plant viruses. *Ann. Rev. Entomol.* **34:** 503-529.

Pathipanowat, W., Jones, R.A.C. and Sivasithamparam, K. (1995). Studies on seed and pollen transmission of Alfalfa Mosaic, Cucumber Mosaic and Bean Yellow Mosaic Viruses in cultivars and accesions of annual *Medicago* species. Australian *Journal of Agricultural Research* 46 **(1):** 153-165.

Perring, T.M. (2001). The *Bemisia tabaci* species complex. *Crop Prot.* 20 **(9):** 725–737.

Pirone, T.P. (1977). Accessory factors in nonpersistent virus transmission. In: K.F. Harris and K. Maramorosch (eds.) *Aphis as virus vectors*, pp. 221-235. Academic Press New York.

Pirone, T.P. and Blanc, S. (1996). Helper-dependent vector transmission of plant viruses. *Annu. Rev. Phytopathol.* **34:** 227-247.

Rawlins, T.E. and Tompkins, C.M. (1936). *Phytopathology* **26:** 578-587.

Roberts, M., Wang, D., Thomas, C.L. and Maule, A.J. (2003). Pea seed-borne mosaic virus seed transmission exploits novel symplastic pathways to infect the pea embryo and is, in part, dependent upon chance. *Protoplasma* **222:** 31–43.

Rochon, D'Ann (2009). Fungal transmission of plant viruses. *Curr. Protoc. Microbiol.* **12:** 16B.4.1-16B.17.

Ryder, E.J. (1964). Transmission ofcommon lettuce mosaic virus through the gametes of the lettuce plant. *Plant Dis. Rep.* **48:** 522-523.

Samuel, G. (1931). *Ann. appl. Biol.* **18:** 494-507.

Seal, S.E. Van den Bosch, F. and Jeger, M.J. (2006). Factors influencing begomovirus evolution and their increasing global significance: implications for sustainable control. *Crit Rev Plant Sci.* 25 **(1):** 23–46.

Sherwood, J.L., German, T.L., Moyer, J.W. and Ullman, D.E. (2009). Tomato spotted wilt virus. The Plant health instructor.

Stout, M.J., Thaler, J.S. and Thomma, B.P.H.J. (2006). Plant-mediated interactions between pathogenic microorganisms and herbivorous arthropods. *Ann Rev Entomol.* 51 **(1):** 663–689.

Tian, T., Rubio, L., Yeh, H.H., Crawford, B. and Falk, B.W. (1999). Lettuce infectious yellows virus: in vitro acquisition analysis using partially purified virions and the whitefly *Bemisia tabaci. J. Gen. Virol.* **80:** 1111-1117.

Varma, A. and Malathi, V.G. (2003). Emerging geminivirus problems: a serious threat to crop production. *Ann Appl Biol.* 142 **(2):** 145–464.

Vaughn, K.C. (2003). Dodder hyphae invade the host: a structural and immunocytochemical characterization. *Protoplasma* **220:** 189–200.

Visser, P.B. (2000). Role of RNA 2 Encoded proteins in Nematode transmission of Tobacco Rattle Virus. Ph.D. Thesis, University of Leiden, Netherlands.

Watson, M.A. and Roberts, F.M. (1939). A comparative study of the transmission of *Hyocyamus* virus3, potato virus Y and cucumber mosaic virus by the vector *Myzus persicae* (Sulz), *M. circumflexus* (Buckton) and *Macrosiphum gei* (Koch). *Proc. R. Soc.* London B **127:** 543-576.

Wijkamp, I., Almarza, N., Goldbach, R. and Peters, D. (1995). Distinct levels of specificity in thrips transmission of tospoviruses. *Phytopathology* **85:** 1069-1074.

Yarwood, C.E. (1955). *Virology* **1:** 268-85.

Chapter 22

Detection and Purification of Viruses

Seweta Srivastava, Saurabh Singh and Asha Sinha*

Department of Mycology and Plant Pathology, Institute of Agricultural Sciences, Banaras Hindu University, Varanasi – 221 005, Uttar Pradesh

Detection

Serology is a traditional technique for virus detection, and based on the use of antibodies, proteins of the immunoglobulin type, raised in animals and capable of specific binding to antigens. Early serological research was to great importance in identificating and classifying viruses, but, in some cases, they lacked the sensitivity for routine diagnosis. A tremendous improvement in sensitivity was achieved with the development of immunoenzymatic techniques, *i.e.*, ELISA (enzyme linked immunosorbent assay), which employs antibodies conjugated to an enzyme, to greatly amplify and signal the presence of amounts of viral antigens.

The potential application of the serological diagnosis was limited for which viruses specific antibodies were produced. Consequently only known viruses, not disease unknown origin can, can be detected by serology. Moreover, since the antigenic properties reside in the coat proteins, viroids cannot be detected by this means.

Antisera may be produced by immunizing rabbits or other suitable animals purified virus. Several injections are administered intravenously or intramuscularly

* Corresponding Author E-mail: shalu.bhu2008@gmail.com

or in combinations of both over a period of several days. Test bleeds to check the level of antibody are made, 1-3 weeks after the immunization. When the antibody level is satisfactory, several larger bleeds can be made without harming the immunized animal. Such blood samples are allowed to coagulate, then, after over-night storage in a refrigerator, the serum may be decanted and clarified by low speed centrifugation. Reviews of antiserum production are available (Berks *et al.*, 1972 and Regenmortel, 1982).

The specific component of antisera is made up of globulins, and the type and amount of these determines the quality of the antiserum. Three main classes of immunoglobulin (gamma globulin) are known, IgG, IgA and IgM (Regenmortel, 1982). Each is made up of heavy and light chains linked together by disulphide bridges. The simplest is IgG which exists as a Y-shaped molecule, of which the arms of the Y are the combining sites which are specific for each antiserum (Figure 22.1).

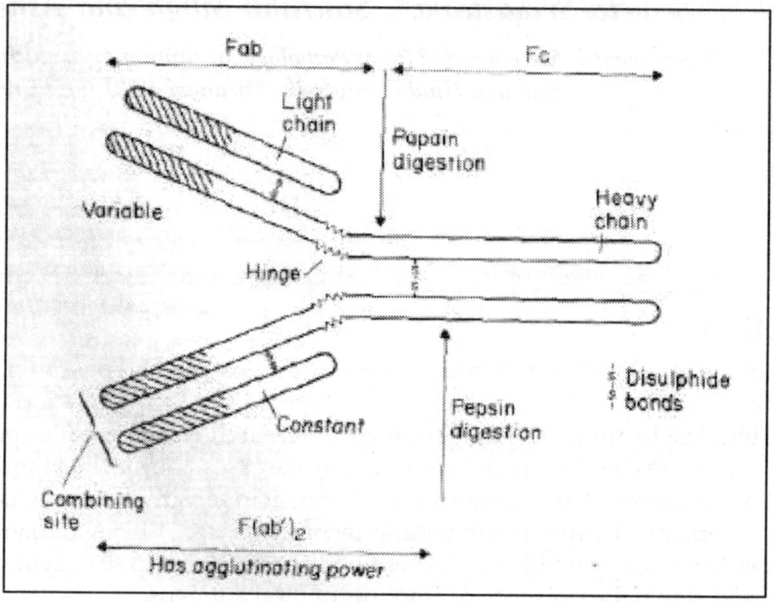

Figure 22.1: Diagrammatic Structure of the IgG Molecule

Serological Terms

Epitopes

Specific regions on antigens that induce and interact with specific antibodies are termed as epitopes.

Monoclonal Antisera

Contain antibodies to one epitope, *i.e.*, more specific and can be used to differentiate strains of many pathogens.

Polyclonal Antisera

Contain antibodies to all available epitopes on the antigen, *i.e.*, less specific.

Virus End Point

Dilutions are always prepared with saline (NaCl 8.5g/L). The highest dilution of the antiserum reacting with antigen (virus) is called antiserum titre and the highest dilution of virus which gives a visible precipitate is called as virus end point (or virus titre).

Serological Tests

Tube Precipitation Test

Whilst infrequently used directly as a diagnostic test, knowledge of the principles behind the tube precipitin test is important, since it is by this method that the titre of an antiserum can be determined for use in further serological investigations. The test depends on the formation of a visible precipitate in small test tubes when antigen and antibody are mixed in suitable conditions. Reactants are usually diluted in neutral 0.85 per cent saline because a precipitate will only form in the presence of electrolytes and most viruses are stable at neutral pH.

Procedure

Various dilutions of clarified virus suspension and antiserum are mixed in glass tubes

⇩

Incubate at 37°C in a water bath

⇩

Precipitation is observed in a mixture

⇩

The type of precipitate formed depends on the shape of the antigen (virus particles)

Observations

Elongated virus particles–Floccular precipitate

Spherical particles–Dense granular precipitate

Drawback

Relatively high concentrations of reagents are needed in this test.

Ring Interface Test

Procedure

The virus containing suspension is carefully layered on top of the viscous, undiluted antiserum suspension (antiserum to which glycerol or sucrose had been added)

⇓

With time, antibody diffuses into the virus solution and virus diffuses into the antiserum

⇓

At interface a ring or band rapidly becomes visible

Drawback

It is a useful quick test, but somewhat insensitive.

Micro-Precipitin Test

A disadvantage of the tube precipitin test is its requirement for large quantities of antiserum. The microprecipitin test (Slogteren, 1955) is a modification of the tube precipitin test which retains its sensitivity and versatility but is much more economical. The principles of the test are the same.

Varying dilutions of antigen preparations and of antiserum are mixed, and the quantity of precipitate formed and the rate at which it is formed are recorded. However, since only single drops of the component antigen and antiserum dilutions are mixed, the quantities used are minimal. A grid titration of the same kind as described for tube precipitation may be set up within a Petri dish (Figure 22.2). Drops of mixed constituents, accommodated within the spaces of the grid, are then covered with a layer of mineral oil to avoid drying out during incubation.

Procedure

Drops of series of dilution mixtures (antiserum and clarified virus suspension) are mixed at the bottom of Petri-dish

⇓

Precipitates produced are observed with a microscope with dark-ground illumination

⇓

The precipitate varies, depending on the ratio of concentrations of antigen and antibodies

Note: This test is a miniature version of the precipitation test.

This test is done on a micro-scale to economize on antiserum.

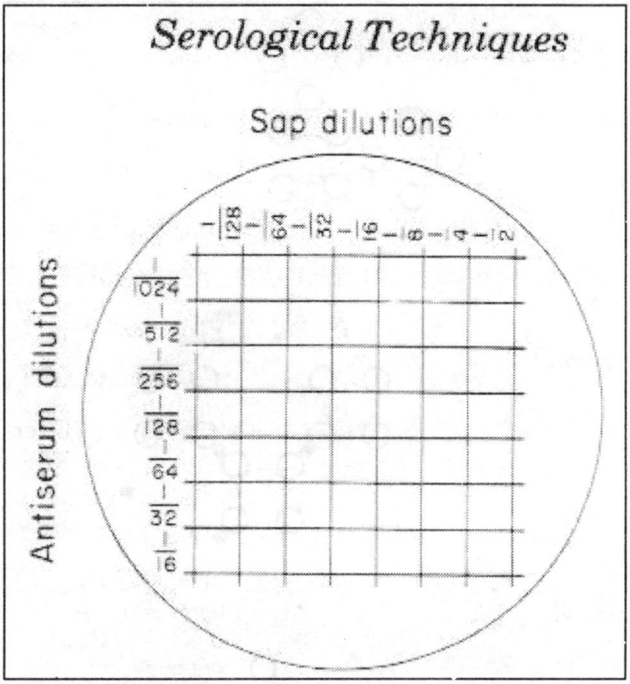

Figure 22.2: Grid for Microprecipitin Test

Agglutination Test

Chloroplast Agglutination Test

Procedure

Few drops of crude freshly expressed plant leaf sap containing high concentration of virus is mixed with double amount of diluted antiserum on a microscope slide

⇩

Due to combination of virus present in crude sap and antiserum, chloroplasts and chloroplast fragments along with small particles of host material clump or co precipitate together, which can be observed with the naked eye or with a microscope

Latex Agglutination Test

Procedure

An antibody is adsorbed onto polystyrene latex beads

⇩

Antibody linked sensitized latex is mixed with antigen in round-bottom microtitre plates (Figure 22.3)

⇩

Flocculation of the latex indicates antigen-antibody combination

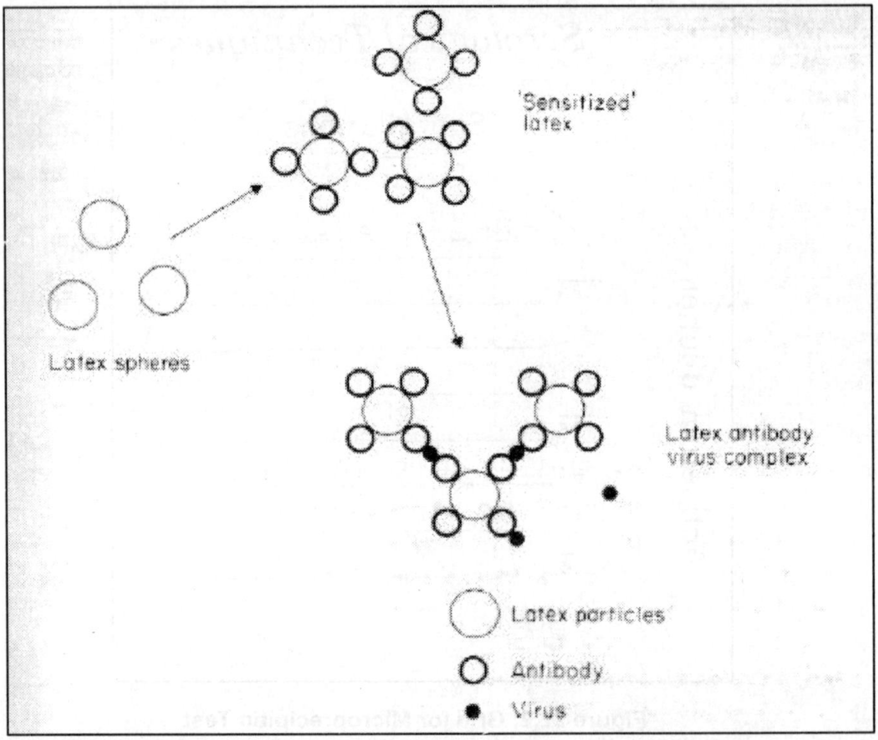

Figure 22.3: The Latex Test has Several Advantages: (*i*) Sensitized latex can be stored or distributed to outstations, and it is (*ii*) 10-100 times more sensitive than slide agglutination, (*iii*) Independent of virus morphology, (*iv*) Moderately economic of antiserum and (*v*) Quick and simple to carry out.

Immunoelectrophoresis

Procedure

Components of the antigen mixtures are first separated in an electric field in an agar gel containing an appropriate buffer

⇩

Antiserum is put into a long trough at one side of the well parallel to the direction of the electrophoretic migration

⇩

Antigen and antibodies then diffuse together and produce bands, which can be observed

⇩

Closely related strains may then be distinguished

Infectivity Neutralization Test

When Viruses are mixed with a specific antiserum, infectivity is reduced to a greater extent than when mixed with a non-immune serum.

Procedure

Antigen (viruses) are mixed with the antiserum

⇓

Loss in infectivity of the virus in the incubated mixture (tested on indicator host)

Note: Sometimes heterologous sera can also inhibit virus infectivity.

Gel Diffusion Test

This test is done in Petri dishes, and precipitation in the form of a visible band or spur is seen in solid agar media (Figure 22.4).

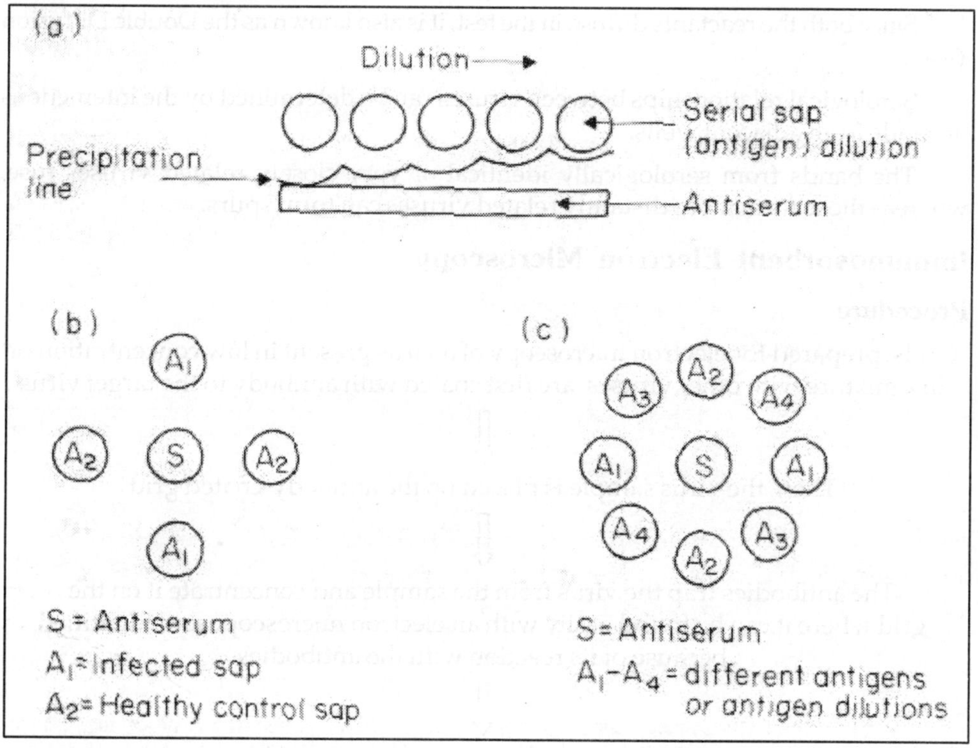

Figure 22.4: Arrangement of Well for Gel Diffusion Tests
(a) Useful configuration to determine antigen dilutions (which will give clear precipitation lines) before proceeding to comparative tests. (b) A simple confirmatory arrangement which can be repeated several times within a Petri dish using different antiserum and sap dilutions. (c) A more complex array of antigen wells around a single antiserum well for comparison of relationships between antigens.

Procedure

The reactants, antiserum and virus solutions, are placed in wells cut in the agar (1 per cent agar containing 0.85 per cent NaCl and 0.02 per cent sodium azide) on the bottom of a Petri dish

⇩

In the Ouchterlony test the antibodies and virus diffuse into the agar towards each other from the adjacent wells

⇩

Precipitation zones in the form of white bands are formed

⇩

Bands can be recorded by direct observation with appropriate lightening

Note: There is no spur or band formation with healthy sap.

Since both the reactants diffuse in the test, it is also known as the Double Diffusion Test.

Serological relationships between viruses can be determined by the interactions of bands from adjacent wells.

The bands from serologically identical or very closely related viruses fuse, whereas those form more distantly related viruses can form spurs.

Immunosorbent Electron Microscopy

Procedure

Grids, prepared for electron microscopy of a virus present in low concentration or in a mixture with other viruses, are first coated with antibody to the target virus

⇩

Then, the virus sample is placed on the antibody-coated grid

⇩

The antibodies trap the virus from the sample and concentrate it on the grid where it can be found easily with an electron microscope and identified because of its reaction with the antibodies

⇩

Identification of the virus is facilitated further by coating the virus particles already on the grid with antibodies (decoration) that make them appear quite distinctive under an electron microscope

Immunofluorescent Staining

Procedure

In this technique, parts of a plant leaf, whole cells, or cell section are first "fixed", *i.e.*, killed with acetone or other organic compounds

⇩

The fixed leaf tissues are then treated with antibodies to a virus that had been labeled previously with a compound, such as fluorescein isothiocyanate (FITC), which fluoresces under ultraviolet light

⇩

If the treated cells are infected with the virus, the virus traps the antibodies and the attached fluorescent compound

⇩

When such cells, in tissues or as protoplasts, are viewed with a microscope supplied with UV light, cells or cell parts that contain virus appear fluorescent while the rest of the cells or cell areas appear dark

Monoclonal Antibodies (MAbs)

For the absolute majority of viruses that have been isolated, purified, and characterized as causal agents of a disease, serology was performed reagents and more recently, monoclonal antibodies. Beside obvious advantages of monoclonal antibodies (*e.g.* specificity, unlimited production, reproducibility of results, easier immunization, possible utilization of mix infection virus sources) these reagents should not be used exclusively. The reason for this is because of its extreme specificity and some false negatives may be realized.

Artificial Polyvalent Antisera

To overcome limitations of strain specific antibodies, polyclonal antisera to several different strains may be mixed (polyvalent antisera) (Koenig *et al.*, 1979 and Uyemoto, 1980). Thus, several viruses will detected simultaneously by using polyvalent antisera (Cambra *et al.*, 1983), and its sensitivity was not compromised, if similar conditions were required for the reliable detection of the different viruses (James, 1997).

ELISA (Enzyme Linked Immunosorbent Assay)

ELISA is a diagnostic techniques utilized for identifying plant viruses. The presence of viral specific antigens in infected sap is detected through a colorimetric reaction, that develops because of the reaction of an enzyme (alkaline phosphatase or horseradish peroxidase) conjugated to antibodies in the presence of an appropriate substrate (paranitrophenylphosphate or tetramethylbenzidine, respectively). Among several variants of ELISA, DAS (double antibody sandwich) ELISA is a standard procedure. Schematically, viral antigens are first trapped by virus-specific antibodies coating the internal surfaces of polystyrene wells, and, then, covered by enzyme-conjugated virus antibodies. Finally, the addition of the substrate induces a colorimetric reaction in the presence of the antigen-enzyme antibody-conjugate complex.

The success of ELISA, adopted in most diagnostic laboratories, is due to the numerous advantages that this technique offers in comparison with others (Clark and Bar Joseph, 1984). For example:

- ✰ Sensitivity for detecting very small amounts of virus, *i.e.*, antigen concentrations of 1-10 ng/ml;
- ✰ Speed of reaction–results are usually available within 6-24 hrs;
- ✰ Scale of operation–several hundred samples can be readily handled, either individually or in groups;
- ✰ Use with plant extracts and purified virus preparations;
- ✰ Specificity for differentiating serotypes;
- ✰ Suitability for both intact and fragmented virions of different size or morphology;
- ✰ Possibility of obtaining quantitative measurements;
- ✰ Possibility of automation and of standardizing tests by the production and availability of commercial kits;
- ✰ Low cost and relatively long shelf life of reagents;
- ✰ Basic requirement for accessory equipment and supplies;
- ✰ Economical technique.

ELISA has been applied to viruses of stone fruit trees since its first introduction into plant virology in 1976. The first approach was with arabis mosaic virus (ArMV) and plum pox virus (PPV), representatives of isometric and filamentous viruses, respectively (Voller *et al.*, 1976 and Clark *et al.*, 1976). The technique was applied later to the majority of the viruses for which antisera were already available.

The use of the ELISA technique for plant virus detection has been reviewed extensively (Clark, 1981 and Torrance and Jones, 1981) and a number of modifications and variations have been described.

DAS-ELISA

The direct double antibody sandwich (DAS) ELISA (Figure 22.5) technique is most frequently employed for the detection of plant viruses and, in most instances, the enzyme used to label the specific antibody has been alkaline phosphatase.

The technique requires (a) purification of gamma globulin and (b) gamma globulin-enzyme conjugation prior to beginning the test process. Prepared globulins and conjugates specific to a small range of viruses can be purchased ready prepared. However, the purification and conjugation procedures are relatively straightforward and once these are carried out within the laboratory most antisera can be used. At its most basic, the ELISA test, including gamma globulin purification and enzyme conjugation can be achieved with simple equipment. However, estimation of gamma

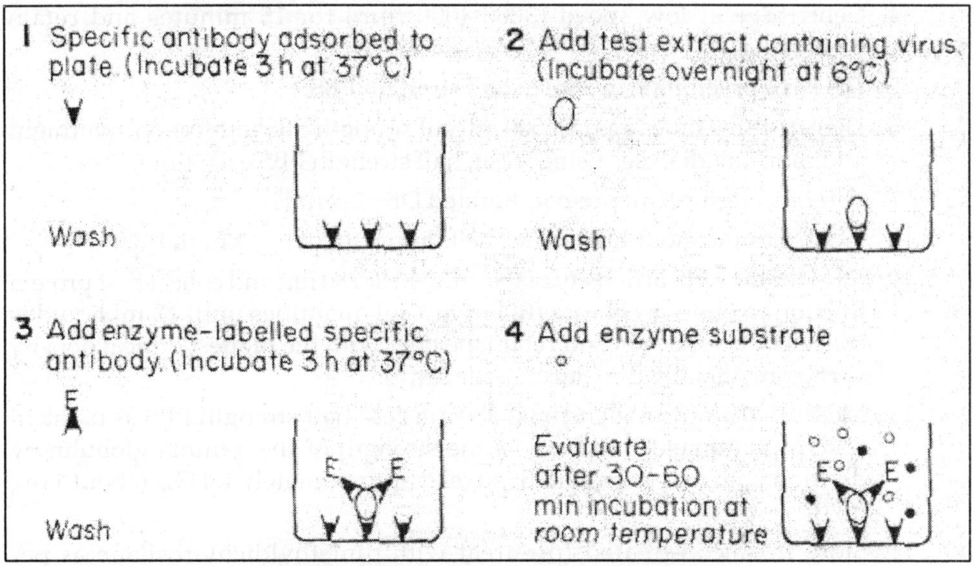

Figure 22.5: The Principles of DAS-ELISA.
Colour intensity in the final evaluation is proportional to virus concentration.

globulin concentration during purification requires access to a.u.v. spectrophotometer, and accurate evaluation of the end-point reactions of ELISA tests may be limited by visual estimation.

For Preparation of Gamma Globulin

Materials

1. Antiserum.
2. Neutral saturated ammonium sulphate.
3. Phosphate buffered saline (PBS).
4. Dialysis tubing (7 mm).
5. Pre-equilibrated DE 52 cellulose column.
6. Dimethyldichlorosilane solution (BDH Chemicals).
7. Suitable tubes for mixing, centrifugation and final storage.
8. Bench centrifuge.
9. Pipettes (1 ml and 10 ml).

Procedure

1. To 1 ml antiserum adds 9 ml distilled water.
2. Add 10 ml saturated ammonium sulphate and mix.
3. Leave to precipitate for 30-60 minutes at room temperature.

4. Centrifuge at low speed (3000-4000 rpm) for 15 minutes and retain precipitate.
5. Dissolve precipitate in 2 ml of half strength PBS.
6. Dialyze three times against 500 ml half strength PBS (afternoon, overnight and morning dialyses using fresh, half strength PBS each time).
7. Filter through 3-5 ml pre-equilibrated DE 52 cellulose.
8. Wash gamma globulin through cellulose with half strength PBS.
9. Monitor the optical density of the effluent at 280 nm and collect first protein fraction to elute *or* collect effluent in 1 ml quantities until 12 ml liquid is collected. Check each and retain the ones with the highest optical density at 280 nm (usually the fourth to the sixth).
10. Measure the optical density at 280 nm (use half strength PBS as blank in spectrophotometer) and adjust the strength of the gamma globulin by dilution with half strength PBS to read approximately 1.4 OD (about 1 mg ml^{-1}).
11. Store in silicon-treated (pre-treat with dimethyldichlorosilane as per manufacturers recommendation) glass tubes at -18°C (but avoid freezing and thawing repeatedly) *or* freeze dry.

Preparation of Enzyme Conjugate

The method most widely used for preparation of antibody alkaline phosphatase conjugates is the one-step glutaraldehyde method.

Materials

1. Alkaline phosphatase. This can be purchased as a suspension in ammonium sulphate (Sigma London, Chemical Co. Ltd, Type VII) or as a solution in sodium chloride (BCL). The activity of each such preparation is defined by the supplier in enzyme units per mg. Both these variables are important in preparation of enzyme conjugates: see below.
2. Gamma globulin (1 mg ml^{-1} concentration).
3. Glutaraldehyde solution (25 per cent as prepared for electron microscopy) (BDH Chemicals).
4. Dialysis tubing.
5. Phosphate buffered saline (PSB).
6. Bovine serum albumin powder (Sigma London, Chemical Co. Ltd; Fraction V).

Procedure

(a) For Alkaline Phosphatase Suspensions in Ammonium Sulphate

1. Centrifuge 1 ml (equivalent to 5 mg or 5000 units) enzyme precipitate (3000-4000 rpm). Discard supernatant liquid.

2. Dissolve precipitate directly in 2 ml (= 2 mg) purified gamma globulin.
3. Dialyse against three changes of 500 ml PBS.
4. Add fresh glutaraldehyde solution to 0.05 per cent final concentration, mix well.
5. Leave for 4 hours at room temperature during which time a very faint yellow brown colour should develop.
6. Dialyse three times (afternoon, overnight and morning) against 500 ml PBS to remove glutaraldehyde.
7. Add bovine serum albumin to give a concentration of 5 mg ml^{-1} and store at 4°C in the refrigerator.

(b) For Alkaline Phosphatase Solutions in Sodium Chloride

Add 0.2 ml (equivalent to 2 mg or 5000 units) enzyme solution to 2 ml gamma globulin (=2 mg). There is no need for dialysis at this stage in the conjugation process which should follow from stage 4 as above.

DAS-ELISA Test Procedure

Preparatory Evaluation of Coating Gamma Globulin and Conjugate

Before use in routine ELISA testing the optimal dilution for each gamma globulin and conjugate preparation should be determined experimentally in a test plate. For many plant viruses gamma globulin suspension of 1 mg ml^{-1} (OD 1.4) may be further diluted to 1 or 2 μg ml^{-1}. Concentrations of greater than 10 μg ml^{-1} are reported to reduce the strength of the virus-specific reaction and increase intensity of non-specific reaction. Conjugate dilutions may be of the order of 1/500 and 1/1000.

The following scheme may be used adopting the routine procedure for DAS-ELISA. Working dilutions of coating gamma globulin and conjugate should be chosen which give the greatest colour at the end of the process, combined with least reaction in control wells. The test plate may also provide an opportunity to determine the best sample extract dilution to use.

Materials

1. Microtitre plate (polystyrene). A variety of these are available from several manufacturers. The most widely used type, has 8x12 flat-bottomed wells of c. 400μl capacity (see Appendix for list of plate types).
2. Adjustable hand pipettes with disposable tips
3. Purified gamma globulin.
4. Phosphate buffered saline (PSB)
5. Coating buffer (carbonate buffer pH 9.6).
6. Substrate buffer (diethanolamine buffer pH 9.8).
7. Tween 20 (Polyoxyethylene sorbitan monolaurate). (Sigma London, Chemical Co. Ltd.)

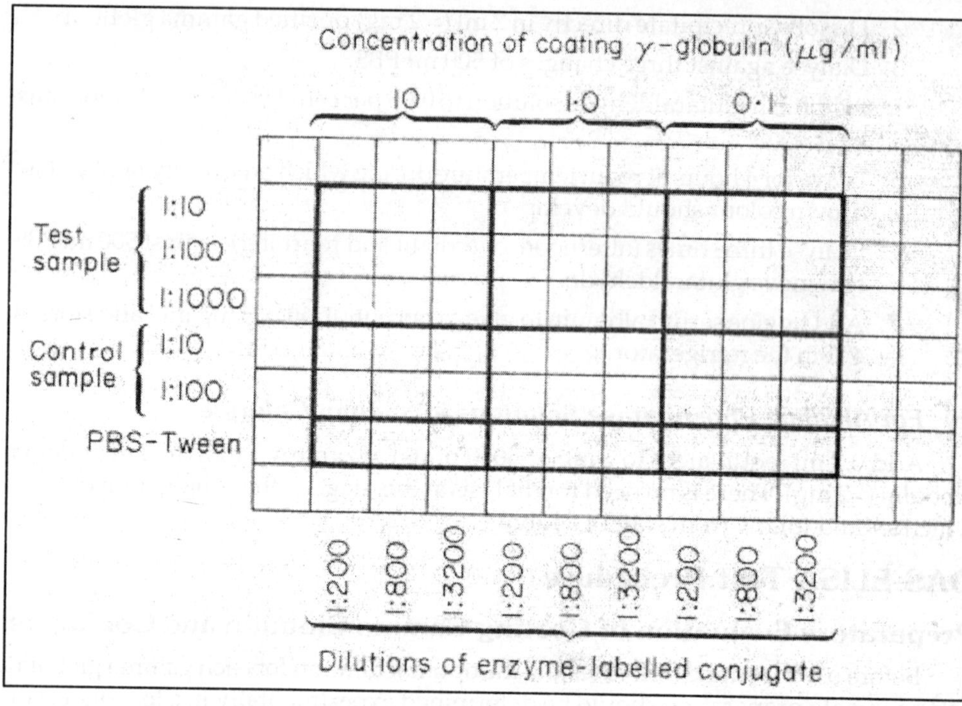

8. Polyvinyl pyrrolidone (PVP) (mol. Wt 44,000). (BDH Chemicals.)
9. Ovalbumen (BDH Chemicals).
10. Enzyme-labelled globulin (conjugate).
11. P-nitrophenyl phosphate (Sigma London, Chemical Co. Ltd). (Tablets each of 5 mg chemical are convenient.)
12. Cling film to cover plates.
13. Glassware for dilution of reagents. To avoid loss of specific components this should be scrupulously cleaned and siliconized (using dimethyldichlorosilane solution; see Appendix).
14. Incubator at 37°C.
15. 3M NaOH.

Procedure for DAS-ELISA

1. Prepare a dilution (experimentally predetermined as above) of coating gamma globulin in carbonate buffer.
2. Add 200 μl of diluted coating globulin to each well of the microtitre plate.
3. Cover plate with cling film and incubate for 2-4 hours at 37°C.

4. Discard well contents and wash by flooding wells with PSB containing 0.5 ml Tween 20 per litre (PSB-Tween). Adopt a standard washing procedure throughout the test: flood empty wells with PSB-Tween and stand for3 min, empty plate and shake dry; repeat three times.
5. Add separate 200 µl aliquots of dilute test sample extract to pairs of wells using a hand pipette and discarding pipette tip between each sample. Samples should be extracted using a pestle and mortar and diluted in PSB-Tween containing 2 per cent PVP plus 0.2 per cent ovalbumin (this should be freshly prepared on the day of use). Each microtitre plate should also contain healthy and known infected control samples and wells filled with sample extracted buffer only.
6. Cover with cling film and incubate overnight at 4°C in a refrigerator, or if speed is essential, at 37°C for 4 hours.
7. Wash plate following procedure outlined in (4) above.
8. Add diluted (experimentally predetermined as above) enzyme-labelled gamma globulin, 200 µl per well, filling all the wells in the plates.
9. Cover with cling film and incubate at 37°C for 3-6 hours.
10. Wash plate following prescribed procedure.
11. Add 300 µl aliquots of freshly prepared substrate (p-nitrophenyl phosphate 0.6 mg ml^{-1} *or* 4x5 mg susbstrate buffer tablets in 30 ml) to all wells.
12. Incubate at room temperature until reactions have progressed sufficiently to allow visualization; usually 30 min is adequate.
13. Reactions may be stopped by adding 50 µl 3M NaOH to each well, whilst shaking the plate to mix.
14. Assess results by *either* visual estimation, *or* measurement of absorbance at 405 nm using a colourimeter.

Direct ELISA	*Indirect ELISA*
Steps Involved (a) Antibodies to the virus. (b) Virus preparation or sap. (c) Antibodies to the virus to which molecules of a particular enzyme have been attached. (d) A substrate for the enzyme, *i.e.*, a substance that the enzyme can break down and cause change in its colour.	Steps Involved (a) Virus preparation or sap. (b) Antibodies to the virus. (c) Antibodies against the antibody proteins of the animal in which the virus antibody was produced. (d) A substrate for the enzyme.
Antibody or antigen is immobilized on a solid surface.	Antigen is bound to bottom of a microtitre plate (Figure 22.6).
Amount of bound antigen, or by antibody, respectively is detected by colour changes in the case of enzyme-linked detection.	Specific antibody (primary antibody) is added that binds to the antigen. An enzyme-linked secondary antibody is added. Intensity of colour produced is proportional to the amount of bound primary antibody.

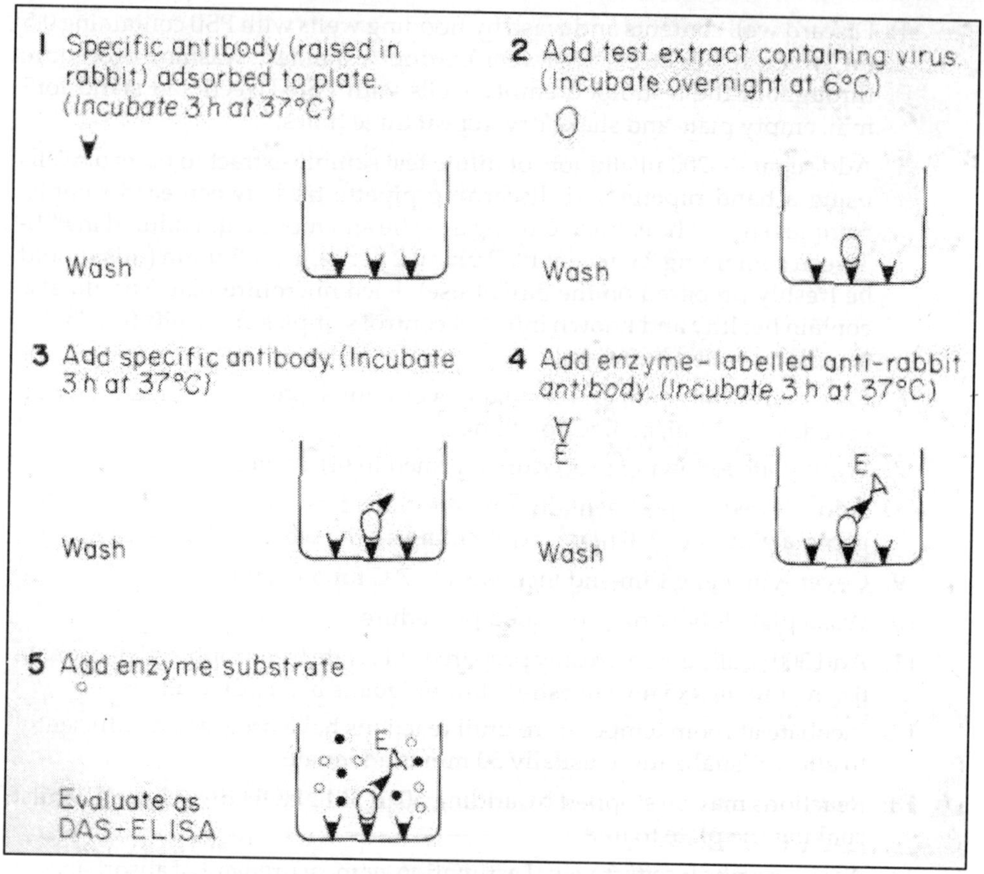

Figure 22.6: Indirect ELISA Sample Protocol

Modifications of the ELISA Process

A number of modifications of the double antibody sandwich ELISA technique have been described and used for detection of plant viruses, each proponent claiming some advantage from their technique. Indirect ELISA systems follow a double antibody sandwich process except that the second layer of specific antibody raised in rabbit is not enzyme labeled, the label being introduced as a conjugated anti-rabbit gamma globulin in a further step (Figure 22.6). This, it is claimed, allows the full binding property of the specific gamma globulin to be used, giving greater sensitivity and overcoming the extreme specificity of DAS-ELISA.

Another modification of ELISA uses Clq (a component of complement obtained from bovine serum) to trap virus antibody aggregates (Figure 22.7). Plates are first coated with Clq after which a mixture of infected plant sap and virus-specific gamma globulin (raised in rabbit) is incubated overnight. Trapped virus antibody aggregates are detected by the subsequent addition of enzyme-labelled anti-rabbit gamma globulin followed by substrate. The Clq assay not offers the advantages of indirect techniques but also employs non-specific coating and enzyme-labelled components. However,

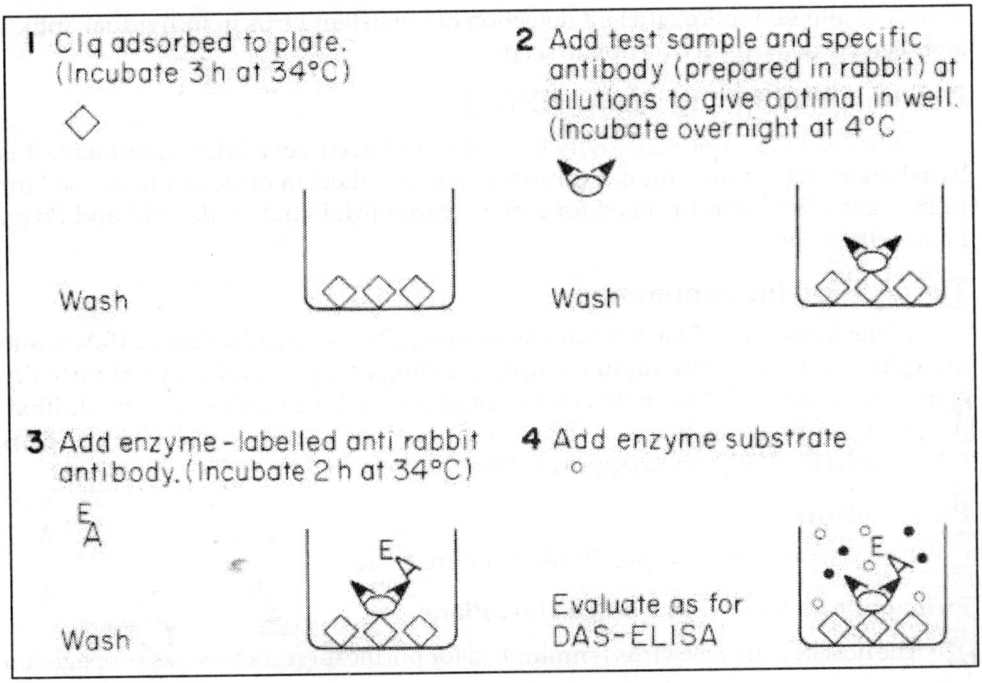

Figure 22.7: Clq-ELISA Protocol

the technique is adversely affected by concentrated sap of certain plant species, and where this occurs sap must be diluted. Clq assay may not therefore be suitable for routine application where such non-specific sap reactions preclude dilution in grouped samples.

A further ELISA modification (Barbara and Clark, 1982) combines the advantages of an indirect assay with those of DAS-ELISA. Antigen is trapped on a solid phase as in DAS but using the $F(ab')_2$ part (Fig. 1) of the gamma globulin molecule (prepared by incubation of gamma globulin with pepsin); trapped virus is detected using an immunoglobulin-based enzyme conjugate specific for the Fc portion of the IgG. Pepsin cleavage of the trapping antibody permits the use of a general purpose enzyme conjugate which discriminates between trapping and detective antibody. Disadvantages of the method are that the specificity of the procedure is dependent on the concentration of the second antibody, higher concentration giving decreased specificity. However, lower background reactions are obtained and the procedure may be useful for investigation where the effort or expense of preparing individual virus specific conjugates is not justified.

The use of a fluorogenic substrate (4 methyl umbelliferyl phosphate) has recently been used in comparison with n-nitrophenyl phosphate for the detection of plant viruses. The fluorogenic substrate allowed increases in sensitivity of two to sixteen times with several viruses in leaf extracts, two to four times in tuber extracts for potato virus and allowed more efficient detection of persistent virus in individual aphids and seed borne virus in true potato seed.

Dot Immunobinding Assay (DIBA)

DIBA has the same sensitivity of ELISA and need very little equipment. It is based on the use of membranes (nitrocellulose or other) in place of plates used for ELISA, and eliminates the need for a plate reader (Makkouk *et al.*, 1993 and Poggi Pollini *et al.*, 1993).

Tissue Blot Immunoassay

While it may not always reach the same sensitivity as ELISA and DIBA, tissue imprinting is remarkably rapid (sample grinding and preparation used virtually eliminated) and, as DIBA, it can be performed with little equipment. In addition, tissue imprinting can provide data on virus localization within plant organs (Makkouk *et al.*, 1993 and Knapp *et al.*, 1995).

Purification

General principles for purification of viruses are:

Propagation of Virus in Choice Host Plant

The host in which the virus is multiplied for purification is known as propagation host.

Preparation of Crude Sap

The prepared sap is centrifuged at law–speed (5,000- 10,000rpm for 15 min) to remove intact cells. Phosphate buffer with ionic strengths 0.05 -0.1 at ph 65 have been most widely used, but alternative buffer like borate/ Tris buffer have also been used.

Use of Stabilizing Agents to Prevent the Inactivation of Virus during Extraction

Chelating agents such as diethyl dithiocarbonate (DIECA) also assist in extraction; prevent virus aggregation and oxidation of polyphenols. Non-ionic detergents like triton x-100 or tween-80 in the extraction buffer helps in the release of virus particles from the host constituents and to dissociate cellular membranes that may contaminate virus particles.

Clarification of Crude Sap/Extract

The initial separation of virus from the bulk of host material is called clarification. The homogenate is subjected to low speed centrifugation (3000-5000 rpm) for 15-20 minutes. The supernatant containing the virus and smaller host cell contaminants is further clarified by addition of organic solvents like n-butanol, $CHCl_3$ or CCl_4, etc and subjected again to law speed centrifugation by which homogenate separate into three layers. The topmost lightest layer is the aqueous phase contains the virus. The aqueous layer is carefully removed and retained. The solvent is evaporated.

Partial Purification and Concentration of Virus

Salt precipitation is a valuable method for viruses that are not inactivated by strong salt solutions.

Precipitation with Ammonium Sulphate or with Polyethylene Glycol

(*a*) Ammonium sulphate (1/4 or ½ saturation), after 3-4 hours when the pptn. is complete the mixture is centrifuged. The precipitate is collected as a pellet by law speed centrifugation (3000-5000rpm for 15minutes). The supernatant still contains soluble & low molecular weight host contaminants.

(*b*) The solvent clarified aq. Phase is mixed c additives like polyethylene glycol (PEG) of 6000 molecular weight along with NaCl and EDTA. Concentration of PEG & NaCl is crucial & standardized for each virus. The concentration of PEG may vary from 4-12 per cent and of NaCl 0.02 to 0.2 M.

The mixture of clarified extract and PEG–NaCl is stirred for 30minutes to 2hours at 4°C. It is subjected to low speed centrifugation. The pellets are suspended in small volumes of buffer and centrifuged. The supernatant is bioassayed for infectivity, contains soluble and low molecular weight host contaminants.

High Speed Centrifugation

High speed centrifugation (ultra centrifugation) leads finally to a virus preparation of reasonable purity. The partially purified preparation, from previous step is centrifuged at 35,000 rpm to 60,000 rpm for 2hours. High speed centrifugation is a physically severe porous that may damage some particles or aggregation of virus particles.

Density-Gradient Centrifugation

It offers the possibility of concentrating viruses without pelleting and is one of the most useful procedures for further purification. During centrifugation upon (50,000–70,000 xg) for varying periods depending upon the virus (several hours), the virus and other host contaminants move along gradient at cliff rates to form layers in the tube, depending upon their shape, shape and density (sedimentation coefficient). The layers can be seen against a dark background as light–scattering zones. Sedimentation coefficient of the virus was then estimated graphically by using pictures of Schlieren patterns. The Schlieren pattern obtained from equilibrium banding of virus in cesium chloride ($CsCl_2$).

Final Purification

Gel Chromatography

Gel chromatography using columns containing Sephadex (modified dextran) and Sepharose (gel-beads) may also be used for purification of viruses.

Immunoaffinity Columns

Monoclonal antiviral antibodies can be bound to a support matrix such as agarose to form a column that will specifically bind the virus from a solution passed through the column. Virus can be eluted by lowering the pH.

Testing for Purity

The homogeneity of the purified virus preparation may be checked by spectrophotometry, electrophoresis, and electron microscopy.

Storage of Purified Virus

It can be stored at low temp. (-20°C) by adding a drop of sodium azide and glycerol or thymol to prevent growth of microbes and stabilize the virus.

References

Agrios, G.N. (2006). Plant Pathology. Fifth Edition. ELSEVIER Academic Press, New York, pp: 744-747.

Barabara, D.J. and Clark, M.F. (1982). A simple indirect ELISA using F(ab´)$_2$ fragments of immunoglobulin. *Journal of General Virology* **58:** 315-322.

Berks, R., Koenig, R. and Querfurth, G. (1972). Plant virus serology. In *Principles and Techniques in Plant Virology* (eds Kado, C. I. & Agarwal, H. O.). pp. 466-472. Van Nostrand Reinhold, New York.

Cambra, M., Llacer, G. and Perez de Sanroman, C. (1983). Use of enzyme-linked immunosorbent assay (ELISA) for virus detection on stone fruit trees in Spain. *Acta Horticulturae* **130:** 145-150.

Clark, M.F. (1981). Immunosorbent assays in plant pathology. *Annual Review of Phytopathology* **19:** 83-106.

Clark, M.F. and Bar Joseph, M. (1984). Enzyme immunosorbent assay in plant virology. *In:* Methods in Virology. Acc. Press. New York. Vol. III 51-85.

Clark, M.F., Adams, A.N., Thresh, J.M. and Casper, R. (1976). The detection of plum pox and other viruses in woody plants by enzyme-linked immunosorbent assay (ELISA). *Acta Horticulturae* **67:** 51-57.

Hill, S.A. (9184). Methods in Plant Virology. Published on behalf of the British Society for Plant Pathology by Blackwell Scientific Publications, London. Pp: 91-121.

Knapp, E., Da Camara Machado, A., Puhringer, H., Wang, Q., Hanzer, V., Weiss, H., Weiss, B., Katinger, H. and Laimer Da Camara Machado, M. (1995). Localisation of fruit tree viruses by immune tissue printing in infected shoots of *Mallus* and *Prunus* sp. *Journal of Virological Methods* **55:** 157-173.

Koenig, R., Fribourg, C.E. and Jones, R.A.C. (1979). Symptomatological, serological and electrophoretic diversity of isolates of Andean potato latent virus from different regions of the Andes. *Phytopathology* **69:** 748-752.

Makkouk, K.M., Hsu, H.T. and Kumari, S.G. (1993). Detection of three plant viruses by Dot-Blot and Tissue-Blot immunoassays using chemiluminescent and chromogenic substrates. *Journal of Phytopathology* **139:** 97-102.

Poggi Pollini, C., Giunchedi, L. and Credi, R. (1993). A chemioluminescent immunoassay for the diagnosis of grapevine closteroviruses on nitrocellulose membrane. *Journal of Virological Methods* **42:** 107-116.

Regenmortel, M.H. van (1982). *Serology and Immunochemistry of Plant Viruses.* Academic Press, London.

Slogteren, D.H.M. van (1955). *Serological microreactions with plant viruses under paraffin oil.* Proceedings of the Second Conference on Potato Virus Diseases, Lisse Wageningen, 1954, pp. 51-54.

Torrance, L. and Jones, R.A.C. (1981). Serological methods in testing for plant viruses. *Plant Pathology* **30:** 1-24.

Uyemoto, J.K. (1980). Detection of maize chlorotic mottle virus serotypes by enzyme-linked immunosorbent assay. *Phytopathology* **70:** 290-292.

Verma, H.N. (2007). Basics of Plant Virology. Oxford & IBH Publishing Co. Pvt. Ltd. Pp: 63-86.

Voller, A., Barlett, A., Bidwell, D.E., Clark, D.E. and Adams, A.N. (1976). The detection of the viruses by enzyme-linked immunosorbent assay (ELISA). *Journal of General Virology* **33:** 165-167.

Chapter 23

Recent Advancements in Understanding of Virus-Vector Interaction and Mechanism for Efficient Virus Transmission by Insects

Kajal Kumar Biswas and Susheel Kumar Sharma*

Advanced Centre for Plant Virology, Division of Plant Pathology, Indian Agricultural Research Institute, New Delhi – 110 012

Importance of Plant Virus-Vectors

Plant viruses can cause severe yield losses to the cereal, vegetable, fruit, and floral industries and substantially reduces the quality of crop products worldwide. Plants being sessile organisms, itself can not transmit viruses except for a few instances where seeds or pollens and movement of plant materials resulting from human intervention through which viruses are transmitted. Plant viruses are obligate parasites and must be spread from one susceptible host to other to be introduced in new living cells for their survival. Viruses are not disseminated by wounds or water or by any other physical means. Generally they do not cause infection resulting in

* Corresponding Author E-mail: kkbiswas@mailcity.com; drkkbiswas@yahoo.co in

disease, unless they come into contact with the contents of the wounded living cells, as virus cannot penetrate the intact plant cuticle and the cellulose cell wall.

There are two principal reasons for which most of the plant viruses depend on vectors for their survival. Firstly, plant epidermis contains impermeable cuticle coats that prevent entry of virus particles, although animal viruses can enter readily through natural openings. Most vectors are insects, particularly Hemipterans insects, and are well adapted to their role as vectors by their capacity to pierce the epidermis and delicately deposit the virus in the cytoplasm without disturbing the integrity of the plant cell. Secondly plants are rooted and lack independent mobility. Therefore, many viruses depend on insects for transport among hosts, unlike animals, that by their own mobility transport the virus to new niches.

The great majority of plant viruses are dependent for their spread on efficient transmission from plant to plant by specific vector(s). Plant to plant spread ensures virus survival, often resulting in disease, particularly in agricultural crop plants. The knowledge of ways of plant virus transmission is important for the following reasons: (1) A disease is recognized as a virus disease only if it can be disseminated to healthy host by some means, (2) Mode of virus spread in the field is essential in designing management strategies. Thus the interactions between virus and their vectors are of considerable biological interest.

Insect Vector of Plant Viruses

Vectors are the regular and specific agents that carry the viruses from the source plants and transmit to new plants. Vector transmission is a specific event in the perpetuation of viruses and its life cycle. Virus-encoded specific determinants interact with the specific receptor that facilitate transmission of viruses by specific vectors. Taxonomically vectors of plant viruses are very diverse and complex containing different organisms like fungi, nematodes, arthropods, fungi, and plasmodiophorids, which are recognized as efficient vectors for various plant viruses. The most common vectors of plant viruses are invertebrate vectors belonging to Arthropoda under the insect order Hemiptera. Arthropod vectors which transmit most of the plant viruses includes aphids, whiteflies, leafhoppers, thrips, beetles, mealybugs *etc.* and non insect arthropods are species of mite (*Acarina sp*). More than 200 aphid species have been identified to transmit viral diseases in non-persistent, semi-persistent and persistent manner. As there are specificity in biology of the host, virus and vector in their interaction, the mechanism of virus uptake from infected hosts and preservation in vector body, and then transport and dissemination to new hosts varies significantly between diverse groups of insect vectors.

Efficient Insect Vector for Plant Viruses

Insect vectors transmitting plant viruses belong to seven orders out of total number of 32 orders under the class *Insecta*. The majority of vectors are found in the two orders, Hemiptera (300) and Thysanoptera (6) having piercing-sucking mouthparts. A few insect vector with chewing type mouthparts belonging to another 5 orders; Orthoptera, Dermaptera, Coleoptera, Lepidoptera and Diptera can also transmit viruses. Apparently, the feeding habits of Hemipteran insects play major roles for successful transmission of plant virusues. Of the 697 virus species recognized by the International Committee on Taxonomy of Viruses (ICTV), insects and other vectors transmit 76 per cent and Hemipteran insects transmit the majority of the vectored viruses (55 per cent). Viruses transmitted by 'piercing-sucking' insects are quantitatively predominant and the traits of morphology that contributes to the ability of these insects to transmit plant diseases so efficiently is their piercing-sucking feeding style. Insects in the order Hemiptera (aphids, leafhoppers, whiteflies), and Thysanoptera (thrips) have similar basic morphologies of the head and body.

Some Remarkable Epidemic Caused by Viruses Transmitted by Hemipteran Insect

Insect-borne plant viruses may cause severe losses to many of the annual and perennial crops worldwide. On occasion, insects are responsible for transition from a non spreading form to the epidemic form of diseases. Majorities of the insect vectors are plant pests, and thus their association with plants makes them ideal agents for efficient local and long-distance spread of viruses. For example, of the 25 rice infecting viruses described recently, 14 are transmitted by either rice pest planthoppers or leafhoppers. Many of these viruses have resulted in significant disease outbreaks in many of the rice growing countries of Southeast Asia (Hibino, 1996). Severe yield losses caused by begomoviruses (Family: *Geminiviridae*), the most widespread of the whitefly (*Bemisia tabaci*) transmitted viruses, have been reported from a number of crop plants like pulses, tomato, cassava, cotton, and other vegetable crops (Briddon and Markham, 2000; Czosnek and Laterrot, 1997). Outbreaks of the aphid-transmitted *Citrus tristeza virus* (CTV) have been reported almost wherever citrus is grown, and in the 1930s in South America millions of citrus trees were lost to due to CTV (Bar-Joseph and Dawson, 2008). In addition, the worldwide incidence and effects of aphid transmitted luteoviruses (genera *Luteovirus* and *Polerovirus*) on food and fiber crop plants have been described. Outbreak of *Tomato spotted wilt virus* (TSWV) is attributed to the spread of the thrip vector *Frankliniella occidentalis*.

The virus *Banana bunchy top virus* (BBTV), member of the genus *Babuvirus* (family *Nanoviridae*) causes tremendous economic losses in banana due to the prevalence of efficient aphid vector *Pentalonia nigronervosa* in Asian countries. *Faba bean necrotic yellows virus* (FBNYV), is one of the most economically damaging disease agents of faba bean, causing up to 90 per cent crop losses in Egypt and the nearby regions of Syria and Turkey. The disease spreads because aphid populations can survive the mild winters and provide a continuous inoculum source for FBNYV. Thus, the importance of Hemipteran vectors as the primary means for spreading many of the important plant viruses is clear.

Mode of Transmission of Plant Viruses by Insect Vectors

The transmission of plant viruses has been investigated for over a century with the most common vectors being sap-feeding insects with pierce-sucking mouth parts, particularly aphids, whiteflies, leafhoppers, planthoppers, and mealy bugs. Pioneer studies have demonstrated the complexity and diversity of the interactions between plant viruses and their insect vectors. Watson and Roberts (1939) coined the terms "non-persistent viruses" and "persistent viruses" for the first time in an attempt to categorize the relationships between plant, virus and vector in virus transmission. Non-persistent viruses are plant viruses for which inoculativity by the vector is retained for only a few seconds/minutes after acquisition from plants and is also lost after molting, where as the persistent viruses are those for which inoculativity by the vector is retained for long periods (days to weeks), often throughout the vector's lifespan, and also is retained after molting.

Non-persistent viruses are efficiently transmitted by vectors after relatively brief acquisition access period (AAP) and inoculation access periods (IAP) while persistent viruses required longer AAP and IAP. Non-persistent viruses need not any latent period between AAP and IAP and subsequent ability of the vector to transmit of the virus to the plants, whereas persistent viruses exhibited a delay, *i.e.* they need a distinct latent period for incubation. The non-persistent viruses are experimentally mechanically transmissible to plants, whereas only a few of the persistent viruses are transmitted mechanically. Finally, aphids are the vectors for the non-persistent viruses, whereas, some aphids, leafhoppers, thrips, and whiteflies transmit viruses in a persistent manner (Reviewed by Ng and Falk, 2006).

Several viruses need an intermediate retention period, longer than non-persistent but shorter than persistent categories of transmission, which is termed as semi-persistent viruses. Inoculativity of semi persistent viruses by the vector is retained for a few hours to a few days after acquisition from plants but lost upon molting.

Based on advancement of research of plant virology, "transmission based classification" of viruses has been revised to "quantitative criteria based classification". In this new classification system, the non-persistent and semi-persistent viruses are grouped together to 'non circulative viruses (NCV)', and the persistent viruses are named to 'circulative viruses (CV)'. Further the category 'circulative' has been broken down into the two subcategories: 'propagative' and 'non-propagative' based on their capacity to replicate in the respective insect vectors (Table 23.1). Some important vectors and its transmission modes with their specific viral genera are given in Table 23.2.

Mechanism of Virus Vector Interaction in Transmission of Plant Viruses

In early study it was observed that a non-persistent virus, as example, *Potato virus Y* losses its vector transmissibility by chemical (formaldehyde) or ultraviolet (UV) treatments that affects the stylets of live viruliferous aphids. This indicated that infectious virus particles are retained in stylets. It was believed that the transmission of 'non-persistent' viruses could be assimilated to mechanical transmission due to contamination of the stylets, which was hypothesized as 'flying needle'. Further it was shown that transmissibility of 'non-persistent' viruses are lost upon moulting of the viruliferous vectors. Later on, the hypothesis of virus uptake has been changed from "flying needles" to 'flying syringes', although the virus-vector relationship was still being considered as non-specific.

Table 23.1: Classification of Plant Virus Transmission Based on Quantitative Criteria

Feature	*Non-circulative (External)*		*Circulative (Internal)*	
	Non-Persistent	*Semi-Persistent*	*Persistent-Circulative*	*Persistent-Propagative*
Duration of retention	Brief (few hours)	Intermediate (few days)	Long (days to months)	Long (days to months)
Duration of acquisition and transmission	Brief (seconds)	Intermediate (hours)	Long (hours to days)	Long (hours to days)
Latent period	Not required	Not required	Required	Required
Tissue where virus is acquired and inoculated	Epidermis and parenchyma	Epidermis, parenchyma and phloem	Mostly parenchyma and phloem	Mostly parenchyma and phloem
Pre-acquisition fasting	Increase transmission	No effects	No effect	No effect
Passage through moult	Negative	Negative	Positive	Positive
Insect specificity	Low	Intermediate	High	High
Sequential inoculation	Poor	Intermediate	Good	Good
Replication in vectors	No	No	No	Yes

Courtesy: Raccah and Fereres, 2009.

Table 23.2: Vectors of Plant Viruses and Transmission Mode

Vector	*Non-circulative*	*Circulative-persistent*		
	Non-persistent	*Semi-persistent*	*Non-propgative*	*Propagative*
Aphids	*Alfamo-, Carla-, Cucumo-, Faba-, Maclura- and Potyvirus*	*Caulimo-, Clostero-, Sequi-, Tricho- and Waikavirus*	*Enamo-, Luteovirus Nanovirus Polerovirus Umbravirus*	*Cytorhabdovirus Nucleorhabdovirus*
Beetles	*Machlomovirus*		*Bromovirus* (?) *Carmovirus* (?) *Comovirus* (?) *Sobemovirus* (?) *Tymovirus* (?)	
Leafhoppers		*Tungrovirus Waikavirus*	*Curtovirus Mastrevirus*	*Cytorhabdovirus Fijivirus Marafivirus Nucleorhabdovirus Oryzavirus Phytoreovirus Tenuivirus Phytoreovirus*
Mealybugs		*Ampelovirus Badnavirus Trichovirus Vitivirus*		
Whiteflies		*Crinivirus* *Ipomovirus*	*Begomovirus*	
Thrips	*Machlomovirus*			*Tospovirus*

?: Llimited iformation is avaiable on virus vector interaction

Courtesy: Andret-Link and Fuchs, 2005.

It has been clearly demonstrated by recent studies that the virus-vector association occurs externally on the cuticle lining the food or salivary canal of the stylets. Because semi-persistent viruses are also lost upon vector moulting, their association with the vector was also proposed to be external, likely in the stylets, though a possible location 'upstream', on the cuticle lining the anterior guts of the insect, was also proposed in some cases. In contrast to non-persistent and semi-persistent transmission, the persistent viruses have internal association with the vectors. Such viruses were shown to pass through the gut epithelium into the hemolymph and join the salivary gland to be ejected together with saliva (Figure 23.2). A latent period of hours to days after acquisition, during which the virus cannot be efficiently inoculated, is needed for completing this cycle within the vector body.

Non Circulative Transmission

Non-circulative viruses are suggested to be those, which do not enter the body of their insect vectors. They simply attach to receptor sites located externally on the cuticle lining anterior part of the digestive tract, and most often they attach to the alimentary/salivary canals within the stylets or foregut region, and wait until the vector has move to another plant, where they are released to initiate a new infection (Figure 23.1). Extensive studies have been made in case of many *Cucumo, Poty* and *Caulimoviruses* which are transmitted in non-circulative manner by various aphid species. The mechanisms of transmissions for non-persistent and semi-persistent viruses are being discussed as following:

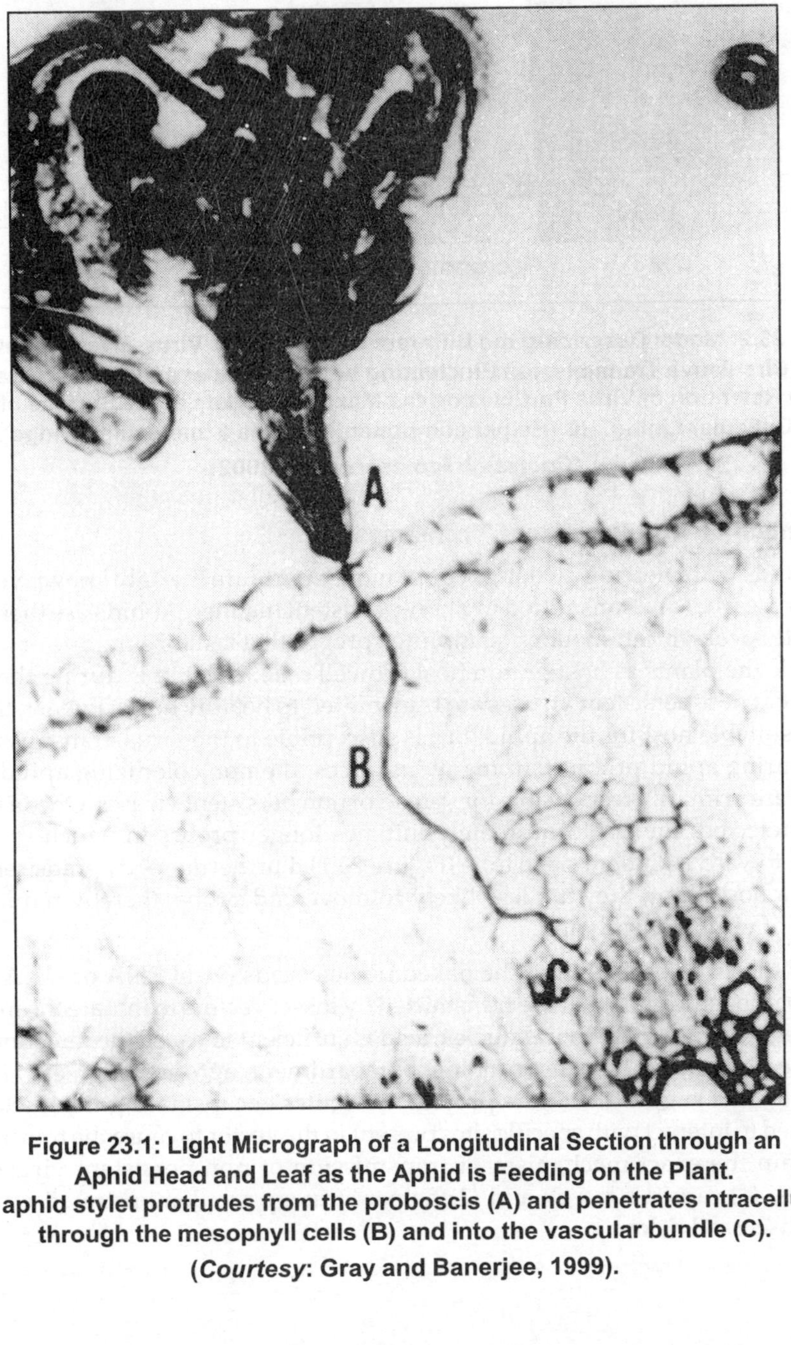

Figure 23.1: Light Micrograph of a Longitudinal Section through an Aphid Head and Leaf as the Aphid is Feeding on the Plant. The aphid stylet protrudes from the proboscis (A) and penetrates ntracellularly through the mesophyll cells (B) and into the vascular bundle (C).

(*Courtesy*: Gray and Banerjee, 1999).

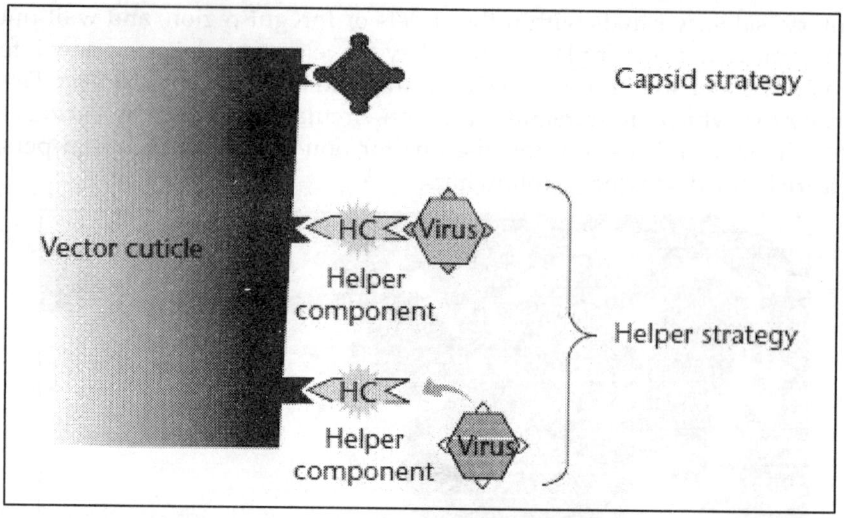

Figure 23.2: Model Describing the Different Strategies for Virus–Vector Interaction in Non Circulative Transmission Elucidating both the Helper and Capsid Strategies; with Retention of Virus Particles on the Maxillary Stylets at the Surface of the Cuticular Lining. HC (Helper component) creates a 'molecular bridge'. (*Courtsey*: Froissart *et al.*, 2002)

Mechanism of Non Persistent Transmission

Aphids with piercing-sucking type of mouth parts are the only known vectors of plant viruses that are transmitted in a non-persistent manner. Aphids use their stylets to initiate several brief, shallow "sampling" probes that last for a minute or less so as to sample the plants as host or non host (Powell *et al.*, 2006). It is during these brief probes that non-persistent viruses are transmitted to healthy hosts. Even if, the plant is not a suitable host for the aphid, but is susceptible to the virus, transmission can result during aphid probing. In many instances, the non colonizing aphids while probing are primarily responsible for spread of non persistent viruses. Once a suitable host is detected, the aphid ultimately initiates longer probes in which stylets are directed toward a phloem sieve tube (Figure 23.1). Furthermore, if aphids select and stay on a host plant, they are less likely to move and probe, thereby reducing the amount of virus transmission.

It is the virus particles not the naked nucleic acids (viral RNA or DNA) which are the pathogenic units that are transmitted by insect vectors to initiate the infection. However, it is known that viral nucleic acid is sufficient to cause the infection when inoculated by artificial means (rubbing, bombardment, agro-infection etc.) in to the plant cell. This suggests that the protein molecules encapsidating the nucleic acid are needed to interact with specific sites present in the vector to cause the transmission. To explain the exact mechanism of transmission of non persistent viruses, two strategies: (1) capsid strategy and (2) helper strategy are demonstrated which are briefly discussed below:

Capsid Strategy (Direct Strategy)

In some aphid transmitted viruses under the genera *Cucumovirus, Alfamovirus,* and *Carlavirus*, purified viral particals are readily acquired and transmitted by the vector. This indicated unequivocally that coat protein (capsid protein) of these viruses must be capable of direct attachment to vector receptor and thus leading to their transmission (Figure 23.2). *Cucumber mosaic virus* (CMV) is the most studied example of virus which is transmitted by non-persistent manner and utilizes the capsid strategy for vector transmission. It has been proved that the CMV encoded coat protein (CP) is a key determinant of aphid transmission. Inoculum composed of CMV RNA 3 from one strain and RNAs 1 and 2 from another strain revealed aphid transmission to be under the control of RNA 3, which encodes the CP. Gera *et al.* (1979) showed that the transmission phenotype of hetero encapsdiated virions was that of the virus from which the CPs was derived. This was further demonstrated by reassembling of free CPs of an aphid transmissible isolate, CMV-T with RNAs from a poorly aphid transmissible isolate CMV-6, or by mixing free CPs from CMV-6 with RNAs from CMV-T. Hetero encapsidation experiment with two cucumoviruses, the aphid (*Myzus persicae*) non transmissible isolate CMV-M and aphid (*M. persicae*) transmissible *Tomato aspermy virus* (TAV) isolate TAV-V, showed that virions having capsids assembled using CPs of TAV-V but not the virions reassembled using CMV-M CPs are aphid transmissible. These experiments confirm the direct role of CP in interaction with aphid vector receptors (Chen and Francki, 1990).

Mutational studies have demonstrated that amono acids at certain position of CP in Cucumoviruses transmitted by aphids, *Aphis gossypii* and *Myzus persicae* are crucial for determination of transmission specificity. The amino acid domain DDALET (positioned from 191 to 197 of CP) which forms the exposed portion of virion particles is crucial for aphid transmission of cucumoviruses.

Helper Strategy (Indirect Strategy)

Helper components are "virus-encoded factors" that are not constituents of the virion, required for vector transmission has been defined (Pirone and Blanc, 1996; Pirone and Megahed, 1996). The natures and origins of helper components and their mechanisms of action for virus transmission may be quite diverse and has been proven to be complex. Later, it has been clearly shown that helper component and the virion can be acquired sequentially by the vector (Froissart *et al.*, 2002)

Early studies showed that *Potato virus Y-C* (PVC) of family *Potyviridae* is not aphid transmissible from plants infected only by PVY-C, but PVY-C is aphid transmissible if plants are co-infected with aphid transmissible PVY (Watson, 1960; Reviewed by Raccah and Fereres, 2009), it is perhaps due to appearance of "some specific protein structure of the virus particles" might determine aphid transmissibility, and that PVY-C gained this characteristic upon co-infecting plants with the normally aphid- transmissible PVY.

Helper phenomenon was reported long back by Kassanis and Govier (1971) showing that mixed infection is not necessary for aphid transmission of PVC. If aphids is first given acquisition access to plants infected with an aphid-transmissible strain of PVY and then to plants infected with non aphid transmissible PVY-C, PVY-C (and PVY) is then aphid transmissible. If the order of acquisition access is reversed (PVY-C followed by PVY), no aphid transmission of PVY-C results, but PVY was still aphid transmissible. Thus, there must be something that aphids acquired from the PVY infected plant before they are given access to PVY-C, and this was sufficient to "help" the aphid transmission of PVY-C. Govier and Kassanis (1974) called this the "helper component" and showed that the helper component (HC) is not the PVY virion, but a separate component from the sap of the PVY infected plants. The biologically active HC could be mixed with purified virions and fed to aphids *in vitro*, thereby making them again aphid transmissible, or the HC could be acquired first, followed by acquisition of PVY virions, and aphid transmission results.

It is clear that in addition to virions, the HC from the *Potyvirus* infected plants was needed for *Potyvirus* aphid transmission. It is also suggested that HC functions to form a "bridge" between the capsid of *Potyvirus* virion and stylet of the aphid, confirming the "bridge hypothesis" for transmission of viruses. The binding of HC to aphid mouthparts on one side and to virions on the other ensures virus retention until it is released into next host. The helper component is thus defined as a non virion and multifunctional protein of viruses that mediates binding between virion and sites within the insect mouthparts during non-persistent transmission. Helper component is a non structural protein encoded by the HC-Pro region of the potyvirus genome. The *Potyvirus* HC is a 50-kD protein, derived from the *Potyvirus* encoded polyprotein by two proteolytic cleavage events (Maia *et al.*, 1996). The helper function in transmission was assigned to the N-terminal and central regions of the HC-Pro. Direct proof was obtained for *Turnip yellow mosaic virus* (TVMV), where loss of HC activity was associated with a mutation from Lys to Glu (K to E) in the highly conserved Lys-Ile-Thr-Cys (KITC) motif of TVMV HC-Pro (Peng *et al.*, 1998). This KITC domain is located near the N terminus (~AA 55) of HC-Pro. In the coat protein of potyviruses, within the `10-15 amino acids from N terminus, is the sequence DAG, which is exposed to the surface of virions. The DAG triplet has been shown to be the important functional domain on the CP for aphid transmissibilty of potyviruses (Atreya *et al.*, 1995; Reviewed by Ng and Falk, 2006).

Mechanism of Semipersistent Transmission

Aphids, whiteflies and leafhoppers are recognized as vectors of many plant viruses under various virus genera and families; those are transmitted in a semi-persistent manner. The acquisition and inoculation of semi-persistently transmitted plant viruses mostly occur from and to phloem tissues of plant. As there is a longer period of acquisition of viruses by vectors, semi-persistent viruses are transmitted successfully compared with non-persistent viruses having a brief acquisition period. It has often been proposed that the difference between non- and semi-persistent viruses, in acquisition and retention time within the vector, was due to differential location of binding sites, the formal being retained in the stylets and the latter higher up in the foregut. Studies have indicated that both the helper and capsid strategies are used for transmission of different viruses transmitted in semi-persistent manner, as described here.

Capsid Strategy (Direct Strategy)

The *Lettuce infectious yellows virus* (LIYV) under genus *Crinivirus*, is transmitted by whitefly (*Bemisia tabaci*) in a semi persistent manner (Duffus *et al.*, 1986). The purified virions of LIYV were shown to be readily transmissible by *B. tabaci* after *in vitro* acquisition (Tian *et al.*, 1999), suggesting that a capsid strategy mediates LIYV transmission by *B. tabaci*. To be sure that the LIYV virions were not contaminated with an HC-like protein, they were purified through various means including cesium sulfate gradients, but virions were still efficiently transmitted by *B. tabaci*. This was the first definitive case wherein semi-persistent vector transmission of a plant virus was shown not to be mediated by non virion (helper) proteins.

Helper Strategy (Indirect Strategy)

Caulimoviruses adopted a HC-dependent transmission strategy similar to those of the helper component-mediated *Potyvirus* transmission (Figure 23.3). The *Cauliflower mosaic virus* (CaMV) requires three-component system: an 18-kDa non virion protein (p2) that is believed to bind to the aphid; a 15-KDa protein p3, that is anchored in the shell of the CaMV virion; and the capsid protein encoded by CaMV virion, p4, (Drucker *et al.*, 2002). The model suggests that p2, the non virion protein, fits the definition given for HC proteins, that can be acquired simultaneously with, or prior to, aphids acquiring the transmissible CaMV virions. Amino acid at position 6 from the N terminus of p2 is shown to be a determinant for facilitating CaMV aphid transmission (Moreno *et al.*, 2005). Further, it was demonstrated that a sequential acquisition of p2 followed by p3 may be natural mode of CaMV acquisition by its aphid vectors (Drucker *et al.*, 2002).

HC-mediated semi-persistent vector transmission has been reported in *Maize cholotic dwarf virus* (genera: *Waikavirus*) transmitted by *Graminella nigrifrons*. In case of *Parsnip yellow fleck virus* (Genera: *Sequivirus*) transmitted by *Cavariella aegopodii* which is not transmitted by aphid unless the aphid first feed on *Anthriscus yellows virus* (genus: *Waikavirus*) infected plant. A HC-dependent transmission relationship are also existed for the leafhopper (*Nephotettix virescens*) transmitted tungro disease complex in rice (Hibino, 1996). Tungro disease is caused by two taxonomically unrelated viruses, : *Rice tungro bacilliform virus* (RTBV, Tungrovirus) and *Rice tungro spherical virus* (RTSV, *Waikavirus*). Both viruses are transmitted by *N. virescens* from doubly infected plants, but RTBV is dependent for its leafhopper transmission on RTSV. RTBV can also be transmitted if leafhoppers are first given access to RTSV-infected plants and then to RTBV-infected plants, providing evidence for involvement of a helper component in transmission of RTBV (Hull, 1996).

Circulative Transmission

Circulative viruses as the name suggest are carried in the interior of the vector body. While feeding on infected plants, vectors ingest circulative viruses from phloem tissues of plants. Viruses reach to the insect vector within minutes to hours depending on species, which explains the long feeding periods required for their acquisition. These viruses cross the mid or hind gut epithelllium, are released into the hemolymph, and can then adopt various pathways to traverse the salivary glands, and be released

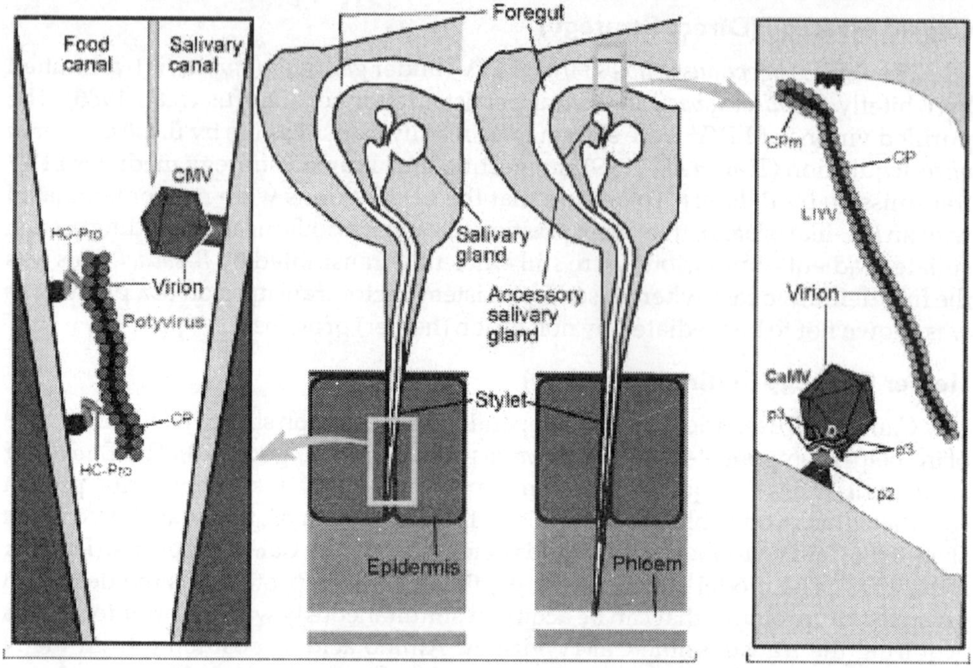

Figure 23.3: Models for Non-persistent and Semi-persistent Virus Transmission (a) Non-persistent transmission. Sampling and virion acquisition (left two panels), and feeding (right panel) for a Hemipteran vector. The left panel shows the tip of the stylet bundle. Upon penetrating the epidermis of an infected source, virions of non-persistent stylet-borne viruses in the genus *Potyvirus* and those in the genus *Cucumovirus* such as *Cucumber mosaic virus* (CMV) can be bound to sites within the stylet. (b) Enlarged view showing the interactions between sites within the insect's foregut region and semi persistently transmitted viruses of *Caulimoviridae* (CaMV), *Closteroviridae* (LIYV). Insect-vector-virus specificity in this case is detemined by helper and capsid strategy, respectively. (Courtesy: Ng and Falk, 2006).

in their lumen, wherefrom they will be inoculated upon salivation into healthy hosts. During this basic cycle, the virus encounters and must overcome diverse cellular barriers, where the existence of specific virus-vector interaction has long been established experimentally, though specific receptors have not been identified so far.

Mechanism of Persistent-Circulative Transmission

The persistent-circulative viruses, also called as persistent non-propagative viruses which move through the gut lumen into the hemolymph or other tissue and finally to salivary glands, but they do not replicate inside insect body, although the insect vector remains infected throughout life in most of the cases.

Virus species under the families of *Luteoviridae*, *Geminiviridae* and *Nanoviridae* are transmitted by insect vectors in the persistent-circulative manner. Luteoviruses and nanoviruses are transmitted solely by aphids, whereas the geminiviruses are by whiteflies or leafhoppers. The association between insect vector and viruses is well understood for luteoviruses. The cycle of *Luteovirus* within their vector body involves specific legand-receptor-like recognition at the cell entry of both the gut epithelium and salivary glands.

Electron microscopy and molecular based studies have determined the route of *Luteovirus* particles within the vectors, across cellular layers (Figure 23.4). Once the virus reaches either the apical membrane of the gut epithelium, or basal membrane of the accessory salivary glands cells, and attaches to the specific receptors, it provokes invagination of the plasmalemma, forming small coated virus-coated vesicles. Soon after budding, the coated vesiclies deliver the virus particles to large uncoated membrane endosomal compartment. Interestingly, luteoviruses mostly escape the route of degradation of internalized material ending in lysosomes. Instead, the virus particle become concentrated in the endosomes, and *de novo* elongated uncoated vesicles are repacked, transporting the virus to the basal or apical membrane, in gut and accessory salivary gland cells, respectively. The elongated vesicles, which contain

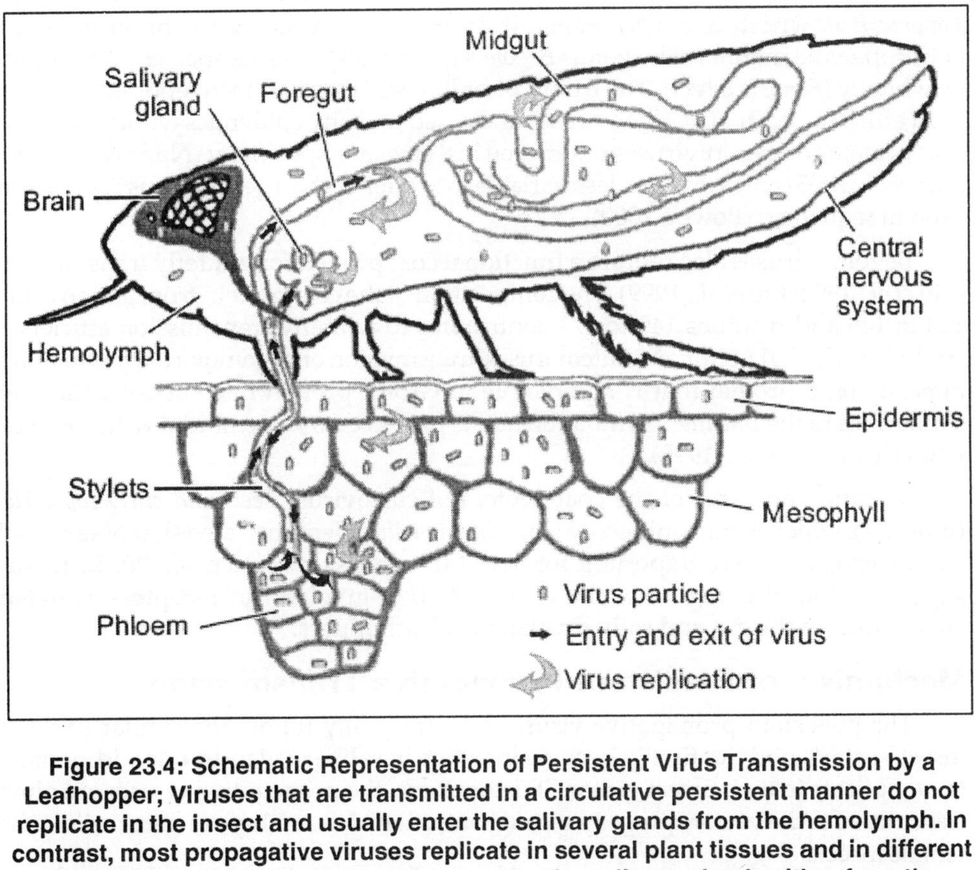

Figure 23.4: Schematic Representation of Persistent Virus Transmission by a Leafhopper; Viruses that are transmitted in a circulative persistent manner do not replicate in the insect and usually enter the salivary glands from the hemolymph. In contrast, most propagative viruses replicate in several plant tissues and in different organs of the insect vectors and may enter the salivary glands either from the hemolymph or from other connecting tissues. (Courtesy: Hogenhout *et al.*, 2008)

rows of virions, finally fuse with plasma membranes and release the virus either into the haemocoel cavity or into the lumen of the salivary ducts.

Virus transfer into the hemolymph is believed to occur by passive diffusion. A study demonstrated that a major protein of the aphid hemolymph, called "symbionin", was mandatory for efficient transmission of luteoviruses. The symbionin, a 52 kDa protein is a homologe to "chaperone", GroEL of the bacterial cell *Escherichia coli*, secreted in aphids by endo-symbiotic bacteria *Buchnera aphidicola*. Elimination of symbiotic bacteria and inactivation of symbionin, by antibiotic treatments, significantly reduces the transmission efficiency of aphid as a vector. Consistently, direct evidence of a physical interaction between symbionin and the *Luteovirus* particles has been detected in several insect species. The virus mutants deficient in symbionin binding are transmitted poorly by insect. Two hypotheses have been proposed to explain the positive action of symbionin: (i) it exhibits protective properties, masking the virus to the immune system and maintaining its integrity during transfer through the hostile hemolymph environment, and (ii) its putative chaperon ensures correct folding facilitating transfer into the salivary glands.

Geminiviruses comprising of genera *Mastrevirus, Curtovirus, Begomovirus* and *Topocuvirus* which are also transmitted by insect vectors in the circulative nonpropagatie manner. The genus *Begomovirus* consist of virus species transmitted by whitefly (*Bemisia tabaci*). The B biotype of *Bemisia tabaci* has been reported to be the most efficient vector of begomoviruses and as led to many epidemics worldwide. The coat proteins of geminiviruses determine insect vector specificity (Noris *et al.*, 1998; Soto *et al.*, 2005) and are much less variable in sequence than geminivirus replication protein sequences (Power, 2000).

Begomoviruses also require a functional coat protein for whitefly transmission (Liu *et al.*, 1992; Liu *et al.*, 1999). The composition of the coat protein from amino acids 123 to 149 and residues 149 to 174 contributes to whitefly transmission efficiency (Hohnle *et al.*, 2001). Similar to luteoviruses, transmission of begomovirus by whiteflies depends on a protein that is homologue to GroEL protein that carries structural similarities to the *Buchnera* symbionin and is produced by a symbiont with coccoid-whitefly (Morin *et al.*, 1999).

Mutation analyses of the coat protein of curtovirus, *Beet mild curly top virus* (BMCTV), which is transmitted by *Circullifer tenellus*, demonstrated that N-terminal amino acids 25-28 are important for insect transmission (Soto *et al.*, 2005). It was suggested that this region of the coat protein is involved in receptor-mediated endocytosis in the gut and salivary glands of leafhoppers.

Mechanism of Persistent-Propagative Transmission

The persistent propagative viruses which usually follow the similar route in insect vector's body to finally become infective in salivary glands, but additionally they replicate in their insect vectors. Furthermore, the propagative viruses are often transmitted to the vectors progeny through infection of the embryo or germ cells in the female insects.

The plant viruses transmitted by persistent-propagative manner includes members of the virus under the families of *Rhabdoviridae, Reoviridae, Bunyaviridae,* and the genus *Marafivirus*. For compatible virus-vector associations, virus particles are released in the hemocoel cavity after infecting the gut epithelium, and then colonize various organ and tissues and ultimately, the salivary glands of the vector. The viruses can either diffuse in the hemolymph and concurrently infect different organs, or spread from organ to organ. This type of virus vector association was demonstrated for rhabdoviruses which move in and spread from the central nervous system.

The relationship between thrips and tospovirus is unique of its kind, as the adult thrip species can transmit *Tomato spotted wilt virus* (TSWV) only if acquisition of viruses occurs in the larval stages (Ullman *et al.*, 1992). Adult thrips that feed on infected plants are unable to transmit virus even if they are allowed lengthy feeding periods on tospovirus-infected plants. The thrip species *Frankliniella occidentalis* is an efficient vector of tospoviruses, transmitting five of the 14 *Tospovirus* species and the TSWV- *F. occidentalis* interaction is the best characterized tospovirus-vector interaction (Whitfield *et al.*, 2005). Tospoviruses encounter multiple tissue systems and membrane barriers along their path from the alimentary canal to the salivary glands in their thrips vectors. Upon ingestion of viral particles, virions travel through the lumen of the foregut into the midgut, the primary site of TSWV-binding and entry into insect cells (Assis Filho *et al.*, 2002). A brush border of microvilli extends into the midgut lumen and forms the first membrane barrier encountered by the virus. Virus particles move across the microvilli into the columnar epithelial cells of the midgut. Following replication in the epithelial cells, virions exit and traverse the basement membrane which is the next barrier encountered by the virus. The midgut epithelium is encircled by alternating series of longitudinal and circular muscle cells. TSWV has been observed in these muscle cells, and entry and exit from these cells presumably constitute the third and fourth membrane barriers that virions must cross on their path to the salivary glands (Nagata *et al.*, 2002).

The primary salivary glands are thought to play a critical role in virus acquisition and transmission. Tospoviruses entering the salivary gland must traverse the basal membrane of this tissue. The lumen of each primary salivary gland lobe is lined with microvilli, and this represents the last membrane the virus must cross for transmission to occur. Once inside the salivary gland lumen, virions can move with saliva into a canal that leads to an efferent salivary canal, a common salivary reservoir, and then a duct that ultimately allows virus-laden saliva to exit the combined salivary-food canal in the maxillary stylets. Studies of TSWV-thrips interactions and of other bunyaviruses provide evidence that the two surface-exposed glycoproteins play an essential role in the infection of insect vectors and animal cells. The two viral membrane glycoproteins, G_N and G_C, are encoded by the M-RNA of virion (Figure 23.5). These two glycoproteins decorate the surface of the virion, and therefore are probably the first viral components that interact with molecules in the thrips midgut. A direct interaction between the G_N glycoprotein and midgets has been demonstrated *in vivo* (Whitfield *et al.*, 2004). Furthermore, it has been demonstrated that G_N serves as a viral ligand that mediates attachment of TSWV to receptors displayed on the epithelial cells of the thrips midgut. It has been further hypothesised that G_N and/or G_C are

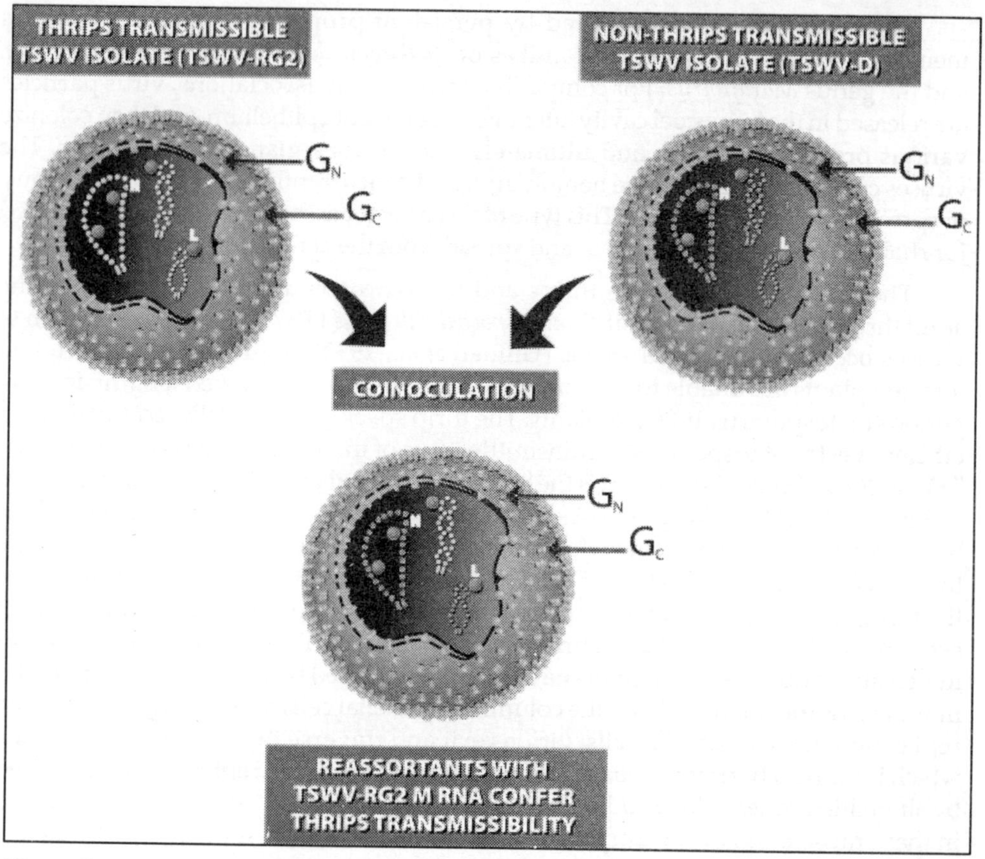

Figure 23.5: Genetic Determinants of Insect Transmission of TSWV Lie on the Viral M RNA in the ORF encoding G_N and G_C

Two parental TSWV isolates were used to coinoculate plants and generate isolates in which the RNA segments reassorted. One parental isolate, TSWV-D, was not transmitted by insects, whereas isolate TSWV-RG2 was transmissible by two insect vector species, *Frankliniella occidentalis* and *F. fusca*. Among the viral reassortants only those containing the M RNA segment from TSWV-RG2 conferred thrips transmissibility. This evidence showing that reassortant viral populations arise from mixed infections with altered traits for insect transmission has important implications for understanding TSWV–thrips coevolution and the emergence of new virus–vector relationships (Courtsey: Sin *et al.*, 2005)

involved in the virus entry and interact with a receptor molecule (50 kDa protein) in thrips.

Conclusion

As plants are sessile organisms, most plant viruses depend on biological vectors for their survival and spread. Although a number of different types of organisms are vectors for different plant viruses, phloem-feeding Hemipterans are the most common and transmit the great majority of plant viruses. Most of the vectors transmit plant viruses in either the circulative or non-circulative manner. Non-circulative viruses (NCV) are carried on the lining cuticle of vectors stylets, while circulative viruses (CV) cross the vectors' gut, move internally to the salivary glands (SG), cross the SG membranes to be ejected upon feeding. The mechanisms of plant virus transmission by insect vectors are highly specific and complex, and the mechanisms used to describe them has undergone many changes.

Recent applications of molecular and cell biology techniques have helped to elucidate the mechanisms underlying the vector transmission of several plant viruses. Two general strategies, the capsid and helper strategies are found for plant viruses that are transmitted by aphids in a non-persistent manner. Transmissibility of NCVs depends on helper proteins (encoded by the virus) in addition to the motifs of coat protein for potyviruses whereas it depends on capsid proteins only in case of cucumoviruses. Evidence from caulimoviruses and criniviruses transmission suggests that both helper and capsid strategies are found for viruses transmitted in a semi-persistent manner.

Transmissibility of CVs depends on proteins comprising the virus capsid (coat protein) and on symbionin (produced by vectors' symbionts). Passage of CV through the gut has been also associated with vectors' proteins, thus making the insect vector-virus association highly specific in case of circulative viruses. These highly complex interactions have been extensively studied in geminiviruses and tospoviruses. Recent understandings in the transmission of plant viruses and their specific interactions with insect vectors has elucidated the novel strategies in the control of plant virus epidemics by targeting the vector landing and feeding behavior.

References

Andret-Link, P. and Fuchs, M. (2005). Transmission specificity of plant viruses by vectors. *Journal of Plant Pathology*. **87**: 153-165.

Assis Filho, F.Md., Naidu, R.A., Deom, C.M. and Sherwood, J.L. (2002). Dynamics of tomato spotted wilt virus replication in the alimentary canal of two thrips species. *Phytopathology*. **92**: 729-733.

Atreya, P.L., Lopez-Moya, J.J., Chu, M., Atreya, C.D. and Pirone, T.P. (1995). Mutational analysis of the coat protein N-terminal amino acids involved in potyvirus transmission by aphids. *J. Gen. Virol.* **76**: 265-270.

Bar-Joseph, M. and Dawson, W.O. (2008). *Citrus tristeza virus. Encyclopedia of Virology* pp 520-525.

Briddon, R.W. and Markham, P.G. (2000). *Cotton leaf curl virus* disease. *Virus Res.* **71**: 151-159.

Chen, B. and Francki, R.I.B. (1990). *Cucumovirus* transmission by the aphid *Myzus persicae* is determined solely by the viral coat protein. *J. Gen. Virol.* **71**: 939-944.

Czosnek, H. and Laterrot,H. (1997). A worldwide survey of *Tomato yellow leaf curl viruses*. *Arch.Virol.* **142**:1391-406.

Drucker, M., Froissart, R., Hebrard, E., Uzest, M. and Ravallec, M., *et al.* (2002). Intracellular distribution of viral gene products regulates a complex mechanism of *Cauliflower mosaic virus* acquisition by its aphid vector. *Proc. Natl. Acad. Sci. USA*. **99**: 2422-27.

Duffus, J., Larsen, R. and Liu, H. (1986). *Lettuce infectious yellows virus*–a new type of whiteflytransmitted virus. *Phytopathology*. **76**: 97-100.

Froissart, R., Michalakis, Y. and Blanc, S. (2002). Helper component in the vector transmission of plant viruses. *Phytopathology*. **92**: 576-579.

Gera, A., Loebenstein, G. and Raccah, B. (1979). Protein coats of two strains of *Cucumber mosaic virus* affect transmission by *Aphis gossypii*. *Phytopathology* **69**: 396-399.

Govier, D.A. and Kassanis, B. (1974). A virus-induced component of plant sap needed when aphids acquire *Potato virus Y* from purified preparations. *Virology* **61**: 420-426.

Gray, S.M. and Banerjee, N. (1999). Mechanisms of arthropod transmission of plant and animal viruses. *Microbiology and Molecular Biology Reviews*. pp 128-148

Hibino, H. (1996). Biology and epidemiology of rice viruses. *Annu. Rev. Phytopathol.* **34**: 249-274.

Hogenhout, S.A., Ammar, El-D., Whitfield, A.E. and Redinbaugh, M.G. (2008). Insect vector interactions with persistently transmitted viruses. *Annu. Rev. Phytopathol.* **46**: 327-359.

Hohnle, M., Hofer, P., Bedford, I.D., Briddon, R.W., Markham, P.G. and Frischmuth, T. (2001). Exchange of three amino acids in the coat protein results in efficient whitefly transmission of a nontransmissible *Abutilon mosaic virus* isolate. *Virology*. **290**:164-171.

Hull, R. (1996). Molecular biology of rice tungro viruses. *Annu. Rev. Phytopathol.* **34:** 275-297.

Kassanis, B. and **Govier, D.A.** (1971). New evidence on the mechanism of aphid transmission of *Potato C* and *Potato aucuba mosaic viruses*. *J. Gen. Virol.* **10**:99-101.

Liu, S., Bedford, I.D., Briddon, R.W. and Markham, P.G. (1997). Efficient whitefly transmission of African cassava mosaic geminivirus requires sequences from both genomic components. *J. Gen. Virol.* **78**:1791-1794.

Liu, S., Briddon, R.W., Bedford, I.D., Pinner, M.S. and Markham, P.G. (1999). Identification of genes directly and indirectly involved in the insect transmission of African cassava mosaic geminivirus by *Bemisia tabaci*. *Virus Genes*. **18**: 5-11.

Maia, I.G., Haenni, A. and Bernardi, F. (1996). Potyviral HC-Pro: a multifunctional protein. *J. Gen. Virol.* 77:1335-1341.

Moreno, A., H´ebrard, E., Uzest, M., Blanc, S. and Fereres, A. (2005). A single amino acid position in the helper component of *Cauliflower mosaic virus* can change the spectrum of transmitting vector species. *J. Virol.* **79**:13587-13593.

Morin, S., Ghanim, M., Zeidan, M., Czosnek, H., Verbeek, M. and van den Heuvel, J.F. (1999). A GroEL homologue from endosymbiotic bacteria of the whitefly *Bemisia tabaci* is implicated in the circulative transmission of *Tomato yellow leaf curl virus*. *Virology*. **256**: 75-84.

Nagata, T., Inoue-Nagata, A.K., van Lent, J., Goldbach, R. and Peters, D. (2002). Factors determining vector competence and specificity for transmission of *Tomato spotted wilt virus*. *J. Gen. Virol.* **83**: 663-671.

Ng, J.C.K. and Falk, B.W. (2006). Virus-vector interactions mediating nonpersistent and semipersistent transmission of plant viruses. *Annu. Rev. Phytopathol.* **44**: 183-212.

Noris, E., Vaira, A.M., Caciagli, P., Masenga, V., Gronenborn, B. and Accotto, G.P. (1998). Amino acids in the capsid protein of tomato yellow leaf curl virus that are crucial for systemic infection, particle formation, and insect transmission. *J. Virol.* **72**: 10050-10057.

Peng, Y.H., Kadoury, D., Gal-On, A., Huet, H., Wang, Y. and Raccah, B. (1998). Mutations in the HC-Pro gene of *Zucchini yellow mosaic potyvirus*: effects on aphid transmission and binding to purified virions. *J. Gen. Virol.* **79**: 897-904.

Pirone, T.P. and Blanc, S. (1996). Helper-dependent vector transmission of plant viruses. *Annu. Rev. Phytopathol.* **34**: 227-247.

Pirone, T.P. and Megahed, S. (1966). Aphid transmissibility of some purified viruses and viral RNA's. *Virology*. **30**: 631-637.

Powell, G., Tosh, C.R. and Hardie, J. (2006). Host plant selection by aphids: behavioral, evolutionary, and applied perspectives. *Annu. Rev. Entomol.* **51**: 309-330.

Power, A.G. (2000). Insect transmission of plant viruses: a constraint on virus variability. *Curr. Opin. Plant Biol.* **3**: 336-340.

Raccah, B. and Fereres, A. (2009). Plant virus transmission by insects. In: *Encyclopedia of Life Sciences (ELS)*. John Wiley & Sons, Ltd: Chichester. DOI:10.1002/ 9780470015902. A0021525. a0000760.pub2.

Sin, S.-H., McNulty, B.C., Kennedy, G.G. and Moyer, J.W. (2005). Viral genetic determinants for thrips transmission ofTomato spotted wilt virus. *Proc. Natl. Acad. Sci. USA* **102**: 5168-5173.

Soto, M.J., Chen, L.F., Seo, Y.S. and Gilbertson, R.L. (2005). Identification of regions of the *Beet mild curly top virus* (family *Geminiviridae*) capsid protein involved in systemic infection, virion formation and leafhopper transmission. *Virology*. **341**: 257-270.

Tian, T., Rubio, L., Yeh, H.H., Crawford, B. and Falk, B.W. (1999). *Lettuce infectious yellows virus*: in vitro acquisition analysis using partially purified virions and the whitefly Bemisia tabaci. *J. Gen. Virol.* **80**:1111-1117.

Ullman, D.E., Cho, J.J., Mau, R.F.L.,Wescot, D.M. and Custer, D.M. (1992). A midgut barrier to tomato spotted wilt virus acquisition by adult western flower thrips. *Phytopathology*. **82:** 1333-1342.

Watson, M.A. (1960). Evidence for interaction or genetic recombination between *Potato viruses Y* and C in infected plants. *Virology.* **10:** 211-232.

Watson, M.A. and Roberts, F.M. (1939). A comparative study of the transmission of *Hyoscyamus virus 3, Potato virus Y* and *Cucumber virus 1* by the vectors *Myzus persicae* (Sulz), *M. circumflex* (Buckton), and *Macrosiphum gei* (Koch). *Proc. R. Soc. London Ser. B.* **127:** 543-576.

Whitfield, A.E., Ullman, D.E. and German, T.L. (2004). Expression and characterization of a soluble form of tomato spotted wilt virus glycoprotein G_N. *J. Virol.* 78:13197-13206.

Whitfield, A.E., Ullman, D.E. and German, T.L. (2005). Tospovirus-thrips interactions. *Annu. Rev. Phytopathol.* **43:** 459-489.

Chapter 24

Grain Amaranth: An Underutilized Crop with Tremendous Potential

Shephalika Amrapali and S.K. Soam

National Academy of Agricultural Research Management, Hyderabad – 500 407, A.P.

Presently the global food security depends on only a hand full of crops. As per the facts by the Bioversity International, over 50 per cent of the global requirement for proteins and calories are met by just three food crops *i.e.* maize, wheat and rice. Only 150 crops are traded on a significant global scale. Yet, it is estimated that there are over 7000 plant species across the world that are cultivated or harvested from the wild for food (Wilson E. O. 1992). The bulk of the agro-biodiversity developed over centuries by generations of farmers and millions of users today are highly marginalized by Research & Development. These limitations have resulted in thousands of species with local relevance becoming neglected and underutilized. Moreover, there is little effort to conserve them in spite of their important role in food and nutrition security, income generation, environmental health and other livelihood benefits.

Neglected and Underutilized Species (NUS) normally exhibit the following features;

- ☆ Locally produced and consumed
- ☆ Highly adapted to agro-ecological niches and marginal areas

* Corresponding Author E-mail: sangeethaau@hotmail.com

- ✰ Ignored by policy makers and excluded from research and development agendas
- ✰ Represented by ecotypes or landraces
- ✰ Harvested in the wild or cultivated and used using indigenous knowledge
- ✰ Not well-represented in *ex situ* gene banks
- ✰ Fragile or non-existent seed supply systems

Since it is very difficult to feed the rapidly rising population with handful of crops we should focus on diversifying our food base by including more food crops. There are many food crops with tremendous potential but still underutilized. One such crop is *Amaranthus* species. Amaranth, a pseudo-cereal has good potential for use as a vegetable as well as a grain crop. The grains are rich in protein, fat and carbohydrate content and are comparable to wheat, rice and oats. These are milled into flour and used as a staple food in the entire Himalayan region.

Historical Perspectives of Amaranths

The word "amaranth" in Greek means "everlasting" and in fact, the crop has endured. To assure a small annual supply for this specialty crop, traditional farmers have continued to grow small plots of the grain each year. Furthermore, the distinctly beautiful appearance of amaranth has helped to prevent the crop from slipping into obscurity. The enchanting beauty of the vividly coloured leaves, stems and seed-heads in an amaranth field is a sight which evokes emotions that other crops cannot stir.

The grain amaranths (*Amaranthus* spp.) are native to the New World. Pre-Columbian civilizations grew thousands of hectares of this pseudo-cereal. Some indigenous populations are said to have used grain amaranth, along with maize and beans, as an integral part of their cropping schemes. The Aztecs relied on amaranth seeds (or "grain") as an important staple food (National Academy of Sciences 1984).

By the middle of this century, the cultivation of grain amaranth had declined to the point where it was grown only in small plots in Mexico, the Andean highlands, and in the Himalayan foothills of India and Nepal. Even now, there is evidence that some traditional farmers are abandoning the cultivation of local landraces of amaranth as they devote more of their land to high-yielding "modern" crops. In an effort to explain the decrease in grain amaranth cultivation, fanciful myths have arisen. The mystery is especially intriguing when one considers that maize, with which amaranth co-evolved, was selected and developed into a major world food crop. The small seed size of amaranth may have been a partial cause for the reduction in amaranth cultivation. Being a small seeded crop, it requires greater attention to detail in the early parts of the growing season than does a larger seeded crop, such as maize.

Botanical Aspects of Amaranths

There are approximately 25 species of *Amaranthus* available in the Asian region. Four *Sanskrit* names (*Rahadri, Rajagiri, Rajashakini* and *Tandulya*) are available in the old literature. The fossil records of pollen have been documented in several excavations in India from Holocene and late Pliestocene periods. Enormous genetic diversity and concentrations of the crop in the Himalayan region unlike peninsular India, its uses and prohibition in different rituals, large number of vernacular names of medicinal and culinary preparations and the amount of genetic diversity in grain amaranths as observed in Himalayas would indicate the greater likelihood of spread of the crop in India from that region in eighth century. Their direct and indirect evidences of the antiquity of the crop in India suggested that possibly grain amaranths were prevalent in south Asia from time immemorial.

As discussed earlier, the Amaranths are used for vegetable and grain purposes. While grain amaranth species may be difficult to distinguish from one another on the basis of morphology, the features they share in common separate them from other amaranths. The vegetable amaranths have smooth leaves and exhibit an indeterminate growth habit. The grain amaranths are annuals and have a main stem axis with a large branched inflorescence at the apex (Stallknecht and Schulz-Schaeffer 1993). The grain species usually range from 0.4 to 3.0m in height. The grain amaranths are dicotyledonous, and, therefore, are not true cereals.

Taxonomy

The grain amaranths belong to the family Amaranthaceae, which contains 169 genera and approximately 2400 species. The most abundant species in the Amaranthaceae family are herbs that colonize shorelines and other open habitats. A few of the genera are cultivated as ornamentals such as *Celosia*, *Iresine*, and *Gomphrena*, known by the common names cockscomb, bloodleaf, and globe amaranth respectively. The grain amaranths are found within the genus *Amaranthus*. Other relatives in the Amaranthanaceae family that are cultivated as crops are from the group of plants formerly known as the family Chenopodiaceae–such as beets, sugar beets (*Beta vulgaris* L.) and quinoa. *Amaranthus genus*. There are 60 to 75 species in the genus *Amaranthus*, sixty of which are native to the Americas while another 15 are indigenous to Africa, Asia, Australia and Africa. These are found mainly in the world's temperate and tropical climates (Sauer 1967). The genus is generally separated into three subgenera: 1) *Albersia*, 2) *Acnida*, the dioecious amaranths and 3) *Amaranthus*, which includes the *A. hybridus* complex. The *A. hybridus* complex consists of the three grain types and their 3 putative progenitors. The majority of the species are wild or weedy. *Amaranthus* species grow best in desert washes, lakeshores, marshes, ocean beaches, and stream banks. Their seeds are naturally dispersed to these habitats by migratory birds that feed on them (Sauer 1967, 1993). *A. hybridus* is also known as smooth pigweed and is considered a particular notorious weed, along with several other members of the *Amaranthus* genus such as waterhemp (*A. tuberculatus* (Moq.) Sauer), redroot pigweed (*A. retroflexus*), and Powell amaranth (*A. powellii* S. Wats) (Wassom and Tranel 2005). Many of these species are rapidly evolving herbicide resistance (Patzoldt *et al.*, 2006).

Leaves, Inflorescences and Flowers

Grain amaranth leaves are petiolate and oval to ovulate-oblong and lanceolate in shape with acute apices. The inflorescence is a dichasial cyme with unisexual flowers, which develop in a variety of colors, including red, purple, orange, or gold

(Iturbide and Gispert 1994, Tapia 1994). The first flower of each of the numerous cymes is staminate followed by an indefinite number of pistillate flowers, frequently over a hundred (Pal and Khoshoo 1974, Sauer 1993). Some pistillate flowers on the cyme develop early before the staminate flower has opened, while others become receptive following the abscission of the male flower. However, because cymes at different developmental stages are present on each indeterminate inflorescence branch, self-pollination is more likely than outcrossing, although both types of fertilization are possible (Sauer 1976).

Fruit

Unlike other cereals, grain amaranths have retained the dehiscent fruits of their wild progenitors (Sauer 1993). The fruits are pyxides, meaning that they house their seeds in circumscissile capsules, which are subtended by colourful bracts and sepals (Tapia 1994). The top half of the papery utricle surrounding each seed acts as a lid-like section, which pops off at the equator of the utricle to reveal the enclosed seed. Thus, although the majority of seeds remain in the densely packed inflorescences, some seeds are lost during the harvest (Sauer 1993). However, in recent years non-shattering grain amaranth populations have been developed (Brenner 2002).

Seeds

The seeds of the grain amaranths are lens-shaped and approximately 1 to 1.5mm in diameter. The seeds come in a variety of colours, ranging from white to yellow to red to black (Iturbide and Gispert 1994, Tapia 1994, Sauer 1993). These colours are governed by simple Mendelian recessive alleles. All three grains produce both dark- and light-coloured seed. Although the dark grains, which are dominant to the light-coloured grains, are edible and were eaten by prehistoric hunter-gathers, the lighter grains have been selected for due to their improved flavour and popping. Furthermore, the pale colour also seems to be linked to a loss of dormancy in the seed (Sauer 1976, 1993). The seeds exhibit epigeal germination, in which the cotyledons emerge above ground as in common beans (*Phaseolus vulgaris* L.). Seedlings emerge three to four days following sowing and after about two and half months the panicle appears and flowering occurs. The seeds maintain viability for over five years at ambient temperature and <5 per cent humidity (Iturbide and Gispert 1994).

Food and Nutrition Aspects of Amaranths

Vegetable Species

No clear separation between vegetable and grain species exists, because the leaves of young grain varieties may be used as potherbs. *A. cruentus*, is cultivated both as vegetable and grain. Its relatives *A. tricolor*, *A. dubius* and *A. lividus* are also grown as vegetables (Stallknecht and Schulz-Schaeffer 1993). Species are mostly grown as potherbs in India, the East India, Southeast Asia, and the Far East. In English, they are known by the common names Chinese spinach, Malabar spinach and tampala (Sauer 1993).

Grain Species

The three amaranths principally grown as grains consist of *A. cruentus*, *A. caudatus*, and *A. hypochondriacus*. In the literature, *A. edulis*, which is grown in the northern Andes of Argentina, is also sometimes cited as a grain species. However, *A. edulis* may be more appropriately considered *A. caudatus* spp. *Mantegazzianus*–a mutant of *A. caudatus* with the phenotype of club-shaped inflorescent branches and determinant growth habit (Sauer 1976). The wild putative progenitor species of the grains include *A. powellli*, *A. hybridus* and *A. quitensis* (Sauer 1950, 1967, 1976). *A. hybridus* is also sometimes cultivated as a grain. Some of the wild relatives of the grains are fairly tall with large inflorescences; however the cultivated species are taller and more robust, with enormous inflorescences. Unlike true cereals, grain amaranths were selected for their high seed production rather than for increased seed size (Sauer 1993).

Nutrients

Amaranth seeds as a grain have been praised for their nutrient content. Amaranths are 50 to 60 per cent starch, with higher protein (15 to 16 per cent) and more fat (7 to 8 per cent) than most cereals (Breene 1991). They also have nutritionally significant levels of vitamins A and C, as well as a higher mineral content than wheat (Becker *et al.*, 1981). Amaranths also have high dietary fiber content reported to be about 8 per cent for pale-seeded types, while the black-seeded grain types may have twice that (Pedersen *et al.*, 1987).

Starches

In amaranth, 78 to 100 per cent of the starch content is found in the branched-chain amylopectin form, while the remaining 0 to 22 per cent of starch content is in the amylose or unbranched form (Tomita *et al.*, 1981, Okuno and Sakaguki 1984). Overall, amaranth's starch composition shows a low gelatinization temperature and good stability during freezing and thawing (Yanez *et al.*, 1986). Amaranth starch is observed in granules that are approximately 1-3 µm in diameter (Irving and Becker 1985)–much smaller than most other commercial cereals. Rice (*Oryza sativa* L.), for example, has starch granules of about 3 to 8µm, while potato's (*Solanum tuberosum* L.) are 100µm in diameter. It has thus been suggested that the small granule size might make amaranth starch useful as a food thickener, a dusting powder in foods and cosmetics, a laundry starch, etc (Yanez *et al.*, 1986).

Proteins

The protein content of the grains has been extensively studied. Amaranth is one of a handful of plants whose protein content approaches animal protein quality on the basis of bioavailability and amino acid content (Bressani 1989). Other examples of plants with essential amino acid patterns that come close to satisfying the needs of the human diet include soybean, high-quality protein maize and quinoa (Bressani 1989). Crude protein content from pale-seeded grain types has been reported to range from 12.5 to 22.5 per cent, with an average of about 15 per cent (Becker *et al.*, 1981, Saunders and Becker 1984, Teutonics and Knorr 1985, Correa *et al.*, 1986, Bressani *et al.*, 1987, Pedersen *et al.*, 1987, Bressani 1989). Furthermore, amaranth is relatively rich in the essential amino acid lysine, which is usually limiting in other cereal crops.. Lysine content ranges from 0.73 to 0.84 per cent of amaranth's total protein content (Bressani *et al.*, 1987). Seed storage proteins from amaranth have been introduced successfully through transgenics into other crop species. Species such as potato and maize that have been modified to express amaranth seed proteins show improved amino acid composition (Chakraborty *et al.*, 2000, Sinagawa-Garcia *et al.*, 2004).

Oils

The 7-8 per cent oil content found in amaranth seeds may be too low and expensive to compete with other oils commercially available, although it is similar in content to corn and cotton seed oils (Bressani *et al.*, 1987). The ranges of fatty-acids observed for the oil content based on Breene's (1991) summary of various studies (Fernando and Bean 1984, 1985; Saunders and Becker 1984; Lorenz and Hwang 1985; Sanchez-Marroquin *et al.*, 1986; Lyon and Becker 1987, Bressani *et al.*, 1987). The saturated/ unsaturated fatty acid ratio has been observed to range from 0.29 to 0.43; this ratio is favorable from a nutritional standpoint because unsaturated fatty acids are predominant in amaranth oil (Breene 1991). High levels of tocopherols (vitamin E) and tocotrienols have been reported in amaranth oil as well (Lehmann *et al.*, 1994). Amaranth oil has been noted for its relatively high concentration of squalene (7-8 per cent) (Bressani *et al.*1987). Squalene is a lucrative ingredient used in cosmetics, skin penetrants, lubricants and is a precursor to cholesterol. The traditional source of squalene for commercial use is liver oil extracted from threatened sea animals such as whales (*Physeter macrocephalus*) and sharks (*Centrophorus squamosus*). Therefore, there is interest in other potential alternative sources. The use of amaranth oil as a squalene source may further its commercialization (Brenner *et al.*, 2000). Recent studies have also shown that amaranth oil may be effective in reducing cholesterol levels in mammals, including humans (Berger *et al.*, 2003, Martirosyan *et al.*, 2007).

Antinutrients

Unlike its relative quinoa, amaranth does not contain high amounts of bitter saponins that must be washed away before consumption (Tapia 1994). Low levels of saponin–around 0.1 per cent of total seed dry weight–that have been observed for *A. cruentus* showed low toxicity in animal tests (Oleszek *et al.*, 1999). Furthermore, amaranth grain shows low levels of some other antinutrients. For example, Lorenz and Wright (1984) studied the tannin and phytate content of *A. hypochondriacus*, *A. cruentus*, *A. hybridus* and some interspecific crosses and found that tannins were localized in seed coat and were present at 0.04-0.12 per cent, while phytates dispersed throughout the kernel were observed at 0.5- 0.6 per cent. However, amaranth seeds and leaves are known to accumulate high levels of trypsin inhibitors as well as α-amylase inhibitors (Sanchez-Hernadez *et al.*, 2004). These antinutritional inhibitors are well documented and the DNA as well as protein sequences are available for some (Valdes-Rodriguez *et al.*, 1993, 1999).

Miscellaneous Uses

Traditionally these nutritious crops continue to be used in Latin America much like they were during pre-Columbian times although to a much lesser extent. In Mexico, the preparation of the sacred Aztec dough, *tzoalli*, by mixing amaranth flour with maguey honey led to the current use of amaranth for preparing *alegria*, a sweet snack. *Alegria* is also made with popped amaranth seeds instead of flour (Iturbide and Gispert 1994). The popped form is also used in cereals. The seed is milled into flour to make a variety of foods, while the leaves are used as a vegetable, particularly in soups. The stems are useful as animal feed (Iturbide and Gispert 1994). In India amaranth leaves are used as pot herb, and also as forage, while the grain amaranth commonly known as Rajgeera or Ramdana is used in the form of popped, flacked grain or milled to make flour called satoo. Flour is used to make chapattis or *tortillas*. Popped grains are used to make laddoos with sugar syrup, which is similar to *alegria*. It is also used to make porridge.

Amaranth, like most grains, has potential for use in industrial products. The oil fraction of the grain is unusually high in squalene, a chemical that sells for thousands of dollars per ton. However, the percent of squalene in the grain is still small, and may not be economical to extract. Greater promise lies with the starch fraction of the grain. Amaranth, like quinoa (*Chenopodium quinoa*), has very small, micro-crystalline starch granules, about one-tenth the diameter of maize (*Zea maize*) starch. The physical characteristics of the starch grains have been cited as being of potential value for both industrial and food product uses, though none has been commercialized to date. Amaranth crop also has potential to be used for the extraction of natural dye and as medicinal plant. Amaranth has also been used as an ingredient in beverages, baby formula, atole, snacks, breakfast cereals, and as a textured vegetable protein (Breene 1991).

The grain amaranths exhibit C4 photosynthesis. Thus, they grow rapidly in bright sunlight, high temperatures, and low moisture conditions. Other cultivated crops that exhibit C4 photosynthesis include maize, sorghum (*Sorghum* spp. L.) and sugarcane (*Saccharum officinarum* L.). Amaranth is better adapted to semiarid environments than these plants, however, because it can make osmotic adjustments that allow it to tolerate dry conditions without wilting or drying (Tucker 1986). Amaranths can also tolerate a variety of unfavorable soil conditions such as high salinity, acidity, or alkalinity (Tucker 1986). Grain amaranths have also been reported to adapt readily to new environments, including some that are inhospitable to traditional grain crops (Gupta and Gudu 1991). All theses adaptations make this crop very suitable for the changing climatic conditions.

Puente a la salud Communitaria: *Amaranthus* Based Food and Nutritional Security in Mexico

Guided by the principles of food sovereignty, Puente seeks to empower communities in Oaxaca, Mexico, to produce their own food in a manner that does not damage the local environment; allowing families to sustainably combat malnutrition using their own resources. Puente a la Salud Comunitaria contributes to food sovereignty and advances the health and well-being of rural communities in Mexico by promoting the cultivation, consumption and commercialization of amaranth. Puente promotes the ecological production of amaranth, combined with other traditional and strategic crops, to encourage the diversification of local agriculture and diet. 57 per cent of rural oaxacan children under five have experienced, stunting, indicating the presence of, long term chronic malnutrition.

Amaranth proves to be one of the best suited crops to address certain health problems in rural Oaxaca. Seeds are 13 to 15 percent protein, among the highest for any grain. Amaranth seeds are also high in fibre, calcium, iron, potassium, phosphorus, zinc, and vitamins A and C. In addition, combining amaranth seeds with corn, a major component of the local diet, forms a complete protein. Leaves are also edible, containing more calcium, phosphorus, and vitamin C than spinach, in addition to the high levels of folate and other nutrients present in the seeds

The high quantities of both micro- and macronutrients in the seeds and in the leaves prove very important in addressing many easily preventable health problems in rural Oaxaca. Studies show that the integration of only a small amount of amaranth grain into the daily diet will help children recover from states of malnutrition. In a study performed by *San Miguel de Proyectos Agropecuarios*, 1,000 children eating the equivalent only 20 grams of amaranth grain daily for one year recovered at a rate of 61.70 percent while the control group only recovered at a rate of 15.33 percent. Such results prove amaranth's viability in fighting malnutrition around the world. Source: http://puentemexico.org.

Conclusion

An all India coordinated project on underutilised crops started in 1982 under which grain amaranth is one of the mandate crops. A great deal of variation is present in the Himalayan region of India and Nepal. In India an extensive collection of over 5450 accessions has been built up from different sources and the germplasm suitable for hill regions and the plains are being maintained at NBPGR Regional Stations at Shimla and Akola, respectively. The characterization and evaluation of these accessions are continuously going on and a wide range of genetic variation was observed for days to flowering, days to maturity, plant height, inflorescence length, spikelet number, 1000 seed weight and seed yield. Based on multi-locational trials involving promising entries, selection IC 42258–1 was identified as the best and was released as "Annapurna". Based on the characterization and evaluation, some accessions have recently been selected and recommended for release for the plains, particularly the states of Gujarat and Maharashtra. However, in view of the increased use of crop, there is a need to develop high yielding varieties for these areas. Undoubtedly, there is a growing awareness of utilising the newer plant resources all over the world. The progress made under the national coordinated research programme on under-utilised and under-exploited plants utilizing the diversity collected but concerted efforts are required to speed up the work on identifying such useful accessions, develop better varieties and standardise their agronomic practices.

References

Barba de la Rosa, A.P., Fomsgaard, I.S., Laursen, B., Mortensen, A.G., Martinez, L.O., Sanchez, C.S., Herrera, A.M., Gonzalez, J., Rodriguez, A.D.L. (2009). Amaranth (Amaranthus hypochondriacus) as an alternative crop for sustainable food production: Phenolic acids and flavonoids with potential impact on its nutraceutical quality. *J. of Cereal Sc.* **49:** 117–121

Bressani, R. (1989). The proteins of grain amaranths. *Food Res. Int.* **5:** 213–238.

Camacho, M.L., Diego, L. Aparicio, G.G.R (2001). A detailed and comprehensive study of amaranth (Amaranthus cruentus L.) oil fatty profile. *Eur. Food. Res. Tech.* **213**: 349–355

Kauffman, C.S. and Webber, L.E. (1990). Grain amaranth, p. 127-139. In: J. Janick and J.E. Simon (eds.). Advances in new crops. Timber Press, Portland,

Pal, M. and Khooso, T.N. (1974). Grain amaranths. *In:* Hutchinson JB (ed.) Evolutionary studies in wold crops: diversity and change in the Indian subcontinent. Cambridge Univ. Press, UK, pp. 129–137

Sauer, J.D. (1950). The grain amaranths: A survey of their history and classification. *Ann. Missouri Bot. Gard.* **37**: 561-632.

Sauer, J.D. (1967). The grain amaranths and their relatives: A revised taxonomic and geographic survey. *Ann. Missouri Bot. Gard.* **54**(2):103-37.

Sauer, J.D. (1976). Grain amaranths(Amaranthus spp, amaranthaceae): In *evolution of crop plants* (Ed.)simmonds, N.W. Chap.2:4-6.

Sauer, J.D. (1977). The history of grain amaranths and their use and cultivation around the world. Proc. First Amaranth Seminar. Emmaus, PA.

Schaeffer, J.R., C.F. McGuire, and Stallknecht, G.F. (1990b). Grain amaranth—research and potential. Proc. First Int. Conf. New Industrial Crops and Products. Oct. 8-12, Riverside, CA.

Stallknecht, G.F. and Schulz-Schaeffer, J.R. (1993). Amaranth rediscovered. P. 211-218. In J. Janick and J.E Simon (eds), New crops, Wiley, New York

Yanez, G.A., Messinger, J.K., Walker, C.E. and Rupnow, J.H. (1986). *Amaranthus hypochondriacus:* starch isolation and partial characterization. *Cereal Chem.* **63**: 273-276.

Chapter 25

Distribution, Host Range, Symptomatology, Biology, Disease Cycle and Management of Devastating Fungus *Sclerotium rolfsii* Sacc.

Manoj Kumar Pandey[1] *and A.B. Rai*[2]

[1]*Krishi Vigyan Kendra (IIVR, Varanasi), Malhana, Deoria, U.P.*

[2]*Indian Institute of Vegetable Research, Varanasi, U.P.*

India is a developing country and moving towards developed nation. Its economy is mainly based on agriculture. Agriculture is the largest private enterprise contributing about 26 per cent in gross domestic product (GDP) of India. About 60 per cent of population depends on agriculture and allied sectors for their employment in rural India and about 13 per cent of the total Indian exports comes from agriculture trade (Babu, 2004). In recent days production of crop yield became constant, but our population increasing is day by day. Total foodgrains production in India reached 182.57 million tonnes (Venkataramani, 2004). It is estimated that India needs to attain the food grain production target of 337 million tonnes to feed an estimated population of 1200 million people by 2011-2012 (Awasthi, 2004). For providing food security to our population, the role of protection technology would be crucial and it is speculated that only this protection technology may fulfill our need and combat with hunger.

Plants suffer from many biotic and abiotic factors, which cause considerable loss in the field. Crop yields are also lost in storage due to different store pest. Total crop loss cause by weeds, diseases, insect-pests, rodents and birds is 10-30 per cent (Chandurkar, 2004). Diseases play an important role in reducing the crop yield. Singh, 2000 described that 10 per cent loss in yield/year is only due to plant diseases. Soil borne diseases are one of the most prominent one. *Sclerotium rolfsii* Sacc. (Teleomorph: *Athelia rolfsii* (Curzi) Tu & Kimbrough) is one of the devastating, cosmopolitan, ubiquitous, noxious, notorious, serious soil-borne and omnivorous pathogen with a diversified host range including both monocotyledonous and dicotyledonous plants encompassing more than 500 host species (Aycock, 1966; Punja, 1985). Blclcl *et al.*, 1994 reported that *S. rolfsii* prevail at 95 per cent in the pea nut fields and produced on average 8.6-21.9 sclerotia/700 g soil samples, and 2.8-3.5 loci/30 mitre (m) peanut rows in Adana province of Turky. Sclerotium (pl. sclerotia), similar word 'sclerosis' is used in medical sciences, which is of greek origin word 'sklerosis' means 'abnormal hardening of body tissues'. Sclerotium is a hard, compact, dark coloured resting body resistant to unfavourable environmental conditions, which may remain dormant for long periods of time and germinate on the return of favourable conditions. Sclerotium is formed by many fungi like, *Rhizoctonia, Sclerotinia* and *Claviceps* etc., whereas formation of Sclerotium (pl. sclerotia) is a typical/basic character in *Sclerotium* for which the same word has been given to this genus. The major difference between sclerotia of *Rhizoctonia solani* and *S. rolfsii* is undifferentiation of *R. solani* into a rind and a medulla while, differentiation of sclerotia of *S. rolfsii* into a rind a cortex and a medulla (RS Singh, 1982). This fungus is mostly facultative parasite, growing saprophytically after killing their host. The pathogen has white thread like mycelium and tan to dark brown sclerotia produced on the colony. The pathogen does not produce inoculum for secondary infection in the neighboring plants in the same season (Fry, 1982). It become severe only when large inoculum is present in the soil. Root diseases caused by *S. rolfsii* generally become severe in moist soil at warm temperature (Cook and Papendick, 1972). Due to wide range, crop rotation is not an adaptable cultural practice that is usually recommended for other soil-borne diseases. Sclerotia formed on undecomposed tissues in the field are capable for initiating infection and serve as the primary inoculum of the disease in the field (Punja and Grogan, 1981).

Peter Henry Rolfs in 1892 first time reported this fungus on tomato in Florida causing more than 70 per cent loss (Aycock, 1966). While, the binomial of Rolfs fungus (*Sclerotium rolfsii*) was first time ascribed by Saccardo (1913) to include those fungi with no known sexual state but formed small, tan to dark-brown spherical sclerotia (0.5 to 2 mm diameter) comprised of rind, cortex and medulla. The large number of sclerotia produced by *S. rolfsii* and their ability to persist in the soil for several years associated with the prolific growth rate of the fungus (2-3 cm per day in culture) make it well-suited facultative and a pathogen of major importance throughout the world. Despite work on several aspects of this fungus for over hundred years, many basic facts about the pathogen remained to be understood such as biology and role of basidiospores of this fungus in disease development, role of different parts of *Cyperus rotundus* and compound from *C. rotundus* which is responsible for basidial stage induction. Recently, some workers (Harlton *et al.*, 1995; Nalim *et al.*, 1995; Okabe *et al.*, 1998) have studied the variability in this fungus taking mycelia as the experimental tool, but involvement of basidiospores in variability of this fungus has not yet been thoroughly explored. In addition to collar rot infection, report on this fungus causing spotted leaf rot disease in many taxonomically unrelated plant species is also available in literature (Singh and Pavgi, 1965). Several hypothesis exit on the possible functions and use of sclerotial exudate during sclerotial development without conclusive reasons. Polyphenoloxidase (PPO) production by the pathogen is reported earlier (Punja and Damiani, 1996). But its involvement in pathogenicity affecting the other plant phenolic is not clearly understood which unequivocally needs further study.

Distribution

S. rolfsii continues to be a problem in a variety of crops when conditions are warm, humid and rainy. This disease occurs around the world in the equatorial zone between the 45° North and South latitudes. Diseases caused by *S. rolfsii* are rampant in the tropics, subtropics and other warm temperate regions, especially in areas of the southern & southeastern United States, Central & South America, the West Indies, Southern European countries bordering the Mediterranean sea, Africa, India, Japan, Australia, New Zealand, Philippines and Hawaii, where temperature are sufficiently high to permit the growth and survival of the fungus (Punja, 1985; Ferreira and Boley, 2004-internet). There has been limited report from Siberia and North China. The pathogen rarely occurs where average winter temperatures fall bellow 0° C. In India, this pathogen has been reported from Uttar Pradesh, Madhya Pradesh, Bihar, West Bengal, Assam, Uttranchal, Jharkhand, Gujrat, Tamilnadu, Rajsthan and almost all states.

Host Range

S. rolfsii has an extensive host range; at least 500 plant species in 100 families are susceptible. The most common hosts are the legumes, crucifers and cucurbits (Ferreira and Boley, 2004-internet; Punja, 1985). French bean (*Phaseolus vulgaris*), chickpea (*Cicer arietinum*), cowpea (*Vigna sinensis*), lentil (*Lens esculentum*), mustard (*Brassica compestris*) and pea (*Pisum sativum*) etc. are important host for *S. rolfsii*.

Diseases Caused by *S. rolfsii*

S. rolfsii causes different types of diseases in plants, i. e., collar rot, leaf spots, southern blight, southern stem rot, southern wilt, basal rot of mango seedling, peanut pod break down, root rot, seedling blight, damping off seedlings, white mould, crown rot, stem rot, foot rot and fruit rot etc. (Singh and Pavgi, 1965, Punja, 1985).

Symptomatology

Sclerotium rolfsii Sacc. causes various types of diseases in different plants. The symptoms appears on plants parts varies on types of infection. It infects seedlings, herbaceous plants, woody plants, roots, bulbs and fruits etc. It infects plant parts near or at the soil surface more frequently, but this fungus may also infect any part of susceptible plant as long as favourable environmental conditions exits. Following symptoms are produce by *S. rolfsii*.

Leaf Spot

Generally low growing plants, their leaves touching the soil may be infected with *S. rolfsii*. Usually the high humidity near the soil surface is an important factor in disease development. Main source for leaf spot development in the plants in nature may be landing of sclerotia with rain splash. The infection of *S. rolfsii* usually starts as a chlorotic translucent spot on the upper surface of the lower leaves closer to the ground (about 3-7 cm above the ground level) but it also infect the leaves which is 30 to 50 cm above the ground level. The spot gradually becomes necrotic measuring 5 to 10 mm in diameter. The spots are circular, medium brown to light brown or straw-coloured with narrow borders. On the basis of distance of leaves from the soil, number of infected leaves in a plants may vary. Humid weather is favourable for development of infection while dry weather check the development of infection. Normally only one leaf spot is formed on a leaf but some times two or more spots develop closely coalesce to each other and form bigger spots. Concentric rings like a target-board invariably appear in the necrotic tissues and small clumps of mycelium on them often develop into tiny pin head-like sclerotia. The sclerotia are pinkish to mustard brown in colour. Concentric rings are dark in colour with dull grey appearance in the center. Drooping do not takes place in infected leaves but occasionally shot holes are appears in the dead tissues. The disease was observed in many hosts while, in *Sphaeranthus indicus, Spathodia compoleta, Blepharis boerhaviaefolia, Ficus infectoria, Morus nigra, Artrica* sp., *Solanum melongena, Bombax malabaricum, Rauvolfia serpentina* and *Ficus religiosa* hosts are reported for the first time.

Collar, Crown, Root and Stem Rot

S. rolfsii infects the collar region of many plants. It cause total yield loss in the infected plants because whole plants dies before flowering due to infection. Disease begins with a small, water-soaked lesion on lower stem/collar region at or near soil surface. The brown lesion spreads rapidly to girdle the stem. The girdling lesion will quickly cause the herbaceous plant to wilt and fall over. After stem rot of herbaceous plant, a white mat of mycelium develops at the lesion site and later spread onto nearby soil surface. After some time white, round and fuzzy mycelial bodies begin to appear which later turn tan or brown sclerotia resembling mustard seeds. The fungus is sometimes referred to as the "mustard seed fungus". Sclerotia develop on infected stems, roots and the surrounding soil help in the identification of the disease. During rotting of lower part of the stem, plants usually remain erect and wilting of foliage start. Wilted leaves gradually become brown and remain hanging on the plant after infection in many host plants. In some plants many branches may wilt and leaves become slightly faded and later turn brown. In some infected plants cankers appear below 2-3 cm of soil surface in dry weather. In case of partial girdling of stems leaves become small and brown without wilting. In wet and hot weather stem become totally rotted excluding the xylem. Sclerotia are serve as primary source of inoculum present on diseased tissues, soil surfaces or in soil crevices. Generally, white thread-like mycelia radiate from a mycelogenous germination of sclerotium infect collar region of the plants under favourable conditions. The collar gets constricted and turns grey to brown in colour. The fungus grows upwards (2-5 cm) on the dried stem depending on the soil moisture. Occasionally it also moves downwards and kills the roots. In case of root rot the leaves eventually die, and branch dieback develops. The lower stem (trunk) above the girdling lesion usually appears normal for a long time and will be the last part of the plant to die. In case of woody plants infection of *S. rolfsii* begin as a crown rot. Characteristic mycelial mat and sclerotia appear at the infection site of crown. In monocots symptoms are totally different, wilting of foliage and dieback develop after rotting of the crown tissues. Brown lesions occur at the crown and lower parts of the culm. Lesions are often small, but they may extend into the hollow part of the culm. Strands of mycelium grow inside the lower internodes. Seed heads may appear normal, but they are devoid of grain. Premature ripening may also occur. The fungus secretes several enzymes, which dissolve various constituents of the host tissues before deep penetration of the hyphae. It takes about 3-5 days for infecting the plants. The affected tissues decay and the mycelium grows very rapidly on such rotten tissue. Infected plants topple down as the constricted collar region becomes weak, hence, unable to bear the weight of the foliage. Sometimes, the plants do not fall on the ground even after complete drying. In case of severe infection patches of grayish brown wilted plants are seen in the field.

Tuber, Bulb and Fleshy Organ Rot

S. rolfsii causes soft rot of fleshy organs of many plant. Potato tubers show small slightly sunken lesions at lenticels later develop into yellow-tan coloured lesions. Infected tissues become soft and collapsed. Invasion of mycelia and sclerotia takes place on rotting tissues. After entry of bacteria on rotten tissues, a sour odor is noticed. When narcissus bulbs become infected, several layers of scales become reddish brown with white streaks. Sclerotia form on rotted tissues. Previously rotten tissues are moist while later become dry and smell like rotten wood.

Fruit Rot

Fruits of many crops plant present at or near soil surface may infected with *S. rolfsii.* Fruit rot symptoms differ with plant species. Fruit of tomato show soft, water-soaked, sunken yellow lesions. These lesions spread quickly to almost all fruits and mycelium and sclerotia develop on lesions.

Biology of the Pathogen

Systemic Position

Sclerotial stage of *S. rolfsii* belong to Kingdom-Mycota, Division-Eumycota, Sub-division-Deuteromycotina, Class-Hyphomycetes, Order-Agnomycetales (Mycelia sterilia), Family-Agnomycetaceae (Hawksworth *et al.*, 1983, Alexopolus, 1962).

Sexual stage *Athelia rolfsii* belongs to higher Basidiomycotina (having true basidia; eubasidia), Class-Hymenomycetes, Sub-class-Holobasidiomycetidae, Order-Aphyllophorales, Family-Corticiaceae (Talbot, 1968, Punja, 1985, Singh, 1982). Generic description of both stages are given bellow.

Asexual Stage (Anamorph)

Mycelium

In early stage of growth in pure culture the fungus mycelium is at first silky-white but gradually loses its lustre and becomes somewhat dull in appearance. The mycelium radiating with abundant aerial hyphae appear as dense tufts dispersed all over the culture medium. Individual hyphae are hyaline, thin walled, branching at acute angles, sparsely septate when young, X-ray diffraction pattern indicate the presence of chitin and b-(1-3) glucan in both hyphal and sclerotial wall (Chet and Henis, 1968). The cells have been reported to be 60 to 350 micron (m) long and 2 to 8 m wide. The broader hyphae often may have clamp connections at the septa. The slender hyphae often lack clamp connections. Number of nuclei per cell is variable but in the secondary and tertiary branches it is mostly two although as many as 50-60 nuclei have been reported in all the cells.

Sclerotia

Sclerotial initials are formed from hyphal strands, which consist of 3-12 hyphae lying parallel. A spherical mycelial mass is soon assumed even if it is only a loose mass of hyphal network. After hardening sclerotium shows an outer layer of polyhedral cells. Rind, cortex and medulla are differentiated as the sclerotium darkens. The rind consists of thickened cells, 2-4 cells broad and the cortex consists of thin-walled cells with densely staining contents 6-8 cells broad. The medulla (central core) contains loosely arranged filamentous hyphae. Mature sclerotia are reddish to dark brown in colour and usually round in shape, surface smooth or pitted. In the beginning they are white but soon become dark and internal tissues of the sclerotium is white. Mature sclerotia germinate through mycelogenous germination. Melanin like pigment is reported to present in sclerotial wall (Chet and Henis, 1968).

Sexual Stage (Teleomorph)

The mechanism of sexual stage formation in many plant pathogenic fungi, especially facultative parasites, are not well understood. Goto (1930) described the perfect stage of *S. rofsii* for the first time and named it provisionally as *Corticium centrifugus* (Lev.) Bres. Curzi (1931) erected a new species *Corticium rolfsii* Curzi. West (1947) transferred these fungi to *Pellicularia*. Venkatarayan (1950) followed another change by keeping them in the genus *Botryobasidium*. Talbot (1973) suggested that the perfect stage of *S. rolfsii* belong to the genus *Athelia*, and then Tu and Kimbrough (1978) described *Athelia rolfsii* as the perfect stage of *S. rolfsii* while studying *Rhizoctonia-Sclerotium* complex. The currently accepted binomial for the sexual stage of *S. rolfsii* is *Athelia rolfsii* (Curzi) Tu & Kimbrough (Punja *et al.*, 1982).

Various types of media were used to induce athelial stage by many workers. Goto (1930) used onion agar, Mundkur (1934) onion-asparagine-peptone and Chet *et al.* (1966) Joham's agar medium containing amino acids. Punja *et al.* (1982) used 8 different media including 2 per cent activated charcoal in potato dextrose agar (C-PDA) medium for inducing athelial stage. They found nutritional or environmental conditions that restricted optimal growth and sclerotial production by the fungus favoured hymenial formation in general and found no apparent correlation between host or area of isolation and ability of an isolate to form hymenia. They also found more extensive hymenial formation in cultures incubated at reduced light intensities (about 80-100 lux) than at higher light intensities (about 1200 lux). Singh *et al.* (1996) used *Cyperous rotundus* rhizome meal agar (CRMA) medium which is easiest and quickest for athelial stage induction.

Basidiocarp

Basidiocarp resupinate, effused, in patches usually 1-2 cm in diameter (sometimes upto 6 cm), pellicular, usually 50-100 um thick and loosely adherent to the substrate. Subicular hyphae small, hyaline, binucleate, loosely arranged, simple septate with rare clamp connections. Hymenophore cariaceous memberanous, meruloid, but becoming smooth on drying, whitish to creamy-white.

The hyphae from which basidia developed initially appeared slender with pro basidial swellings. As the terminal portions enlarged, clavate basidia are discernable. Sterigmata developed as the basidia matured and basidiospores subsequently formed. The athelial stage grows as a spreading white, yellow, or buff-coloured granular or encrusted area with a slightly wavy surface of hymenium (unprotected layer of basidia) on the host surface while in culture it appears as dense white crustiform masses in small irregular area. The hymenium closely adhere to the substrate and always gymnocarpous. In appearance the hymenium may be coarsely aerolate with clusters of basidia arranged on a tenuous subiculum. The hymenial thickness may be upto 30-40 micron and almost white in culture. The hymenia may be grey, yellow or buff coloured on the host. The sub-hymenial layer is composed of loosely interwoven hyphae which branch monopodially. Branches are septate and later develop into basidia (Singh, 1982). Basidiocarp usually well developed, effused and mostly saprobe, some times parasite (facultative parasite). Cystidia and rhizomorphs are absent (Punja *et al.*, 1982).

Basidia

Basidia originate as the terminal cell of a dicaryotic hyphae. Basidia are typical holobasidia (aseptate basidia). Basidia are in cluster, relatively short, clavate, mostly 4-spored, without a basal clamp, 15-19X5-6 um in size. Basidium bears 4 parallel or divergent, slender, tapering sterigmata, which are 2.5-4 micron long (Punja *et al.*, 1982).

Basidiospores

Basidiospores are formed on sterigmata, and are ballistospores. Basidiospores are haiploid, unicellular, elliptical to obovate, sometimes rounded or pyriform, smooth, thin walled, 5-7X3-5 um in size, hyaline and apiculate at base. Germinating basidiospores produce 1-3 germ tubes and subsequently an extensive mycelium was produced (Punja *et al.*, 1982).

Disease Cycle and Epidemiology

In the rainy season, hyphal growth resumes from infected tissues and germinating sclerotia. Hyphae and germinated sclerotia come into contact with susceptible crown, root, bulb, fruit tissues and penetrate directly but wounds facilitate infection. infection of stem, roots, bulbs, fruits and leaves may occur in susceptible tissue if temperature, humidity and other environmental conditions are favourable. Hyphae may be intracellular or intercellular. The fungus produces a variety of biochemicals like, polygalacturonase and oxalic acid, which help in pathogenicity. Secondary cycles occur and disease spreads as hyphae contact new healthy susceptible tissues. Also, basidiospore production may contribute to secondary cycles. The fungus overwinters as sclerotia and mycelia in infected plants and plant debris and some times as developing hymenial layers. It has been reported that sclerotia may pass through the digestive tract of cattle or sheep and still be viable.

The pathogen survives as mycelium on dead organic matter when living plants are not available. Fungal infection is enhance when dead organic matters are available around susceptible plants. Optimum growth and sclerotia production of the fungus occurs between 27-35°C. In addition to temperature, hyphal growth and sclerotia germination require a water-saturated soil, high humidity also favoures fungal development. Formation of sclerotia and its survival favoured by soil pH below 7, light, well-aerated and well-drained soil.

Management

Despite continuous research over the past one century, this pathogen continues to plague growers causing considerable economic loss. Management efforts have often met with limited success, due to the extensive host range, prolific growth and ability to produce large number of sclerotia that may persist in the soil for several years. Further more, control measures effective for a particular crop or area may not be adaptable elsewhere due to regulatory or economic constraints. Several fungicides effectively controlled this pathogen on various crops in the field (Blackman and Rodridguez-Kabana, 1975; Brown and Hendrix, 1980; Haas, 1976; Punja *et al.*, 1982), but the major limitations to their widespread use to control this pathogen are requirement of large amounts of chemicals, variable effects on different crops, and variation in results in different seasons.

References

Chandurkar, P.S. (2004). Dynamic approach helps. *The Hindu Survey of Indian Agriculture 2004*. pp 145-146.

Awasthi, U.S. (2004). Pragmatic views, concerns. *The Hindu Survey of Indian Agriculture 2004*. pp 147-148.

Venkataramani, G. (2004). Third time in a row. *The Hindu Survey of Indian Agriculture 2004*. pp 5-7.

Babu, S.C. (2004). More supportive policies. *The Hindu Survey of Indian Agriculture 2004*. pp 13-14.

Singh, R.S. (2000). Plant disease. Oxford & IBH Publishing Co. Pvt. Ltd., New Delhi pp 4-8.

Singh, R.S. (1982). Plant pathogen: The fungi. Oxford & IBH Publishing Co. Pvt. Ltd., New Delhi pp 324-325.

Chapter 26

Insect Pests of Pigeonpea and their Management

Ram Keval and G.M. Narasimha Murthy

Department of Entomology & Agricultural Zoology, Institute of Agricultural Sciences, Banaras Hindu University, Varanasi – 221 005, U.P.

Pigeonpea, *Cajanus cajan* (L) Millsp., is the second most important pulse crop in India with an area of 3.53 million ha and production of 2.43 million tones. The mean realized yield of pigeonpea (nearly 690 kg ha^{-1}) is quite low in comparison to the potential yields of 1500-1800 kg ha^{-1} in short duration cultivars and 2500-3500 kg ha^{-1} in medium and long duration cultivars. In most of the areas, biotic stresses mainly insect pests have been recognized as the major constraints in realizing potential yields of pigeonpea. Pigeonpea plants and seeds attract over 200 species of insects that damage their roots, shoots, flowers, and seeds. Most of these insect species are sporadic in their distribution and therefore may not all be regarded as pests. Pigeonpea plant can compensate for most insect damage during its vegetative phase. However, the reproductive parts of the plant are most attractive to pests and recovery from damage at the reproductive phase is slow and dependent on plant type, soil moisture, and climatic conditions. Thus, most of the economically important insect pests attack pigeonpea at the reproductive phase and in storage, when they damage buds, flowers, pods, and seeds. Broadly, the key insect pests of pigeonpea at the reproductive phase are grouped into three. The flower- and pod-feeding Lepidoptera larvae (mainly *Helicoverpa armigera, Maruca vitrata, Etiella zinkenella*), the pod sucking Hemiptera (mainly *Clavigralla* spp.), and seed-feeding Diptera (*Melanagromyza* sp.).

Insects feed on all parts of the pigeonpea plant. The most serious pests, and the primary focus of pigeonpea pest management research, are those that attack reproductive structures, including buds, flowers, and pods. Pigeonpea has a great capacity to tolerate and recover from early season losses of flowers and young pods, provided the general health of the plant is good and that sufficient soil moisture is available. Removal of all flowers and pods for up to 5 weeks after flower initiation did not reduce seed yields in 10 short- and medium-duration pigeonpea cultivars. Thus, only pests that are continuously present or that attack at the middle or end of the crop cycle are economically important. The key pests of pigeonpea can be grouped into three categories: flower- and pod-feeding Lepidoptera, pod-sucking Hemiptera, and seed-feeding Diptera and Hymenoptera.

Pest Complex

The key insect pests of pigeonpea can be grouped in to three categories:

1. Flower and pod feeding Lepidoptera,
2. Pod-sucking Hemiptera, and
3. Seed feeding Diptera and Hymenoptera

Flower and Pod Feeding Lepidoptera

Nearly 30 species of Lepidoptera in six families feed on the reproductive structures of pigeonpea. Among them the gram pod borer, *Helicoverpa armigera* (Noctuidae), spotted pod borer *Maruca virtata* (Pyralidae), Green pod borer *Etiella zinckenella* (Pyralidae), Plume moth *Exelastis atomosa* (Pyralidae) are of considerable importance.

Helicoverpa armigera

It is a major biotic constraint to increasing pigeonpea production throughout India. Annual pigeonpea losses due to *H. armigera* have been estimated at $317 million worldwide. In general, moths prefer to oviposit on plants in the reproductive growth stage and are attracted to flowering crops, perhaps by the nectar, which is a carbohydrate source for adults. On pigeonpea, more than 80 per cent of eggs are laid on calyxes and pods. In addition to its preference for feeding on reproductive structures, four features of *H. armigera* life history make it one of the most serious and widespread insect pests in the Old World: high fecundity, extensive polyphagy, strong flying ability, and a facultative diapause. In India, *H. armigera* has been recorded on at least 181 plant species from 45 families. Though highly polyphagous, *H. armigera* prefers maize and sorghum to most other host plants.

Identification and Monitoring

Adult is a stout moth with dark yellow-olive forewings and pale hind wings. Eggs are laid single and in all parts of plant and yellowish, shiny. Full grown larva is 40 mm long and hairy and varied in color. Use pheromone traps for monitoring. Visual observations at weekly intervals at all stages. Setting of light traps (1 light trap/5 acre) to know the range of pest incidence.

Damage Symptom

Small larvae eat up the green portion of the plant. The larger larvae eat up the floral parts, flowers, leaves and pods. The larvae eat up the bulged portion of the pods. The larva eats up the floral buds. By eating the buds holes are formed and the larvae proceed by further eating up the grains.

Biology and Life Cycle

The biology and ecology of *H. armigera* have been extensively reviewed, and the general features do not differ when pigeonpea is used as a host. Females oviposit at night and fecundity is high, with up to 3000 eggs reported from a single female. The eggs are white and nearly spherical when freshly laid and darken with age. Eggs hatch in 3–5 days, and the number of instars, from 5 to 7, varies with temperature and host plant. The generation time of *H. armigera* is highly variable. In tropical regions it can be as short as 28 days, with up to 11 generations per year. Mean development time on six short-duration pigeonpea genotypes was approximately 21 days for larvae and 15 days for pupae. Pupation occurs in a pupal cell 2–18 cm below ground. The prepupal stage lasts for 1–4 days. The pupal stage requires 10–14 days for nondiapausing individuals but may last several months during diapause. The variable development time on different host plants, varying number of generations per year, strong migratory ability, and co occurrence of diapausing and nondiapausing individuals all contribute to produce overlapping generations in the field.

Maruca virtata

It is distributed throughout tropical and subtropical regions worldwide. It has a wide host range but is restricted to legumes. *M. vitrata* is a serious pest of pigeonpea in India, Sri Lanka, and Africa, with annual losses estimated at US$ 30 million worldwide. During the dry season, when crop host plants are not available, *M. vitrata* feeds on wild leguminous shrubs and trees.

Identification and Monitoring

The adult are brown forewings with white club shaped marking and hind wings are white with irregular blotch. The larva is whitish-green. The eggs are oval and yellow laid in small batches commonly on terminals.

Damage Symptom

The larva webs together the leaves, buds and pods and feeds inside these webs. The caterpillar also bores into pods and eat up the ripening seeds. A crown mass of excrement is seen at the entrance into the larval burrow.

Biology and Life Cycle

The biology and life cycle of *M. vitrata* appear to be similar on the two host plants. Regardless of the host plant, eggs are primarily laid on buds and flowers. Fecundity of more than 400 eggs per female has been reported from laboratory studies. Eggs are usually laid in groups of 4–6, though up to 16 eggs have been found in some groups. Eggs hatch in 2–5 days and larvae pass through five instars over 8–14 days.

The prepupal period lasts for about 2 days and the pupal period 6 to 9 days. Pupation may occur either on the plant or in the soil in a pupal cell made of silk and covered with debris. Generation time is typically 18–25 days, but can be as long as 57 days. Larvae feed from inside a webbed mass of leaves, flowers, and pods. This concealed feeding complicates control as pesticides and natural enemies have difficulty penetrating the shelter to reach the larvae. Pigeonpea genotypes with determinate growth habit, where pods are bunched together at the top of the plant, are more prone to damage than genotypes with indeterminate growth habit, in which the pods are arranged along the fruiting branches.

Exelastis atomosa

It is one of the most destructive and major constraint in the successful cultivation of pigeonpea particularly in eastern U.P. As the pest damage by boring various reproductive parts, timely control of this pod borer is more important for realizing better yield.

Identification and Monitoring

Adult is small moth with yellowish brown wings, forewings are cut into 2 plumes and hind wings into 3. The green oval eggs are laid singly on buds and pods. Larvae are green or brown, spindle shaped and covered with spines and hair. Peak populations are during Nov to March. Caterpillars are more rampant during the post rainy season than during the rainy season.

Damage Symptom

The larvae bore into the buds, flowers and pods and feed on the developing grains. By eating the buds holes are formed and the larvae proceed by further eating up the grains. The larvae excrete inside the damaged grains and pods and due to this, fungus develops in it.

Biology and Life Cycle

The mean longevity adult was 6.59±0.38 days. The average number of eggs laid bay an adult female was 93 to 101 eggs hatched in 2.92 to 3.02 days. There were five larval instars witch took 23.12+-0.93 days to enter into pupal stage. Pupation took place on pod surface or entrance of hole or in the burrow of infested pods and the pupal period lasted for about 7.97+-0.33 days. The life cycle of *E. atomosa* was completed in 40to 42 days.

Pod-sucking Hemiptera

A large number of Hemiptera, mainly in the families Alydidae, Coreidae, and Pentatomidae, feed on pigeonpea and are commonly referred to as pod-sucking bugs. Relatively few are serious pests; the most important are the coreids *Anoplecnemis*, *Clavigralla* (*Acanthomia*), and *Riptortus*. Research has focused on three *Clavigralla* species; *Clavigralla tomentosicollis* is widespread in sub- Saharan Africa, while *Clavigralla scutellaris* is found from Kenya through Yemen, Oman, Pakistan, and India. The third species, *Clavigralla gibbosa*, is restricted to India and Sri Lanka.

Both adults and nymphs of *Clavigralla* spp. feed on pigeonpea by piercing the pod wall and extracting nutrients from the developing seeds. Damaged seeds are dark and shriveled, and they are difficult to distinguish from those damaged by drought. For this reason, bug damage is frequently underestimated. Damaged seeds do not germinate and are not acceptable as human food. Combined losses due to *C. gibbosa* and *C. scutellaris* in India vary among regions and occasionally exceed 50 per cent.

Biology and Life Cycle

The egg to adult development of *Clavigralla* spp. reared on pigeonpea is completed in 15–40 days under ambient temperatures. Both *C. gibbosa* and *C. tomentosicollis* pass through five nymphal instars before reaching the adult stage. Adult *Clavigralla* spp. can live more than 150 days and females occasionally lay as many as 430 eggs in clusters of varying size. Field collections of *Clavigralla* egg clusters in India showed that cluster size ranges from 2 to 62 eggs (mean D 18 eggs), with more than 70 per cent of clusters containing 7–24 eggs (81). Similar ranges in egg cluster size have also been reported for African *Clavigralla* spp. Longevity and fecundity of adults and egg cluster size of laboratory-reared bugs may differ significantly from field-collected bugs.

Seed Feeding Diptera and Hymenoptera

Two Diptera and one Hymenoptera feed on developing seeds within the pigeonpea pod. The most important is *Melanagromyza obtusa* (Diptera: Agromyzidae), the pigeonpea pod fly, which appears to be restricted to Asia. Its biology, ecology, and management have been extensively studied. A second agromyzid species, *Melanagromyza chalcosoma*, is a pest of pigeonpea in eastern and southern Africa. Though less well studied, it seems to occupy a similar ecological niche. Both species feed only on pigeonpea and closely related species within the subtribe *Cajaninae.*

Melanagromyza obtusa

Identification and Monitoring

Adult is small black fly and is about 5 mm in length. Eggs are laid in the wall of an immature pod. Maggot is milky white, legless and about 3 mm in size. Five brownish strips runs along the entire mid dorsal line of the body.

Damage Symptom

The maggot feeds on the developing grain. The infested pods do not show external evidence of damage until the fully grown larvae chew holes in the pod walls. These bore the grains and make the tunnel in them. This hole provides an emergence "window" through which the adults exit the pod. Damaged grains do not mature and due to excreta fungus may develop in the grain. The infested grains loose their viability.

Pod fly damage has been reported from several countries. In India, the pod fly is a more serious pest in northern and central areas than in other parts of the country. Damage levels in farmers' fields range from 10 to 50 per cent. In Vietnam, *M. obtusa* is the key pest of pigeonpea, causing seed losses of more than 90 per cent, while damage of 43 per cent is reported from Taiwan .

Biology and Life Cycle

M. obtusa females produce up to 80 eggs and lay them individually into developing pigeonpea pods. Development of the immature stages under field conditions includes 3–5 days for the egg stage, 6–11 days for the three larval instars, and 9–23 days for the pupal stage. Adults live up to 12 days when fed with honey and about half as long without food. The population dynamics of *M. obtusa* are governed by its narrow host range and feeding niche. In India, pigeonpea pods are available in the field from approximately October to April, and infestations increase rapidly over a relatively short period. Fewer eggs are laid in December and January when temperatures are low, and populations increase as temperatures rise. Long-duration pigeonpea crops mature in March or April and can be heavily damaged. *M. obtusa* may survive the offseason on alternate hosts such as *Rhyncosia minima*, which have been found to be infested with eggs, larvae, and/or pupae between April and November.

Tanaostigmodes cajaninae

The larvae of Hymenopteran borer, *Tanaostigmodes cajaninae* (Hymenoptera: Tanaostigmatidae) also feed on developing pigeonpea seeds. In addition to pigeonpea, *T. cajaninae* has been reported feeding on 13 non crop legumes. Female wasps lay individual eggs on the surface of flowers or young pods. The larva bores into the pod and feeds for 8–10 days on the developing seeds. Pupation, which occurs within the pod, lasts 5–7 days, and the adult lifespan is 7–9 days. Low infestation levels (<2.4 per cent pods) have historically been reported in farmers' fields in southern India, but higher infestation levels, up to 58 per cent of pods, were found on research stations, especially in pesticide-treated plots.

Pest Management

Pigeonpea pest management is complicated by several factors. The crop is attacked by at least three key pest groups with very different biologies. These differences include host range, apparency (feeding on the plant surface versus concealed feeding), and feeding mode (chewing versus piercing and sucking). The pests also have highly variable population dynamics between years and locations, and at least one, *H. armigera*, has developed high levels of resistance to several insecticides. The key pests are all direct pests, feeding on the portion of the crop most valued by humans, and each is capable of completely destroying a crop. Economic thresholds have not been developed for any pest of pigeonpea. Given the variety of pests, the long reproductive phase and compensatory ability of the crop, and the socioeconomic constraints of farmers in most pigeonpea-producing countries, it is doubtful if useful or practical economic thresholds could be developed. Another obstacle to progress in pigeonpea pest management is that pigeonpea has been considered a marginal crop or is the neglected component of a mixed cropping system and is thus given less attention by farmers, crop protection specialists, and policy makers.

The primary focus of pigeonpea pest management has been on *H. armigera* and *M. obtusa*, with emphasis on chemical control and host plant resistance. In India, calendar sprays are recommended and followed, with the first application at 50 per cent flowering and second and third applications at 15-day intervals. The rapid increase in pesticide use on pigeonpea is alarming and emphasizes farmers' concern with insect pests. The trend also highlights the need for safe and effective management strategies.

The difficulty in managing insecticide-resistant populations of *H. armigera* has given impetus to the development and use of alternative insecticides such as plant-derived products [*e.g.* neem (*Azadiracta indica*)] and insect pathogens, particularly the *Helicoverpa* nuclear polyhedrosis virus (NPV). These products are generally considered to be safer for humans and the environment and have less negative impact on beneficial organisms than conventional insecticides. Both neem and NPV products suffer from poor and highly variable quality and a more limited distribution network than conventional insecticides. These problems must overcome before these products can be considered effective and practical alternative control methods.

Significant work for different components of Integrated Pest Management has been done during the last one decade both under All India Coordinated Research Project on Pigeonpea and at ICRISAT, India. The progress made on different aspects is briefly discussed below.

Host Plant Resistance

Pigeonpea lines with resistance to either or both *H. armigera* and *M. obtuse* have been reported, but little progress has been made in incorporating resistance in cultivars that are acceptable to farmers. No insect-resistant pigeonpea genotypes are widely cultivated by farmers. Frequently, the resistant lines are less preferred in terms of taste, seed color, and/or size and are often susceptible to wilt, sterility mosaic virus, or other diseases. Traditional pigeonpea landraces are medium-to-long–duration and may have been selected to avoid peak pest attack. Delaying planting to avoid high pest populations has been an effective strategy in research station trials but has not been widely adopted. Selecting companion crops or cultivars has also been investigated as a means of minimizing pest damage. The widespread practice of intercropping the longer-duration pigeonpea genotypes with one or more companion crops may have evolved through farmers' desire to reduce the risks of insect or other losses. But the companion crop(s) is usually harvested before pigeonpea flowers when medium- and long-duration pigeonpea cultivars are used. Medium duration varieties JA-4 or JKM-7 are resistance to pod bug.

Cultural Control

- ☆ Deep summer ploughing.
- ☆ Sowing should be done by the end of June to avoid pod borer attack.
- ☆ Remove the weeds from the field.
- ☆ Intercropping of early maturing pigeon pea with mung bean in alternate and paired row results in low infestation of pod borer. Intercropping with jowar, maize or groundnut etc.
- ☆ The varieties bahar and sharad should be grown in pod borer endemic areas of northern India.
- ☆ Short duration varieties escapes from the attack of pod borer.

- ✰ Maintaining complete field sanitation.
- ✰ Do not repeat sowing of pigeon pea crop in same field.

Mechanical Control

- ✰ The population of pod borer can be regulated by using pheromone traps.
- ✰ In cases of heavy infestation, physical shaking of pigeon pea plants to dislodge larvae is favored.
- ✰ Bird perches placed just above the crop canopy will also help to reduce the population of the pest.
- ✰ Remove and destroy the damage plant parts.

Biological Control

- ✰ Conserve predators like spiders and wasp etc.
- ✰ Use of NPV at 500 LE with adjuvant like teepol, tinopal and jaggery etc.
- ✰ Applying HNPV at a rate of 500 larval equivalents (LE) per ha.
- ✰ This application can be repeated at 15-20 days intervals.
- ✰ Female moths can be deterred from ovipositing by the spraying of 5 per cent neem kernel suspension.
- ✰ Conserve *Ormyrus sp* (parasite of pod fly).
- ✰ In nature these are prayed by cocinellid beetles and chrysoperia. So conserve lady beetles, green lace wing, diaretiella rapae, menochiles sexamaculatus.

Chemical Control

- ✰ Apply chemical insecticide only if the pest population crosses ETL.
- ✰ Spraying of endosulfan 35 EC 0.07 per cent (2ml of 35 EC/lit. of water) or monocrotophos 36SL 0.04 per cent (1 ml of 36 SL /litre of water) or chlorpyriphos 20 EC @ 3.5 ml/lit. of water at 600-1000 lit. of spray material per ha. with hand sprayer.

Future Prospects

Pigeonpea farmers in some parts of India and Africa have rapidly adopted the use of pesticides as the primary means of pest management. Past experience in developing countries has shown that pesticide use is often inappropriate and unsafe and that farmers frequently fall into a cycle of increasing the amounts and/or frequency of pesticide applications. To avoid this "pesticide treadmill," pigeonpea farmers need effective alternative pest management practices. There is no shortcut or magic bullet to reduce losses due to insect pests immediately. Progress will be incremental, and in the short term, the greatest impact may come from improving insecticide application. This would involve enhancing the skills needed to scout fields and properly mix and apply insecticides and providing unbiased information on the relative risks and benefits of different insecticides. A strategy for the medium term should concentrate on developing improved cultivars that combine high yield and disease and insect-resistance into backgrounds with consumer-preferred agronomic characters. The identification of specific resistance mechanisms, such as increasing the density of nonglandular trichomes on pods, would be a good start. A longer-term solution to insect pest problems in pigeonpea must focus on ways to enhance natural control processes, either by the introduction of exotic natural enemy species or by enhancing the effectiveness of endemic species.

References

Banu, M.R., Muthiah, A.R. and Ashok, S. (2005). Evaluation of pigeonpea (*Cajanus cajan* L.) genotypes against gram-pod borer (*Helicoverpa armigera*). Abstract in 4th *International Food Legume Research Conference on Food Legumes for Nutritional Security and Sustainable Agriculture*, Oct. 18-22, 2005, New Delhi, India. pp. 317.

Fitt, GP. (1989). The ecology of *Heliothis* in relation to agroecosystems. *Annu. Rev. Entomol.* **34**:17–52.

Lateef, S.S., Reed, W. and LaSalle, J. (1985). *Tanaostigmodes cajaninae* LaSalle sp. n. (Hymenoptera: Tanaostigmatidae), a potential pest of pigeonpea in India. *Bull. Entomol. Res.* **75**:305–13.

Majumder, N.D. and Singh, F. (2005). Pigeonpea improvement in India. Souvenir, 4th International Food Legume Research Conference on Food Legumes for Nutritional Security and Sustainable Agriculture, Oct. 18-22, 2005, New Delhi, India. pp. 53-65.

Sachan, J.N. (1992). Present status of *Helicoverpa armigera* in pulses and strategies for its management. In Helicoverpa *Management: Current Status and Future Strategies, Proc. First Natl. Workshop*, ed. JN Sachan, pp. 7–23. Kanpur, Uttar Pradesh, India: Dir. Pulses Res. 158 pp.

Reed, W. and Lateef, S.S. (1990). Pigeonpea: pest management. See Ref. 56, pp. 349–74.

Shanower, T.G. and Romeis J. (1999). Insect Pests of Pigeonpea and their Management. *Annu. Rev. Entomol.* **44**: 77–96.

Chpater 27

Entomopathogenic Fungi as a Tool for Sustainable Pest Management: An Overview

Vibha[1], Rakesh Pandey[2], P.K. Jha[1] and R.C. Rai[1]

[1]*Department of Plant Pathology, Rajendra Agricultural University, Pusa (Samastipur), Bihar,*

[2]*Krishi Vigyan Kendra, (Indian Institute of Vegetable Research) Sant Ravidas Nagar, U.P.*

Invertebrate Pathology is acquiring a new dimension under sustainable agriculture, as enotomopathogenic fungi (EPF) are biological controlling agents and well known potential alternative to chemical pesticides for the control of insect-pests and also available commercially for the pest control purposes in numerous countries world- wide. Its roots, however, can be traced to ancient history with reference to solution for preventing diseases in honey bees and silk worms (Steinhaus, 1956, 1975). Since the late 1940, insect pest control has relied mostly on chemical pesticides, although in many industrialized nation, pest management strategies have now been shifting to use of transgenic plants expressing particular traits such as resistance to insects, fungi, herbicides or viruses. However, the replacement of chemicals with transgenic plants does not represent a fundamental change in approach. Therefore, the application of micro organisms for control of insect pests was proposed by early notable pioneers in invertebrate pathology such as Agostino Bassi, Louis Pasteur, and Elie Metchnikoff (Steinhaus, 1956, 1975). It was not until the development of the bacterium *Bacillus thuringiensis* Berliner that the uses of microbes for the control of insects become widespread (Lacey *et at.*, 2001). Today, a variety of entomopathogens are used for the control of invertebrate pests in glasshouse and row crops, orchards ornamentals etc. (Burges, 1981, Tanada and Kaya, 1993).

The soil is an excellent reservoir for a diversity of entomopathogenic fungi *viz. Beauveria* spp, *Metarhizium anisopliae* and *Paecilomyces* spp. which contribute significantly to regulate the insect population (Keller and Zimmerman, 1989). Key components of population dynamics of the entomopathogenic fungi are the buildup of the population, the infection of hosts, and the survival and dispersal in the environment (Anderson and May, 1981). Dispersal of infective stages of a pathogen is an important factor in disease development (Anderson and May, 1981). Infective propagules of entomopathogenic fungi in the Hypocreales are passively dispersed, and this is mainly considered to occur through the action of weather components like wind and rain (Hajek, 1997; Inglis *et al.*, 2001 and Shah and Pell, 2003). Entomopathogenic fungi infect insects primarily by secreting extracellular enzymes, whether of mycelia or spore origins which are the hallmark of infection process (Qazi and Khachatourian, 2007). As the infection proceeds, the fungi produce large array of toxic metabolites that can target different internal organs in insects (Clarkson and Charnely, 1996). Strategies for the use of entomopathogenic organisms for insect pest control are basically the same as that for other biological control agent (Harper, 1987). These may be used to augment naturally occurring pathogens, conserve or activated in nature, introduced into pest population as classical biological control agents to become established and exert long- term regulation of the pest (Lacey *et al.*, 2001).

Ecology

Adaptation to insect parasitism among the entomopathogenic fungi varies owing to diversity among insect host, which regulate the insect population by breaking the insect defense system and differential absorption of nutrient from host cell. Some fungi alter host behavior (*e.g.* summit disease in which infected insect- exhibit climbing behavior), but there are considerably fewer examples with hypocrealen infected insects than in entomophthoralean infected ones (Roy *et al.*, 2006). However, behavioral avoidance of entomopathogenic fungi has been reported for various insects like *E. bassiana* is avoided by *Anthrocoris nemorus* (Meyling and Pell, 2006) and *Coccinella septumpunctata,* while *Coptotermes lacteus* avoids *M. anisopoliae* (Staples and Milner, 2000). Avoidance indicates recognition of the fungus by the insect, although the specific mechanism for avoidance is not known (Vega *et al.*, 2009). Entomopathogenic fungi built up their population up to infective stage in the presence of arthropod hosts. The relationships by which different species of fungi obtain energy form their insect host include biotrophy, necrotrophy, and hemibiotrophy (Vega *et al.*, 2009). The outbreaks of disease among insect population referred as epizootics, the development of epizootics rely on host population dynamics, the number of infective stages in the pathogen population and the viabilities of these, infection efficiency and development (Anderson and May, 1981) and a complex set of environment factors and timing (Inglis *et al.*, 2001).

Distribution

Most terrestrial ecosystem harbours soil inhabiting entomopathogenic fungi which play pivotal role in regulating insect population, particularly soil- dwelling insect pests (Keller and Zimmerman, 1989 and Jackson *et al.*, 2000). The entomopathogenic fungi such as *Beauveria bassiana, Nomuraea rileyi, Metarhizium anisopliae, Paecilomyces fumosoroseus* and *P. furinosus*, have been considered as a source of biocontrol agents (Goettel *et al.*, 2000) and more recently these microbes have also been considered to be a rich source of natural bio-active compound (Lee *et al.*, 2005). Knowledge of indigenous entomopathogenic fungi provide an insight into naturally occurring fungal biodiversity, which are highly pathogenic to arthropod pests, and provide a pool of potential biological control agents for developing eco-friendly pest management tool. The richness of fungi in subtropical regions are similar to those recorded in other temperate regions (Vanninen 1995), but lower to that recorded for tropical habitat (Evans, 1982). Furthermore, fungal biological control agents frequently perform inconsistently in the soil due to a lack of environmental competence (Jackson *et al.*, 2000). It is important to understand the best suited environmental conditions for improving biological control potential of entomopathogenic fungi.

Factors Affecting the Distribution

Successful development of mycosis is feasible only in the presence of optimum temperature, moisture, nutritional and physiological state of host accompanied with soil management practices.

Temperature and Humidity

It is well established that development of the fungus on the cadaver, and their sporulation is associated with optimum temperature and humidity close to saturation point for fungal auto multiplication of the fungal inoculum and subsequent development of mycosis results in heavier contamination of healthy insects. Sivasankaran and his co-worker (1998) reported that the high ambient humidity was essential for entomopathogenic fungi for germination, infection and subsequent sporulation on insect cadavers. A major impediment to the use of fungi in biological control is their sensitivity to condition of low humidity (Matewele *et al.*, 1994). Contrary reports indicate that infection from an initial application of conidia is unaffected by low ambient relative humidity when the microclimate on the insect cuticle is suitable for infection (Barson *et al.*, 1994 and Selman *et al.*, 1997). Selection of fungal pathogens tolerant to the temperature range in the ecosystem in which they are able to be used is imperative for their effectiveness as mycopesticides (Ferron *et al.*, 1991). Doberski (1981) selected fungal strain with pathogenic activity below 15°C for insect pests in temperate region, McClatchie *et al.* (1994) selected strain active at temperature below 30°C for use against pest in tropical regions. Genetically based resistance to desiccation and extreme temperatures would be a distinct advantage, both during infection and during mycopesticide formulation and storage (St. Leger and Screen, 2001). Since these traits are under polygenic control, strain modification through genetic transformation does not seem possible; immediate advances are more likely from strain selection (St. Leger and Screen, 2001). Conidia of entomopathogenic fungi must retain high viability and virulence for effective biological control (McClatchie *et al.*, 1994). The conidia must, therefore, have predictable shelf life. The prediction of conidial survival is now possible using equations driven by temperature, even if, temperature fluctuates (Hong *et al.*, 1997; Hong *et al.*, 1998 and Hong *et al.*, 1999). Moreover, this approach can be applied over wide geographic regions when combined with spatial metrological date and models which quantities the effect of ambient temperature and relative humidity on equilibrium conidial moisture content (Hong *et al.*, 2002). The viability of conidia of entomopathogenic and plant pathogenic fungi has already been developed (Hong *et al.*, 1997, 1998, 1999)

Physiology and Nutritional Status

Important information has been generated by Magan's group on the amount of endogenous polyols and trehalose accumulated in conidia and the effect of these compounds on conidial performance. Conidia of insect pathogenic fungi *Beauveria bassiana*, *Metarhizium anisopliae* and *Isaria* (Syn. *Paecilomyces fumosororeus*) produced on media with low water activity or with high concentration of glycerol had increased accumulation of polyolys and trehalose in conidia and they are more virulent than conidia produced on a rich media (Sabouraud dextrose agar) without stress (Anderson *et al.*, 2006; Hallsworth and Magan, 1994 and Magan, 2001). Trehalose hydrolysis is a major event during early conidial germination and presumably provides glucose for energy (Elbein *et al.*, 2003) and, high concentration of trehalose and polyols in conidia have also been related to increased stress tolerance (Thevelein, 1984). Both trehalose and polyols serve as easily mobilized energy reserves for rapid conidial germination; and this may explain, at least partially, why conidia produced under one culture condition showed improved germination speed in relation to conidia from some other culture conditions when germination of conidia obtained from both the conditions tested on same medium (Anderson *et al.*, 2006; Rangel *et al.*, 2005 and Shah *et al.*, 2005). Furthermore, intracellular trehalose and mannitol immobilization can contribute to some intent to the energy requirements of endogenous respiration and provide energy for spore germination (d'Enfert *et al.*, 1999; Elbein *et al.*, 2003 and Thevelein, 1984.). Conidia of *Metarhizium anisopliae* produced under stress condition had increased virulence, increased germination speed and conidial adhesion to host cuticle (Rangel *et al.*, 2008). Nutrient deprivation triggers increased Prl transcription and rapid secretion of this enzyme (Freimoser *et al.*, 2005) and this enzyme being extracellular chemoelastase that constitute the major protein synthesis by fungus during penetration through insect proteinaceous cuticles (Goettel *et al.*, 1989 and St. Leger *et al.*, 1987). The subtilisin Pr1is upregulated on minimum medium (Freimoser *et al.*, 2005) and on the insect cuticle (St. Leger *et al.*, 1991).

Influence of Soil Management System

Microbial assemblages in agricultural soil are important for ecosystem services in sustainable agricultural systems, including pest control (Altieri, 1999). Higher populations of beneficial soil borne organisms are characteristic of healthy soil (Magdoff, 2001). The soil environment constitutes an important reservoir of diversity of entomopathonogenic fungi which can contribute significantly to the regulation of insect population (Keller and Zimmerman, 1980). Many species of hypocreales inhabit the soil, of these, *Beauveria* spp. *Metarhizium anisopliae* and *Paecilomyces* spp. are especially common (Keller and Zimmerman, 1989). Entomopathogenic fungi contribute in lowering the arthropod population, whether due to naturally occurring conidia that persists in soil or application of EPF in high value crops (Hajek, 2004). Higher EPF detection occurred in soil on organic compared with conventionally managed farms (Klingen *et al.*, 2002) and on farms relying on biological, rather than chemical inputs (Hummel *et al.*, 2002) suggesting conservation of EPF in the absence of pesticides and synthetic fertilizers (Jabbour and Barbercheck, 2009). Higher detection of EPF has been shown in conservation tillage system compared to conventional tillage system compared to conventional tillage (Hummel *et al.*, 2002), although response may differ by fungal taxa (Sosa-Gomez *et al.*, 2001). Similar detection of *B. bassiana* occurred in no till and conventionally tilled soybean systems (Bing and Lewis, 1993); however, *Metarhizium* and *Isaria* were conserved in no-till soybeans as compared to conventionally tilled systems (Sosa–Gomez *et al.*, 2001). Effect of cover crops on EPF have not been explored although the use and type of cover crops have been shown to affect the abundance and community composition of other soil invertebrate and microbes (Makamoto and Tsukamoto, 2006 and Peachey *et al.*, 2002). Furthermore, laboratory experiments have shown decreased survival of *M. anisopliae* (Li and Holdom, 1993) and *B. bassiana* (Studdert *et al.*, 1990) conidia in wet soil and fallowing, addition of carbon and nitrogen sources (Lingg and Donaldson, 1981). However, full tillage systems with less crop residue may allow increased spread of fungal spores via rainfall (Bruck and Lewis, 2002) or could provide habitat for arthropods. Arthropods serves as host or vector to promote EPF population, but can also limit entomopathogenic fungi by competing for hosts and feeding on spores (Broza *et al.*, 2001and Roy and Pell, 2000). Collembolans can serve as vectors of EPF, transmitting sufficient spores to cause infection of a host (Dromph, 2003).

Mode of Action

Among the biocontrol agents, parasitic fungi can penetrate directly their target through the gut, spiracles and also through the surface of the integument without getting much affected by the adverse climatic conditions. Mycoparasitic fungi are particularly well suited since they share the same lifestyle and in many instances even the same habitat of their hosts (Adams, 1990 and Chet *et al.*, 1997). Infection of insects does not require any specialized mode of entry and begins with attachment of fungal spores to the target host. In response to cuticle surface clues, the fungus germination, and the emerging germ tubes produce a variety of enzymes that combine with mechanical pressure and begin the process of cuticle penetration (Fan *et al.*, 2011). Bioassays with entoronopathogenic fungi can be used to determine and quantify host pathogen relationship and to assess the effect of biotic and abiotic factors in insect parasitism (Asensio *et al.*, 2005). These studies include (i) determination of virulence, (ii) Comparison of strain virulence, (iii) determination of host range, (iv) determination of epizootic potential, (v) study of effects of biotic and abiotic factors (host age, host plant temperature, humidity and formulation) on parasitism, and study of the mode of action of entomopathogenic fungi on their insect host (Asensio *et al.*, 2005). Fungi do not exploit all its possible ways of infection against a given insect because of the barriers to infection (wart layer, cuticle etc.), differential production of enzymes, toxins and insects life. For instance, extracellular proteases production by *Metarhizium anisopliae* had immobilized over 70 per cent of soluble enzymes activity which would influence the extent and nature of cuticle degradation (St. Leger *et al.*, 1986).

Toxins

During the infection process fungi produce large array of toxic metabolites that can target different internal organs in insects (Clarkson and Charnley, 1996). Destruxins are one of the most potent toxins which can be synthesized by different species of entomopathogenic fungi such *Metarhizium anisopliae, Lecanicillium langisporum,* and *Aschersonia* sp. particularly, in *M. anisopliae,* the production of destruxins has been found to be an important virulence factor that accelerates the death of infected insects (Kershawa *et al.*, 1999). Structurally, destruxins are cyclic hexadepsipeptides composed of an α- hydroxy acid and five amino acid residues have been isolated (Pedras *et al.*, 2002). Insecticidal properties are exhibited by destruxins when applied to the external surface of the cuticle, ingested, injected or by any combination of these administered to *Spodoptera litura* (Sawjayna *et al.*, 2008), *Bemisia tabaci* nympns (Hu *et al.*, 2009), *Empoasca vitis* nymphs (Poprawski *et al.*, 1994) and *Agrotis segetum* and *Pieris brassicae* early larval stage (Thomsen and Eilenberg, 2000). Destruxins also target some internal organs like epithetial cells of malpighian tubules and gut tissues of *Galleria mellonella*, pycnosis of the nucleus and changes to mitochondria shape and density (Dumas *et al.*, 1996). Furthermore, this toxin induces oxidative stress, cause detachment and damage of microvilli, epithelial cell vacuolization and disruption of the epithelial cell membranes (Sowjanya and Padmaja, 2008). Ruiz-Sanchez *et al.* (2010) reported that the target sites for this peptide toxin might be associated with inhibition of the V-type H^+ ATPase of malpighian tubule cells of *Rhodnius prolixus*. Likewise, beauvericin is produced by *Beauveria bassiana, Paecilomyces fumosoroseus* (Hamil *et al.*, 1969) and also by *Fusarium subglutinans* and *Fusarium semitectum* (Gupta *et al.*, 1991), a toxic hexadepsipeptide (Hamil *et al.*, 1969), triggers the activation of the phenoloxidase system and then the production of melanin (Ganassis *et al.*, 2002) leads to death in insects such as mosquito larvae, blowflies, and Colorado potato beetle (Grove and Pople, 1980 and Gupta *et al.*, 1991). However, bassiacridin produced by *B. bassiana* is a monomer with low molecular weight of 60KDa exhibited β-glucosidase, β-galactosidase and N-acetyl glucosaminidase activities, and the insecticidal protein showed a specific activity toward locusts (Moraga and Vey, 2004).

Enzymes

Entomopathogenic fungi infect insects primarily by breaking the cuticular barrier. The protected barrier of insect cuticle is comprised of protein and chitin and there is strong agreement between the chitin embedded in protein fibrils. Moreover, cuticle contains more protein than the chitin (75-80 per cent). Hence, fungi overcome this constrain by producing a variety of extracellular enzymes involved in the degradation of protein, chitin and lipids (Charnley, 1984; St. Leger *et al.*, 1988; Khachatourians, 1991 and St. Leger, 1993). Extracellular enzymes whether of mycelial or spore origin are the hallmark of infection process. The successful infection relies upon efficient release of exoenzymes which have potential to degrade insect cuticle (Khachatourians, 1996). Among these, proteases are important, as they are among the first to appear during the infection process (St. Leger *et al.*, 1986 and St. Leger, 1995). Moreover, proteolytic enzymes (proteinases and peptidases) are thought to attach the cuticle before chitinolytic enzymes (chitinases and β-N acetyl- glucosaminidase) as protein masks the chitin microfibers (Fukamizo and Kramer, 1985 and Smith *et al.*, 1981). Three major proteinases have so for been purified and characterized from the culture filterates of *M. anisoliae*, Prl, Pr2 (St. Leger *et al.*, 1987), and Pr4 (Cole *et al.*, 1993). Pr1 is a chemoelastase with an active site serine residue which hydrolyses range of substrate, Pr 2 is a serine proteinase with an active site arginine and lysine residues having ability to degrade casein and albumin and Pr4 is a cysteine proteinase exhibited trypsin- like specificities (Samuels and Paterson, 1995) and was also more effective in hydrolyzing insect cuticle than Pr2. Proteinases with similar properties to both Pr1 and Pr2 have also been isolated from culture filterates of the entomopathogenic Deuteromycetes, *B. bassiana, M. anisopliae, Verticillium lecanni, Nomuraea rileyi* and *Aschersonia aleyrodris* (St. Leger *et al.*, 1987). In contrast to the five species of Deutromycetes described by St. Leger *et al.* (1987), in which Pr1 and Pr2 like activities are found as separate enzymes, the entomopathogenic Entomophthoraceae, *E. rhizospora, E. dipterigena and E. neoaphidis*, produce single endoprotease with activity against tryptic and chemotryptic substrates (Samuels *et al.*, 1990).The regulation of production of Pr1 and Pr2 is controlled by carbon and nitrogen (Paterson *et al.*, 1994). Under carbon and nitrogen starvation condition, Pr2 is induced by a range of proteinaceaus substrates; whereas Pr1 is only induced by insect cuticle (Paterson *et al.*, 1994) and specifically by cuticular protein (Paterson *et al.*, 1994). Pr4 is not produce under same condition as either Pr1 or Pr2, Pr4 are produced during the infection process (Samules and Paterson, 1995).

Biological Control

Entomopathogenic fungi can be used as component of integrated pest management (IPM) of many insect pests. Often, these pathogens cause natural mortality in insect population due to various differential traits associated with them and attacking host. After the successful use of *Neozygite fresenii* against *Aphis gossypii*, the classical biological control gained the importance. However, inundative applications of entomopathogens as biopesticide have surpassed the inoculative introductions due to various regulatory regulations associated with the former.

Classical Biological Control

The first programme of classical biological control that brought widespread attention to the great potential success possible with this method after targeted control of the cottony cushion scale (*Icerya purchasi* Maskell) insect. This pest was introduced to California around 1868 and, by 1886, the new and growing California citrus industry was being decimated by damage caused by this scale (DeBach, 1974). The successful introduction of vedalia beetle, *Rodolia cardinalis* (Mulsant), a coccinellid from Australia in 1889 (Caltagirone and Doutt, 1989), which marked the initiation of extensive use of introduced arthropod predators and parasitoids to control arthropod pests and phytophagous insects and plant pathogens to control weed (Clause, 1978 ; Greathead and Greathead, 1992 and Julien and Griffiths, 1998). Hajek *et al.* (2007) prepared the criteria for including programs (refers as release of one species of agent in one geographic area) in the catalogue (refers as summary of documented introductions of pathogens and nematodes for classical biological control) and evaluating their success as discussed in the catalogue. The used criteria for including programs in the catalogue and its analysis are given below:

1 The target pest was an insect or mite
2. The species, strain or biotype of the microbial pathogen or nematode that was introduced was exotic (non–native) to the area of introduction.
3. The intent of the program was to establish the pathogen in the release area, hopefully resulting is long term (and not temporary) control.

The five most frequently species used in classical biocontrol programs were the *O. rhinoceros* virus, *Metarhizium anisopliae, E. maimaiga, R. culicivorax* (nematode) and *Deladenus siricidicola*. Entomophthorales possess good attributes as classical biological control agent to Ascomycetes as they are host specific, have ability to cause epizootics and form specialized resting spores for persistence. They are not preferred as they fail to establish in introduced area and not easily mass produced (Hajek *et al.*, 2007). Major constraint is variation in environmental conditions over the different geographical locations. Denoth *et al.* (2002) suggested that in the past, multiple agents have been released not to increase the cumulative impact on target host, rather to increase the likehood so that the right control species can be released. Such reasoning was replaced by International standards established by FAO/IPPC (1996) that the importation of entomopathogens for small scale (?4ha) experimental purposes should address the concerns of the environmental community. Regulatory restrictions on introduction of exotic pathogens have nearly eliminated the classical biological control.

Entomopathogenic Fungi as Biopesticides

Biodiversity in agro-ecosystem and global interdependence of markets for agriculture produce have brought to forefront the need to develop the practices that mitigate adverse effects on the environment and that result in products that are safe for human consumption (Vega *et al.*, 2009). Despite being an estimated 700 species belonging to 90 genera (Roberts and Humber, 1981), only 12 species have been used for development of the commercially produce fungi are species of *Beauveria, Lecanicillium* and *Isaria* that are relatively easy to mass produce (Vega *et al.*, 2009). The entomopathogenic fungi presents the wide range of biologies, varies from obligate parasitism to opportunistic pathogens. The life cycles of obligate parasites, such as those species in the genus Ceolomycetes, may have ability to involve and include intermediate hosts (Couch and Bland, 1985). The fungi Imperfecti (Deuteromycotina: Hyphomeycetes) on the other hand, have simpler life cycles and lack sexual reproduction, but many have considerably broader insect host range (Lacey *et al.*,

2001). Entomophthorales are the major group of entomopathogenic fungi which are responsible for epizootics, but are not preferred for commercial production as they are difficult to produce and their primary conidia are short lived. Entomopathogenic Hyphomycetes are the main contenders for commercial production, and use against a broad range of insect pests, including whiteflies, aphids, thrips, termites, grasshoppers and locusts, beetles, and others (Keller *et al.*, 1997, Milner, 1997 and Goettel *et al.*, 1995). Furthermore, development of epizootics caused by entomopathogenic fungi depends on interaction processes, both environmental and biotic. These include sensitivity to solar radiation, microbial antagonists, host behavior, physiological condition and age, pathogen vigour and age, presence of pesticides, and appropriate temperature, humidity and inoculums thresholds (Lacey *et al.*, 2001). Commercial products of entomopathogenic fungi *viz.*, *Metarhizium anisopliae* (Metschnikoff) Sorokin, *Beauveria bassiana* (Balasomo) Vuillemin, *Verticillium lecanii* (Zimmermann) Viegas and *Paecilomyces fumosoroseus* (Wize) Brown and Smith are currently in use; their success depends on the use of virulent propagules, susceptible host range, favorable environmental conditions and formulation technique. Further improvement in the microbial control activity of entomopathogenic fungi can be expected by their combination with other inventions and technologies, and use of other biological control agents, environmental manipulation to favour the infection process, and use of targeted pests to aid in the dissemination of fungus (Lacey *et al.*, 2001). Entomopathogenic fungi effective against phylloplane insect pests should have discrete, infective propagules provided by spore forms to satisfy the requirement for complete coverage of the foliar surface to ensure contact and infection of the host (Vega *et al.*, 2009). In rhizosphere, the rihzosphere competence of fungal strain is dependent on the plant community and not necessarily the presence of an insect host (Hu and St. Leger, 2002).

Mass Production and Formulation

Effective non- chemical control measures are needed for both pest control to reduce harmful side effects on public health and the environment. Several biocontrol agents such as predators, parasitic Hymenoptera, and microbial bioagents are currently being considered for use against these pests (Batta, 2003). There is a particular interest in microbial control agents especially entomopathogenic fungi which include *Verticillum lecanii, Paecilomyes fumosoroseus, Aschersonia aleyrodis, Paecilomyces farinosus* and *Beauveria bassiana* (Fransen, 1992; Osborne and Landa, 1992; Lacey *et al.*, 1996 and Lacey *et al.*, 1999). The entomopthogenic fungi *Metarhizium anisopliae* has been reported to infect more than one hundred species of insects belonging to variety of insect orders (McCoy *et al.*, 1988 and Zimmermann, 1993). However, use of the fungus to actively control pest infestations has been tested against relatively few insect species those reportedly controlled by the pathogen include wheat grain beetles, termites, black field cricket, wax moth cockroaches, locusts, grasshopper and western flower thrips (Kucera, 1980; Ferson 1981, Prior and Greathead, 1989, Rath, 1992; Farrow *et al.*, 1993; Quarles, 1995; Milner *et al.*, 1996; Lomer and Kooyam 1997and Ludwing and Oething, 1998). A considerable number of mycoinsecticides have gradually reached the market place and millions of hectares are treated annually with entomopathogenic fungi worldwide (Faria and Wraight,

2007). Development of simple and reliable production system follows the basic multiplication procedure of blastospore, which are short lived, and hydrophilic (Romback, 1989) or solid state fermentation (Rousson *et al.*, 1983) for production of aerial conidia. However, the most viable mass production technologies include making use of a diaphasic strategy in which the fungal inoculum is produced in liquid culture, which is further utilized for inoculating the solid substrate(s) for conidia production (Burges and Hussey, 1981). The most suitable liquid medium tested for higher spore production of *B. bassiana, P. fumosoroseus, V. lecanni* and *M. anisopliae* was coconut water (Sahayaraj and Namasivayam, 2008 and Danger *et al.*, 1991). Among the solid substrate, rice grain provides the suitable substrate for mass multiplication of *B. bassiana* (Sharma *et al.*, 2002) and, sorghum for *Paecilomyces farinosus and V. lecanii* (Gopalkrishnan *et al.*, 1999 and Lakshmi *et al.*, 2001).

Despite recent advances, use of mycopesticides is proportionally limited when compared to their counterparts, even in countries where many mycopesticides are produced. The growing adoption of these microbial bioagent depends on factors such as (1) development of better products, (Faria *et al.*, 2010) (2) development and implementation of truly integrated pest management strategies in which biological options are emphasize (Lomer *et al.*, 1999; Thomas,1999 and Lacey *et al.*, 2001), (3) the capacity of biopesticide manufactures/ retailers to maintain marketing and product support teams, (4) cultural changes (acceptance by farmers of slow acting, narrow host range products), and (5) sound knowledge-based recommendation for product use. However, better product development includes increased concentrations of active ingredients, more predictable shelf life under non-refrigerated conditions (Hong *et al.*, 1997and Waright *et al.*, 2001), improved shipping and handling characteristics, greater UV tolerance (Inglis *et al.*, 2001),and ultimately greater efficacy and reliability under field conditions (Lacey *et al.*, 2001). Besides, physical properties of the fungus formulations, especially viscosity and stability of the emulsion, remain constant over time (Batta, 2003).

Scope and Challenges

Shrinking agricultural land, ever changing climate, threat to emergence of new pest species are raising a big question mark on eco-friendly production of agri-produce. Hence, future of sustainable farming and protecting the ecosystem in India depends on scientific diligence of researchers to develop and introduce a sound IPM-EPF based alternative. Several governments are taking political decision in the favour of sustainable agriculture which will require rebuilt of present pest management strategy. Therefore, the global demand for new biocontrol agents, their mass production and other related technologies will have to take the momentum. What need to be done for the coming generation has to start now. Void of phases-wise removal of chemicals can filled by efficient entomopathogenic fungi. Era of IPM-EPF based alternative will require the following advances; improvement in pathogens, their production, and formulation; better understanding of how they will fit into the integrated system and their interaction with the environment and other IPM components; greater appreciation for their full advantages (efficacy, safety, selectivity etc.), not simply their comparison with chemical pesticides; and acceptance by grower and the general public. However, molecular characterization of several properties including growth, sporulation, virulence and environmental stress response for fungus remains the major challenge. The current knowledge of applied mycology and biotechnology could be used for improved understanding of the complex traits and properties of the EPF and use of such knowledge in their production formulation, and hence performance.

References

Adams, P.B. (1990). The potential of mycoparasites for biological control of plant diseases. *Annual Review of Phytopathology*. **28:** 59-72.

Altieri, M.A. (1999). The ecological role of biodiversity in agroecosystems. *Agriculture Ecosystem and Environment*. **74:**19-31.

Anderson, M., Magan, N., Meed, A. and Chandler, D. (2006). Development of population based threshold model of conidial germination for analyzing the effect of physiological manipulation of the stress tolerance and infectivity of insect pathogenic fungi. *Environmental Microbiology*. **8:** 1625- 1634.

Anderson, R.M. and May, R.M. (1981). the population dynamics of microparasites and their invertebrate host. *Philosophical Transactions of the Royal Society B*. **291:** 451-524.

Asensio, L, Lopez-Llorca, L.V. and Lopez–Jimenez, J.A. (2005). Use of light scanning electron microscopy and bioassays to evaluate parasitism by entomopathogenic fungi of the red scale insect of palms (*Phoenicococcus marlatti* Ckll., 1899). *Micron* **36:** 169-175.

Barson, G., Renn, N. and Bywater, A.F. (1994). Laboratory evaluation of fine species of entomopathogenic fungi for the control of the house fly (*Musca domestica* L.), a pest of animal units. *Journal of Invertebrate Pathology*. **64:** 107-113.

Batta, Y.A. (2003). Production and testing of novel formulations of the entomopathogenic fungus *Metarhizium anisopliae* (Metschinkoff) Sorokin (Deuteromycotina; Hyphomycetes). *Crop Protection*. **22 :** 415-422.

Bing, L.A. and Lewis, L.C. (1993). Occurrence of entomopathogenic *Beauveria bassiana* (Balsamo) Vuillemin in different tillage regimes and in *Zea mays* L. and virulence towards *Ostrinia nubilalis* (Hubres). *Agriculture Ecosystems and Environment*. **45:** 147-156.

Broza, M., Pereira, R.M. and Stimae, J.H. (2001). The non-susceptibility of soil collembole to insect pathogens and their potential as scavengers of microbial pesticides. *Pedobiologia*. **45:** 523-534.

Bruck, D.J. and lewis, L.C. (2002). Rainfall and crop residue effect son soil dispersion and *Beauveria bassiana* spread to corn. *Applied soil Ecology*. **20:** 183-190.

Burges, A.D. and Hussey, N.W. (1981). Microbial Control of Insect Pests and Mite, Academic Press, London, pp. 161-167.

Burges, H.D. (1981). Strategy for the microbial control of pests in 1980 and beyond. In: Burges, H.D. (Ed.), Microbial Control of Pests and Plant Diseases 1970–1980. Academic Press, London, pp. 797–836.

Caltagirone, L.E. and Doutt, R.L. (1989). The history of the vedalia beetle importation to California and its impact on the development of biological control. *Annual Review of Entomology*. **34:** 1–16.

Charnley, A.K. (1984). Physiological aspects of destructive pathogenesis in insect by fungi: A speculative review. In: Symposium Series, British Mycological Society. Cambridge University Press, Cambridge. pp., 85-125.

Chet, I., Inbar, J. and Hadar, Y. (1997). Fungal antagonism and mycoparasites. In: Wicklow DT, Soderstrom B (eds.). The Mycota IV, Environmental and Microbial Relationships, 4. Springer, Berlin- Heidelberg. Pp., 165-184.

Clarkson, J.M. and Charnley, A.K. (1996). New insights into the mechanisms of fungal pathogenesis in insects. *Transactions Microbiol.* **4:** 197-203.

Clausen, C.P. (Ed.) (1978). Introduced Parasites and Predators of Arthropod Pests and Weeds: A World Review. USDA, Agriculture Handbook No. 480, Washington, DC.

Cole, S.C.J., Charnley, A.K. and Cooper, R.M. (1993). Purification and partial characterization of a novel trypsin-like cysteine protease from *Metarhizium anisopliae. FEMS Microbiology Letter.***113:**189-196.

Couch, J.N. and Bland, C.E. (1985). "The Genus *Coelomomyces.*" Academic Press, New York.

d'Enfert, C., Bonini, B.M., Lepella, P.D.A., Fontaine, T., da Silva, A.M. and Terenzi, H.F. (1999). Natural trehalase catalyzes intracellular trehalose breakdown in the filamentous fungi *Aspergillus nidulans* and *Neurospora crassa. Molecular Microbiology*. **32:** 471- 483.

Dangar, T.K., Geetha, L., Jayapal, S.D. and Pillai, G.B. (1999). Mass Production of the Entomopathogens *Metarhizium anisopliae* in Coconut Water. *Journal of Plantation Crop* **19:** 54-59.

DeBach, P. (1974). Biological Control by Natural Enemies. Cambridge University Press, London, UK.

Denoth, M., Frid, L. and Myers, H. (2002). Multiple agents in biological control: improving the odds? *Biol. Control* **24:** 20–30.

Doberski, J.W. (1981). Comparative laboratory studies on three fungal pathogens of the elm bark beetle, *Scloytus scolytus*: effect of temperature and humidity on infection of *Beauveria bassiana, Metarhizium anisopliae* and *Paecilomyces farinosus. Journal of Invertebrate Pathology*. **37:**195-200.

Dromph, K.M. (2003). Collembolans as vectors of entomopathogenic fungi. *Pedobiologia*. **47:** 245-256.

Dumas, C., Ravallec, M., Matha, V. and Vey, A. (1996). Comparative study of the cytological aspect of the mode of action of destruxins and other peptidic fungal metabolite on target epithelial cells. *Journal of Invertebrate Pathology*. **67:** 137-146.

Elbein, A.D., Pan, Y.T., Pastuszak, I. and Carroll, D. (2003). New insights on trehalose; a multifunctional molecule. *Glycobiology*. **13:** 17R- 27R.

Evans, H.C. (1982). Entomogenous fungi in tropical forest ecosystems: an appraisal. *Ecological Entomology.* **7:** 47-60.

Fan, Y., Zhang, S., Kruer, N. and Keyhani, N.O. (2011). high-throughput insertion mutagenesis and functional screening in the entomopathogenic fungus *Beauveria bassiana. Journal of Invertebrate Pathology.* **106:** 274-279.

FAO/IPPC (Food and Agriculture Organization, International Plant Protection Convention). (1996). Code of Conduct for the Import and Release of Exotic Biological Control Agents. ISPM No.3, IPPC Secretariat, FAO, Rome, Italy.

Faria, M. and Wraight, S.P. (2007). Mycoinsecticides and Mycoacaricides; a comprehensive list with worldwide coverage and international classification of formulation types. *Biological Control.* **51:** 346-354.

Faria, M., Hotchkiss, J.H., Hajek, A.E. and Waright, S.P. (2010). Debilitation in conidia of the entomopathogeinc fungi *Beauveria bassiana* and *Metarhizium anisopliae* and implication with respect to viability determinations and mycopesticide quality assessments. *Journal of Invertebrate Pathology.* **105:** 74-83.

Farrow, R.A., Mc Donald, C.S. and Grandstahle, P.D. (1993). Potential impact of rapid climate change through the green house effect on pests of pastures in south eastern Australia. In: Delfosse ES (ed.), Pests of pastures. CSIR, Melbourne. pp., 142-151.

Ferron, P., Fargues, J. and Riba, D. (1991). Fungi as microbial insecticides against pests. In: Arora DK, Mukherji KG and Drouhet E. (eds.), Handbook of applied mycology; humans, animal and insects, Vol.2, Marcel Dekker, New York, pp., 665-705.

Ferson, P. (1981). Pest control by the fungi: *Beauveria* and *Metarhizium.* In: Burges HD (ed.), Microbiol control of pests and Plant diseases. Academic Press, New York. pp., 465-482.

Fransen, J.J. (1992). Natural enemies: Fungi. In: Gerling D. (ed.), Whiteflies–Their Bionomics, Pest status and Management. Intercept Ltd., Andover UK. pp., 186-210.

Freimoser, F.M., Hu, G. and St. Leger, R.J. (2005). Variation in gene expression patterns as the insect pathogens *Metahizium anisopliae* adapts to different host cuticles or nutrient deprivation in- vitro. *Microbiology.* **151:** 361-371.

Fukamizo, T. and Kramer, K.J. (1985). mechnization of chitin hydrolysis by the binary chitinase system in insect molting fluid. *Insect Biochemistry.* **15:** 141-145.

Ganassi, S., Moretti, A., Pagliai, A.M.B., Logrieco, A. and Sabatini, M.A. (2002). Effect of beauvericin on *Schizaphis graminum*(Aphidiae). *Journal of Invertebrate Pathology.***80:** 90-96.

Goettel, M.S., Inglis, G.D. and Wraight, S.P. (2000). Special considerations for the application and evaluation of pathogens fungi. In: Lacey LA, Kaya HK (eds.), Field Manual of Technologies in Invertebrate Pathology. Kluwer Academic Press, Netherlands, pp. 255-585.

Goettlel, M.S., St. Leger, R.J., Rizzo, N.W., Staples, R.C. and Roberts, D.W. (1989). Ultrastructural localization of a cuticle degrading protease product by the entomopathogenic fungus during penetration of host (*Manduca sexta*) cuticle. *Journal of General Microbiology.* **135:** 2233-2239.

Gopalakrishnan, C., Anusuya, D. and Narayanan, K. (1999). Occurrence of entomopathogenic fungi *Paecilomyces farinosus* (Holmskiold) Brown and Smith and *Zoophthora radicans Plutella xylostella* L. on cabbage. *Entomon* **24:** 363–369.

Greathead, D.J. and Greathead, A.H. (1992). Biological control of insect pestsby insect parasitoids and predators: The BIOCAT database. Biocontrol News Inf. 13. 61N–68N.

Grove, S. and Pople, M. (1980). the insecticidal activity of beauvericin and the enniantin complex. *Mycopathologia.***70:**103-105.

Gupta, S.C., Leathers, T.D., El-Sayed, G.N. and Ignoffo, C.M. (1991). Production of degradative enzymes by *Metarhizium anisopliae* during growth on defined media and insect cuticle. *Experimental Microbiology.* **15:**310-315.

Hajek, A.E., Michael, L.M. and Italo, D.J. (2007). A review of introductions of pathogens and nematodes for classical biological control of insects and mites. *Biological Control.* **41:** 1–13.

Hajek, A.E. (1997). Ecology of terrestrial fungal entomopathogens. *Advances in Microbial Ecology.* **15:**193-249.

Hajek, A.E. (2004). Natural enemies an introduction to biological control. Cambridge University Press, Cambridge, UK, New York.

Hallsworth, J.E. and Magan, N. (1994). Effect of carbohydrate type and concentration of palyhydroxy alcohol and trehalose content of conidia of three entomopathogenic fungi. *Microbiology.* **140:** 2705-2713.

Hummel, R.L., Walgenbach, J.F., Barbercheck, M.E., Kennedy, G.G., Hoyt, G.D. and Arellano, C. (2002). Effects of production practices on soil-borne entomopathogens in western North Carolina vegetable systems. *Environmental Entomology.* **31:** 84–91.

Hamill, R.I., Higgens, C.E., Boza, H.E. and Gorman, M. (1969). The structure of beauvericin, a new depsipeptide antibiotic toxic to *Artemia salina. Tetrahedron Letter.* **49:** 4255-4258.

Harper, J.D. (1987). Applied epizootiology: Microbial control of insects.*In* "Epizootiology of Insect Diseases" (J. R. Fuxa and Y.Tanada, Eds.), pp. 473–496. Wiley, New York.

Hong, T.D., Ellis, R.H. and Moore, D. (1997). Development of model to predict the effect of temperature and moisture on fungal spore on longevity. *Annals of Botany.* **79:** 121-128.

Hong, T.D., Ellis, R.H., Gunn, J. and Moore, D. (2002). Relative humidity temperature and the equilibrium moisture content of conidia of *Beauveria bassiana* (Balsamo) Vuillenun; a quantitative approach. *Journal of Stored Produces Research.* **38:** 33-41.

Hong, T.H., Ellis, R.H. and Moore, D. (1998). Limits to the negative logarithmic relationship between moisture content and longevity in conidia of *Metarhizium flavoviride. Annals of Botany.* **81:** 625-630.

Hong, T.H., Ellis, R.H. and Moore, D. (1999). Fluctuating temperature and the longevity of conidia of *Metarihizium flavoviride* in storage. *Biocontrol Science and Technology.* **9:**165-176.

Hu, G. and St.Leger, J. (2002). field studies using a recombinant mycoinsecticide (*Metarhizium anisopliae*) reveal that it is rhizosphere competent. *Applied Environmental Microbiology.* **68:** 6383-6387.

Hu, Q.B., An, X.C., Jin, F.L., Freed, S. and Ren, S.X. (2009). Toxicities of destruxins against *Bemicia tabaci* and its natural enemies, *Serangium japonicum. Toxicon.* **53:**115-121.

Inglis, G.D., Goethel, M.S., Butt, T.M. and Strasser, H. (2001). Use of hyphomycetous fungi for managing insect pest. In: Butt TM, Jackson C, Magan N (eds.), Fungi as biocontrol agent: Progress Problems and Potential. CABI Publishing pp. 23-69.

Jabbour, R. and Barbercheck, M.E. (2009). Soil management effects on entomopathogenic fungi during the transition to organic agriculture in a feed grain rotation. *Biological Control.* **51:** 435-443.

Jackson, T.A., Alves, S.B. and Pereira, R.M. (2000). Success in biological control of soil-dwelling insects by pathogens and nematodes.In: Gurr G, Wratten S (eds), Biological Control: measures of success. Kluwer Academic Press, London, pp. 271–296.

Julien, M.H. and Griffiths, M.W. (eds.). (1998). Biological Control of Weeds–A World Catalogue of Agents and their Target Weeds. CABI Publisher, Wallingford Oxon, UK.

Keller, S., Schweizer, C., Keller, E. and Brenner, H. (1997). Control of white grubs (*Melolontha melolontha* L) by treating adults with the fungus *Beauveria brongniartii. Biocontrol Science Technology.* **7:** 105–116.

Keller, S. and Zimmerman, G. (1989). Mycopathogens of soil insects. In: Wilding N, Collins NM, Hammond PM and Webber JF (eds.), Insect-funus interactions. Academic Press, London, UK. pp., 240-270.

Kershawa, M.J., Moorhousea, E.R., Batemand, R., Reynoldsa, S.E. and Charnley, A.K. (1999). The role of destruxins is the pathogenicity of *Metarhizium anisopliae* for three species of insect. *Journal of Invertebrate Pathology.***74:** 213-223.

Khachatourians, G.G. (1991). physilogy and genetics of entomopathogenic fungi. In: Arora DK, Ajello L and Mukerji KG (eds.), Handbook of Applied Mycology, Humans, Animals and Insects, Vol.2, Marcel Dekkar, New York.pp.,613-661.

Khachatourians, G.G. (1996). Biochemistry and molecular biology of entomopathogenic fungi. In: Howard DH and Milner JD (eds.). The Mycota VI Human and animal relationships. Springer-Verlas. Berlin Heidelberg, Germany. pp., 331-363.

Klingen, I., Eilenberg, J. and Meadow, R. (2002). Effects of farming systems, field margins and bait insects on the occurrence of insect pathogenic fungi in soils. *Agriculture, Ecosystem and Environment*. **91:** 191-198.

Krasnoff, S.B. and Gibson, D.M. (1996). New destruxins from the entomopathogenic fungus *Aschersonia* sp. *Journal Nat. Prod.* **59:** 485-489.

Kucera, M. (1980). Proteases from the fungus *Metarhizium anisopliae* toxic for *Galleria melonella* larvae. *Journal of Invertebrate Pathology*.**35:** 304-310.

Lacey, L.A., Fransen, J.J. and Carruthers, R. (1996). Global distribution of naturally occurring fungi of *Bemisia*, their biology and use as biological control agents. In: Gerling D and Mayer R (eds.), *Bemisia* 1995; Taxonomy, Biology, Damage, and Management. Intercept Ltd., Andover UK. pp., 401-433.

Lacey, L.A., Fratos, R., Kaya, H.K. and Vail, P. (2001). Insect pathogens as biological control agents: Do they have a future. *Biological Control.* **21:** 230-248

Lacey, L.A., Krik, A.A., Millar, L., Mercandiers, G. and Vidal, C. (1999). Ovicidal and larvicidal activity of conidia and blastospores of *Paecilomyces fumorsoroseus* (Ductromycotina: Hyphomycetes) against *Bemisia argentifolii* (Homoptera: Aleyrodiadae) with a description of bioassay system allowing prolonged survival of control insects. *Biocontrol Science Technology.* **9:** 9-18

Lakshmi, S.M., Alagammai, P.L. and Jayaraj, K. (2001). Studies on Mass Culturing of the Entomopathogen Whitehalo Fungus *Verticillium lecanii* on Three Grain Media and Its Inefficacy on *Helicoverpa armigera*, In: Ignachimuthu S, Sen S (eds.), Microbials In Insect Pest Management, Oxford and IBH publishing Pvt. Ltd., New Delhi, pp. 23-27.

Lee, S.Y., Nakaiima, I. Ihara, F., Kinoshita, H. and Nihira, T. (2005). Cultivation of entomopathogenic fungi for the search of antimicrobial compounds. *Mycopathologia.* **160:** 321-325.

Li, D.P. and Holdom, D.G. (1993). Effect of soil metric potential on sporulation and conidial survival of *Metarhizium anisopliae* (Deutomycotina: Hyphomycetes). *Journal of Invertebrate Pathology*. **62:** 273-277.

Lingg, A.J. and Donaldson, M.D. (1981). Biotic and abiotic factors affecting stability of *Beauveria bassiana* conidia is soil. *Journal of Invertebrate Pathology*. **38:**191-200.

Lomer,G.J., Prior, C. and Kooyman, C. (1997). Development of *Metarhizium* spp. for the control of grasshoppers and locusts. In: Goettlel MS and Johnson DL (eds.), Microbial control of Grasshoppers and locusts. *Memoirs of the Entomological Society of Canada.* **171:** 265-286.

Lomer, C.J., Bateman, R.P., Dent, D., De-groote, H., Douro-Kpindou, O.K., Kooyman, C., Langewald, J., Ouambama, Z., Peveling, R. and Thomas, M. (1999). Development of strategies for the incorporation of biological pesticides into the integrated management of locusts andgrasshoppers. *Agric. Bur. Entomol.* **1:** 71–88.

Ludwing, S.W. and Oetting, R.D. (1998). Control of thrips (*Frankliniella occidentalis*) and the fungus (*Bradysia* spp.) in potting media. *SNA Research Conference Proceeding*. **43:** 166-167.

Magan, N. (2001). Physiological approaches to improving the ecological fitness of fungal biocontrol agents. In: Butt TM, Jackson CW and Magan N(eds.). Fungi as Biocontrol Agents. CABI Publishing, Wallingford. pp., 239-251.

Magdoff, F. (2001). Concept, component and strategies of soil health in agroecosystems. *Journal of Nematology*. **33:** 169-172.

Makanmoto, T. and Tsukamoto, M. (2006). Abundance and activity of soil organisms in fields of maize grown with a white clover living mulch. *Agriculture, Ecosystem and Environment*. **115:** 34-42.

McClatchie, G., Moore, D., Bateman, R.P. and Prior, C. (1994). Effect of temperature on the viability of the conidia of *Metarhizium flavoviridae* in oil formulations. *Mycological Research*. **98:** 749-756.

McCoy, C.W., Samson, R.A. and Boucias, D.G. (1988). Entomogenous fungi. In: Ignoffo CM and Mandava NB (eds.), Handbook of natural pesticides; Microbial Insectidides, Vol.V, Part A. CRC Press, Inc., Boca Roton, Fl. Pp.,151-236.

Metewele, P., Trinci, A.P.J. and Gillespies, A.T. (1994). Mutants of entomopathogenic fungi that germinate and grow at reduced water activity and reduced relative humidities are more virulent to *Nephotettix virescens* (green hopper) than the potential strains. *Mycological Research*. **98:** 1329-1333.

Meyling, N. and Pell, J.K. (2006). Detection and avoidance of an entomopathogenic fungus by a generalist insect predator. *Ecological Entomology*. **31:**162-171.

Milner, R.J., Miller, L., Lutton, G.G. and Driver, F. (1996). Biological control of black field cricket *Teleogryllus commodus* Walker (Orthoptera: Gryllidae) using fungal pathogen *Metarhizium anisopliae* (Metsch.) Sorokin (Deuteromycotina: Hyphomyceters). *Plant Protection*. **11:** 9-13.

Milner, R.J. (1997). Prospects for biopesticides for aphid control. *Entomophaga* **42:** 227–239.

Moraga, Q. and Vey, A. (2004). Bassiacridin, a protein toxic for locust secreted by the entomopathogenic fungus *Beauveria bassiana*. Mycological Research.441-452.

Osborne, L.S. and Landa, Z. (1992). Biological control of whiteflies with entomophagus fungi. *Florida Entomology*. **75:** 456-471.

Paterson, I.C., Charnley, A.K., Cooper, R.M. and Clarkson, J.M. (1994). Specific induction of the cuticle- degrading protease of the insect pathogenic fungus *Metarhizium anisopliae*. *Microbiology*. **140:**185-189.

Peachey, R.E., Moldenke, A., William, R.D., Berry, R., Ingham, E. and Groth, E. (2002). Effect of cover crops and tillage system on *Symphylan* (Symphlya: *Scutigerella immacualta,* Newport) and *Pergamasus equisquiliarum cenestrini* (Acari: Mesostigmata) population and other soil organisms in agricultural soils. *Applied Soil Ecology*. **21:** 59-70.

Pedras, M.S.C., Zaharia, L.I. and Ward, D.E. (2002). The destruxins; synthesis, biosynthesis biotransformation and biological activity. *Phytochemistry.* **59:** 579-596.

Poprawski, T.J., Robert, P.H. and Maniania, N.K. (1994). Contact toxicity of the mycotoxin destruxin E to *Empoasca vitis* (Go¨ the) (Hom., Cicadellidae). Zeitschrift Fur Angewandte Entomologie-*Journal of Applied Entomology.* **117:** 135–14.

Prior, C. and Greathead, D.J. (1989). Biological control of locust: The potential for the exploitation of pathogens. *Plant Protection Bulletin.* **37:** 37-48.

Qazi, S.S. and Khachatourians, G.G. (2007). Hydrated conidia of *Metarhizium anisopliae* release a family of metalloproteases. *Journal of Invertebrate Pathology.* **95:** 48-59.

Quarles, W. (1995). New Technologies for termite control. *IPM Practitioner.* **17:** 1-9.

Rangel, D.E.N., Alston, D.G. and Roberts, D.W. (2008). Effect of physical and nutritional stress conditions during mycelia growth on conidial germination speed, adhesion to host cuticle and virulence of *Metarhizium anisopliae*, an entomopathogenic fungus. *Mycological Research.* **112:** 1355-1361.

Rangel, E.N., Barga, G.U.L., Anderosn, A.J. and Roberts, D.W. (2005). Influence of growth environment on tolerance to UV–B radiation, germination speed, and morphology of *Metarhizium anisopliae* var. *acridum* connida. *Journal of Invertebrate Pathology.* **90:** 55-58

Rath, A.C. (1992). *Metarhizium anisopliae* for control of Tasmania pasture scarab *Adoryphorus couloni.* In: Jackson TA and Glare TR (eds.), Use of pathogens in scarab pest management. Intercept Ltd., Andover, UK. pp., 217-228.

Roberts, D.W. and Humber, R.A. (1981). Entomogenous fungi. In: Cole GT and Kendrick B(eds.), Biology of conidial fungi.Academic Press, New York. Pp.,201-236.

Romback, M.C. (1989). Production of *Beauveria bassinaa* Conidia in Submerged Culture, *Entomophaga,* **5:** 45-52.

Rousson, S., Rainbautt, M., and Lonsane, B.K. (1983). Zymotics a Large Scale Fermenter Design and Evaluation, *Appl. Biochem. Biotechnol.,* **42:**161-167.

Roy, H.E. and Pell, J.K. (2000). Interaction between entomopathogenic fungi and other natural enemies; implication for biological control. *Biological Science and Technology.* **10:** 737-752.

Roy, H.E., Steinkraus, D.C., Eilenberg, J., Hajek, A.E., and Pell, J.K. (2006). Bizarre interactions and endgames : entomopathogenic fungi and their hosts. *Annual Review of Entomology.* **51:** 331-357.

Ruiz-Sanchez, E., Orchard, I. and Lange, A.B. (2010). Effect of the cyclopeptide mycotoxin destruxin A on the Malpighian tubules of *Rhodinus prolixa* (Stal). *Toxicon.* **55:**1162-1170.

Sahayaraj, K. and Namasivayam, S.K.R. (2008). Mass production of entomopathogenic fungi using agriculture product and by product. *African Journal of Biotechnology.***7:**1907-1911.

Samuel, R.I. and Paterson, I.C. (1995). Cuticle degrading proteases from insect moulting fluid and culture filterates of entomopathogenic fungi. *Compendium of Biochemical Physiology,* **110B:** 661-669.

Samuel, R.I., Charnley, A.K. and St.Leger, R.J. (1990). the partial characterization of endoproteases from three species of entomopathogenic Entomophthorales and two species of Deuteromycetes. *Mycopathologia.* **110:** 145-152.

Selman, B.J., Dayer, M.S. and Hasan, M. (1997). Pathogenicity of *Beauveria Bassiana* (Bals.) Vuill. to the larvae of diamondback moth *Plutella xylostella* L. (Lep., Yponomeutidae). *Journal of Applied Entomology.* **121:** 47-49.

Shah, F.A., Wang, C.S. and Butt, T.M. (2005). Nutrition influences growth and virulence of the insect pathogenic fungus *Metarhizium anisopliae.* FEMS *Microbiology Letters.* **251:** 259-266.

Shah, P.A. and Pell, J.K. (2003). Entomopathogenic fungi as biological control agents. Applied *Microbial Biotechnology.* **61:** 413-423.

Sharma, S.P., Gupta, R.B.L. and Yadava, C.P.S. (2002). Selection of a Suitable Medium for Mass Multiplication of Entomofungal pathogens, *Indian Journal of Entomology.* **14:** 255-261.

Sivasankaran, P., Eswaramoorthy, S. and David, H. (1998). Influence of temperature and relative humidity on the growth, sporulation and pathogenecity of *Beauveria* Bassiana. *Biological Control.* **12:** 71-75.

Smith, R.J., Perkrul, S. and Grula, E.A. (1981). requrirementr of sequential enzymatic activities for penetration of the integument of the corn earthworm (*Heliothis zea*). *Journal of Invertebrate Pathology.* **38:** 335-344.

Sosa-Gomez, D.R., Delpin, K.E., Moscardi, F. and Farias, J.R.B. (2001). Natural occurrence of the entomopathogenic fungi *Metarhizium, Beauveria* and *Paecilomyces* in soybean under till and no- till cultivation systems. *Neotrophical Entomology.* **30:** 407-410.

Sowjanya, S.K., Padmaja, V. and Murthy, Y.L. (2008). Insecticidal activity of destruxin, a mycotoxin from Metarhizium anisopliae (Hypocreales), against *Spodoptera litura* (Lepidoptera: Noctuidae) larval stages. *Pest Management Science.* **64:** 119–12.

Sowjanya, S.K., Padmaja, V. and Murthy, Y.L. (2008). Insecticidal activity of destruxin, a mycotoxin from *Metarhizium anisopliae* (Hypocreales), against *Spodoptera litura* (Lepidoptera: Noctuidae) larval stages. *Pest Management Science.* **64:** 119–125.

St. Leger, R.J. (1995). The role of cuticle-degrading proteases in fungal pathogenesis of insects. *Canadian Journal of Botany.* **73:** S1119–S1125.

St. Leger, R.J. and Screen, S. (2001). Prospect for strain improvement and fungal pathogens of insects weed. In: Butt TM, Jackson C and Magan N (eds.), Fungi as biocontrol agents: Progress Problems and Potential. CAB International, London, UK. pp., 219-237.

St. Leger, R.J., Charnley, A.K. and Cooper, R.M. (1986). Cuticle degrading enzymes of entomopathogenic fungi: Mechanisms of interaction between pathogen enzymes and insect cuticle. *Journal of Invertebrate Pathology.* **47:** 295- 305.

St. Leger, R.J., Cooper, R.M. and Charnley, A.K. (1988). The effect of melanization of *Manduca sexta* cuticle on growth and infection by *Metarhizium anasopliae. Journal of Invertebrate Pathology.* **52:** 459-470.

St. Leger, R.J., Cooper, R.M. and Charnley, A.K. (1987). Production of cuticle degrading enzymes by the entomopathogen *Metarhizium anisopliae* during infection of culticle form *Calliphore vomitoria* and *Manduca sexta*. Journal of General Microbiology. **133:** 137-1382.

St. Leger, R.J., Goettel, M., Roberts, D.W. and Staples, R.C. (1991). Prepenetration events during infection of host cuticle by *Metarhizium anasopliae. Journal of Invertebrate Pathology.* **58:** 168-179.

St. Leger, R.J., Joshi, L., Bidochka, M.J., and Roberts, D.W. (1996). Construction of an improved mycoinsecticide over expressing a toxic protease. *Proceeding of the National Academy of Science,*UAS. **93:** 6349-6354.

St. Leger, R.J. (1993). Biology and mechanism of insect-cuticle invasion by Deutromycete fungal pathogens. In: Becage NE, Thompson SN and Federici BA (eds.), Parasites and Pathogens of Insects, Vol.2 : Pathogens. Academic Press, San Diego. pp., 211-229.

Staples, J.A. and Milner, R.J. (2000). A laboratory evaluation of the repellence of *Metarhizium anisopliae* conidia to *Coptotermes lacteus* (Isoptera : Rhinotermitidae). *Sciobiology.* **36:** 133-148.

Steinhaus, E.A. (1956). Microbial control: the emergence of an idea. *Hilgardia* **26:** 107-160.

Steinhaus, E.A. (1975). Disease in a minor cord. Ohio State University Press. Columbus, OH.

Studdert, J.P., Kaya, H.K. and Duniway, J.M. (1990). Effect of water potential, temperature and clay coating on survival of *Beauveria bassiana* conidia is a loam and peat soil. *Journal of Invertebrate Pathology.* 417–427.

Tanada, Y. and Kaya, H.K. (1993). "Insect Pathology." Academic Press, New York.

Thevelein, J.M. (1984). Regulation of trehalose metabolization in fungi. *Microbiology Review.* **48:** 42- 59.

Thomas, M.B. (1999). Ecological approaches and the development of "truly integrated" pest management. *Proc. Natl. Acad. Sci.* USA **96:** 5944–5951.

Thomsen, L. and Eilenberg, J. (2000). Time-concentration mortality of *Pieris brassicae* (Lepidoptera : Pieridae) and *Agrotis segetum* (Lepidoptera : Noctuidae) larvae from different destruxins. *Environmental Entomology.* **29:**1041–1047

Vanninen, I. (1995). Distribution and occurrence of four entomopathogenic fungi in Finland; effect of geographical location, habitat type and soil type. *Mycological Research.* **100:** 93-107.

Vega, F.E., Goettel, M.S., Blackwell, M., Chandler, D., Jackson, M.A., Keller, S., Koike, M., Manianina, N.K., Monzon, A., Ownley, B.H., Pell, J.K., Rangel, D.E.N. and Roy, H.E. (2009). Fungal entomopathogens : new insights on their ecology. *Fungal Ecology*. **2:** 149-159.

Wraight, S.P., Jackson, M.A. and De Kock, SL. (2001). Production, stabilization and formulation of fungal biocontrol agents. In: Butt TM, Jackson C, Magan N. (Eds.), Fungi as Biocontrol Agents. CAB International, Wallingford, UK, pp. 253–287.

Zimmermann, G. (1993). The entomopathogenic fungus *Metarhizium anisopliae* and its potential as biocontrol agent. *Pest Science* **37:** 375-379.

Chapter 28

Approaches for Health Management of Planting Material in Ornamentals

M.N. Khare and S.P. Tiwari

Department of Plant Pathology, J.N.K.V.V., Jabalpur – 482 004, Madhya Pradesh

Floriculture was not an organized sector earlier. Flowers were used only for worships as loose flowers or garlands and in other social purposes. Roses, gerbera, anthurium, lily, chrysanthemum etc. were used as cut flowers for home decorations. According to Singh (2009) the picture has changed and the flower industry is advancing 10-20 per cent per year. The area under floriculture has increased to around 1.26 lac hectares in Maharashtra, Karnataka, Tamil Nadu, Andhra Pradesh, West Bengal, Uttar Pradesh, Delhi, Sikkim and attracted growers of other states also like Bihar, Haryana, Madhya Pradesh, Rajasthan, Gujarat, Himachal Pradesh, Uttarakhand. The present production of cut flower is 1,952 million (Singh, 2009). Six Agri Export Zones concerning floriculture are set up one each in Sikkim, Uttarakhand, Karnataka, Maharashtra and two in Tamil Nadu. For commercialization it is essential to have new attractive varieties with longer durability as loose flowers, cut flowers, potted plants, dry flowers and other floricultural products like essential oils, dyes, pot pourri. Besides Indian markets, export potential has tremendously, increased. The hitech protected cultivation of flowers is increasing fast. Seeds of flowers are produced in the country specially Punjab, Haryana, Karnataka etc. Tissue culture is used for fast production of disease free promising material. The plants are grown through seeds, cuttings, tubers, bulbs, rhizomes, corms, cormlets and other plant parts. The planting material has to be of good quality and pathogen free. The seeds carry the pathogens internally, externally, intra-embryal, extra-embryal as contaminant, inert matter and associated with inert matter. Pathogens are also associated with plant parts used as seed. Necessary steps are essential for health management of planting material of ornamental plants. The seed material must be obtained from reliable source, grown in disease free areas following scientific approach. Pathogens can be detected in propagating source by visual inspection, culture indexing, virus indexing, serology, biochemical methods and biotechnological approaches (Hudson *et al.*, 1997). The quality certified seed is essential for better outcome as they are properly cleaned and graded, have higher germination and vigour, genetically and physically pure, free from pathogens and weed seeds and provide uniform stand and higher returns. The treatment of planting material depends upon the type of association of the pathogen with seed and the mode of seed to plant transmission. Suitable chemicals, microflora antagonistic to the pathogens and botanicals are selected for seed treatment. Mode of plant to seed transmission and time of infection must be explored to check the seed infection in order to get pathogen free seed. It is essential to follow seed rules, certification procedure and sanitary and phytosanitary measures as per WTO requirements for import and export of such materials.

International Level

Globally more than 145 countries are involved in the cultivation of ornamental crops and the area is increasing fast. World wide trade in floriculture products was estimated at over US$ 7.9 billion in 2001. Cut flowers account for 50 per cent of sales, plants were 41 per cent, bulbs 9 per cent and cut foliage about 9 per cent. Seven countries export 73 per cent of the value of the world's floriculture crops, the Netherlands, Columbia, Italy, Belgium, Denmark, the United States and Ecuador. The Netherlands continues to dominate the world floricultural industry. It was estimated that in 2004 almost 51 per cent (US$ 6.27 billion) of world floriculture exports were from the Netherlands. Columbia was the second largest exporter at 7.5 per cent, Italy, Belgium, Denmark, the United States, Ecuador and Germany followed with approximately 3 per cent each of exported products. Kenya, Costa Rica, Israel and Spain produced about 2 per cent each. Major markets are Germany, the United States, Britain, France and the Netherlands. These five countries account for almost 70 per cent of all imports of floriculture products with an 8 per cent annual growth rate, world exports are expected to reach US$ 16–18 billion by 2015 (Jain, 2010).

Government Role

Floricultural export from India during 1999–2000 was Rs.105.15 crore and Rs.190.63 crore in 2000–2001. The main importing countries of Indian floricultural products were the Netherlands, the USA, Japan, Germany, Italy, Denmark, Egypt, Singapore, Switzerland, France, Australia, UAE, Belgium and Sri Lanka. Looking to the importance of floriculture and export potential the floriculture sector has been recognized as 'Extreme Focus Thrust Area'. The Agricultural and Processed Food Products Export Development Authority (APEDA) has taken steps to facilitate export of ornamentals. The National Horticulture Board (NHB) is providing soft loans for establishing infrastructure facilities like pre-cooling units, cold storage, packaging and grading sheds, refrigerated transport and green houses.

Ornamental Plants

The flower plants are annuals and perennials. Annuals are grown according to season like in winter season antirrhinum, aster, carnation, calendula, cineraria, larkspur, lupin, nasturtium, nigellia, pansy, petunia, phlox, salvia, sweet pea, viola, verbena, marigold, cosmos, linum, poppy, hollyhock etc. are preferred. In summer gaillardia, portulaca, sunflower, zinnia etc. and in rainy season balsam, gomphrena, zinnia, cosmos, cockscomb etc. are grown. These ornamentals are grown through seeds. They suffer due to biotic and abiotic stresses. Among biotic stress fungi, bacteria, nematodes, viruses are seed borne as well as soil borne. The important diseases and pathogens of some commonly grown ornamentals are mentioned herewith.

Sunflower–stem rot (*Sclerotinia sclerotiorum*), powdery mildew (*Erysiphe cichoracearum*), leaf spots (*Cercospora helianthi, Septoria helianthi, Colletotrichum helianthi*), Downy mildew (*Plasmopara halstedii*), Snap dragon-anthracnose (*Colletotrichum antirrhini*), Phyllosticta blight (*Phyllosticta antirrhini*).

Geranium–leaf blight (*Alternaria tenuis*), grey mould, blossom blight (*Botrytis cinerea*), rust (*Puccinia pelargonii–zonalis*), black stem rot (*Pythium splendens*), leaf spots (*Cercospora brunkii*), bacterial leaf spot (*Xanthomonas pelargonii*).

Larkspur–crown rot (*Deplodinia delphinii*), stem canker (*Fusarium oxysporum delphini*), powdery mildew (*E. polygoni, E. cichoracearum*), Diaporthe blight (*Diaporthe arctii*), black blotch (*Pseudomonas delphinii*).

Petunia–root and foot rot (*Pythium sp., Phytophthora sp.*), root rot (*Rhizoctonia solani*), leaf blight (*Alternaria alternata*), Phytophthora crown rot (*Phytophthora parasitica*), Leaf blight (*Cercospora petuniae*), leaf spot (*Ascochyta petuniae*), mosaic (TMV, CMV).

Phlox–leaf spot (*Septoria drummondii*), powdery mildew (*Sphaerotheca fuliginea*).

Pyrethrum–grey mould (*B. cinerea*), leaf and bud nematode (*Aphelenchoides ritzemabosi*).

Zinnia–damping off, root rot, foot rot (*Pythium sp., Phytophthora sp.*), grey mould (*B. cinerea*), leaf spot (*Alternaria zinniae*), root rot (*R. solani*), foot rot (*Sclerotium rolfsii*), powdery mildew (*E. cichoracearum*), leaf spot (*Cercospora zinniae*).

Aster–wilt (*Phialophora asteris*), downy mildew (*Basidiophora entospora*), powdery mildew (*E. cichoracearum*), leaf spots (*A. alternata, Cercospora asterata, Septoria ateris*).

Cosmos–leaf spots (*Cercospora sp., Septoria sp.*), powdery mildew (*E. cichoracearum*).

Sweet pea–powdery mildew (*E. polygoni*).

Balsam–foot rot (*Pythium sp.*).

Pansy–stem rot (*S. sclerotiorum*).

Poppy–stem rot (*S. sclerotiorum*), downy mildew (*Peronospora arborescens*), leaf smut (*Entyloma fusum*).

Hollyhock–anthracnose (*Colletotrichum malvacearum*), rust (*P. malvacearum*), leaf spots (Species of *Cercospora, Ascochyta, Phyllosticta, Alternaria*).

Anthurium–anthracnose (*Gloeosporium minutum, Colletotrichum sp.*).

Marigold–leaf spots (Species of *Alternaria, Cercospora, Septoria*), root rot (*R. solani*), collar rot (*S. rolfsii*), blight (*B. cinerea*), charcoal rot (*R. bataticola*), flower bud rot (*Alternaria alternata, A. dianthi*).

Carnation- Streak mosaic (Streak mosaic virus), leaf spot (*Septoria dianthi*), rust (*Uromyces dianthi*), blight (*Alternaria dianthi*), stem rot (*R.solani*).

Antirrhinum–downy mildew (*Peronospora antirrhini*), Phyllosticta blight (*P. antirrhini*), anthracnose (*C. antirrhini*), rust (*P. antirrhini*), grey mould (*B. cinerea*), powdery mildew (*Oidium sp.*), stem rot (*S. sclerotiorum*), root rot (*R. solani*).

Calendula–powdery mildew (*S. fuliginea*), leaf spot (*Cercospora calendulae*), rust (*P. flaveriae*), stem rot (*S. sclerotiorum*), leaf blight (*B. cinerea*), smut (*Entyloma calendulae*).

Gaillardia–leaf spot (*Septoria gillardiae*), powdery mildew (*E. cichoracearum, S. humuli*).

Coreopsis–leaf spots (*Cercospora choreopsidis, Septoria coreopsidis, Phyllosticta choreopsidis*), blight (*B. cinerea*), stem rot (*R. solani*).

Linum–root rot (*R. solani*), stem rot (*S. sclerotiorum*), root knot (*Meloidogyne incognita*).

Twenty four fungi were reported from 22 samples of 13 species of flowering crops, *Alternaria carthami* (*Zinnia*), *Colletotrichum dematium* (*Celosia* and globe amaranth), *Curvularia lunata* (*Tagetes patula* and *Gomphrena globosa*), *Drechslera rostrata* (*T. patula*) and *Phoma* sp. (*G. globosa*). These seeds were imported from other countries to Taiwan (Chou and Wu, 1995).

The seed borne pathogens result in seed rot, seedling rots and diseases at later stages of plant growth. The seed to plant infection is systemic, local or both. Several pathogens dwell in soil and attack the new plant when the conditions are congenial. Both the seed and soil borne pathogens are responsible for recurrence of the disease, subsequently spread of the disease occurs. Sclerotia of *S. sclerotiorum* and *Sclerotium rolfsii* are mixed with the seed as inert matter. Soil solarization is necessary at least in nurseries. Several pathogens are transmitted through the infected plant parts which are mixed with the seed as inert matter. Cleaning of seed is very essential. Detailed investigations have been made on *Myrothesium roridum* (Barapete, 1986) and *Alternaria alernata* (Pagnis, 1987) associated with seeds of hollyhock at J.N.K.V.V., Jabalpur, M.P.

Some flower plants are grown through vegetative parts like tubers, corms cormels, bulbs, rhizomes etc. They harbour pathogens which cause rots, hence one has to be careful while selecting such material for planting. Some examples are given herewith.

1. **Dahlia** (Seed, cuttings, tubers)–Rots (*Sclerotinia sclerotiorum, R. bataticola;* Wilt (*Verticillium alboatrum, Fusarium* sp.), Crown gall (*Agrobacterium tumefaciens*), tuber rot nematode (*Ditylenchus destructor*), root knot (*Meloidogyne hapla, M. incognita*)
2. **Gladiolus** (corms, cormels)–Dry rot (*Stromatinia gladioli, F. oxysporum* f.sp. *gladioli*); rot (*Rhizopus arrhizus*); storage rot (*Penicillium gladioli*); hard rot (*Septoria gladioli*), Botrytis dry rot (*Bortytis cinerea, B. gladiolorum*).
3. **Tulip** (Bulb)–Basal rot (*F. oxysporum f.sp. tulipae;* root rot (*Phythium* spp.) blight (*Botrytis cinerea*); grey bulb rot (*Rhizoctonia solani*), bulb nematode (*Ditylenchus dipsaci*), soft rot (*Erwinia carotovora*), Crown rot (*S. rolfsii*).
4. **Lily** (Bulb)–Fusarium bulb and scale rot (*F. oxysporum* f.sp. *lilii*); foot rot (*Phytophthora* spp.); root rot (*Pythium* spp.) bulb soft rot (*Rhizopus stolonifer*), blue mold bulb rot (*Penicillium cyclopium, P. corymbiferum*) bacterial soft rot (*Erwinia carotovora*), leaf spot (*Botrytis cinerea*), bulb and leaf nematode (*Aphelenchoides fragariae*), root lesion nematode (*Pratylenchus pratensis, P. penetrans*).
5. **Canna** (Rhizome)–Bud rot (*Xanthomonas cannae*), Rhizome rot (*R. solani*), rust (*Puccinia thaliae*).
6. **Chrysanthemum**–(Cuttings, root, suckers)–Rots (*R. solani, S. rolfsii, Sclerotinia sclerotiorum, Fusarium* spp.). Crown rot (*Erwinia tumefaciens*), wilt (*Verticillium alboatrum*), leaf spots, blotch (*Septoria chrysanthemella*), grey mold (*Botrytis cinerea*), leaf and bud nematode (*Aphelenchoides ritzemabosi*).
7. **Iris** (Rhizome, bulb, offsets)–Rhizome rot (*Botryotinia convoluta*), black rot (*S. sclerotiorum*), crown rot (*Sclerotium rolfsii*), ink spot (*Mystrosporium adustum*), blue mold (*Penicillium* sp.).
8. **Narcissus** (Bulb)–Blue mold rot (*Penicillium* spp.), Crown rot (*R. solani, S. rolfsii*), dry rot (*S. sclerotiorum*), basal rot (*F. oxysporum* f.sp. *narcissi*).
9. **Cactus** (Pseudostem)–Wilt (*F. oxysporum*), pad decay (*Aspergillus alliaceus*), soft rot (*Botrytis cinerea*), crown gall (*Erwinia tumefaciens*).
10. **Orchids**–(Bulbs, tubers)–Black rot–(*Phytophthora palmivora, P. nicotianae var. parasitica*), wilt (*Fusarium oxysporum*), rot (*S. rolfsii*), white rot (*S. sclerotiorum*), leaf rot (*Phythium* spp.), brown rot (*Phytomonas cattleyae*), basal rot (*R. solani*), wilt (*F. oxysporum* f.sp. *cattleyae*), Soft rot (*Erwinia carotovora*).
11. **Roses**–cuttings grafting, budding–Black spot (*Diplocarpon rosae*), wilt (*Verticillium dahliae*), rust (*Phragmidium mucronatum*), die back (*Diplodia rosarum*), nematodes (*Meloidogyne hapla, M. incognita, Pratylenchus penetrans*).
12. **Tuberose** (Bulbs)–Stem rot, Root knot nematode.
13. **Cuttings**–Several plants are propagated through cuttings like carnation, roses, bougainvillea etc. The cuttings should be used from disease free plants. It is better to treat them with contact fungicides before planting.

It is observed that different planting materials of various crops are commonly infected by the species of *Pythium, Phytophthora, Fusarium, Rhizoctonia, Sclerotium, Sclerotinia, Verticillium, Stromatinia, Rhizopus, Penicillium, Botrytis, Septoria, Erwinia,*

Xanthomonas, Ditylenchus, Aphelenchoides, Meloidogyne etc. Some foliage pathogens also survive in the planting materials like *Alternaria, Collectotrichum* etc. It is necessary to keep a strict watch on the mother plants to keep them disease free.

Management

The planting material needs proper checking for the associated microflora, as they should be free from pathogens. They should be treated with suitable chemicals like Benomyl, Carbendazim, Captan, Mancozeb, Copper oxychloride etc. *Trichoderma* species are effective biocontrol agents against disease causing pathogens. It is essential to search for proper eradicants. Management of pathogens in plant propagation has been nicely dealt by Hudson *et al.* (1997) which start prior to propagation by strict watch of mother plants from which the plant parts are used for propagation.

It is essential to control pathogens associated with seeds for which seed treatment is done with thiram, captan, Mancozeb, Bavistin, Kavach, Phaltan etc. 2-3 g per kg seed or thiram + Bavistin (2:1) 3 g per kg (Khare and Bhale, 2006). For checking soil borne pathogens soil treatment with suitable chemicals, biocontrol agents like *Trichoderma viride, T. harzianum* and solar treatment should be done more specifically in nurseries.

It is necessary to check the diseases in field as soon as they appear more specially when the crops are for use as seed. In rose powdery mildew and black spot disease is very important and can be controlled by spraying 0.1 per cent Bavistin, 0.1 per cent Karathane, 0.2 per cent sulfex or wettable sulphur (0.3 per cent) and 0.2 per cent captan, mancozeb (0.2 per cent), respectively. Kavach 0.2 per cent is used against grey mould caused by *Botrytis cinerea*. In Chrysanthemum for the control of leaf spots caused by different pathogens Phaltan, Mancozeb, Captan sprays 0.15-0.20 per cent are recommended. Powdery mildew is checked by the application of Karathane 0.2 per cent. Grey mould infection is controlled by Zineb, Mancozeb, Maneb 0.2 per cent spray. The Gladiolus suffers due to Botrytis grey mould, leaf spots due to *Septoria gladioli, Stemphylium* sp., *B. cinerea, Curvularia lunata* for which Kavach, Mancozeb, Zineb, copper oxychloride 0.2 per cent are highly promising. For corm rot captaf 0.3 per cent is used, besides *Trichoderma harzianum* has proved to be a good biocontrol agent. In carnation, wilt is caused by *F. oxysporum*. f.sp. *dianthi* and *Verticillium alboatrum*. It is necessary to take cuttings from green house grown healthy plants. The stock plants should be sprayed with captan (0.2 per cent). Leaf spots are caused by *Septoria dianthi, A. alternata, B. cinerea* and can be checked by the spray of the same fungicides used in other ornamentals. In Tuberose leaf spots and blights can be controlled by spraying Iprodione 0.025 per cent, Difencanozole 0.05 per cent. Collar rot due to *S. rolfsii* is checked by *T. viride* 20 g/m^2, carbendazim 0.1 per cent + captan 0.2 per cent. The Gerbera plants are attacked by *B. cinerea, Gloeosporium* sp., *Erysiphe cichoracearum, Phytophthora* spp. and can be checked by the use of Benomyl 0.1 per cent, Kavach 0.2 per cent, difencanazole 0.05 per cent, Mancozeb (0.2 per cent), *T. viride* 20 g/m^2. In Dahlia powdery mildew and Botrytis blight are important besides common leaf spot disease. Use of wettable sulphur, Karathane, Zineb, Mancozeb etc. can check the diseases. In orchids anthracnose, black rot, leaf spots due to the species of *Cercospora, Diplodia, Laptothyrium, Phyllosticta* occur which can be controlled by the spray of copper oxychloride, Mancozeb, Zineb etc.

Nematode Diseases and their Management

Nematodes cause enormous damage to ornamental plants like species of *Meloidogyne, Rotylenchulus, Aphelenchoides, Ditylenchus, Pratylenchus, Tylenchus, Helicotylenchus* etc. Root-knot nematodes are most serious and obnoxious endoparasite to limit ornamental crop productivity (Zarina and Abid, 1995). *Meloidogyne* spp. have the potential to damage many important nurseries of ornamental crops (Benson and Barker, 1985) and form disease complexes with certain soil-borne fungal pathogens, thus increasing their hosts susceptibility (Nigh, 1972; Santamour and Riedel, 1993; Walker and Melin, 1998). Certain ornamental crops *viz.*, begonia, boxwood, camellia, daylily, gardenia, gladiolus, gerbera, daisy, hibiscus, liriope and numerous annual and perennial flower plants suffer due to root-knot nematode. Symptoms vary with the host such as *Sansevieria cylindrical* develop leaf discoloration and tip necrosis in 4 to 5 months after infection with *M.incognita* (Mishra and Mishra, 1997), *Philodendron selloum*, exhibits a reduction in leaf size when infected with *M. incognita* (Mishra and Mishra, 1993), and *Juniperus horizontalis* var. *plumosa* exhibit thickened roots and slight galling at post-infection stage with *Meloidogyne* spp. (Nemec and Morrison, 1972). *Gladiolus hortulanus* plants infected with *M. incognita* race 2 exhibited leaf drying, reduction in floral stalk height and girth, and reduced number of florets (Khanna *et al.*, 1998). Some ornamental plants exhibit minute galls following infection with *Meloidogyne* spp. In such conditions, root-knot nematode females can be seen protruding from the infected roots. Other plants, such as *Rheum* spp., *Begonia* spp., and *Thunbergia* spp. produce large galls, measuring up to 0.6 mm in the latter case (Bird, 1974).

Foliar nematodes like *Aphelenchoides fragariae, A. ritzemabosi, A. besseyi* cause problem in *lilies, Iris* etc. Tuberose are infected by *A. besseyi*, Dahlia by *A. ritzemabosi.* Other ornamental plants *viz., Amaranthus tricolor, Anemone* hybrid, *Baptista australis, Fragaria ananassa, Papaver orientale, Anthurium andraeanum, Hibiscus rosa-sinensis, Cyclamen persicum, Ocimum basilicum, Pelargonium hortorum, Saintpaulia ionantha, Asplenium nidus, Athyrium goeringianum* and species of Hepatica, Heuchera, Hosta, Hypericum, Ipomoea, Iris, Ligularia, Ligustrum, Lilium, Malva, Narcissus, Paeonia, Phlox, Polygonatum, Rhododendron, Ageratum, Begonia, Coleus, Ficus, Impatiens, Lilium, Orchids, Peperomia, Salvia, Blechnum, Dryopteris, Nephrolepis, Osmunda, Polypodium, Polystichum, Pteris etc. are host of *Aphelenchoides* spp.

Ditylenchus destructor, D.dipsaci and *D. angustus* attack Narcissus Aster, Begonia, Carnation, Chrysanthemum, Dahlia, Hyacinths, Iris, Phlox, Roses and Tulip (Siripong, 1962.).

The lesion nematode, *Pratylenchus penetrans* and *P. pratensis* damage bulbous flower plants. Feeding of lesion nematode (*Pratylenchus vulnus*) observed on cortical tissues causes numerous small dark lesions on fibrous roots. Darkened and stunted roots and reduction in the size of the root system may also be seen on severely damaged plants. The nematode moves through root tissues and makes thin galleries in root tissues. This nematode has been noticed on Boxwood, Carnation, Chrysanthemum, Iris, Narcissus, Phlox, Rose, Violet, Forsythia, Pine, Willow and even some annual flowers.

The dagger nematode (*Xiphinema* spp.) feed on fibrous (smaller) roots and root tips making them swollen and discolored. Several lateral roots may appear above the damaged root tips. Galls may be confused with those of root-knot. The nematode migrates along roots. Occurrence of nematode is more common in Ash, Azalea, Maple, Oak, and Sycamore.

Belonolaimus spp. commonly known as sting nematode free-living in most of the soil infest roots which turn darker and rot. Root systems are sparse, stunted, or stubby. Camellia, Holly, Juniper, Magnolia, Oak, are major ornamental plants, attacked by the nematode. Snap dragon and Carnation are susceptible to *Heterodera* spp. and Begonia to *Rotylenchulus reniformis*.

The planting material can be soaked in 0.2-0.4 per cent Triazophos or Carbosulfan to kill the foliar nematodes and carbofuran for the management of root associated nematodes. Carbosulfan can also be sprayed (Chawla and Singh, 2009).

In ornamental crops where plant parts are used as seed all possible care should be taken to keep them free from pathogens. In case of seed production of ornamentals plant to seed transmission of pathogens must be checked. The time of infection is very important based on which fungicide spray may be given at the appropriate time.

Seed Certification

Like other crops seed certification is a must for ornamental crops. Seeds and plant parts used as seed are also governed by ISTA rules (Anonymous, 1985). Seed Certification Agencies get the inspections made by Inspectors for field standards and seed standards. Samples are drawn as per rules and required quantity is sent as submitted sample for testing. Quantity of seed of submitted sample of flower plants is decided by the weight and size of seed as given in ISTA rules. Vegetatively grown plant parts are sent 250 by number. The seeds are tested for genetic purity, inert matter, moisture percentage, germination percentage, association of pathogens and diseases. The designated seed borne diseases and permissible limits for the seed crop of ornamental flower plants at field stage are given in Table 28.1.

According to the gazette notification No. So. 1165 (E) dated 24th November, 1999 the amendments have been made by the Technical Committee of Central Seed Certification Board for the required germination per cent and physical purity per cent in some ornamentals is given below:

Crop	Germination (per cent)	Physical Purity (per cent)
African marigold (seed)	70	97
Annual carnation (seed)	75	97
Annual chrysanthemum (seed)	50	98
Marigold (seed)	70	97
Ornamental sunflower (seed)	70	98
Periwinkle (seed)	60	95
Petunia (seed)	75	98
Snapdragon (seed)	70	98

Table 28.1: Designated Seed Borne Diseases and Permissible Limits for Seed Crop of Ornamental Flower Plants at Field Stage

Crop Group	*Disease/Causal Organism*	*Field Standard (per cent)*	
		Foundation	*Certified*
Annual carnation (Magurite or Chaubad type) (*Dianthus caryophyllus*)	Streak mosaic virus (at third and final inspection before harvesting capsule start opening (shattering)	0.10	0.20
Snap dragon (*Antirrhinum* spp.) and hybrid	Anthracnose (*Colletotrichum antirrhini* and *Colletotrichum falcatum*) Blight (*Phyllosticta antirrhinum*) at final stage of inspection (III) prior to harvesting)	0.10	0.20
Annual chrysanthemum (*Chrysanthemum* spp.)	Grey moulds (*Botrytis cinerea*)	0.10	0.20
	Blotch (Septoria chrysanthmella) at final stage of inspection (III) prior to harvesting	0.10	0.20
Petunia (*Petunia* spp.)	Leaf blight (*Cercospora petuniae*) Leaf spot (*Ascochyta petuniae*) Leaf spot (*Phyllosticta petuniae*) Leaf blight (*Alternaria alternata)* Crown rot (*Phytophthora parasitica*) Tobacco mosaic virus (TMV) Cucumber mosaic virus (CMV) at third and final inspection before harvesting)	0.10	0.20
Ornamental sunflower (*Helianthus* spp.)	Downy mildew (*Plasmopara halstedii*) at each three inspection	0.050	0.050
	Plant infected with *Orobanchy cumana* (at final third inspection) (First inspection at the stage of 6-7 pair of leaves, second inspection during flowering and third at maturity and prior to harvesting)	None	None
Ornamental sunflower (*Helianthus* spp.) hybrids	Downy mildew (*Plasmopara halstedii*) at each four inspection	0.050	0.050
	Plant infested with *Orobanchy cumana* (at fourth inspection)	None	None
Marigold (*Tagetes* spp.) and hybrids	Leaf spot (*Alternaria tagetics*) Flower bud rot (*Alternaria alternata Alternata dianthi*) Collar rot (*Rhizoctonia solani*) (At final (III) inspection I marigold and at final (IV) inspection in hybrids crop maturity stage)	0.10	0.20

For clonally propagated crop through tubers, bulbs, bulblets, corms, cormels, rhizomes of the following crops, the minimum limit of survival is 90 per cent and the minimum limit for pure seed living clones is 95 per cent.

Defodil (bulbs,bulbils), Narcissus (bulbs, bulbils), Tube rose (bulbs), Tuberous Begonia (tubers), Tulip (bulbs).

For clonally propagated crop through planting stakes, root slips, self rooted plantlets produced by tissue culture the minimum limit of survival is 90 per cent and the minimum limit for pure living clones is 98 per cent.

Antirrhinum (off shoots), Chrysanthemum (suckers and planting stakes), French Jasmine (root slips),Gerbera (sister clumps), Orchids (off shoots, pseudo-bulbs), Perennial Carnation (planting stakes), Rose (planting stakes), Rose-geranium (planting stakes), Rose marium (self rooted plants).

New policy on Seed Development on 16th Sept. 1988 liberalized import of quality seed and planting material by Indian farmers. Testing of the material is done at 26 quarantine and fumigation stations. Post entry quarantine facilities are made available.

Seed Acts

The Destructive Insects and Pests Act 1914 was followed in procuring seed and planting material of all agricultural or horticultural crops and all trees, bushes or plants from other countries. Seed Act 1966 came in force from 2nd October 1969. Seed Rules 1968 deal with the functions of different committees, Seed Inspectors, Seed Analysts, procedure of tagging etc. The Seed Order 1983 made the registration of seed dealers compulsory. License is granted for three years which is renewed every three years.

According to Plants, Fruits and Seed Order 1989 (regulation of import into India) horticultural and fruit seeds, plant parts used as seed are included. Cuttings, saplings and bud woods of flowers or ornamental plants, seeds and plant material of fruits are imported subject to special conditions prescribed for them in schedule II, according to which exotic pests and diseases are to be specified in import permit. The consignments shall be grown in approved post-entry quarantine. No consignment can be imported into India without a valid permit. Post entry inspection is done and steps are taken at the point of entry and fumigation, disinfection, disinfestation, destruction or return is suggested as per need. An official phytosanitary certificate of exporting country is necessary. Provided that cut flowers, garlands, bouquets, fruits and vegetables weighing less than two kg imported for personal consumption may be allowed to be imported without a Phytosanitary Certificate or an import permit. This order has surpassed Order 1984 (Anonymous, 2000).

Seeds Bill 2004

The Bill was tabled in the parliament in December 2004 for approval. The requirements and conditions laid down in earlier passed Seed Acts and Seed Rules were incorporated in the new Bill along with the recent developments in agriculture like Genetically Modified Crops. Besides compulsory labeling, the concept of misbranding and selling of fictitious company name is treated as an offence.

Transgenic seed, extant varieties and existing varieties are protected. Compulsory registration of all varieties sold in India is essential; such registration is valid for 15 years for annual and biennial crops and 18 years for perennials. All transgenic varieties will be registered only after approval under Environmental Protection Act, 1986.

Seed dealers and horticultural nurseries are needed to be registered. No seed of any kind or variety shall, for the purpose of sowing or planting by any person, be sold unless such seed is registered. It is essential for a registration holder of horticultural nursery to keep a complete record of the origin of source of every planting material and performance record of mother trees in the nursery. A layout plan showing the position of the root stocks and scions used in raising the horticulture plants must be kept. The nursery plants as well as the parent trees used for the production or propagation of horticulture plants should be kept free from infectious or contagious insect pests and diseases affecting plants. It is necessary to furnish such information to the State Government on the production, stocks, sales and price of planting material in the nursery as may be prescribed. The State Government shall establish a State Seed Committee to advise the government on registration of seed producing units, seed processing units, seed dealers and horticultural nurseries, maintain in each district, a list of seed dealers, seed producers, seed processing units and horticulture nurseries.

National Horticulture Board (NHB), Ministry of Agriculture, Government of India is the Nodal Agency for rating for Horticulture Nurseries and for Awarding Accreditation Certificate. The Seed Act and the Nursery Registration Act have been in operation since December 1966 but presently it has been enforced in eight States–Punjab, Maharashtra, Himachal Pradesh, Uttar Pradesh, Uttarakhand, Jammu and Kashmir, Orissa and Tamil Nadu; in nine States some system of registration/ monitoring exists–Andhra Pradesh, Assam, Bihar, Goa, Haryana, Karnataka and Kerala but in thirteen States no Horticulture Nursery Act exists–Arunachal Pradesh, Chattisgarh, Jharkhand, Madhya Pradesh, Manipur, Meghalaya, Mizorum, Nagaland, Rajasthan, Sikkim, Tripura and West Bengal.

Plant Quarantine

The Directorate of Plant Protection, Quarantine and storage (DPPQS) under the Ministry of Agriculture is to enforce quarantine regulations and conduct quarantine inspection. Disinfestation, disinfection, destruction of the commodities is done through Plant Quarantine Stations. Post-entry quarantine inspection is also done. Heads Department of Plant Pathology of Agricultural Universities and some Agricultural Institutes are made inspection authorities (Anonymous, 2006).

The National Bureau of Plant Genetic Resources (NBPGR), New Delhi is authorized to handle quarantine processing of germplasm and transgenic planting material imported for research purposes into the country by public and private sectors.

WTO

Sanitary and Phyto-sanitary Measures are essential under WTO. So far 32 International Standards have been developed by International Plant Protection

Convention (IPPC) of Food and Agriculture Organization (FAO) of the United Nations. They provide guidelines on insect pest and disease prevention, detection and eradication. Procedures for undertaking survey for pest and disease free areas, locations under less damage, pest risk analysis etc.

Penalties

According to the Seeds Act 1966, any person who contravenes any provision of the Act or prevents a Seed Inspector from taking samples or prevents Seed Inspector from exercising any other power shall be punished for the first offence with a fine up Rs 500/-. If the offence is repeated he may be imprisoned up to six months and/or fined up to Rs 1,000/-.

When any person has been convicted under this Act for the contravention of any of the provisions of this Act or the rules made there under, the seed in respect of which the contravention has been committed may be forfeited to the Government.

Where an offence under this Act has been committed by a company, every person who at the time the offence was committed was in charge of, and was responsible to, the company, shall be deemed to be guilty of the offence and shall be liable to be proceeded against and punished accordingly.

Where an offence under this Act has been committed with the consent or connivance of, or is attributable to any neglect on the part of, any director, manager, secretary or other officer of the company, such director, manager, secretary or other officer shall also be deemed to be guilty of that offence and shall be liable to be proceeded against and punished accordingly.

According to the Seeds Bill 2004, which has yet to be passed by the parliament the punishments proposed are–any person who contravenes any provisions of the Act, prevents a Seed Inspector from taking samples etc. shall be punished for the first offence with a fine up to Rs 500/-. If the offence is repeated he may be imprisoned up to 6 months and/or fined up to Rs 1000/-. Any person who contravenes any provisions of the Act or imports, sells or stocks seed derived to be misbranded or not registered can be punishable by a fine of Rs 5000/- to Rs 25000/-. The penalty for giving false information is prison term up to 6 months and/or a fine up to Rs 50000. According to PPVFR Act, 2001 penalty for applying false demonination to a variety is imprisonment up to two years and/or a fine between Rs. 50,000 and Rs five lakh penalty for falsely representing a variety as registered is imprisonment up to three years and /or a fine between Rupees one lakh and Rupees five lakh or both. Penalty for subsequent offence is imprisonment up to three years and /or a fine between Rs. 2 and 20 lakh.

Farmer can claim compensation under the Consumer Protection Act.

Conclusion

Ornamentals have gained great importance. Newer plant materials are imported and exported. Each state should develop a floriculture information centre to provide all information about floriculture and its marketing to dealers who are usually ignorant about production technology of floricultural crops and newly developed novel floral varieties developed at research institutes and universities. A continuous supply of flowers can be managed by staggered plantings. India occupies two mega biodiversity centers in Eastern Himalayas and Western Ghats and nine phyto-geographical regions due to which great diversity exists and over 17,500 species of plants are available. Most of the floricultural crops can be grown in the country at one or the other region (Datta *et al., 2007*). All the rules and regulations must be followed in order to receive and supply insect pests and disease free material. List of insect pests and diseases prevailing in the country must be made along with the global information country wise. Timely inspections to check the source of seed and planting material, diseases and insect pests are essential. No dealer is permitted to bring or import seed or planting material from other countries directly. All out efforts are needed to manage the health of planting material of ornamental plants.

References

Anonymous, (1985). International Rules for Seed Testing. Seed Science and Technology 13 **(2)** : 520.

Anonymous, (2000). Plants, Fruits and Seeds (Regulation to Import into India) Order 1989. Seed Tech News 30 **(3-4)** : 35-53.

Anonymous, (2006). Plant Quarantine (Regulations of Import into India) (Second Amendment) Order 2006. GOI Notification Sept. 2006, pp 25.

Barapete, R.D. (1986). Studies on Myrothesium leaf spot of Hollyhock. M.Sc. Ag. Thesis. JNKVV, Jabalpur, M.P. pp. 62.

Benson, D.N. and Barker, K.R. (1985). Nematode a threat to ornamental plants in the nursery and landscape.Plant Disease **69:** 97-100.

Bird, A.F. (1974). Plant response to root-knot nematode. Annual Review of Phytopathology **12:** 69-85.

Chawla,G. and Singh, K.P. (2009). Grow disease free flowers. Indian Horticulture **54:** 36-37.

Chou, J.K. and Wu, W.S. (1995). Seed borne fungal pathogens of ornamental flowering plants. Seed Science & Technology **23:** 201-209.

Datta, S.K., Goel, A.K. and Roy, R.K. (2007). Floriculture expanding in a big way. The Hindu Survey of Indian Agriculture: 172-175.

Hudson, T.H., Dale, E.K., Fred, T.D. and Robert, L.G. (1977). Plant Propagation Principles and Practices. Prentice Hall of India Pvt. Ltd. New Delhi. pp. 770.

Jain, P.K. (2010). Seed Production Technologies of Floriculture Plants. Seed Business Training, JNKVV, Jabalpur, M.P. pp. 15.

Khanna, A.S., Chandel, S.S. and Malhotra, R. (1988).Evaluation of varietal resistance in gladiolus to *Meloidogyne incognita* (race 2). Pest Management and Economic Zoology **6:** 27-30.

Mishra, S.D. and Misra, R.L. (1997). New record on the occurrence of root-knot nematode, *Meloidogyne incognita* and *Philodendran sellarum,* K.Koch. Current Nematology **4:** 245-246.

Misra, R.L. and Mishra, S.D. (1997). *Senseviera cylindrica* Rojer-a new host of root-knot nematode (*Meloidogyne incognita*). Progressive Horticulture. **29:** 196.

Khare, M.N. and Bhale, M.S. (2006). Management of seed borne pathogens. In Plant Protection in New Millenium Vol I. Eds. Gadewar, A.V. and Singh, B.P. Satish Serial Publishing House, Delhi: 141–163.

Nemec, S. and Morrison, L.S. (1972). Histopathology of *Thuja orientalis* and *Juniperus plumose* infected with *Meloidogyne incognita.* Journal of Nematology **4:** 72-74.

Nigh, E.L.Jr. (1972). Susceptibility of Arizona grown ornamentals to attack several nematode species. Plant Disease Reporter **56:** 914-918.

Pagnis, S. (1987). Studies on Alternaria leaf spot of Hollyhock (*Althea rosea* Can.). M.Sc. Ag. Thesis.JNKVV, Jabalpur M.P.pp. 50.

Saha, L.R. (1990). Handbook of Plant Protection. Kalyani Publishers, New Delhi. pp. 740-804.

Santamour, F.S. Jr. and Riedal, L.G.H. (1993). Susceptibility of various landscape trees to root-knot nematode. Journal of Arboriculture **19:** 257-259.

Siriphong, I. (1962). Nematode diseases of some flowering ornamental plants. The Kasetsart Journal 2 **(1):** 1-16

Singh, H.P. (2009). Floriculture Industry in India, the bright future ahead. Indian Horticulture 54**(1):** 3-9.

Walker, J.T and Melin, J.B. (1998). Host status of herbaqceous perennials to *Meloidogyne incognita* and M. *arenaria*. Journal of Nematology **30:** 607-610.

Zarina,B. and Abid, M. (1995). New host records of root-knot nematode (Meloidogyne spp.) in Pakistan. Pakistan Journal of Nematology **13:** 49-50.

Chapter 29

Role of Seed-borne Fungi in Seed Health and their Management through Plant Products

Jai Prakash Rai[1]* *and Asha Sinha*[12]

[1]*Department of Mycology and Plant Pathology (KVK),*
[2]*Department of Mycology and Plant Pathology,*
Institute of Agricultural Sciences, Banaras Hindu University, Varanasi – 221 005, U.P.

Diversity is perhaps the greatest feature of nature. In the biotic systems operating in the nature we witness a plethora of diverse organisms occupying their respective niches in complete harmony with one another. Almost all the organisms have certain kind of association with another kind of organisms. This phenomenon of association of organisms with one another is precious tool of nature in order to create variations and maintain an element well within its boundaries.

Association of mankind with plants is a landmark in the evolution of the former. Man learnt to cultivate plants to meet its requirements of food, shelter, clothing, medicines and a number of other necessities. Agriculture is the biggest venture based on this association and it has come to such a point that existence of mankind has become dependent on the existence of crop plants. In somewhat similar context, association of microorganisms with plants is also a much studied topic with a pile of information and increase in this knowledge of interactions between microorganisms and plants has paved the way for improved crop cultivation practices in agriculture.

* Corresponding Author E-mail: drjaibhu@gmail.com

Seeds are perhaps the most important input of crop production and we have witnessed manifolds increase in the yields of various crops through improvement in the seed quality. Association of microorganisms with seeds of plants has given rise to the discipline of seed pathology which studies seed health and effects of microbial association on this important component. Seeds carry a wide range of microorganisms either externally of internally and those microorganisms become active in favourable conditions causing considerable damage to the seed and several diseases on the crop raised from such seeds in various ways. The microorganisms thrive on the seeds at the expense of easily digestible components. The successful invasion or colonization, however, depends largely upon the efficiency of microorganisms to degrade complex molecules into simpler forms (Bilgrami and Verma, 1978). Fungi are perhaps the most important microorganisms which have pronounced effects on seed quality and therefore, on the performance of the resultant plant. That some of the seed-borne microflora might reduce the germinability of the seeds when planted, or result in diseases in the growing plants, has been recognized long back. To check the spread of such microorganisms seed health testing procedure was considered to be introduced and in accordance, plant pathologists all over the world developed seed health testing procedures in 1920. Later on, Dr. Paul Neergaard took up initiatives specifically on this problem and in 1966 after a meeting, the International Rules for Seed Health Testing were recommended. The two methods of assessment of seed microflora, *viz.* agar plate method and standard blotter technique as recommended by International Seed Testing Association (ISTA) are used worldwide for seed health testing. Some important organizations related with the process of seed health testing include (*i*) FAO (Food and Agriculture Organization), (*ii*) ISTA (International Seed Testing Association), (*iii*) IAPSC (Inter African Phytosanitary Commission), (*iv*) SEAPRPC (Plant Protection Committee for South East Asia and Pacific Region), (*v*) EPPO (European Plant Protection Organization) and (*vi*) NEPPC (Near East Plant Protection Commission).

In order to meet season to season requirements of the seeds and other purposes, they are needed to be stored. Storage of seeds and food grains has been an age old practice in agriculture. Storage of seeds is done mainly for the following three purposes:

1. Seeds are stored for about six months from one harvest to the next planting season.
2. Seeds may also be stored for 18-36 months to insure against a following crop of poor yield and low quality, fluctuation in price and market demand against shortages, during outbreaks of war and when the produce could not be sold in the present year.
3. Long term storage of germplasm.

The conditions of storage do affect the quality of seeds stored and based on the two environments (*viz.* field and storage) which are altogether different from each other, seeds are involved in associations with fungal microorganisms of two kinds-one which are most active under the environmental factors as presented in the field and play their major role under such environments, known as *field fungi* and the other, which mostly get associated with the seed in the field itself but do not play their active role under field conditions. Instead, they remain silently associated with the seed and start their activity only when the environment under the storage conditions is developed to suit them. These are aptly called, *storage fungi*.

Among the first to recognize that the fungi might be involved in the deterioration of stored grains and seeds were Drs. P.E. Ramstad and W.F. Geddes in the Department of Agricultural Biochemistry of the University of Minnesota in the late 1930s (cited by Christensen and Kaufmann, 1969). Later many workers such as Nagel and Semeniuk (1947) and Christensen and Kaufmann (1965; 1969) studied the biodeterioration of grains due to storage fungi.

Field Fungi

These are the fungi that invade seeds developing on the plants in the field after the seeds have matured and the plants are either still standing or are cut and swathed, awaiting threshing (Christensen, 1972). The fungi require a moisture content in equilibrium with a relative humidity of at least 90-95 per cent to grow (Koehler 1938). In some regions, in some years a combination of wet weather during harvest, lack of drying facilities or lack of sufficient drying capacity and lack of transport may result in much grain of high moisture content being piled on the ground. If such high moisture content is coupled with a temperature that allows microflora to grow, it will deteriorate rapidly. (Christensen, 1972).

Pepper (1960) lists approximately 180 species of filamentous fungi and about 20 species of yeasts that have been reported from barley kernels. Christensen and Kaufmann (1969) state, 'from a single gram of malting barley about 25 kernels, we have isolated tens of thousands of colonies of yeasts and several million colonies of bacteria.' Malone and Muskett (1964) describe 77 species of seed-borne fungi isolated from seeds of soybean, 32 species from those of maize, 34 species from those of rice, 29 species from those of sorghum and 28 species from seeds of wheat. The same publication lists *Alternaria* from seeds of more than 100 species and *Fusarium* from seeds of 200 species of plants.

The fungal species associated with crop seeds play an important role in seed deterioration during storage (Singh *et al.*, 2001). Concerning the effects of field fungi, Christensen and Kaufmann (1969) state, 'Field fungi may discolour seeds, cause death of the ovules, shriveling of the seeds or kernels, weakening or death of the embryos and development of compounds toxic to man and to other animals. In grading of some grains, this discolouration is referred to as "weathering"- a misnomer, since the discolouration is a product of growth of microflora.

Reduction in seed germinability resulting from invasion of seeds by field fungi has been reported by several workers, Christensen and Stakman (1935) found a high correlation between increasing percentage of seeds infected with *Helminthosporium* and *Fusarium* and decreasing germination percentage. Christensen (1936) confirmed the finding for barley seeds. Machacek and Greaney (1938) carried investigations over 'black point' or 'kernel smudge' of wheat and observed that blackpoint caused by *Alternaria* did not result in decreased germinability but that caused by *Helminthosporium* did. Even if kernels with heavily discoloured pericarps as a result of invasion by *Alternaria* did not usually have the embryo invaded by the fungus. *Fusarium* may invade and kill the developing or mature embryo without causing any noticeable discolouration of either the pericarps or embryo-the seeds appear sound but actually are diseased or even dead.

Storage Fungi

Storage fungi comprise chiefly several 'group species' of the genus *Aspergillus* and about an equal number of less well-defined species of *Penicillium*. The group species of *Aspergillus* that are involved, *viz. Aspergillus restrictus, Aspergillus glaucus, A. candidus, A. versicolor, A. ochraceous* and *Aspergillus flavus* (listed in the order of increasing moisture required for growth) are quite distinct from one another and are relatively easy to identify with a high degree of certainty. *Aspergillus restrictus* and *Aspergillus glaucus* are considered to be associated with incipient deterioration. The other common species of storage fungi-*Aspergillus candidus, A. ochraceous, A. versicolor, A. flavus* and *Peniciliium* develop only later, after the growth of *A. restrictus* and *A. glaucus* has increased the moisture content of the grain mass or of a portion of it, sufficiently to permit the growth of high-moisture requiring or less xerophytic species (Christensen, 1972).

As far as harmful effects of storage fungi on the stored grains and seeds are concerned, extensive work has been done in this field. The major harmful effects can be listed as- discolouration of seed and embryo, reduction in seed germinability, biochemical changes in the seed, increase in the moisture content and heating of the seed or grain mass, reduction in processing quality, enzymatic activities of seeds and production of mycotoxins.

Invasion of seeds by storage fungi causes blackening of the germ. This invasion takes place normally in the storage, but the contamination of seeds and, in certain cases, invasion of seeds (wheat, cotton and groundnut) by some species of *Aspergillus* (*Aspergillus flavus, A. candidus*) and *Penicillium* takes place in the field itself before harvesting of the crop. The invasion causes discolouration, shriveling of the seeds, prolonged dormancy and quality losses. Such types of seeds are termed as 'sick' or damaged seeds.

Reduction in the germinability of the seeds has been attributed to the death of the embryo and reduced seedling emergence to the pre-emergence mortality of germlings resulting from the invasion by storage fungi. The whole seed lot may be rendered useless for sowing and consumption purposes. Decrease in germinability of seeds is higher and rapid following their storage at high moisture levels.

The biochemical changes in stored seeds brought about by storage fungi include changes in the three major nutritional components of seeds, *viz.* carbohydrates, proteins and lipids (Prasad *et al.*, 1988; Saxena and Karan, 1991) the last incurring losses in the oil quality and quantity of oilseeds. Loss in the quality of oilseeds through biochemical changes has been reported to be caused by fungi associated with them. Fungi like *Aspergillus niger, Aspergillus flavus, Alternaria dianthicola, Curvularia lunata, Curvularia pellescens, Fusarium oxysporum, Fusarium equiseti, Macrophomina phaseolina, Rhizopus stolonifer, Penicillium digitatum* and *Penicillium chrysogenum* cause discoloration, rotting, shrinking, seed necrosis, loss in germination capacity and toxification to oilseeds (Chavan and Kakde, 2008). Since oilseeds are rich in oil content, which causes a boost in the vigor of pathogenic fungi biodeterioration of oilseeds in brought about by production of lipase. (Umatale, 1995; Waghmare, 1996; Kakde and Chavan, 2011a; Kakde and Chavan, 2011b). Fungi growing on stored grains can reduce the germination rate along with loss in the quantum of carbohydrate, protein and total oil content, induce increase in moisture content and free fatty acid content enhancing other biochemical changes of grains (Bhattacharya, 2002). Such seeds are not fit for human consumption and are also rejected at the industrial level.

Development of storage fungi increases moisture content and temperature of seed lots due to increased respiration of seeds. The moisture so produced in one corner can diffuse to the other areas and render it vulnerable to spoilage. By the time when the fungi have grown enough in the seed lot to have raised the moisture content high enough so that seed respiration becomes measurable, the seeds are dead and decayed.

Storage fungi reduce the processing quality of grains for milling as in wheat, corn and many other cereals. Musty odour often produced during spoilage of grains persists even in the preparation of food and beverages from such grains. The storage fungi deteriorate the quality of seeds used for oil production and reduce the oil content in the affected seeds. Oil from infected samples is discoloured with acrid smell.

Stimulation of activities of pectic enzyme complex, C_x, amylase, invertase and protease by storage fungi in finger millet has been reported (Prasad *et al.*, 1988). Storage fungi are known to stimulate hydrolysis of starch and protein producing extracellular amylase and protease besides cellulose and lipase (Prasad, 1979).

Fungi growing on stored seeds are well known to produce metabolites toxic to other organisms including domestic animals and man (mycotoxins). The diseases caused by consumption of mycotoxins are cumulatively known as 'mycotoxicoses.' Aflatoxins are the most important of the known mycotoxins. These are produced chiefly by *Aspergillus flavus*, a common storage fungus. The problem of aflatoxin production is confounded by the fact that *A. flavus* is cosmopolitan and worldwide in distribution and can attack a wide variety of substrates under storage conditions. Aflatoxins may cause serious disorders in the human beings or animals, when consumed, including liver damage and even cancer. Some other fungi producing mycotoxins include *Penicillium rubrum, P. perpurogenum* (rubratoxins); *Aspergillus ochraceous, Penicillium viridicatum, P. cyclopium* (ochratoxins); *Aspergillus clavatus, A. terreus, A. patulum, Penicillium cyclopium* (Patulin); *P. palitans, Aspergillus flavus, Penicillium cyclopium* (tremortins); *Fusarium oxysporum, F. moniliforme, F. roseum* (Zearalenone); *Penicillium citreoviride, P. toxicarinum* (citreoviridin); *P. citrinum* (citrinin) and *P. cyclopium* (penicillic acid).

Plant Products for Management of the Losses

Although, chemical fungicides have been used since long for treatment of the seeds so infected by fungi that deteriorate health and quality of the seed, they are well known for their nontarget effects over the beneficial organisms or even the mankind and its properties. Chemical hazards have forced the scientists of the era to rethink of their use particularly with the much advocated theme of sustainable agriculture. Consequently, people associated with plant protection sciences tried to search for a suitable alternative of the agrochemicals, particularly fungicides or fungicidal principles, in the higher plants and their products. Due to high potency of antimicrobial properties, nonphytotoxic nature and easy availability with virtually no side effects, the higher plants and herbs are rapidly becoming popular alternatives to control several diseases of human being and animals. Their applications against phytopathogenic mycoflora in general and in storage are given much attention now-a-days to control these organisms causing severe damage to the crop and food commodities. The volatile and nonvolatile fractions of higher plants have been given much emphasis due to their high fungitoxic properties against parasitic, saprophytic and pathogenic fungi. The volatiles do not leave any residue; hence there is least chance of residual toxicity in treatment of food commodities.

In comparison to synthetic chemical compound, the microbicidal compounds of plant origin are more durable for their antimicrobial properties and have little or no side effects on human beings (Kumar *et al.*, 1995). Green plants appear to be reservoir of biotoxicants and constitute inexhaustible source of a large number of pesticides (Swaminathan, 1978). Hooda and Srivastava (1998) have mentioned that natural fungicides are free from environmental toxicity as compared to synthetic compounds. Natural compounds are less phytotoxic, easily biodegradable and more systematic (Saxena *et al.*, 2005).

A large number of plants belonging to angiosperm and gymnosperms have been screened for their fungitoxic properties by several workers. Manoharachary and Gourinath (1991) found that aqueous leaf extract of *Eucalyptus lanceolatus* was inhibitory for the germination and growth of *Curvularia lunata, Cylindrocarpon lichenicola* and *Fusarium solani*. Singh and Prasada (1993) found that leaf extract of *Azadirachta indica* and *Ocimum sanctum* inhibited the growth of *Fusarium oxysporum*. Aqueous extracts of plants have been used mostly to evaluate their fungitoxic properties (Thapliyal *et al.*, 2000; Algesaboopathi and Balu, 2002). Nguefack *et al.* (2007) tested the effect of seed treatment of rice with ethanolic extracts and essential oil of *Callistemon citrinus* and *Ocimum gratissimum* to manage the seed-borne fungus, *Bipolaris oryzae*. They reported that the incidence of the target fungus was reduced at a range of 42 to 100 per cent and also, that seedling emergence of variety Tox was doubled to the non-treated control following seed treatment with these botanicals when compared. Based on the results of an experiment conducted by Shafique *et al.* (2007), the investigators recommended that aqueous extracts of allelopathic trees especially those of *A. indica* and *M. indica* can be used to treat the wheat grains for 10 minutes before sowing or storage to reduce the fungal incidence. Similarly, Meena *et al.* (2010) tested leaf extract of ten medicinal plants against *Alternaria cucumerina*. They reported that, leaf extract of *Azadirachta indica, Calotropis gigantia* and *Aloe barbadensis* were found to be effective in controlling spread of *Alternaria cucumerina*. Aqueous extract of *Eucalyptus angophoroides* showed retardation of growth of storage fungi which reveals that it is most fungitoxic and can be used in biopesticide formulations. Rashid *et al.* (2010) has determined that garlic tablets at a dose of 1:3 (w/v) show better performance in increasing seed germination and reducing prevalence of fungal pathogens. Simultaneously, Kakade and Chavan (2011c) reported antifungal activity of aqueous extracts of leaves of several medicinal plants including *Azadirachta indica, Polyalthia longifolia* (both showed activity against *Macrophomina phaseolina, Rhizopus stolonifer* and *Penicillium digitatum*), *Eucalyptus angophoroides* (fungitoxic to the growth of *Alternaria dianthicola, Curvularia, pallescens, Fusarium oxysporum, Macrophomina phaseolina, Rhizopus stolonifer, Penicillium digitatum* and *P. chrysogenum*), *Vitex nigundo* (reduced the growth of *Alternaria dianthicola* and *Penicillium digitatum*), *Annona squamosa* (caused inhibition of the growth of *Penicillium digitatum* and *Fusarium equiseti*), *Murraya koeningii, Jatropha curcus, Withania somnifera* and *Datura strominum* with varying range of activity against several species of seedborne fungi tested.

References

Algesaboopathi, C. and Balu, S. (2002). Antifungal activity of some species of *Andrographis wallichex* Nees on *Helminthosporium oryzae* Breda deHann. *J. Economic and Taxonomic Botany* **24:** 705-707.

Bhattacharya, K. and Raha, S. (2002). Deteriorative changes of maize, groundnut and soybean seeds by fungi in storage. *Mycopathologia* **155:** 135–141.

Bilgrami, K.S, and Veram, R.N. (1978). Physiology of Fungi. Vikas Publishing House Pvt. Ltd., New Delhi. pp 597.

Chavan, A.M. and Kakde, R.B. (2008). Studies on abnormal oilseeds mycoflora from Marathwada region. *Bionano Frontier* **2 (2):** 101-104.

Christensen, C.M and Kaufmann, H.H. (1965). Deterioration of stored grain by fungi. *Annu. Rev. Phytopathol.* **3:** 69-84.

Christensen, C.M. (1972). Microflora and Seed Deterioration. *In:* Viability of Seeds (Roberts, E.H., *Ed.*). Chapman and Hall, London, pp. 59-93.

Christensen, C.M. and Kaufmann, H.H. (1969). *Grain Storage- The Role of Fungi in Quality Loss.* Univ. Minnesota Press, Minneapolis. 153pp.

Christensen, J.J. (1936). Association of micro-organisms in relation to seedling injury arising from infected seed. *Phytopathology* **26:** 1091-1105.

Christensen, J.J. and Stakman, E.C. (1935). Relation of *Fusarium* and *Helminthosporium* in barley seed to seedling blight and yield. *Phytopathology* **25:** 309-327.

Hooda, K.S. and Srivastava, M.P. (1998). Biochemical response of scented rice as influenced by fungitoxicant and neem products in relation to rice blast. *Indian J. Pl. Pathol.* **16:** 64- 66.

Kakde, R.B. and Chavan, A.M. (2011a). Extracellular lipase enzyme production by seed-borne fungi under the influence of physical factors. *Intern. J. of Biology* **3(1):** 94-100.

Kakde, R.B. and Chavan, A.M. (2011b). Effect of carbon, nitrogen, sulphur, phosphorus, antibiotic and vitamin sources on hydrolytic enzyme production by storage fungi. *Recent Research in Science and Technology* **3:** 20-28.

Kakde, R.B. and Chavan, A.M. (2011c). Deteriorative changes in oilseeds due to storage fungi and efficacy of botanicals. *Curr. Bot.* **2(1):** 17-22.

Koehler, B. (1938). Fungus growth in shelled corn as affected by moisture. *J. Agric. Res.* **56:** 291-307.

Kumar, A., Roy, S.K., Saxena, D.C. and Saxena, A.R. (1995). *In vitro* control of *E. coli* by herbal treatment. *Neo Botanica* **3:** 1-2.

Machacek, J.E. and Greaney, F.J. (1938). The 'black point' or 'kernel smudge' disease of cereals. *Can J. Res., C,* **16:** 84-113.

Malone, J.P. and Muskett, A.E. (1964). Seed-borne fungi-description of 77 fungus species. *Proc.Int. Seed Test. Ass.* **29:** 179-384.

Manoharachary, C. and Gourinath (1991). Antifungal properties of *Eucalyptus lanceolatus* extracts. *Ind. Bot. Reptr.* **10:** 37-39.

Meena, A.K., Godara, S.L. and Gangopadhaya, S. (2010). Efficacy of fungicide and plant extract against Alternaria blight of Clusterbean. *J. Mycol. Pl. Pathol.* **40 (2):** 272-275.

Nagel, C.M. and Semeniuk, G. (1947). Some mould induced changes in shelled corn. *Pl. Physiol.* **22:**20-23.

Nguefack, J., Torp, J., Leth, V., Lekane, Dongmo, J.B., Fotio, D. and Amvam Zollo, P.H. (2007). Effects of plant extracts and chemical fungicide in controlling a rice seed-borne fungus under laboratory and in irrigated cropping system in Ndop-Cameroon. *African Crop Science Conference Proceedings* **Vol. 8:** pp. 791-796.

Pepper, E.H. (1960). The microflora of barley kernels: their isolation, characterization, etiology and effect on barley, malt and malt products. *Ph.D. Thesis,* Dept. of Botany and Plant Pathology, Michigan State Univ., East Lansing (unpublished).

Prasad, B.K. (1979). Enzymatic studies of seedborne fungi of coriander. *Indian Phytopath.* **32:** 92-94.

Prasad, B.K., Shanker, U., Narayan, N., Dayal, S. and Kishor, A. (1988). Physico-chemical changes in food reserve of coriander seed due to storage moulds. *Indian Phytopath.* **41:** 386-388.

Rashid, M.M., Ruhul Amin, A.B.M. and Rahman, F. (2010). Determination of effective dose of garlic for controlling seedborne fungal diseases of tomato. *J. Yeast and Fungal Res.* **1 (9):** 183-187.

Saxena, A.R., Sahni, R.K., Yadav, H.L., Upadhyay, S.K. and Saxena, M. (2005). Antifungal activity of some higher plants against *Fusarium oxysporum* f.sp. *pisi. J. Liv. World* **12:** 32- 39.

Saxena, N. and Karan, D. (1991). Effect of seedborne fungi on protein and carbohydrate contents of sesame and sunflower seeds. *Indian Phytopath.* **44:** 134-136.

Shafique, S., Javaid, A., Bajwa, R. and Shafique, S. (2007). Effect of aqueous leaf extracts of allelopathic trees on germination and seed-borne mycoflora of wheat. *Pak. J. Bot.* **39(7):** 2619-2624.

Singh, H.N.P. and Prasada, M.M. (1993). Efficacy of leaf extracts of some medicinal plants against disease development in banana. *Lett. Appl. Microbiol.* **17:** 269-271.

Singh, M.T., Singh, M.S. and Pandey, R.R. (2001). Effect of storage fungi on electrolyte leakage from rice seeds. *Indian Phytopathol.* **54:**123-125.

Swaminathan, M.S. (1978). Inaugural address, *Fruit, Bot. conference,* Meerut, India, pp. 1-31.

Thapliyal, M., Ghosh, M., and Benne, S.S.R. (2000). Screening of six medicinal plants for their antifungal protein activity. *Asian J. Microbiol., Biotechnol. Environmental Sci.* **2:** 215- 218.

Umatale, M.V. (1995). Studies on fungal enzymes and toxins in biodeteriorations of oilseeds. *Ph.D. Thesis,* Dr. Babasaheb Ambedkar Marathwada University, Aurangabad (MS), India pp- 227.

Waghmare, B.M. (1996). Studies on seed-borne species of *Fusarium* (Link) from different Plant seeds. *Ph.D. Thesis,* Dr. Babasaheb Ambedkar Marathwada University, Aurangabad (MS), India pp- 217.

Chapter 30

Botanicals in Crop Disease Control

M.N. Khare[1], *S.P. Tiwari*[1] *and Roopa V. Sangvikar*[2]

[1]*Jawaharlal Nehru Agricultural University, Jabalpur, M.P.*

[2]*Department of Botany, Yeshwant Mahavidyalaya, Nanded, M.S.*

Crops are attacked by various pathogens from seed germination to adult plants and their produce resulting in enormous losses in quantity and quality. On the contrary due to ever increasing population attempts are being made to develop high yielding varieties and hybrids to get increased production per unit area, time and expenditure. The diseases are caused by fungi, bacteria, viruses, nematodes and phytoplasma. The chemicals are usually used to combat these maladies but they are hazardous to human and animal life. Besides they have nontarget effects influencing the beneficial microbes and disturbing the ecological relationship of various organisms in nature. The pathogens develop resistance to such chemicals when used in lesser dosage unknowingly by the farmers. The residues of chemicals in agricultural commodities are harmful to consumers and is of great concern at global level. The used chemicals cause pollution of water, soil and air as well.

Biological control has attracted the attention to check the pathogens either by the use of microbes antagonistic to pathogens or through botanicals. The plants possess antipathogenic chemicals in various parts, which are applied in different ways as seed treatment, soil treatment, pastes, spray and dusts. Their use after proper evaluation, experimentation *in vitro* and *in vivo* lead to an ecofriendly approach to manage crop diseases.

Historical

References of plants in use of plant disease control are made in 'Atharveda' and 'Rigveda'. The science of medicine for plants is narrated in 'Vraksayurveda' written by Surapala (Raychaudhuri, 1964; Sadhale, 1996). Accordingly the diseases are external and internal, caused by vata, pitta, kafa. The diseases of vata are soil borne; kafa types appear in winter and spring and of pitta type occur at the end of summer. The kafa type of diseases are controlled by bitter, strong and astringent decoctions made of panchmula *i.e.* roots of five plant species- sriphala, sarvatobhadra, patala, ganikarika and syonaka with fragrant water. The plants with pitta type diseases are controlled by watering the decoction of milk, honey, yastimadhu and madhuka. The vata type diseases are cured by fumigation of the mixture of fat of hog, ghee, hemp, hairs of horses and cows horn. In this way different types of diseases were treated.

Forsyth (1802) recommended decoctions of tobacco and elder buds with lime and sulphur for plant disease control. The decoction was used to wash young tender shoots infected with downy mildew. Fawcett and Spencer (1970) reviewed the earlier work on antifungal natural products from plants like carboxylic acid, aminoacids, phenolic compounds, quinones which usually induce resistance in host plants against pathogens. For last many decades use of plants in the control of plant pathogens and diseases caused by them is in practice. The active principle involved in mode of action has been isolated and identified. Khare and Shukla (1998) have reviewed the work on plants used in crop disease control. Mawar and Lodha (2008) have listed some plants whose extracts have been used in the control of plant pathogens.

Plant Part and Product Selection

The plant parts in controlling pathogens and diseases are bark, stem, root, kernel, flowers, pollen grains, bulbs, rhizomes, corms, seeds etc. The commonly used plant products in this endeavor are oils, oilcakes, resins, latex etc. The commonly used formulations are given below:

Cold Water Extract

The plant parts are pre-treated with chlorax with 7 per cent chlorine, ethanol or mercuric chloride 0.2 per cent,washed with sterile distilled water turned into pieces and ground. Sterile cool distilled water is added in the ratio of one part ground material and one part water, churned in a mixer grinder and filtered through cotton wool, muslin cloth, Whatman No.1 filter paper and finally with Seitz filter. This extract is considered of 100 per cent strength. Dilutions are made with sterilized distilled water.

Hot Water Extract

In place of cold water the temperature is raised to 70° C for extraction. At higher temperature the extract looses the antipathogenic property.

Use of Organic Solvents

Organic solvents like ethanol, ethyl acetate, dichloromethane, chloroform, methanol, acetone, hexane, petroleum ether etc. are used for extraction.

Two Fold Broth Dilution Method

The collected plant material is shade dried and grinded to powder form in a grinder and kept in air tight containers for further use. For extraction of bioactive components 100 g of the powder is subjected to continuous soxhlet extraction using 100 per cent ethanol as solvent and the obtained extracts are concentrated under reduced pressure using rotary vacuum evaporator. The crude extracts are kept in bottles, sealed and stored in refrigerator (4°C). The working solutions are prepared in dimethyl sulfoxide (DMSO) (Maharjan *et al.*, 2010). The antifungal activity can be interpreted in the form of minimum inhibitory concentration (MIC) and minimum fungicidal concentration (MFC).

Vapour Evaluation

Mishra and Dixit (1979) have devised a simple method to determine the fungitoxicity of vapours of plant extracts.

Oilcake Extract

Oilcakes are used as such as soil application for the control of plant pathogens including nematodes. Their extracts are also applied for disease control. The required quantity of oilcake is powdered, soaked in distilled water @ 1 g/ml, kept overnight, processed in mixer grinder and filtered. This is 100 per cent standard extract.

Oils

Plant oils are used as such with Teepol 0.1 per cent or as emulsion in desired concentration. Essential oil is extracted by hydrodistillation technique. A quantity of 500 g plant part is pre-treated with 0.2 per cent mercuric chloride solution for five minutes, thoroughly washed with double distilled water and pulverised in a sterilized grinder. This pulp is hydrodistilled in Clevenger's apparatus to collect aroma as essential oil. The oil is dried over anhydrous sodium sulphate to remove traces of moisture.

Leaf Powder

Leaves are dried in shade and powdered. This is used as such or diluted with talc or any other suitable inert matter.

Resins

These are dried plant exudates called resins or gums and used diluted in sterilized distilled water.

Plant Latex

Some plants exude latex from stem, leaves, fruits which are collected and used after dilution with distilled water.

Screening Methods

The plant extracts and other products are evaluated for their efficacy against pathogens *in vitro* by various techniques using several dosages. Those exhibiting positive performance are tested *in vivo* in pots and field. The testing methods are the same as used for testing the fungicides against pathogens and diseases (Dhingra and Sinclair, 1995).

In vitro Testing

Several techniques are available which are usually used in testing fungicides against pathogens in laboratory.

Filter Paper Disc Method

Filter paper discs of normal size are used in testing. They are dipped in the extract of known strength, dried and placed in the centre of petriplates on the medium seeded with the test pathogen. Potato dextrose agar, corn meal agar or any other medium preferred by the pathogen is used. Filter paper discs dipped in sterilized distilled water are used as control. The zone of inhibition is measured and the efficacy of the test material is calculated.

Poisoned Food Technique

Potato-dextrose–agar or any other suitable medium is prepared in a flask, sterilized and the required quantity of plant extract or any other material under test is added after cooling the medium. On proper mixing 20 ml of the medium is poured in each 9 cm diameter petriplate and solidified. A 0.7 cm disc is cut from a seven day old culture of the fungus grown on PDA and transferred in the centre of the petriplate upside down. Control is kept by transferring a disc without culture. The growth of the fungus is recorded after every 24 hours. The colony diameter compared with check exhibits the toxicity of the plant material.

Growth in Liquid Medium

Richard's or Czapek's liquid medium is generally used. A definite quantity of plant extract is mixed with the liquid medium in 250 ml conical flasks. One 7 mm disc of the test fungus is added in each flask. In the control flask the disc is added without the fungus culture. The incubation is done at the suitable temperature. The mycelial mat is harvested and fresh and dry weight is recorded. The decrease in dry weight due to the extract is calculated.

Cavity Slide Method

The plant extract is taken in test tubes in different dilutions and fungal spores are added. The suspension is transferred in cavity slides and incubated at the required temperature in moist chamber for germination. Sterile distilled water without plant extract is used as control. Effect on germ tube elongation is also observed.

One drop of plant extract is placed in cavity slide and spore suspension in 1 per cent sucrose is placed over the dried plant extract in the cavity and incubated in moist chamber. The spore germination count is made after regular interval.

Hanging Drop Method

Spore suspension is prepared in different plant extracts in water with definite number of spores per ml or per microscopic field. Suitable control without extract is maintained. A drop is transferred on a cover glass which is inverted in a cavity of the cavity slide and kept in moist chamber. The spore germination is recorded after a definite period and inhibition percentage of germination is calculated.

Detached Leaf Technique

Leaves are detached, washed with sterile water and dried to remove extra water. Plant products are sprayed or dusted on both the leaf surfaces and again dried. Spore suspension is either dropped or sprayed on leaf surfaces and kept for incubation in moist chamber. The leaves are examined for fungal activity after definite time intervals under a suitable microscope (Tuite, 1969).

Leaf Disc Test

In order to examine the movement of plant extract of product through the leaf, detached leaf is inoculated by applying dry spores of the fungus or their suspension on the leaf surface and discs of uniform size (1 cm or more) cut by cork borer. The test plant product as solution in different doses is taken in petriplate and the discs are floated. The discs can also be cut first followed by inoculation. Incubation is done until the disease symptoms develop on control discs in sterile water. Disease intensity and phytotoxicity is assessed (Fawcett and Spencer, 1970).

In vivo Testing

Pot Culture Studies

The seeds are soaked in plant extract for 30 minutes, dried in shade for two hours and sown in pots containing sterilized soil infested with the pathogen. The proportion of the inoculum depends on the type of the pathogen. Similarly number of seeds per pot depends on the size of the plant type. Pots with uninfested sterilized soil and untreated seeds are used as control. Data are recorded for seedling emergence and post emergence diseases. The plant materials or their products are added in soil and seeds infested with the pathogen are sown. Data are recorded as in the previous one.

Seed Treatment with Leaf Powder

Leaves are dried, powdered and seeds are treated @ 4 g/kg seed before sowing. The treated seeds can be stored to check their efficacy under storage.

Plant Extracts as Spray and Plant Powders as Dust

Based on the results of *in vitro* testing, specific concentrations of plant extracts are sprayed on crop plants at the required growth stage. Artificial inoculation of the pathogen is done at the proper time and sprays are given prior to artificial inoculation and after the inoculation. The phytotoxicity, if any, is also observed. Powders are dusted and evaluated in controlling the pathogens and diseases.

Field Trials

In case of annuals, the plants are raised in field or in pots with suitable replications and design. The plant product is sprayed or dusted at the proper time. Suitable control is maintained with spray of sterile distilled water. The percent disease incidence is recorded.

Mode of Activity of Plant Products

When the extracts of plant parts and other products are applied to check the activity of pathogens and the diseases on crops they inhibit the spore germination, the growth of the germ tube, formation of appressoria, induce resistance in host plants and act as barrier in penetration of the pathogen into the host system. The spores get deformed and distorted. Some plant materials act as antisporulant and disturb the physiological activities interfering with the life processes of the pathogen. Antipathogen compounds are produced by the plant parts.

Efficacy of Plant Extracts and Products

The effectiveness of the plant products against the pathogens has been tested *in vitro* and *in vivo* by several workers. The preliminary screening of a large number of plants is done by *in vitro* method and the promising ones are tested *in vivo*.

Dubey and Dwivedi (1991) reported inhibitory effect of extract of leaf and bulb of *Allium cepa* and *A. sativum* and fruit and bark extract of *Acacia arabica* on the growth and sclerotial variability of *Macrophomina phaseolina*. Kalaichelvan and Nagarajan (1992) extracted an alkaloid monocrotaline from seeds of *Crotalaria paleda* which at 500 µg inhibited 58 per cent spore germination of *Curvularia lunata*. The action is fungistatic. Leaf extracts of *Azadirachta indica, Lantana camera and Cathranthus roseus* checked the mycelial growth and spore germination of *C. lunata, Fusarium moniliforme, Aspergillus flavus and Rhizopus stolonifer* (Meena and Mariappan, 1993). They have also reported similar activity in flower extracts of *C.roseus*. According to Patil *et al.* (1992) 10 per cent extract of *Ocimum sanctum* leaves inhibited the spore germination of *R.arrhizus* by 69.5 per cent and of *Botryodiplodia theobromae* by 65.6 per cent and also checked activities of pectinolytic and cellulolytic enzymes. Jariwala *et al.* (1992) reported effectivity of dry root powder of *Asparagus adscendens* in inhibiting the radial growth of *Dechslera oryzae* and *Alternaria solani*. Bhowmick and Choudhary (1982) found antifungal activity in leaf extracts of medicinal plants against *A. alternata* and Bhowmick and Varadhan (1981) and Upadhyay and Gupta (1990) against *C. lunata*.

The napthoquinones isolated from dichloromethane extract of *Newbouldia laevis* roots was effective against *Cladosporium cucumarium,* I benzofuran. 2 lignans. 2 dammarane triterpenoids and limonoid from the bark of *Aglaria elaegnoidea* acted against *C. cucumarium*. Potassium dichromate oxidised cashew nut shell liquid was most active against *C. cladosporioides*. Aqueous extract of *Ranunculus asiaticus* completely inhibited the growth and sporulation of *A. solani, Rhizoctonia solani* and *Helminthosporium sativum*.

Ethanolic extracts of plants tested by filter paper disc method exhibited very strong activity of *Piper betle* leaf extract against *A. alternata, B theobromae, Phomopsis caricae papayae*. Pollen from *Xanthium strumarium* inhibited spore germination of *D.oryzae* (Tripathi *et al.*, 1982). Iridodial β- monoenol acetate isolated from essential oils of *Nepata leucophylla* was most effective against *Sclerotium rolfsii*. Actinidine obtained from *N. clarkei* was highly active against *M. phaseolina* (Saxena and Mathela, 1966). Methnol extract of *Aegle marmelos* leaves at 150 ppm completely checked formation of sclerotia in *S.rolfsii* (Prithviraj *et al.*, 1996). Sivasithambaram *et al.* (1981) checked *Phytophthora cinnamomi* by adding fresh saw dust and composted tree barks.

Prithviraj *et al.* (1977) isolated anacardic acid. They reported antifunfal activity of anacardic acid and xanthone against *Colletotrichum capsici, A. brassicae, A. alternata, A. carthami and C. lunata.*

Rai *ct al.* (1995) reported that power density memory of water has synergistic effect on the inhibitory activity of ajoene, disturbing the enzymatic activity of spores of fungi like *A. alternata, A tenuissima, Fusarium udum, F. lini, H.turcium and D. oryzae.* Several pathogens and diseases caused by them in different crops have been tested for the efficacy of plant products for their control.

Plant Products

Plant products like oil cakes, resins, latex and oils have been used in controlling plant pathogens and diseases.

Oil Cakes

Rhizome rot of ginger caused by *Pythium amphanidermatum* and *F. solani* was checked by soil amendment with cakes of neem, mustard, *Pongamia glabra* and *Callophylum inophyllum* (Thakore *et al.*, 1987). Ekka and Prasad (2010) also reported cakes of neem and *P. glabra* 20 q/ha to reduce the incidence of rhizome rot due to *P. aphanidermatum* and *F.oxysporum* f.sp. *zingiberi* to 54.40 per cent and 49.60 per cent and increased yield by 20.44 per cent and 19.83 per cent respectively. Neem cake 150 kg/ha effectively controlled sheath blight of rice when applied in soil (Senapoty, 2010). Pre and post emergence rots caused by *R.solani* in cotton were highly reduced by adding neem cake in soil. The treatment also reduced root rot of soybean (*M.phaseolina*), wilt of coconut (*Ganoderma lucidum*), betle vine (*Phytophthora capsici*) and root rot of fenugreek (*R. solani*) (Rajan *et al.*, 1991). Thanjavur wilt of coconut was controlled by neem cake application. Ratnoo and Bhatnagar (1993) controlled ashy stem blight caused by *M. phaseolina* by adding straw and oil cakes. The extracts of neem cake, groundnut cake, sesame cake and coconut cake at 1,2 and 3 per cent concentrations w/v stimulated the germination and germ tube elongation followed by lysis of sclerotia of *M.phaseolina* (Muthusamy and Mariappan,1992). Groundnut and mustard oil cakes at 2 per cent concentration of soil w/w reduced the population and activity and F. *oxysporum* f.sp. *lycopersici* propagules and disease incidence (Raj and Kapoor,1996).

Resins

The mycelial growth of *R.solani* and *S.sclerotiorum* was completely inhibited at 7000 and 2000 ppm concentrations of asafoetida in Czapek's Dox broth. Formation of sclerotia was completely inhibited at 1000 ppm and above and to both the pathogens (Chaurasia and Dayal, 1990). Asfoetida is obtained as resin from *Ferula foetida* and it contains ferulic acid and certain volatile substances which are the active components. The application of ferulic acid to rice plants induced resistance to *P.oryzae* by neutralizing the toxin pyricularin.

Latex

The latex of *Euphorbia hirta* (100 per cent) controlled damage of tomato fruits against *A. niger* and insect (*Drosophila busckii*) when dipped in latex. Dilutions were not so effective (Sinha and Saxena, 1990). Latex of *Calotropis procera* controlled seed mycoflora of wheat (Abul-Fazal *et al.*, 1987). Khare and Dhingra (1974) reported fungistatic nature of papaya latex against *Gloeosporium papayae*. The fruits exhibited resistance to pathogen till they are rich in latex. The latex contains phenolic compounds like euphosterol, phytosterol, phytosterolin, organic acids like gallic, mellissic, palmic, oleic, lenoleic acids and alkaloid xanthorhamnin, which adversely affect both the pathogen and insect. Sengupta *et al.* (2008) reported seed treatment of paddy with aqueous plant extract 400 µg/ml of *Plumeria actifolia* for six hours and dried before sowing resulted in reduced disease incidence due to *R solani*. The foliar spray had 30 per cent reduction in the disease. Several alkaloids and glycoside were found in the latex extract.

Oils

Oils from various plants have been found useful in controlling plant pathogens. Banerjee *et al.* (1989) reported inhibition of germination of sclerotia of *S. rolfsii*, *S. hydrophillum* and *Rhizoctonia oryzae sativae* when soaked in 5 per cent emulsion of oil obtained from *A. indica* and *Cymbopogon nardus*. Senthilnathan and Narasimhan (1994) ovserved maximum inhibition of mycelial growth and spore germination of *A. tenuissima* causing onion leaf blight. Singh and Singh (1982) reported inhibition of conidial germination of *Erysiphe polygoni* causing powdery mildew of pea by oils. Neem oil (100 per cent) resulted in 100 per cent inhibition of *F. moniliforme* and *M. phaseolina* (Vir and Sharma, 1985). Jagannathan and Narsimhan (1988) reported spray of neem oil to be effective against blast and blight of finger-millet. Spray of 3 per cent neem oil controlled powdery mildew of blackgram (Mariappan and Narsimhan, 1987) and sheath rot of paddy (Narsimhan *et al.*, 1994). Six plant oils were tested against *S. rolfsii* and ten other soil fungi, neem leaf oil was most effective followed by that of *Eucalyptus globulus* and *Ocimum canum*. Neem oil also checked the spore germination and growth of *Pyricularia grisea* and *H. nodulosum* (Jagannathan and Narsimhan, 1988).

Essential oil of *Ocimum canum* had most toxic effect and checked germination of sclerotia of *M. phaseolina* followed by *Citrus medica* and *Pinus roxburghii*. Both volatile and non volatile constituents are responsible for the inhibition of sclerotial germination (Dubey, 1991). According to Guenther (1966) essential oil of *C.medica* contained citral, limonens and ditentene. *O. canum* contained citral, citronellal, linalool, methyl cinnamats, eugenol, d-camphor, traces of phenols and acetic acid. *P.roxburghii* has α and β-pipene, I-limonen, candinene, cryptone,phellandrene, fatty acids and phenols (Guenther, 1972). Singh and Singh (1980) observed inhibition of growth and sclerotium formation in *R. solani* by the treatment with garlic oil. Essential oils obtained from *Apium graveolens, Cuminum cyminum and Zanthoxylem alatum* are reported fungistatic at lower and fungicidal at higher doses against aflatoxin producing strains of *A.flavus* and *A. parasiticus* (Dubey *et al.*, 1991).

Cereals

Paddy

Paddy diseases have attracted more attention. Mycelial growth of *D. oryzae* was checked by the leaf extract of *Adenocalymma allicea* by Chaturvedi *et al.* (1987) and

Lawsonia inermis by Natrajan and Lalithakumari (1987). Inhibition of spore germination of *D. oryzae* was obtained by leaf extracts of *Eleveise caracona* and *Sapium japonicum* (Oigashi *et al.*, 1972). Leaf extract of *Juglans regia, Allium sativum, Origanum vulgare* and *A.wallichii* resulted in the inhibition of mycelial growth in terms of dry weight by 64.11, 59.47. 56.22 and 52.77 per cent respectively (Bisht and Khulbe, 1995). The leaf extracts of *Tridax procumbens* and young stem extract of *Cylindropuntia ramissima* exhibited antisporulent activity (Selvaraj and Narayanasamy, 1994 b). the extracts of *Ipomoea cornea* and *Cascabela thevetia* reduced brown spot in field to a considerable level (Selvaraj and Narayanasamy, 1994a). Neemazal 3 ml/l and Wanis 5 ml/l provided 26 per cent reduction in leaf spot phase and Neemgold, Achook, Thuja leaf and garlic clove extract reduced stalk rot incidence by 19.3 per cent (Sunder *et al.*, 2010).

The spore germination of *Sarocladium oryzae* was checked by pepper seed extract and garlic bulb extract (Kanagarajan, 1974). The young stem extract of *Euphorbia tirucalli* and matured leaf extract of *Urginea indica* exhibited antisporulant activity (Selvaraj and Narayanasamy, 1994 b). Extracts of *Tribulus terrestris, Cartharanthus roseus* and *Ocimum tenuiflorum* reduced both brown spot and sheath rot of paddy when applied as spray (Selvaraj and Narayanasamy, 1994 a).

Seed extract of Ajwain (*Trachispermum ammi*) (1:20) did not allow any growth of *R. solani* causing sheath blight of paddy. It was followed by *Ocimum* sp. and *L.inermis*. In pot experiment Ajwain seed extract reduced the disease by 72.25 per cent when the sclerotia of the fungus were inoculated between leaf sheath (Ansari, 1995). Maximum inhibition of germination was observed by leaf and oil cake extracts of *Thevatia peruviana, Prosopis juliflora* and *Eucalyptus globulus* (Ezhilan *et al.*, 1994).

Ethanolic leaf extracts of *Calotropis procera, A. indica* and *Datura stramonium* were effective against *Pyricularia oryzae, R.solani* and *F.moniliforme* (Mishra *et al.*, 1990). Garlic extract was inhibitory to *P.oryzae, D.oryzae* and *Cortcium sasakii* also (Tewari and Dath,1984). Two benzopyran derivatives, 6 acetyl-2,2-dimethyl 1,2-benzopyran and 6-acetyl -7-4 hydroxy 2,2-dimethyl 1,2-benzopyran were isolated from ethanol extract of sunflower receptacles which were highly effective against *P. oryzae*. Behura *et al.* (2000) reported leaf extracts of *Curcuma longa* highly inhibitory to *R. solani, D. oryzae* and *F. moniliforme* and lesser effective against *Trichoconis padwickii* and *C. lunata*. *Adhatoda vasica* leaf extract (5 per cent) spray effectively controlled bacterial blight caused by *Xanthomonas oryzae* pv. *oryzae*. This activity is due to high levels of glycoprotein and tannin contained in the extract (Madhiazhagan *et al.*, 2002). Kandhari *et al.* (2010) tested spraying of essential oils, aroma compounds and plant extracts against sheath blight of rice and found among essential oils winter green (*Gaulteria procumbens*) had 82.3 per cent disease reduction followed by Patchouli. Out of aroma compounds used eugenol had 70.7 per cent disease reduction. Root dipping of seedlings before transplantation also had disease reduction at adult stage of crop growth. *Piper betle* extract showed 44.4 per cent reduction in disease incidence. Khandari (2007) found neem formulations Achook (0.15 per cent), neemark (0.5 per cent), Tricure excellent in controlling sheath blight. Khan and Sinha (2006) have reported FYM+ *T. harzianum*, neem cake+ *T harzianum* and sawdust + *T. harzianum* controlled sheath blight and gave higher yields and grain weight. Murlidharan *et al.*

(2003) found application of Achook. Neemazal,Neemgold and Wanis to reduce the severity of both leaf and neck blast. Tricure was the best in controlling sheath blight tiller infection. Achook, Neemazal, Neemgold, Spictaf and Wanis also reduced sheath blight. They all gave higher yields.

Wheat

Extracts of *Lawsonia inermis, Datura stramonium* and neem were highly effective against black point disease of wheat caused by *A. alternata* (Abdul *et al.*, 2005).

Sorghum

Neem seed extract (0.2 per cent) seed treatment plus spray resulted in 17.5 per cent increase in grain yield and 17.9 per cent increase in fodder yield while the efficacy of disease control of leaf blight caused by *Exserohilum turcicum* was 50.7 per cent (Bunker and Mathur, 2008).Singh and Nair (2000) have reported control of ergot of sorghum by the spray of crude aqueous garlic extract (12 per cent). The conidial germination was completely inhibited at 6 per cent concentration.

Maize

Drechslera maydis causes leaf spot disease in maize. Extract of *Portulaca oleracea* controls the disease as it possesses therapeutic property against the pathogen (Noriel and Rables, 1990).

Barley

Yadav and Gour (2007) found aqueous extract of leaves of neem provided control of leaf stripe caused by *Drechslera graminea*. The treated leaves exhibited significantly high activity of enzymes phenylalanine ammonia–lyase (PAL) and tyrosine ammonia-lyase (TAL) with accumulation of phenolic compounds.

Pulses

Pea

Garlic bulb extract (3000 ppm), garlic oil (750 ppm), neem leaf extract (3000 ppm), neem stem extract (3000 ppm), neem pulp (2000 ppm), neem oil (3000 ppm), turmeric rhizome extract (3000 ppm) and ginger rhizome extract (3000 ppm) completely inhibited the germination of conidia of *Erysiphe polygoni* causing powdery mildew of pea (Singh and Singh,1983). In field, ginger extract (3000 ppm) had the minimum powdery mildew inoulum intensity when tested as spray followed by garlic bulb extract (3000 ppm), garlic oil and neem leaf extracts (Singh *et al.*, 1984, 1991).

For obtaining the extract of ginger rhizome, they were crushed in a blender and centrifuged at 500C rpm for 30 minutes. The supernatant was taken as 100 per cent. The dilution was made with sterile distilled water. Singh *et al.* (1983) observed the active principle to be 3, 7-dimethyl-2, 6 octadieneol (citral).

Chickpea

Singh *et al.* (1979) tested the effect of aqueous garlic leaf extract in liquid medium on the growth of *Fusarium oxysporum* f.sp. *ciceri* and *Sclerotinia sclerotiorum* causing wilt and rot in chickpea. Their growth was greatly checked at 7000 and 5000 ppm respectively. The germinability of sclerotia of *S.sclerotiorum* was also checked. Extract treated seed, sown in infested soil, produced wilt free seedlings. Induction of resistance in chickpea against *S. sclerotiorum* by the use of extract of *A. marmelos* leaves has been reported by Singh *et al.* (1990). Ratnoo and Bhatnagar (1993) reported control of grey stem blight by soil amendment with neem cake. Allicin a putative antimicrobial agent was isolated from garlic extract as well as an antithrombotic compound called ajoene.

Pigeonpea

Singh *et al.* (1990) tested the antifungal activity of ajoene which checked germination of conidia of *F. oxysporum, F.udum* and *Colletotrichum* sp. Chouhan and Singh (1991) and Singh and Chouhan (1992) examined the effect of ajoene by mixing in potato dextrose broth. It checked the growth of *Phytophthora drechsleri* f.sp. *cajani* sporangium formation, its germination and zoospore germination at 20 ppm concentration. Singh and Singh (1983) reported inhibition in the growth of *F.udum* when infested soil was amended with ether distillate of Margosa cake.

Lentil

Singh *et al.* (1995) reported efficacy of extracts of *Ranunculus scleratus* at different growth stages. The leaves harvested just before flowering were most fugitoxic to *F.oxysporum* f.sp. *lentis* followed by stage after flowering. Singh *et al.* (1992, 1995 b) screened 40 species of higher plants against the fungus of which extract from *Impatience balsamina, Lawsonia inermis* and *Adena calymma allicea* completely inhibited the mycelial growth. They have also reported that *Mentha spicata* leaf extract (1:2 w/v) inhibited the growth of the wilt fungus completely. The extract was active after six days storage and after autoclaving for 30 minutes at 1.05 kg/cm2 (Singh *et al.*, 1994). Alkaline extracts of pine bark were more effective than neutral or acidic extracts in checking the growth of *R.solani, S. rolfsii, F.oxysporum, Phytophthora parasitica, A. solani* and *S. sclerotiorum*. The emergence of lentil was sifnificantly increased by the addition of fresh or composted Pine bark powder to soil infested with *R.solani* and *S. rolfsii*.

Phaseolus mungo

A protein mungin was isolated from beans which had inhibiting property against *R. solani, Botrytis cinerea, F. oxysporum* and *Mycosphaerella arachidicola* (Ye and Neg,2000).The application of extract of *A.indica* reduced severity of powdery mildew of urd and mungbean (*Erysiphe polygoni*) (Lakpale *et al.*, 2008). Dubey (2003) reported the efficacy of Karanj leaf extract (10 per cent) as spray in controlling the web blight caused by *R. solani* on both blackgram and greengram with higher yields. The soil application of Karanj cake and foliar spray of Karanj leaf extract resulted in minimum disease intensity and maximum grain yield.

Oilseeds

Groundnut

Ganapathy and Narayanasamy (1994) tested products of 43 plant species against late leaf spot and rust of groundnut and found that extract of *A. indica* young leaf and seed, *Nerium odorum* leaf, *Vinca rosea* leaf and *Allium cepa* bulb inhibited spore germination upto 90 percent. Neem based oil, neem oil (1:1 with water); castor seed oil and extract of neem cake also checked the spore germination of both the fungi to some extent. Conidia of the two fungi did not germinate on leaves when sprayed with the extract of young leaf and seed of *A. indica, Eucalyptus* sp., red periwinkle, nerium, coir pith and liquorice leaf. In neem oil the conidia germinated but did not form appressoria due to which the fungus did not penetrate the leaf surface. In nerium leaf extract the conidia of *P. personata* swelled but failed to germinate. Malformation of conidia of *Curvularia pallescens* was observed in the extracts of *Eucalyptus* sp. and Jasmine by Rajiv Kumar and Sachan (1979). Neem oil caused distortion of rust spores which did not germinate. Chandrasekar *et al.* (1994) reported neem seed kernel extract useful as spray in checking Tikka disease when applied with K_2O. The disease severity of rust (*Puccinia arachidis*) was reduced by the application of extracts of *A.indica, L. camera, Parthenium* sp. and *Ipomoea cornea* (Lakpale *et al.*, 2008). Aqueous leaf extracts of *Tinospora cordifolia, Solanum nigram, Acalypha indica, Vitex trifolia* and six more plants reduced seed mycoflora of sesame, groundnut and castor and increased seedling growth (Tippeswamy *et al.*, 2003).

Mustard

White rust (*Albugo candida*) severity was much less when the extracts of *Euphorbia hirta, L. camera, S nigram* and *I.cornea* were applied as spray (Lakpale *et al.*, 2008). Yadav (2009) found three sprays of 1 per cent garlic clove extract to check white rust, Alternaria blight and Sclerotinia rot. It gave higher yields. However, Kumar (2009) reported least incidence of white rust and staghead infection by spraying extracts of Eucalyptus, neem and garlic.

Soybean

Seed treatment followed by foliar spray of *Lawsonia inermis* (1 per cent) + alum (0.1 per cent) was effective in reducing leaf anthracnose and pod blight incidence (Chandrasekaran and Rajappan, 2002). Patni *et al.* (2005) tested leaf extract of six plants to control Alternaria blight (*A. brassicae*) by spray of which *Eucalyptus globulus* leaf extract resulted in least disease with higher yield.

Vegetables

Brinjal

Jacob and Sivaprakasam (1994) found the leaf extracts of *Eucalyptus terticornis* to give best control of *Pythium aphanidermatum* causing damping off in brinjal. Narayan Bhat and Sivaprakasam (1994) tested cold and hot water leaf extracts of 25 crop plants and 30 forest trees *in vitro*. The cold water exract of *Polyalthia longifolia* exhibited 56.6 per cent inhibition of mycelial growth. Whereas hot water extract of *E.microtheca* resulted in 90 per cent inhibition. Leaf extract of some plants were more effective in cold water, whereas in some cases hot water was more effective. Raja (2010) found 10 per cent garlic extract spray highly toxic to *Alternaria tenuissima* which causes leaf spot and fruit rot. It was followed by neem extract.

Tomato

Seedling treatment with a mixture of asafoetida 1g + turmeric powder 5 g in 10 l water as spray or soil drenching at 15,30 and 45 days after transplanting controlled bacterial wilt caused by *Rolstonia solanacearum* (Sharma and Kumar, 2009). They have also found that soil treatment with Karanj (*Deris indica*) 10 q/ha reduced the inoculum load of the bacterium in soil and disease incidence.

Chilli

Muthulaxmi and Seetharaman (1994) tested leaf extracts of five plants *A.marmelos, P.juliflora, I. cornea, O.sanctum* and *Bougainvillea spectabilis* against *A. tenuis* causing fruit rot of chilli (*Capsicum annum*). In the *in vitro* studies poisoned food technique *A. marmelos* exhibited maximum inhibition of mycelial growth (87.56 per cent) followed by *P.juliflora* (83.72 per cent) and *I. cornea* (70.12 per cent). In pot studies both pre-inoculation and post-inoculation spray of leaf extracts significantly reduced fruit rot. *A.marmelos* leaf extract (10 per cent) spray had least fruit rot incidence 21.70 and 13.48 percent respectively followed by *P. juliflora* leaf extract (26.8 and 18.77 per cent) as compared to control (84.62 per cent under pre and post inoculation spray respectively. The yield was maximum (78g/pot) in case of *A.marmelos* leaf extract followed by *P. juliflora* leaf extract (72 g) as compared to control (55g). Bohra *et al.* (2006) successfully managed damping off of brinjal and chilli by application of neem oil (50 per cent) and azadirachtin (1500 ppm) as soil drench and seed treatment.

Sponge Gourd

Ahmad and Prasad (1995) reported that the leaf extract of *Datura fistulosa* and *Ocimum sanctum* reduced soft rot of Sponge gourd fruits caused by *Fusarium scirpi.*

Watermelon

Garlic clove juice was effective in controlling Fusarium wilt of watermelon (El-shami *et al.*, 1986).

Cowpea

Seed treatment of cowpea with fresh leaf extract of *Moringa olifera* controlled seed borne *Colletotrichum destructum* casual agent of anthracnose disease (Akinbode and Ikotun, 2008).

Clusterbean

Jain and Jaiman (2005) reported reduction on the incidence of *M.phaseolina* due to seed treatment with neem leaf powder as well as turmeric powder with higher percentage of seed germination. Bhatnagar *et al.* (2009) found garlic extract (1:1) effective against stem blight caused by *M.phaseolina*. It resulted in three times yield. The crop suffers due to root rot complex for which *R. solani* and *F.solani* are responsible. Jatav and Mathur (2005) tested four neem based formulations, neem seed extract, neem oil, azadiractin and achook 0.2 per cent and found achook to be best in controlling the disease.

Mushroom

In *Agaricus bisporus* white button mushroom wet bubble (*Mycogone perniciosa*), dry bubble (*Verticillium fungicola*) and wilt disease (*Fusarium moniliforme*) are important diseases. Singh and Singh (2005) evaluated extracts of 26 plants species of which six plants showed promise under *in vitro* testing. *Erigeron karvinskianus* was the best in controlling the three pathogens *in vivo*.

Onion

Stemphylium blight of onion (*S. botryosum*) is seed as well as soil borne disease. Leaf extracts of six plants (20 per cent) were tested in field. The lower leaf disease intensity was recorded in case of neem leaf extract followed by *Datura metel* extract. They gave better yield also.

Potato

Maharjan *et al.* (2010) reported antifungal activity of crude ethanol extract of *Brassica nigra* by agar well diffusion technique against *Phytophthora infestans*. Ghorbani *et al.* (2005) found compost tea extract to reduce the late blight of potato 30 to 40 per cent. Khair and Wafaa (2007) reported reduction in late blight disease severity by the application of extracts of lemon grass and chilli fruits. Somani (2009) tested fresh leaf extracts of 20 plants in water 1:5 w/v as 30 min dip treatment of naturally infected tubers with black scurf (*R. solani*) in naturally infested field and found extracts of *Vinca rosea, Wethania somnifera* and *Ocimum canum* very effective in controlling the disease. According to him these ecofriendly botanicals have the potential to control black scurf and their continuous use may help in better management of the disease. Dhaliwal *et al.* (2003) controlled black scurf of potato by treating the infected tubers with essential oils from *Cyperus scariousus, Mentha piperita* and *Cymbopogon citratus* by dip method.

Fruits

Mulberry

Sarvamangala *et al.* (1993) reported 92 per cent inhibition of germination of uredospores of *Cerotelium fici* by leaf extract of *A.indica* followed by 79.1 per cent by *Eucalyptus* sp. Leaf spot disease caused by *Cercospora moricola* was effectively controlled by leaf extract of *Eucalyptus* sp. According to Biswas *et al.* (1995) whole plant extract (excluding root) of *Adhatoda zeylanica* was most effective against powdery mildew (*Phyllactinia corylea*), leaf spot (*Psuedocercospora mori*) and rust when used as spray. It was followed by leaf and stem extract of *A.indica.*

Banana

In banana fruit rot caused by *B. theobromae, H. spiciferum, Trichothecium roseum, A. flavus* is controlled by fruit dip in the extract of *A. indica* and *O. sanctum*. The symptoms were delayed in treated fruits and they had minimum loss in weight (Singh *et al.*, 1993).

Mango

Fruit rot due to *R.arrhizus* and *B. theobromae* can be minimised by dipping the fruits in 10 per cent extract of *O. sanctum* leaves (Patil *et al.*, 1992). Reduction in rots under post harvest condition in Alphanso mango fruit by dipping in garlic bulb extract has been reported by Hasabins and D'souza (1987).

Lemon

Babu and Reddy (1986) reported control of lemon rot caused by *Colletotrichum gloeosporioides* by the application of leaf extract of *Lawsonia inermis.*

Grapes

Geotrichum candidum causes severe fruit rot of grapes on vines, storage and transit. Extract of garlic (10 per cent) and onion (50 per cent) checked the rot development of grape fruits completely (Mehta and Mehta, 2005).

Pomegranate

Extracts of garlic cloves and turmeric rhizome (5 per cent) application resulted in good control of fruit rot due to *Alternaria alternata* (Singh and Majumdar, 2001).

Kinnow

Foliar spray of garlic plus green chilli extract 2.5 g and 250 mg respectively per lit water reduced canker caused by *Xanthomonas axonopodis* pv. *citri* (Gaur and Sharma, 2010).

Other Crops

Sugarcane

High inhibitory effect of *Polyalthia longifolia* against *Colletotrichum falcatum* causing red rot of sugarcane has been reported by Kishore *et al.* (1982).

Coconut

Anish Kumar *et al.* (2004) evaluated extracts from 17 plant species against three major pathogens of leaf rot *Colletotrichum gloeosporioides, Exserohilum rostralum* and *Fusarium solani* and found acetone extracted extracts of *Adenocalymina allicea, Lawsonia inermis, Azadirachta indica* were highly promising.

Spices

Gangopadhyay *et al.* (2010) have reported two sprays of aqueous extracts (10 per cent) of *Calotropis procera, Azadirachta indica* seed kernel and leaf suppressed Alternaria blight of Cumin and enhanced seed yield.

Cashew

The inflorescence blight of cashew is controlled by the spray of extract of *Piper guineense* 5 per cent and 10 per cent and in combination of garlic extract. The combined extracts of *P. guineense* 5 per cent, *Ocimum gratissimum* 7.5 per cent and *Chromalaena odorata* 10 per cent reduced the incidence of the disease and increased yield (Adejumo and Otuonye,2002).

Neem

Die-back of neem is caused by *Phomopsis azadirachtae*, it results in 100 per cent loss of fruit production in severely diseased trees. Out of 25 plants tested *Lawsonia inermis* gave promising results (Sateesh, 2009).

Tea

Aqueous extract of *Catharanthus roseus* application reduced foliar blight caused by *Alternaria alternata*. The defence enzymes β-1,3 glucanase, chitinase, phenylalanine ammonia lyase and phenols were induced to a high level by the extract (Chakraborty *et al.*, 2007).

Betelvine

Anthracnose caused by *Colletotrichum capsici* and bacterial leaf spot caused by *Xanthomonas campestris* pv. *betlicola* are important diseases of betelvine. Daka *et al.* (2008) found extract of *Polygonum hydropiper* and *Tagetes erecta* inhibited the growth of both the pathogens *in vitro*. In field sanitation plus *P. hydropiper* extract spray resulted in 9.7 per cent disease index and was quite effective in controlling the leaf spot complex over control.

Seed Treatment

Several plant products have been tried for seed treatment and have been found useful in controlling seed borne pathogens. The treatment also helps in safe storage of seeds.

Wheat

Shah *et al.* (1993) tested stem, leaf and flower extracts of *Chrysanthemum roseum, C.coronarium* and *C. cinerariafollum* at 100 per cent concentration as seed treatment of wheat for 30 and 60 minutes. The treatment checked 12 seed borne fungi completely but *A. alternata* and *D. graminea* partially and the seed germination was improved. Seed treatment for 60 minutes was better. Khan (1989a) reported antimicrobial action of compositae plants on seed microflora of wheat. He found *Nicotiana* spp. to be very effective.

Paddy

Rao and Ratnasudhakar (1992) controlled fungi and bacteria associated with paddy grains by seed treatment with rhizome powder of *Acorus calamus* and storing in steel bins. The treatment did not affect the cooking quality. Leaf extract of *Mentha piperita* reduced seed borne *D. oryzae* significantly (Alice and Rao, 1986). Sagar *et al.* (2002) reported the effect of extracts of *A. indica, Ocimum sanctum, Parthenium hysterophorus, Prosopsis julifora* and *Bougainvillea* to control seed mycoflora of paddy with higher seed germination and seedling vigour.

Setaria italica

Seed treatment with aqueous extract of *Datura alba* (20 per cent) and *Allium sativum* (10 per cent) checked *C. lunata* associated with seed.

Sorghum

Rao (1991) tested leaf bits, leaf powder, fruit powder of *A. indica*, rhizome cut pieces and powder of *Acorus calamus* and *Curcuma longa*, leaf bits and leaf powder of *Eucalyptus* sp. and *N. tabaccum* as pre-storage seed treatment of sorghum to control molds in storage. All the treatments had lesser incidence of fungi, however leaf bits and leaf powder of *N. tabaccum* was most effective. The leaf extracts of *A. indica*, *C. roseum and L. camera* checked seed borne *A. alternata*, *F. moniliforme and A. flavus* (Meena and Mariappan,1993). Seed treatment with leaf powder of *A. marmelos* was also effective in checking molds and resulted in higher germination (Meena and Mariappan, 1994). Lakshmanan *et al.* (1988) effectively controlled molds by seed treatment with leaf extract of *A. indica*. Somda *et al.* (2007) reported oil from *Cymbopogon citratus* to control seed borne *Colletotrichum graminicola* effectively.

Maize

The wood ash of *Vernonica amygdalina*, *Gliricidia sepium*, *and Cassia siamea* as seed treatment controlled seed borne fungi *D. maydis*, *C. lunata*, *F. semitectum*, *F. moniliforme* and species of *Aspergillus* and *Penicillium*. The treatment resulted in increased germination and reduced pre- and post- emergence mortality. Rai *et al.* (2002) found leaf extract of *Melia azedarach* to check seed mycoflora. Out of six medicinal plants tested Kiran *et al.* (2010) found seed extract of *Psoraelea corylifolia* to be highly inhibitory to six seed borne fungi of maize.

Bajra

Plant extracts of *Ocimum sanctum*, *Mentha viridis*, *Ipomoea cornea* and *Abutulon indicum* have been reported inhibitory to seed mycoflora of bajra by Rathore and Kagane (2005).

Pulses

Kotkar *et al.* (2002) found leaf extract of *Annona squamosa* to control seed borne fungi of pulses due to flavonoids.

Lentil

Hashmi *et al.* (1992) controlled the seed borne fungi of lentil by treating the seed with neem bitter extract which included *F. moniliforme*, *F. oxysporum and F. semitectum*.

Tomato

Tomato seeds treated with aqueous extract of garlic (30 g/100 ml water) for 12 hours controlled seed borne *Xanthomonas campestris* pv. *vesicatoria* and reduced the severity of disease (Mangamma and Sreeramulu, 1991).

Cowpea

Seed treatment with fresh leaf extract of *Moringa olifera* controlled *Colletotrichum dertructivum* (Akinbode and Ikotun, 2008).

Sangvikar (2011) tested 36 plants for their antimicrobial activity and selected seven most promising ones. The alcoholic extract of *Hemidesmus indicus* roots was most effective followed by ethyl acetate extract of *Curcuma amada*. Both the extracts at 10 per cent strength checked seed borne fungi associated with sorghum, wheat, maize and cotton. The extracts exhibited stimulatory effect on seed germination, emergence, root and shoot length and inhibitory effect on rots.

Soybean

Dipping of soybean seeds in 100 percent extract of ginger, garlic and neem for 30 minutes controlled seed borne *Colletotrichum dematium* f.sp. *truncatum* (Hossain *et al.*, 1999). Chandrasakaram *et al.* (2000) observed reduced incidence of soybean anthracnose when seeds were treated with 10 per cent leaf extract of *L. inermis*. Arora and Khausik (2003) found dry hot water extract of *Lantana camera, Berberis aristate* to be highly effective against seed borne anthracnose causing pathogen.

Other Uses

Plants are used to check pathogens as well as diseases caused by them as mulches, green manuring, decoy crops and inter and mixed cropping.

Mulches

Plant parts are incorporated in soil as organic matter to act as soil amendment. Stubbles of previous crops like cereals, legumes and weeds are added in soil. They control some pathogens as well as diseases by reducing their population directly or indirectly (Baker and Cook, 1974; Lewis and Papavizas, 1975). Soil amendment with barley straw decreased Verticillium potato wilt (Huber and Watson, 1970) and cotton root rot caused by *M. phaseolina* (Lewis and Papavizas,1975), wheat straw decreased black scurf of potato caused by *R. solani* (Gudmestad *et al.*, 1978), oat straw reduced bean root rot due to *Thielaviopsis basicola*, cruciferous crop refuge checked pea rot caused by *Aphanomyces euteiches* and sugarcane residues decreased banana wilt due to *F. oxysporum cubense* (Seqeira,1962). Singh *et al.* (1990) observed reduction in the incidence of Sclerotinia stem rot of chickpea when decomposed leaves of *A. marmelos* were added in soil 30 days before planting. When finely ground wheat straw was added in soil infested with *M. phaseolina*, it gave 91.4 per cent emergence of chickpea as compared to 69.4 per cent in control.

Green Manuring

Green manuring with rape, pea reduced take all disease of wheat (Grossman, 1967), pea, *Melilotis officinalis* reduced root rot of cotton caused by *Phymatotrichum omnivorum* (Cook *et al.*, 1978) and barley and oats checked black surf of potato caused by *R. solani*.

Decoy Crop

Decoy crops are non-host crops for soil borne pathogens which waste the infection potential. Such crops activate the dormant propagules of fungi in absence of the actual host. *Tagetes minuta* is used as decoy crop for *Verticillium albo-atrum* causing disease in olive (Baker and Cook, 1974). *Datura stramonium* for *Spongospora subterranea* in potato. Rye grass, *Roseda odorata* for *Plasmodiophora brassicae* in brassicas (Garret, 1970).

Inter and Mixed Cropping

It is usual recommendation to have inter and mixed cropping system in crop production programme as the same pathogen does not infect two crops simultaneously and one can be saved. Plant to plant spacing of a crop is increased and one crop acts as a barrier for the other in case of air borne pathogens. The root exudates of one crop may contain chemicals which are toxic to the soil borne pathogens infecting the other crop. When pigeonpea and sorghum are grown together, the root exudates of sorghum check *Fusarium udum* causing wilt. Sarkar (2000) reported that in sunhemp-Jowar intercropping the incidence of wilt in sunhemp was considerably less. When sunhemp and Jowar seed was mixed in 3:1 ratio and sown in line the incidence of wilt was remarkably low.

Crop Residues

Use of organic amendments in the form of crop residues in soil is beneficial in controlling soil borne pathogens. They release a number of biologically active compounds which have antipathogenic property like isothiocynates, glucosinolate, allyl-isothiocyanates. Residues of cruciferous crops like Brassicas had controlled *M. phaseolina* (Lodha,2010), *Thielaviopsis basicola* causing root rot of sesame, *Sclerotium rolfsii* and *Pythium ultimum* (Stapleton and Duncan, 1998), *R solani* in snapbean (Manning and Grossan, 1969).

Viruses

Control of virus diseases is possible by the use of plant extracts which have been found to possess antiviral properties. Verma and Verma (1993) reviewed the antiviral properties of a number of plant species, which act in various ways by complexing chemically modifying or denaturing, dissociating or precipitating viruses. The inhibitory substance is proteinaceous in nature. The antiviral proteins from plant acted locally as in case of *Amaranthus caudatus, Atriplex nitens, Chenopodium album, C. amaranticolor, Dianthus caryophyllus and Phytolacca americana* with well characterised protein. *C amaranticolor, Datura sp., D. caryophyllus, P. americana, Spinacea oleracea* etc. with basic protein and *P. americana* with glycoprotein. Systemic action was reported in root extract of *Boerhavia diffusa* and leaf extract of *Bougainvillea spectabilis, Clerodendron fragrans, Psuedoeranthemum bicolour, P. tricolour, Mirabilis jalapa* (Verma and Awasthi, 1979). Certain alkaloids were isolated from *Clivia miniata* by Leven *et al.* (1983), which were antiviral.

Kurucheve *et al.* (1995) reported complete inhibition of tomato spotted wilt virus in cowpea by the application of *Eucalyptus* and *Geranium* leaf extracts. The plant extracts of *M. jalapa, P. thirsiflora, C amaranticolor* and *B spectabilis* checked tomato spotted wilt tospovirus and increased yields in tomato.

Verma *et al.* (1996) purified a nonphytotoxic systemic resistance inducing specific basic protein CA-SRI which was isolated from leaves of *Clerodendron aculeatum.* Treatment of plants with purified protein preparation induced a very high level of systemic resistance against tobacco mosaic tobamovirus on tobacco and sunhemp rossete virus within 5 to 30 minutes. Verma and Awasthi (1979) also found and isolated a systemic resistance inducer from the roots of *B. diffusa* which induced local as well as systemic resistance in several susceptible hosts.

The induction of resistance in hosts may be due to the prevention of movement of virus from originally infected cells; blocking of infection through interference with the host cells; inhibition of virus replication at the level of RNA or protein synthesis and prevention of assembly of RNA and coat protein into functional virus particles (Verma and Verma, 1993).

Meena (2001) found neem extract to check mosaic disease of blackgram and greengram. Meena *et al.* (2008) found the application of leaf powder extract of *Clerodendron aculeatum* (0.1 per cent) most effective against leaf curl disease of chilli. It decreased the disease incidence by 60-70 per cent, besides it increased seed germination, plant height and number of fruits. Biju *et al.*(2007) used two neem oil formulations azadirachtin and dihydro azadirachtin 2500 ppm and found they adversely affected the survival, feeding behaviour and acquisition of rice tungro viruses by the vector green hopper, *Nephotettix virescens*.

Thirumalaisamy *et al.* (2003) found extracts of nine plants out of 21 tested to be inhibitory to urdbean leaf crinkle virus of which extract of *Zingiber officinale, Prosopis juliflora* and *Piper longum* was better performer.

Nematodes

Nematicidal value of plants against plant parasitic nematodes has been worked out time to time. Many allelopathic compounds in their native or processed forms have potential in plant-parasitic nematode management strategies. Allelo-chemicals have been identified that possess differing levels of activity against a wide range of plant-parasitic nematodes. In general, these compounds are less toxic to nontarget species, and less persistent in soil than chemical nematicides. Operative mechanisms for plant-parasitic nematode control with allelopathic compounds include nematicidal activity, nematostatic activity, and nematode behavior modification (Burelle, 2006). Chemicals produced by plants are a potential source of new chemistry for development of new pesticidal compounds. Nematicidal phytochemicals are generally safe for the environment and humans (Chitwood, 2002). Chinese herbal remedies may be a source of new nematicidal compounds (Zasada *et al.*, 2002). Many nematicidal phytochemicals with great variety of chemical structures have been isolated from numerous plant families (Gommers and Bakker, 1988; Chitwood, 2002). A majority of these isolated nematicidal phytochemicals are from the plant family Asteraceae (Gommers and Baker, 1988). α-Terthienyl and related compounds isolated from *Tagetes* spp. have shown to be nematicidal at low concentrations *in vitro*. These phytochemicals, however, were not effective in nematode control in soil (Gommers and Bakker, 1998).

Polyacetylenes are the chemical group from Asteraceae family with nematicidal activity. For example, nematicidal polyacetylenes have been isolated from flowers of *Carthamus tinctorius* and roots of *Cirsium japonicum* (Kogiso *et al.*, 1976; Kawazu *et al.*, 1980), and dithio-acetylenes have been isolated from *Milleria quinqueflora, Iva xanthiifolia, Ambrosia artemisiifolia, A. trifida, Schkuhria pinnata,* and *Eriophyllym caespitosum*. Thiarubrine C isolated from the roots of *Rudbeckia hirta* has been shown to have nematicidal activity against *M. incognita* and *Pratylenchus penetrans*.

Unfortunately, none of these compounds or their derivatives could be developed into commercial nematicides. Plant essential oils, mainly monoterpenes, have been evaluated for their nematicidal activity, and some were highly effective in nematode suppression (Oka *et al.*, 2000; Oka, 2001). However, use of natural essential oils as nematicides is not cost effective. Various neem tree (*Azadirachta indica*) preparations are well known commercially available nematode control products (Mojumdar, 1995). Elecampane (*Inula viscosa*) (Asteraceae), has been found to have nematicidal activity in the shoot (Oka *et al.*, 2001). Another species, *I. helenium*, has been known to have anthelminthic activity, due to sesquiterpenoid lactones such as alantolactone (Mahajan *et al.*, 1986; Bourrel *et al.*, 1993). Sesquiterpenic acids (costic acid and isocostic acid) from *I. viscosa* leaf extracts were found to be the nematicidal phytochemicals (Oka *et al.*, 2001). A mixture of these compounds was toxic to *M. javanica* at concentration as low as 50 mg/kg in soil.

Linford *et al.* (1938) observed effect of chopped pine apple leaves against *Meloidogyne* spp. Reduction of root knot population by *Melilotus alba* var *annua* and *Sorghum vulgare* was reported by Patel and Desai (1964); chopped Karanj leaves against *M. javanica* in tomato; leaves of *Casia fistula, Crotolaria juncea* and *Sesbania aculata* against *Meloidogyne* infested soil (Singh and Sitaramaiah,1967); antagonistic plant parts of *Crotolaria*, Marigold, Kentucky blue grass in powder form reduced number of galls in *M. hapla* ; chopped leaves of subabool have acted against *M. incognita* in okra.

Application of *Azadirachta indica, Calotropis procera, Datura stramonium, Crotolaria juncea* and *Vitex negundo* were found to be superior and effective in reducing the lesion nematode population and increasing the yield significantly (Sundararaju *et al.*, 2003). Leaf extracts of *Glyricidia maculata, Ricinus communis, Crotalaria juncea, Glycosmis pentaphylla, Azadirachta indica, Kalanchoe pinnata, Piper betle* and *Moringa oleifera* have been reported to be lethal to *Radopholus similis* (Tasy and Koshy, 1992).

The nematicidal efficacy of leaf water extracts of *Tripterygium wilfondii, Nicotiana tabacum, Asarum sieboldii, Nerium indicum, Paederia scandens, Ricinus communis, Tagetes erecta* and *Lantana indica*, had the killing rates of 100, 100,100, 90.0, 87.1, 84.3, 77.3 and 32.4 per cent respectively in 24 h; the killing rates of flower, leaf and stem water extracts of Tagetes erecta were 100, 77.3 and 48.2 per cent respectively (Yang *et al.*, 2003).

Fresh leaf extracts of *Datura stramonium, Calotropis procera, Verbesena enceloides, Parthenium hysterophorus, Morus alba, Phyllanthus amarus, Eichhornea crassipes, Ricinus communis, Jatropha curcas, Azadirachta indica,Tinospora cordifolia, Clerodendron multiflorum, Catharanthus roseus* and *Adhatoda vesica* exhibited nematicidal activity against root-knot nematode, Meloidogyne incognita (Sharma and Trivedi, 2002). Eggmasses or larvae of *Meloidogyne incognita* were exposed to varying concentrations of neem leaf (fresh and dry), *Borelia* sp.,groundnut leaf and garlic bulb. Neem leaf and garlic bulb extracts inhibited the hatching of eggmasses and were lethal to larvae (Agbenin *et al.*, 2005). Emulsifiable concentrate formulations of *Inula viscosa* pastes killed *M. javanica* juveniles in sand at a concentration of 0.01 per cent (paste, w/w) and reduced the galling index of cucumber seedlings as well as the galling index and number of nematode eggs on tomato plants (Yuji *et al.*, 2006).

Extracts from plants of Liliaceae family reduced galling of tomato by *M. javanica* without showing phytotoxic effect on the host. Similarly, isothiocyanate-yielding plants in the Brassicaceae suppressed root galling without phytotoxicity. Other plant extracts, including those from *Azadirachta indica, Nerium oleander*, and *Hedera helix*, suppressed root galling but were phytotoxic at the higher concentrations against host crops (Zasada *et al.*, 2002).The cold and hot aqueous extracts of *Luffa cylindrica, Momordica charantia, Euphorbia hirta, Desmodium scorpiurus and Stachytarpheta cayennensis*, wood ash of *Gmelina aborea* inhibited the hatching of the nematode and reduced galls on the roots of cowpea (Ononuju and Nzenwa, 2011).

Exposure to standard extract of *Punica granatum, Thymus vulgaris* and *Artemisia absinthium* for 72 h reduced the number of active nematodes by 100 per cent for *Meloidogyne incognita* and by 95.7 per cent, 71.4 per cent and 42.9 per cent for *Helicotylenchus dihystera*, respectively (Korayem *et al.*, 1993). The essential oils of four medicinal plants belonging to Lamiaceae were explored for phytonematode control. The four oils inhibited ($P<0.05$) nematode motility but *Mentba spicata* was generally more effective in reducing the numbers of active nematodes followed by *Thymus vulgaris, Majorana bortensis* and *Mentba longifolia*. The main corresponding compound of each oil, determined by GLC analysis, was carvone (58.14 per cent), P-cymene (40.5 per cent), terpinen-4-ol (41.6 per cent) and carvone (70.36 per cent) (Mahfouz *et al.*, 1995).

Plant parts of *Argemone maxicana, Asystacia gangetica, Potytrials amoura, Gliricidia sepium, Leucaena leucocephala, Eucalyptus globarus, Datura metel, Phylanthus niruri*, exhibited nematicidal properties as they inhibited egg hatching, reduction in production of galls and eggmasses of *Meloidogyne javanica* and *Rotylenchulus reniformis*.

Storage Problems

Fungi cause seed deterioration of various crops in storage. The problem is more severe in tropical regions. Plant products are useful in controlling fungi responsible for seed spoilage in storage (Tripathi *et al.*, 2006). Some fungi associated with seeds produce toxins hazardous to humans and animals like aflatoxin from *Aspergillus flavus*. The species of fungal genera responsible for seed spoilage are *Aspergillus, Penicillium, Fusarium, Rhizopus, Mucor, Chaetomium, Cladosporium, Stemphylium, Curvularia, Drechslera, Stachybotrys, Nigrospora, Epicoccum, Mennoniella, Aureobasidium* etc. Temperature, relative humidity, moisture content and nutrient levels of seed are key factors in the spoilage. Extracts of plants successfully used against storage fungi of cereals and pulses by various workers time to time are listed by Tripathi *et al.* (2006). Significant antifungal activity of aqueous extract of *Acacia nilotica, Achras zapota, Datura stramonium, Emblica officinalis, Eucalyptus globulus, Lawsonia inermis, Polyalthia longifolia, Prosopis juliflora, Punica granatum* has been reported against seven species of *Aspergillus* as storage fungi concerning biodeterioration in case of sorghum, maize and paddy (Satish *et al.*, 2007). Vijaya and Mouli (2010) have reported the efficacy of extract of *Costus specious* rhizomes against species of *Penicillium, Curvularia,*

Cladosporium and *Aspergillus*. Reddy *et al.* (2007) demonstrated the ability of crude clove extract and purified eugenol to inhibit Aspergilli and arrest colonization of rice grains. Singh *et al.* (2004) found seed treatment of paddy with leaf extract of *Plectranthus ternifolius* did not allow fungal invasion in storage.

Antipathogen Compounds

The plant and their parts contain certain compounds which are inhibitory to fungi, bacteria, viruses and nematodes and they help in controlling plant diseases. Different methods are applied in their extraction like ammonium sulphate precipitation method (Stumpf, 1955). Mitra *et al.* (1984) have given a list of antifungal compounds produced by higher plants in a review.

Asparagus recemosus: Quercitin (Sangvikar,2011)

Azadirachta indica: Proto-meliacins, meliacins, penta-nortriterpenoids and other nortriterpenoidal group. Azadirachtin (Schmutterer, 1996).

Allium sativum: Allicin, Ajoene

Citrus medica: Citral, limonens, dipentene

Clerodendrum aculeatum (root): protein CA-SRI

Costus specious (Rhizome): methyl 3-(4-hydroxyphenyl)-2(E)-propenoate (Vijaya and Mouli, 2010)

Curcuma amada (mango ginger rhizome): difurocumenonol (Policegoudra *et al.*, 2007)

Curcuma longa: Curcumin (1, 7 bis [4-hydroxy-3-methoxy-phenyl-1] heptan-1, 6-diene-3, 5 dione) (Deepa *et al.*, 2000)

Eucalyptus globulus: Phenolic compounds

Ferula foetida: Ferulic acid and volatile substances

Gingiber officinale: 3,7-dimethyl-2,6-octadiencol (Citral), Citral citronellal, linalool, methyl cinnamats, eugenol, d-camphor, traces of phenols and acetic acid.

Hemidesmus indicus: 2-hydroxy-4-methoxy benzoic acid (Sangvikar, 2011)

Lawsonia inermis (Henna): Triterpinoids, sterols, naphthoquinone derivatives, phenolic constituents coumarins, xanthones and flavonoids (Ali,1996)

Ocimum gratissimum: Alkaloids, cardiac glycosides, flavonoids, glycosides, resins, steroidal terpens and tannins (Mbata and Saikia, 2008)

Parthenium hysterophorus: parthenin–a sesquiterpene lactone (Ganeshan and Jayachandra, 1993).

Plant latex: Phenolic compounds like euphosterol, phytosterol, phytosterolin, organic acids like gallic, mellissic, palmitic oleic, linoleic acid and alkaloid xanthorhamnin.

Pinus roxburghii: LB pipene, 1-limonene, candinene, cryptone, phellandrene, fatty acids and phenols

Trachispermum ammi: Thymol and phenol

Conclusion

It is evident from the experiences of crop disease control by using botanicals that the use of chemicals can be avoided. It is necessary to utilize those plants which are commonly available and cheap. Simple methods for their application in the field must be searched. The process of making formulations of botanicals need to be easy. Specific plants and their parts with higher quantities of active principle should be continuously searched. The plant materials can be further enriched by biotechnological, molecular, biological techniques and genetic engineering. Botanicals have great potential and future in plant disease control and will occupy a privileged position in near future.

References

Abul-fazal, M., Khan, M.I. and Saxena, S.K. (1987). Seed mycoflora of wheat and their control with latex of *Calotropis procera* R. *J.Sci. Res.* **9:** 1-3.

Abdul latif, M., Abu Kaoser, Saleh, M., Ashik Iqbal Khan, M. and Habibur, Rahman (2006). Efficacy of some plant extracts in controlling seed-borne fungal infections of mustard. *Bangladesh J.Microbiol* **23:** 168-170.

Agbenin, N.O., Emechebe, A.M., Marley, P.S. and Akpa, A.D. (2005). Evaluation of Nematicidal Action of Some Botanicals on Meloidogyne incognita *in vivo* and *in vitro*. *Journal of Agriculture and Rural Development in the Tropics and Subtropics:* 106

Alam, M.M. (1993). Neem an ideal component of integrated pest management. In Integrated Pest Management (Eds. Dwivedi, B.K., Alam, M.M.; Pandey,G.) Bioved Research Society, Allahabad, India, pp. 119-130.

Alice, D. and Rao, A.V. (1986). Management of seed borne *Drechslera oryzae* of rice with plant extracts. *Int. Rice Res. Newsl* **11:**19.

Adams, P.B. (1971). Effect of soil temperature and soil amendments on Thielaviopsis root rot of sesame. *Phytopathology* **61:** 93-97.

Adejumo, T.O. and Otuonye, A.H. (2002). The use of botanicals in the control of inflorescence blight disease of cashew, (*Anacardium occidentale*). *Nigr. J.Sci.*36 **(1):** 75-80.

Ahmad, S.K. and Prasad J.S. (1995). Efficacy of foliar extracts against pre- and post-harvest of sponge gourd fruits. *Lett. in Appl. Microbiol.* **21:** 373-375.

Akinode, O.A. and Ikotun T. (2008). Efficacy of certain plant extracts against seed borne infection of *Colletotrichum dertructivum* on cowpea (*Vigna ungiculata*). *African J. Biotechnology* **7:**383-385.

Ali, M. (1996). Chemical and medicinal evaluation of *Lawsonia inermis* (henna). *Hamdard Medicus* **39:** 43-48.

Anishkumar, V.K., Rehna, J.S. and Srinivasan, N. (2004). Evaluation of botanicals against pathogens of coconut leaf rot disease and their antagonistic organisms. *Journal of Plantation Crops* **32** (Sppl.): 321-329.

Ansari, M.M. (1995).Control of sheath blight of rice by plant extracts. *Indian Phytopath.* **48:** 268-270.

Arora, C. and Kaushik, R.D. (2003). Fungicidal activity of plant extracts from Uttaranchal hill against soybean fungal pathogens. *Allelopathy Journal* **11:**217-228.

Babu, K.J. and Reddy, S.M. (2002). Efficacy of some indigenous plant extracts in the control of lemon rot by two pathogenic fungi. *Natl. Acad. Sci. Lr.* **9:**133-134.

Baker, K.F. and Cook, R.J. (1974). Biological Control of Plant Pathogens. W.H. Freeman and Co. San Francisco. pp. 433.

Behura, C., Ray, P., Rath, C.C., Mishra, R.K., Ramchandraih, O.S. and Charyulu, J.K. (2000). Antifungal activity of essential oils of *Curcuma longa* against five rice pathogens *in vitro*. *J.Essential Oil Bearing Plants* **3:** 79-84.

Banerjee, S., Bhattacharya, I. and Mukherjee, N. (1989). Sensitivity of three sclerotial rice pathogens to plant oils. *Int. Rice Res.Newsl.* **14:** 23.

Bhatnagar, K., Tak, S.K., Gurgen, R.B. and Majumdar, V.L. (2009). Management of stem blight disease of cowpea induced by *M.phaseolina* through bioagent and plant products (Abstr.) *J.Mycol Pl Pathol* **39:** 571.

Bhowmick, B.N. and Varadhan, V. (1981). Antifungal activity of some leaf extracts of medicinal plants on *Curvularia lunata* *Indian Phytopath.* **34:**385-386.

Bhowmick, B.N. and Choudhary, B.K. (1982). Antifungal activity of some leaf extracts of medicinal plants on *Alternaria alternata* (Fr.) Keissler. *Indian Botanical Reports* **1:** 164-165.

Biju, C.N., Niazi, F.R. and Singh, J. (2007). Effect of azadirachtin and dihydra azadirachtin on *in vitro* acquisition of rice tungro viruses. *Indian Phytopath.* **60:** 362-365.

Bisht, G.S. and Khulbe, R.D. (1995). In vitro efficacy of leaf extracts of medicinal plants on *Curvularia lunata.Indian Phytopath.* **48:** 480-482.

Biswas, S., Das, N.K., Quadri, S.M.H. and Saratchandra, B. (1995). Evaluating different plant extracts against three major diseases of mulberry. *Indian Phytopath.* **48:** 342-346.

Bohra, B., Vyas, B.N. and Mistry, K.B. (2006). Biocontrol agents and neem formulations for management of damping off in brinjal and chilli. *Indian Phytopath.* **59:** 223-226.

Bunker, R.N. and Mathur, K. (2008). Evaluation of neem based formulations and chemical fungicides for the management of sorghum leaf blight. *Indian Phytopath.* **61:**192-196.

Burelle, N.K. and Rodrigao, Rodri'guez- Kabana. (2006). Allelochemicals: Biological Control of Plant Pathogens and Diseases. *Disease Management of Fruits and Vegetables* **2:** 15-29.

Bourrel, C., Vilarem, G., and Perineau, F. (1993). Chemical analysis, bacteriostatic and fungistatic properties of the essential oil of Elecampane (*Inula helenium* L.) *Journal of Essential Oil Research.* **5:** 411–417.

Chakraborty, B.N., Biswas, R.D. and Sharma, M. (2007). Induction of resistance in tea plants against *Alternaria alternata* by foliar application of leaf extracts. *J. Mycol. Pl.Path.* **37:** 60-64.

Chandrasekar, V., Narayanasami, R. and Ramabadran, R. (1994). Effect of foliar spray of potash and neem seed extract on the Tikka leaf spot of groundnut. *Indian Phytopath.* **47:**188-189.

Chandel, S. and Tomar, M. (2007). Evaluation of fungicides and biopesticides against Fusarium wilt of gladiolus. *Indian Phytopath.* **60:** 115-117.

Chandel, S., and Tomar, M. (2008). Effectiveness of bioagents and neem formulations against Fusarium wilt of carnation. *Indian Phytopath.* **61:**152-154.

Chandrasekaran, A. and Rajappan, K. (2002). Effect of plant extracts, antagonists and chemicals (individual and combined) on foliar anthracnose and pod blight of soybean. *J. Mycol. Pl.Pathol.* **32:** 25-27.

Charles, J.S.K. and Venkitesan, T.S. (1993). Status reports on the nematological investigations on banana in Kerala Agricultura1 University, Publ. Director of Extension. 43 pp.

Chaturvedi, R., Dikshit, A. and Dixit, S.N. (1987). *Adenocalymma allicea* a new source of a natural fungitoxicant. *Trop. Agric.* **64:**318-322.

Chaurasia, S.N.P. and Dayal, R. (1990). Inhibition of mycelial growth and Sclerotium formation in *Rhizoctonia solani*, Kuhn and *Sclerotinia sclerotiorum* (Lib.) deBary by Asafoetida. *Trop. Sci.* **30:**15-19.

Chauhan, V.B. and Singh, U.P. (1991). Effect of volatiles of some plant extracts on germination of zoospores of *Phytophthora dreschleri* f.sp.*cajani. Indian Phytopath* **44:**197-200.

Chitwood, D.J. (2002). Phytochemical-based strategies for nematode control. *Annual Review of Phytopathology* **40:** 221–249.

Clevenger, J.F. (1928). Apparatus for the determination of volatile oil. J.Amer. Pharmn. Assoc. **17:** 346.

Cook, R.J., Boosalis, M.G. and Doupnik, B. (1978). Influence of crop residues on plant diseases. In Crop Residue Management System. *Amer. Soc. Agron. Spec. Pub.* **31:**147-163.

Deepa, S., Shrivastva, Birendra and Garg, S.P. (2000). Isolation and antimicrobial studies of curcumin from *Curcuma aromatica. Current Agriculture* **24:**101-103.

Deka,U.K, Dutta, P.K., Gogoi, R. and Ghosh, P.K. (2008). Management of leaf spot disease complex of betelvine by bioagents and plant extracts. *Indian Phytopath.* **61:**337-342.

Dhaliwal, H.J.S., Thind, T.S. and Mohan, C. (2003). Activity of some essential oil against *Rhizoctonia solani* (Kuhn), the cause of black scurf of potato. *J. Mycol. Pl. Path.* **33**:399-402.

Dhingra, O.D. and Sinclair, J.B. (1995). Basic Plant Pathology Methods, Second Edition. CRC Press Inc. Lewis Publishers, Boca Raton, USA.pp.434.

Dubey, R.C. (1991). Fungicidal effect of essential oils of three higher plants on sclerotia of *Macrophomina phaseolina*. *Indian Phytopath.* **44**: 242-243.

Dubey, R.C. and Dweivedi, R.S. (1991). Fungitoxic properties of some plant extracts against vegetative growth and sclerotial variability of *Macrophomina phaseolina*. *Indian Phytopath.* **44**: 411-413.

Dubey, S., Upadhyay, P.D. and Tripathi, S.C. (1991). Fungitoxic and insect repellent efficacy of some spices. *Indian Phytopath.* **44**:101-105.

Dubey, S.C. (2003). Integrated management of web blight of urd and mungbean. *Indian phytopath.* **56**: 413-417.

Ekka, S. and Prasad, S.M. (2010). Evaluation of oilcakes against rhizome rot of ginger. *Indian Phytopath.* **63**:106-107.

El-Shami, M.A., Fadk, F.A., Tawfick, K.A., Sirry, A.R. and El-Zayat, M.M. (1986). An antifungal property of garlic clove juice compared with fungicidal treatment against Fusarium wilt of watermelon. *Egyptian J. Phytopath.* **17**: 55-62.

Ezhilan, J.G, Chandrasekar, V. and Kurucheve, V. (1994). Effect of six selected plant products and oil cakes on the sclerotial production and germination of *Rhizoctonia solani, Indian Phytopath.* **47**:183-185.

Fawcett, C.H. and Spencer, D.M. (1970). Plant chemotherapy with natural products. *Ann.Rev.Phytopathol.* **8**: 403-418.

Forsyth, W. (1802). A Treatise on the Culture and Management of Fruit Trees. Longman. London. pp. 370.

Ganapathy, T. and Narayanasamy, P. (1994). Screening plant products effective against Tikka and rust diseases of groundnut. Crop Diseases Innovative Techniques and Management (eds. Sivaprakasam,K., Seetharamam, K.) Kalyani Publishers, Ludhiana. India, pp. 347-354.

Ganeshan, G. and Jaychandra. (1993). Antifungal activity of *Parthenium*. *Indian Phytopath.* **46**: 193-194.

Gangopadhyay, S., Bhakar, B.R. and Godara, S.L. (2010). Effect of botanicals on cumin blight caused by *Alternaria burnsii*. *J. Mycol. Pl.Path.* **40**: 570-574.

Garrett, S.D. (1970). Pathogenic Root Infecting Fungi. Cambridge Univ. Press, Cambridge UK. pp.294.

Gommers, F.J. and Bakker, J. (1988). Physiological diseases induced by plant responses or products. In: Diseases of Nematodes. Vol 1., Eds. Poinar G.O., Jr., Jansson, H.B. and Boca, Raton, FL. CRC Press Inc. pp. 3–22.

Gour, R.B. and Sharma, R.N. (2010). Prevalence of canker in kinnow (*Citrus deliciosa* Ten.) and its management. *Indian Phytopath.* **63:** 345-347.

Ghorbani, R., Wilcockson, S. and Leifert, C. (2005). Alternative treatments for late blight control in organic potato: Antagonistic microorganisms and compost extracts for activity against *Phytophthora infestans. Potato Research* **48:**181-189.

Grossman, F. (1967). Grundung als Pflanzenschutzmabnahme. Z. Pflazenkrankh Pflanzenschutz **74:** 143-149.

Gudmested, N.C., Haguelet, J.E. and Zink, R.T. (1978). The effect of cultural practices and straw incorporation into the soil on Rhizoctonia disease of potato. *Plant Dis. Reptr.* **62:** 985-989.

Guenther, E. (1966). The Essential oils Vol.III.D.Van Nostrand Co. Inc. London, pp. 423.

Guenther, E. (1972).The Essential Oils Nostrand Co. Inc. London, pp 201-307.

Hasabnis, S.N. and D'souza, T.F. (1987). Use of natural plant product in the control of the storage rot in Alphanso mango fruit. *J.Maharashtra Agri. Uni.* **12:** 105-106.

Hashmi, R.Y., Saleem, S., Khanzada, A.K., Aslam, M., Kazmi, S.A.R. and Shahzad, S. (1992). Studies on the seed mycoflora of lentil and its control. *Pakistan J. of Sci. & Indust. Res.* **35:** 345-347.

Hossain, I., Suratuzzaman, M. and Khalil, M.I. (1999). Seed health of soybean and control of seed borne fungi with botanicals. Bangladesh J. of Training and Development **12:** 99-105.

Huber, D.M. and Watson, R.D. (1970). Effect of organic amendment on soil borne plant pathogens. *Phytopathology* **60:** 22-26.

Jacob, C.K. and Sivaprakasam, K. (1994). Evaluation of some plant extracts and antagonists for the control of pre-emergence damping off of brinjal (*Solanum melongena*). Crop Diseases Innovative Techniques and Management (eds. Sivaprakasam, K.,Seetharama,K.) Kalyani Publishers, Ludhiana, India, pp 289-294.

Jagannathan, R. and Narsimhan, V. (1988). Effect of plant extracts/products on two fungal pathogens of finger millets. *Indian J. Mycol. Pl.Pathol.* **18:** 250-254.

Jain, S.C. and Jaiman, R.K. (2005). Effect of traditional seed testing materials on the incidence of *Macrophomina phaseolina* (Tassi) Goid in stored seeds of clusterbean (*Cyamopsis tetragonoloba*) Abstr. Second Global Conference on Plant Health Global Wealth. ISMPP, MPUAT, Udaipur:103.

Jasy, T. and Koshy, P.K. (1992). Effect of certain leaf extracts and leaves of *Glyricidia maculata* (H.B & K) Stend as green manure on *Radopholus similis. Indian Journal of Nematology* **22:** 117-121.

Jatav, R.S. and Mathur, K. (2005). Bio-agents and neem based seed treatment for management of root rot complex in clusterbean. *Indian Phytopath.* **58:** 235-236.

Khan, A.L and Sinha, A.P. (2006). Integration of fungal antagonists and organic amendments for the control of rice sheath blight. *Indian Phytopath.* **59:** 363-365.

Khan, M.I. (1989 (a)). Antimicrobial action of some plants of family compositae on seed mycoflora of wheat (*Triticum aestivum* L.) *Indian J. Appl. and Pure Biol.* **4:** 23-26.

Khan, M.I. (1989 (b)). Effect of *Nicotiana* spp. on the seed mycoflora of wheat. *Indian J. Appl. and Pure Biol.* **4:** 69-71.

Kandhari, J. (2007). Management of sheath blight of rice through fungicides and botanicals. *Indian Phytopath.* **60:** 214-217.

Kandhari, J., Devakumar, C., Kumar, R. and Singh, D. (2010). Management of sheath blight of rice through essential oils, aroma compounds and plant extracts. *Indian Phytopath.* **63:** 199-202.

Kalaichelvan, P.T. and Nagarajan,G. (1992). A fungitoxic alkalod from *Crotalaria paleda. Indian Phytopath.* **45:** 252-253.

Kanagarajan, R.S. (1974). Studies on certain aspects of seed borne fungi IV. Seed borne fungi of some starchy seeds. *Acta Bot. Indica* **2:**129-135.

Khair, A.H. and Wafaa, M.H. (2007). Application of some Egyptian medicinal plants extracts against late blight and early blights. *Research Journal of Agricultural and Biological Sciences* 3 **(3):** 166-175.

Khare, M.N. and Dhingra, O.D. (1974). Laboratory studies on preharvest rot of papaya fruits caused by *Gloeosporium papayae. Mysore J. Agric Sci.* **8:**115-120.

Khare, M.N and Shukla, B.N. (1998). Utility of Plants in Crop Disease Control. Vasundhara (International Journal of Environmental Biology) **3:** 1-15.

Kiran, B., Lalitha,V. and Raveesha, K.A. (2010). Screening of seven medicinal plants for antifungal activity against seed borne fungi of maize seeds. *African J. of Basic and Applied Science* **2:** 99-103.

Kishore, N., Dubey, N.K., Tipathi, R.D. and Singh, S.K. (1982). Fungitoxic activity of leaves of some higher plants. *Nat. Acad.Sci.Letters* **5:** 9-10.

Kotkar, H.M., Mendki, P.S., Sadan, S.V.G.S., Jha,S.R., Upasani, S.M. and Maheswari, V.L. (2002). Antimicrobial and pesticidal activity of partially purified flavonoids of *Annona squamosa. Pest Manag. Sci.* **58:** 33-37.

Kumar, A. (2009). Ecofriendly options for the management of white rusts in Indian mustard under mid hill conditions of North-West India. *Indian Phytopath.* **62:** 44-48.

Kurucheve,V., Jeyarajan, R., Doraiswami, S. and Balasubramanian, P. (1995). Spent leaves of *Eucalyptus* and *Geranium*; a source of TSWV inhibitor. *Indian J. Mycol. Pl. Pathol.* **25:** 127.

Kawazu, K., Nishii, Y. and Nakajima, S. (1980). Two nematicidal substances from roots of *Cirsium japonicum. Agricultural and Biological Chemistry.* **44:** 903–906.

Kogiso, S., Wada, K. and Munakata, K. (1976). Isolation of nematicidal polyacetylenes from *Carthamus tinctoris* L. *Agricultural and Biological Chemistry*. **40:** 2085–2089.

Korayem, A.M., Hasabo, S.A. and Ameen, H.H. (1993). Effects and mode of action of some plant extracts on certain plant parasitic nematodes Anzeiger für Schädlingskunde, 66 **(2):** 32-36.

Lakpale, N., Khare, N. and Thrimurthy,V.S. (2008). Evaluation of botanicals of antimicrobial properties. *J. Mycol.Pl.Pathol.* **38:** 614-617.

Lakshmanan, P., Parambaramani, C. and Mohan, S. (1988). Antifungal properties of certain plant extracts on grain mould pathogens of *Sorghum bicolour* L. Moencl. Proc. National Seminar on Management of Crop Diseases with Plant Products TNAU, Madurai. p.16.

Leven, M., Vlletinch, A.J., Vanden Berghe, D.A., Fotte Dommise, R., Esmans, E. and Vanden Berghe, D. (1983). Plant antiviral agents III isolation of alkaloids from *Clivia miniata* Regal (Amaryllidaceae). *Virology* **16:**57.

Lewis, J.A. and Papavizas, G.C. (1975). Survival and multiplication of soil borne plant pathogens as affected by plant tissue amendments.In: Biology and control of Soil borne plant pathogens (ed. Bruchi, G.) *Am. Phytopathol. Soc.* pp.84-89.

Linford, M.B., Yap, F. and Oliveira. (1938). Reduction of soil populations of root knot nematode during decomposition of organic matter. *Soil Science* **45:** 596-609.

Lodha, S. (2010). Potentials of cruiferous residues in suppression of soil borne plant pathogens. *Indian Phytopath.* **63:** 250-260.

Madhiazhagan, K., Ramadoss, N. and Anuradha, R. (2002). Effect of botanicals on bacterial blight of rice. *J.Mycol.Pl.Pathol.* **32:** 68-69.

Maharjan, B.L., Shrestha, K. and Basnyat, S. (2010). Botanical control of late blight of potato. Nepal J. Sci and Technol. **11:** 37-40.

Mahajan, R., Singh, P., Bajaj, K.L. and Kalsi, P.S. (1986). Nematicidal activity of some sesquiterpenoids against root-knot nematode (*Meloidogyne incognita*). *Nematologica* **32:** 119–123.

Mahfouz, M., Abd-Elgawad and Elsayed, A. Omer. (1995). Effect of essential oils of some medicinal plants on phytonematodes Anzeiger fur Schadlingskunde. 68 (4):82-84.

Mangamma, P. and Sreeramulu, A. (1991). Garlic extract inhibitory to growth to *Xanthomonas campestris* pv. *vesicatoria*. *Indian Phytopath.* **44:** 372-373.

Manning, W.J. and Crossan, D.F. (1969). Field and greenhouse studies on the effect of plant amendment on *Rhizoctonia* hypocotyl rot of snapbean. *Plant Dis. Reptr.* **53:** 227-231.

Mariappan, V. and Narsimhan, V. (1987). Field In:Botanical Pest Control IRRI. Philippines. pp.10.

Mawar, R. and Lodha, S. (2008). Yes! Our sick crops can be cured eco-friendly. In Disease Management in Arid Lands Crops. Eds.S. Lodha, R. Mawar and B.S Rathore, Scientific Publishers (India) Jodhpur. pp. 379-430.

Mbata, T.I. and Saikia, A. (2008). Antibacterial activity and phytochemical screening of crude ethanolic extract of leaves of *Ocimum gratissimum* on *Listeria monocytogenes*. *The Internet Journal of Microbiology* **4:** 56-58.

Meena, L.R. (2001). Allelopathic potential of purple nutsedge (*Cyperus rotundus*) on germination and seedling growth of pigeonpea and urdbean *Ann, Agric.Res.* **22:** 491-494.

Meena, R., Patni, V. and Arora, D.K. (2008). Management of leaf curl of chilli with leaf proteins of *Clerodendrum aculeatum J.Mycol Pl.Pathol.* **38:** 178-181.

Meena, S.S. and Mariappan,V. (1993). Effect of plant products on seed borne mycoflora of sorghum.*Madras Agric.J.* **80:** 383-387.

Meena, S.S. and Mariappan, V. (1994). Effect of plant leaf powders and containers on germination of mycoflora infested sorghum seeds. In: Crop Diseases Innovative Techniques and Managemrnt. (Eds. Sivaprakasam, K. Seetharaman,K.) Kalyani Publishers, Ludhiana, India, pp.331-334.

Mehta, A. and Mehta, P. (2005). Antifungal potency of plant stem extract on growth, pectolytic and cellulolytic enzymes production and rot development on grapes by *Geotrichum candidum*. *J.Mycol. Pl.Pathol.* **35:** 156-158.

Mishra, S.B. and Dixit, S.N. (1979). A simple method for determining fungitoxicity of vapours of plant extracts. *Indian J Mycol. Plant Pathol.* **9:** 250-252.

Mishra, M., Tewari, S.N. and Mishra, M. (1990). Ethanolic extract toxicity of three botanicals against five fungal pathogens of rice. *Nat. Acad. Sci. Letters,* **13:** 409-412.

Mitra, S.R., Choudhuri, A. and Aditya Choudhury, N. (1984). Production of antifungal compounds by higher plants a review of recent researches. *Pl. Physiol. & Biochem.* **11:** 53-77

Muralidharan, K., Reddy, C.S., Krishnaveni, D. and Laha, G.S. (2003). Evaluation of plant derived commercial products for blast and sheath blight control in rice. *Indian Phytopath.* **56:** 151-155.

Muthulakshmi, P. and Seetharaman, K. (1994). Use of plant extracts in the management of fruit rot disease of chilli caused by *Alternaria tenuis* Nees. In: Crop Diseases Innovative Techniques and Managemrnt. (Eds. Sivaprakasam, K. Seetharaman,K) Kalyani Publishers, Ludhiana. India, pp.295-302.

Muthusamy, S. and Mariappan, V. (1992). Disintegration of sclerotia of *Macrophomina phaseolina* (Soybean isolate) by oil cake extracts. *Indian Phytopath.* **45:**217-272.

Mojumdar, V. (1995). Effects on nematodes. The neem tree, *Azadirachta indica* A. Juss., and other meliaceous plants: in Source of unique natural products for integrated pest management, industry, and other purposes. Ed. Schmutterer H, Weinheim, Germany VCH. pp. 129–150.

Narayan Bhat, M. and Sivaprakasam, K. (1994). Antifungal activity of some plant extracts. In: Crop Diseases, Innovative Techniques and Management. (Eds.

Sivaprakasam, K. and Seetharaman,K.) Kalyani Publishers, Ludhiana, India, pp.335-339.

Nariel, L.M. and Robles, R.P. (1990). Fungicidal activity of *Portulaca oleracea* extract against *Helminthosporium maydis* Wisik and Miyake in corn (*Zea mays* L.). *Philippine Journal of Weed Science* **17:** 26-32.

Narsimhan, V., Chandrasekaran, A. and Kareem, A.A. (1994). Efficacy of botanicals in the management of sheath rot disease of rice. In Crop Diseases, Innovative Techniques and Management. (Eds. Sivaprakasam, K. Seetharaman,K.) Kalyani Publishers, Ludhiana, India,pp.375-378.

Natrajan, M.R. and Lalithakumari, D. (1987). Antifungal activity of the leaf extract of *Lawsonia inermis* on *Drechslera oryzae*. *Indian Phytopath.* **40:** 390-395.

Ohigashi, H., Kawazu, H., Egaw, H. and Mitsuit. (1972). Antifungal constituent of *Sabium japonicum*. *Agril.Biol.Chem.* **36:**1399-1403.

Oka, Y., Ben-Daniel, B.H. and Cohen, Y. (2001). Nematicidal activity of powder and extracts *of Inula viscosa*. *Nematology* **3:** 735–742.

Oka, Y., Nacar, S., Putievsky, E., Ravid, U., Yaniv, Z. and Spiegel, Y. (2000). Nematicidal activity of essential oils and their components against the root-knot nematode. *Phytopathology* **90:** 710–715.

Oka, Y. (2001). Nematicidal activity of essential oil components against the root-knot nematode *Meloidogyne javanica*. *Nematology* **3:** 159–164.

Ononuju, C.C. and Nzenwa, P.O. (2011). Nematicidal effects of some plant extracts on egg hatchability and control of *Meloidogyne* spp. in cowpea (*Vigna unguiculata* (L.) Walp). *African Journal of Plant Science* 5**(3):** 176-182.

Osman, A.A., Farahat, A.A. and Amin, A.W. (1989). Organic amendments as co-agents with systemic nematicides to control nematodes inducing cowpea. Bulletin of the Zoological Society of Egypt **38:** 49-63.

Patel, R.M. and Desai, M.V. (1964). A possible biological control of root knot nematode. *Plant Dis. Reptr.* **48:** 894

Patni, C.S., Kolte, S.J. and Awasthi, R.P. (2005). Efficacy of botanicals against Alternaria blight (*Alternaria alternata*) of mustard. *Indian Phytopath.* **58:** 426-430.

Patil, R.K., Patil, K.D., Sharma, A. and Pathak, V.N. (1992). Inhibitory effect of *Ocimum sanctum* extract on fruit rot fungi. *Indian J. Mycol. Pl Pathol.* **22:**199-200.

Prithviraj, B., Kishore, S., Ram, D. and Singh, U.P. (1996). Effect of methanol extract of *Aegle marmelos* leaves on *Sclerotium rolfsii*, *Int. Pharmacognosy* **34:** 148-150.

Prithviraj, B., Manickam, M., Singh, U.P. and Roy, A.B. (1977). Antifungal activity of anacardic acid a naturally occurring derivative of salicylic acid. *Can. J.Bot.***75:** 207-211.

Policegoudra, R.S., Divakar, S. and Aradhya, S.M. (2007). Identification of difurocumenonol, a new antimicrobial compound from mango ginger (*Curcuma amada* Roxb.) rhizome. *J. Apply. Microbiol.* **102:** 594-602.

Prasad, S.M. and Sharma, M.K. (2004). Evaluation of plant extracts in management of Stemphylium blight of onion. *Indian Phytopath.* **57:** 110-111.

Raj, H. and Kapoor, I.J. (1996). Effect of oil cake amendment of soil on tomato wilt caused by *Fusarium oxysporum* f.sp. *lycopersici*. *Indian Phytopath.* **49:** 355-361.

Rai, V.R., Lokesh, S. and Ayub Khan. (2002). Occurrence and management of seed-borne fungal pathogens of maize and sorghum *in vitro*. *Seed Research* **30:**112-117.

Rai, S., Singh, U.P., Mishra, G.D., Singh, S.P., Samerketu and Wagner, K.G. (1995). Synergistic effects of ajoene and the microwave power density memories of water on germination inhibition of fungal spores. *Medical and Biological Engineering & Computing* **33:** 313-316.

Rajan, F.S., Vedamuthu, P.G.B., Khader, M.D.A. and Jeyarajan, R. (1991). Management of root disease of fenugreek. *South Indian Horticulture* **39:** 221-223.

Rao,V.K. (1991). Effect of certain indigenous plant materials on grain microflora of sorghum during storage. *Indian J. Mycol. Pl Pathol.* **21:** 268-269.

Rao, V.K. and Ratnesudhakar,T. (1992). Effect of some plant powders on grain microflora of paddy during storage. *Indian Phytopath.* **45:** 55-58.

Rathore, O.S. and Kagne, R.M. (2005). Bioprospecting of some traditional medicinal plants as antifungal against skin infections. *J. Bot. Soc. Univ.* Sagar **40:**149-152.

Ratnoo, R.S. and Bhatnagar, M.K. (1993). Effect of straw, oilcakes on ashy stem blight *Macrophomina phaseolina* (Tassi) Goid of cowpea. *Indian J. Mycol. Pl. Pathol.*, **23:**186-187.

Raya, P. (2010). Fungitoxic properties of plant extracts against *Alternaria tenuissima*: a new sp. infecting egg plant in India. *Indian Phytopath.* **63:** 45-46.

Raychaudhuri, S.P. (1964). Agriculture In Ancient India. ICAR, New Delhi, India, pp.167.

Reddy, C.S., Reddy, K.P.N., Prameela, M., Mangla, U.N. and Murlidharan, K. (2007). Identification of antifungal component in clove that inhibits *Aspergillus* spp. colonizing rice grains. *J.Mycol. Pl.Pathol.* **37:** 87-94.

Sadhale, Nalini. (1996). Surpala's Vrikshayurveda. Asian Agri-History Foundation, Secunderabad, India, pp.104.

Sagar, S.D., Hedge, Y.R., Shrikant, Kulkarni and Rao, M.S.L. (2002). Effect of seed treatment with plant extracts on seed mycoflora in rice. *Plant Pathol. Newslet.* **20:** 37-39.

Sarvamanagala, H.S., Govinaiah and Data, R.K. (1993). Evaluation of plant extracts for the control of fungal diseases of mulberry. *Indian Phytopath.* **46:** 398-401.

Sangvikar Roopa, V. (2011). Studies on Biocontrol of Some Phytopathogens. Ph.D. Thesis. Swami Ramanand Teerth Marathwada University, Nanded (M.S.) India. pp. 215.

Sarkar, S.K. (2000). Effect of intercropping Jowar on the incidence of sunhemp wilt (*Fusarium udum* f.sp.*crotlariae*) *J. Mycol Pl. Pathol.* **30:** 41-43.

Sateesh, M.K. (2009). Bioprospecting of botanicals against *Phomopsis azadiractae* on neem (Abstr.) *J. Mycol. Pl Pathol.* **39**:567.

Satish, S., Mohan, D.C., Raghavendra, M.P. and Raveesha, K.A. (2007). Antifungal activity of some plant extracts against important seed borne pathogens of *Aspergillus* spp. *J. of Agric.Tech.* **3**:109-119.

Saxena, J. and Mathela, C.S. (1996). Antifungal activity of new compounds from *Nepata leucophylla* and *N. clarkei*. *Applied and Environmental Microbiology,* **62:** 702-704.

Schmitz, H. (1930). Poisoned food technique. Indust. Engin. *Chem. Analyst*. 361-363.

Schmutterer, H. (1996). The Neem tree Azadirachta indica, A. Juss. and other meliaceous Plants. Weinheim, New york, pp.89-126.

Selvaraj, C. and Narayanasamy, P. (1994 a).Effect of plant extracts on the incidence of brown spot and sheath rot disease of rice. In: Crop Diseases Innovative Techniques and Management. (Eds. Sivaprakasam, K. Seetharaman,K.) Kalyani Publishers, Ludhiana, India, pp. 325-329.

Selvaraj, C. and Narayansamy, P. (1994 b). Antifungal effect of some of the plant extracts against *Helminthosporium oryzae* and *Sclerotium oryzae* in rice. In: Crop Diseases, Innovative Techniques and Management. (Eds. Sivaprakasam, K. Seetharaman,K) Kalyani Publishers, Ludhiana. India,pp.303-306.

Senapaty, D. (2010). Efficacy of soil amendments for the management of rice sheath blight. *Indian Phytopath.* **63:** 2010.

Sengupta, S., Ghosh, S.N. and Das, A.K. (2008). Antifungal potentiality of the latex of *Plumeria actifolia* Poir. *Indian Phytopath.* **61:** 499-503.

Senthilnathan, V. and Narsimhan, V. (1994). Effect of plant extracts/products on mycelial growth and spore germination of *Alternaria tenuissima* inciting the blight of onion and the nature of the antifungal components. In: Crop Diseases Innovative Techniques and Management. (Eds. Sivaprakasam, K. Seetharaman,K.) Kalyani Publishers, Ludhiana, India, pp.307-313.

Sequeira, L. (1962). Influence of organic amendments on survival of *Fusarium oxysporum* f.sp.*cubense* in the soil. *Phytopathology* **52:** 976-983.

Shah, N.H., Khan, M.I. and Mohd., I. (1993). Plant extracts vs fungicide on mycoflora of wheat seed. In: New Approaches in Agricultural Technology Vol.2. Bioved Research Society, Allahabad, India, pp.471-474.

Sharma, J.P. and Kumar, S. (2009). Management of Ralstonia wilt of tomato through microbes, plant extract and combination of cake and chemicals. *Indian Phytopath.* **62:** 417-423.

Sharma, Nidhi and Trivedi, P.C. (2002).Screening of Leaf Extracts of Some Plants for Their Nematicidal and Fungicidal Properties Against *Meloidogyne incognita* and *Fusarium oxysporum*. *Asian J. Exp. Sci.* **16** : 21-28.

Shtacher, G. and Kashman, Y. (1970). 12-Carboxyeudesma-3,11(13)-diene. A novel sesquiterpenic acid with a narrow antifungal spectrum. *Journal of Medicinal Chemistry* **13:** 1221–1223.

Siddiqui, S., Siddiqui, B.S., Faizi, S. and Mahmood, T. (1988). Tetracyclic triterpinoids and their derivatives from *Azadirachta indica*. *J. Natural Products,* **51:**30-43.

Sinha, P.S. and Saxena, S.K. (1990). Effect of latex of *Euphorbia hirta* on fruit rot of tomato caused by *Aspergillus niger* in the presence of *Drosophila busckii*. *Indian Phytopath.* **43:** 462-463.

Singh, H.B. and Singh, U.P. (1980). Inhibition of growth and sclerotial formation in *Rhizoctonia solani* by garlic oil.*Mycologia,* **72:** 1022-1025.

Singh, H.B. and Singh, U.P. (1982). Effect of volatiles of some plant extracts and oils on conidia of *Erysiphe polygoni* DC. *Aust.Pl. Path.* **10:** 66-67.

Singh, H.B. and Singh, U.P. (1983). Effect of some new sulphur compounds, plant extracts and vegetable oils on conidia of *Erysiphe polygoni* DC. Agric. Res. Rural Dev.,**6:** 5-10.

Singh, J. and Majumdar,V.L. (2001). Efficacy of plant extracts against *Alternaria alternata*–the incitant of fruit rot of pomegranate (*Punica granatum* L.). *J. Mycol Pl Pathol.* **31:**346-349.

Singh, J., Tripathi, N.N. and Singh, J. (1995). Effect of growth stages of *Ranunculus scleratus* L. on its fungitoxicity. *Indian Phytopath.,* **48:** 363-364.

Singh, M. and Singh, R.P. (2005). Management of mushroom pathogens through botanicals. *Indian Pytopath.* **58:** 189-193.

Singh, M., Sobita Devi, Singh, M. and Singh, M.T. (2004). Effect of plant extracts on seed mycoflora of rice during storage. *Indian Phyopath.* **57:** 205-207.

Singh, N. and Singh, R.S. (1983). Inhibition of *Fusarium udum* pigeonpea wilt pathogen by ether distillate of margosa cake amended soil. *Indian J. Mycol. Pl. Pathol.,* **13:**329-330.

Singh, R.S. and Sitaramaiah, K. (1967). Effect of decomposing green leaves,saw dust and urea on the incidence of root knot of okra and tomato. *Indian Phytopath.* **20:** 349-355.

Singh, S.D. and Navi, S.S. (2000). Garlic as a biocontrol agent for sorghum ergot. *J. Mycol. Pl Pathol.* **30:** 350-354.

Singh, U.P., Pathak, K.K., Khare, M.N. and Singh, R.B. (1979). Effect of leaf extract of garlic on *Fusarium oxysporum* f.sp.*ciceri, Sclerotinia sclerotiorum* and on gram seeds. *Mycologia.***71:** 556-564.

Singh, U.P. and Singh, H.B. (1983). Comparative efficacy of some commercial fungicides,plant extracts and oil for the control powdery mildew (*Erysiphe polygoni* DC.) of pea (*Pisum sativum* L.). *Aust. Pl. Path.* **12:**22-24.

Singh, U.P., Singh, H.B. and Chauhan,V.B. (1984). Effect of some fungicides,plant extracts and oil on inoculum density of different nodal leaves of pea by *Erysiphe polygoni*. Zeitschrift fur Pflanzenkrankheiten und Pflanzenschutz.**91:** 20-26.

Singh, U.P., Ram, D. and Tewari,V.P. (1990). Induction of resistance in chickpea (*Cicer arietinum*) by *Aegle marmelos* leaves against *Sclerotinia sclerotiorum*. Zeitschrift fur Pflanzenkrankheiten und Pflanzenschutz. **97:** 439-443.

Singh, U.P., Pandey,V.N., Wagner, K.G. and Singh, K.P. (1990). Antifungal activity of ajoene, a constituent of garlic (*Allium sativum* L.). *Cand. J. Bot.* **68:** 1354-1356.

Singh, U.P., Srivastava, B.P., Singh, K.P. and Mishra, G.D. (1991). Control of pea powdery mildew with ginger extract. *Indian Phytopath.* **44:** 55-59.

Singh, U.P. and Chauhan,V.B. (1992). Effect of ajoene, a compound derived from garlic (*Allium sativum*) on *Phytophthora dreschleri* f.sp. *cajani*. *Mycologia* **84:**105-108.

Singh, U.P., Prithviraj, B., Wagner, K.G. and Schumacher, K.P. (1995). Efeect of ajoene, a constituent of garlic (*Allium sativum)* on powdery mildew (*Erysiphe pisi*) of pea (*Pisum sativum*). Zeitschrift fur Pflanzenkrankheiten und Pflanzenschutz. **102:** 399-406.

Singh, V., Tripathi, R.D., Tripathi, N.N. and Dixit, S.N. (1983). Isolation and properties of fungitoxic principle for *Zingiber officinale*. *Indian J. Plant Path.* **1:** 89-96.

Sinha, P.S. and Saxena, S.K. (1990). Effect of latex of *Euphorbia hirta* on fruit rot of tomato caused by *Aspergillus niger* in the presence of *Drosophila busckii*. *Indian Phytopath.* **43:** 462-463.

Sivasithambaram, K., Smith, L.D.J. and Gross, O.M. (1981).Effect of potting media containing fresh saw dust and composted tree barks on *Phytophthora cinnamomi*. *Aust. Plant Pathol.* **10:** 20-21

Somani, A.K. (2009).Management of black scurf (*Rhizoctonia solani*) of potato through seed treatment with botanicals. *Potato J.* **36:** 155-159.

Somda, I., Leth, V. and Sereme, P. (2007). Antifungal effect of *Cymbopogon citratus*, *Eucalyptus camaldulensis* and *Azadirachta indica* oil extracts on Sorghum seed borne fungi *Asian J of Plant Sciences* **6:**1182-1189.

Stapleton, J.J. and Duncan, R.A. (1998). Soil disinfestations with cruciferous amendments and sub lethal heating:effects on *Meloidogyne incognita*, *Sclerotium rolfsii* and *Phythium ultimum*. *Plant Path.* **47:** 737-742.

Stumpf, P.K. (1955). Enzymes of protein metabolism. In: Methods In Enzymology Vol.II Academic Press Inc.New York. pp. 265-272.

Sundararaju, P., Padmanaban, B. and Sathiamoorthy, S. (2003). Efficacy of certain botanicals against root lesion nematode, *Pratylenchus coffeae* in banana. *Nematol. medit.* **31:** 201-205

Sunder, S., Singh, R. and Dodan, D.S. (2010). Evaluation of fungicides, botanicals and non-conventional chemicals against brown spot of rice. *Indian Phytopath.* **63:**192-194.

Thippeswamy, G., Lokesh, S. and Rai, V.R. (2003). Influence of some indigenous medicinal plants extracts on seed mycoflora and seedling growth of some oilseed crop species. *Adv. Plant Sci.* **16:** 67-74.

Thirumalaisamy, P.P., Rathi, Y.P.S. and Tripathi, H.S. (2003). Screening of some plant extracts inhibitory to urdbean leaf crinckle virus. *Indian Phytopath.* **56:** 233-235.

Tewari, S.N. and Dath, P.A. (1984).Effect of leaf extract media on some plants on the growth of three fungal pathogens of rice. *Indian Phytopath.* **37:** 458-461.

Thakore, B.B.L., Mathur, S., Singh, R.B., Chakravarti, B.P. and Mathur, S. (1987). Soil amendment with oil cakes in field for rhizome rot control. *Korean J. Plant Prot.***26:** 267-268.

Tripathi, N.N., Kumar, N., Singh, P. and Mishra, A.K. (2006). Potential of Plant Extracts against Storage Pests of Cereals and Pulses: A Review. In Plant Protection In New Millennium. Eds. Gadewar, A.V. and Singh,B.P. Satish Serial Publishing House, Delhi. pp 245-282.

Tripathi, R.N., Pandey, D.K., Ttripathi, N.N. and Dixit, S.N. (1982). Antifungal activity in pollens of some higher plants. *Indian Phytopath.* **35:** 346-348.

Tuite, J. (1969). Microscopic Techniques in Plant Pathological Methods : Fungi and Bacteria. Bergers Publishing Co. Minneapolis. U.S.A. pp. 239.

Upadhyaya, M.L. and Gupta, R.C. (1990). Effect of extracts of some medicinal plants on the growth of *Curvularia lunata. Indian J. Mycol. Pl Pathol.,* **20:**144-145.

Verma, A. and Verma, H.N. (1993).Plant virus inhibitors-A perspective. In: Integrated Pest management (eds Dwivedi, B.K., Alam,M.M.,Pandey, G.) Bioved Research Society. Allahabad, India, pp. 217-230.

Verma, H.N. and Awasthi, L.P. (1979). Antiviral activity of *Boerhavia diffusa* root extract and the physical properties of the virus inhibitor. *Can.J.Bot.* **57:**926-932.

Verma, H.N., Srivastava, S., Varsha and Kumar, D. (1996). Induction of systemic resistance in plants against viruses by a basic protein of *Clerodendron aculeatum* leaves. *Phytopathogy* **86:** 485-492.

Vijaya, T. and Mouli, K.C. (2010). Microbial and Botanical biopesticides in plant disease management. In Plant Diseases and its Management. Ed. P.C.Trivedi. Pointer Publishers, Jaipur, India.pp.187-205.

Vir, D. and Sharma, R.K. (1985). Effect of fungicide XXXII. Evaluation of neem oil for control of plant pathogens. *Asian Farm. Chemicals* 1 **(7/8):** 23-24.

Wang, W.Q., Ben-Daniel, B.H. and Cohen, Y. (2004). Control of plant diseases by extracts of *Inula viscosa. Phytopathology* **94:**1042–1047.

Yadav, M.S. (2009). Biopesticidal effect of botanicals on the management of mustard diseases. *Indian Phytopath.* **62:** 488-492.

Yadav, S.C. and Gour, H.N. (2007). Management of stripe disease of barley caused by *Drechslera graminea* using fungicides, botanicals and bioagents. *J. Mycol. Pl Pathol.* **37:** 457-459.

Ye, X.Y. and Neg, T.B. (2000). Mungin a novel cyclophilin like antifungal protein from the mung bean. Biochemical and Biophysical Research Communications.**273:**1111-1115.

Yang, Xiu-juan, He, Yu-xian, Chen Fu-ru, and Tong, R.U. (2002-03). Evaluation on Nematicidal Activity of Different Plant Extracts. Acta Agriculturae Universitis Jiangxiensis pp. 19.

Yuji, Oka, Bat-Hen Ben-Daniel, and Yigal, Cohen. (2006).Control of *Meloidogyne javanica* by formulations of *Inula viscosa* leaf extracts. *J Nematol.* 38 **(1):** 46–51.

Zasada, I.A., Ferris, H. and Zheng, L. (2002). Plant Sources of Chinese Herbal Remedies: Laboratory Efficacy, Suppression of *Meloidogyne javanica* in Soil and Phytotoxicity Assays.*J Nematol.* **34(2):** 124-129.

Chapter 31

Seed Priming in Respect to Disease Resistance

Bandana Bose[1], Sananda Mondal[1], Asha Sinha[2], and Parmanand Trivedi[2]

[1]Department of Plant Physiology,
[2]Department of Mycology and Plant Pathology,
Institute of Agricultural Sciences, Banaras Hindu University,
Varanasi – 221 005, U.P.

Seed is the ripened ovule that consists of an embryo and stored food for germination and contain a protective covering or one can also say that a seed is a small embryonic plant enclosed in a covering called seed coat along with some stored food. The term seed also has a general meaning *i.e.* a seed is any living material that can be sown and which gives rise to a functional plant, eg. seed potato is a part of a tuber and the setts of sugarcane is the parts of the stem. The seed plant (*Spermatophyta*) consists of two major classes: Gymnosperms (800 living species) and Angiosperms (250000 living species). However, a true seed is comprised of three important parts: seed coat, embryo and endosperm.

Seed coat or testa or seed covering protects the embryo and endosperm from the environmental stresses. It may be membranous, thicker, in few cases coloured or sometimes fused with fruit tissue to form a pericarp (eg- cereal grains) or endocarp (mango, coconut). The seed coat is derived from integument; in gymnosperm and angiosperm it contain single and double layers respectively. In mature seeds the outer cell layers of the integument form a dead covering layer, while inner cell layer may remain alive.

Embryo or germ is the second important living part of a seed or can be denoted as future plant (as embryonic plant). It consists of embryonic axis namely plumule (shoot tip), radicle (root tip) and cotyledons (seed leaves). Based on the number of cotyledons present in a seed it is classified into two types: monocotyledonous or monocots (mature seeds consist of single cotyledon, eg- corn) and dicotyledonous or dicots (mature seeds consists of two cotyledons or embryonic leaves, eg- chick pea). The cotyledon in the grass family is highly modified to a plate like or a leafy structure known as scutellum.

Endosperm the third important part of seed surrounds the embryo in flowering plants it also provides nutrition to the embryo at the time of its development as well as at the time of germination of seed and seedling establishment. The seeds having endosperm as main storage tissue are called albuminous seed and eg. are most of the grasses and the dicots are like fenugreek, lucerne castor, *Arabidopsis*, tomato, chilli etc.

Seeds are heterotropic organs totally dependent on nutrient imported from the mother plant or parent plant for their growth, development and storage. So, the nutrients available for the development of the seeds from the mother plant influences or determines the total seed number or individual seed size or seed weight which are the properties of biological and agronomic significance. Except photosynthetic machinery every seed contains all other metabolic activities like carbohydrates, protein and fat metabolisms; storage tissue of seeds also contain different mineral salts and micronutrients (like nitrogen, sulphur, sodium, potassium, ammonium salt, calcium, magnesium, chlorine etc). These nutrients are transported in seeds at the time of seed development through phloem by both symplastic and apoplastic pathways from the mother plant facilated by different transporters and anion channels present in the plasma membrane.

The seed maturation phase starts once the embryo and the endosperm have completed the morphogenesis and patterning stages. The characteristic of this phase is arrested growth followed by synthesis and accumulation of preserves whose degradation upon germination will produce or provide nutrients to the growing seedling before the photosynthetic capacity is fully acquired. In the early and mid phases of maturation are dominated by the action of abscisic acid (ABA). Initially, ABA is synthesized in the maternal tissue and latter on in the embryo and endosperm. The ABA promotes desiccation tolerance in the maturing seed by promoting synthesis of different proteins like Late Embryogenesis Abundant Protein (LEA protein), responsive to ABA (RAB) and Dehydrin proteins (DHN). The ratio of ABA to GA controls seed maturation seed dormancy and seed germination (Bewley and Black 1978).

Germination represents the start of the dynamic phase in plants life, from where the embryo is considered to be awaken. During this time period the metabolic activities in the seeds increase greatly which toggles the whole process which causes a new plant to come into existence.The main mechanism that triggers this process is simply liquid water. When water gets into the embryo and hydrates its cells, it speeds up metabolic activities which allow to increase the process of cell division as a result

growth increases in rapidity. However it is not always so simple to start this process since several impairments–both chemical and physical–necessary for successful germination.

Germination includes a number of events that starts with the uptake of water by the quicent dry seed and terminate with the elongation of the embryonic axis. The germination starts with the rupturing of the seed coat surrounding the embryo and coming out of radicle. The uptake of water by a mature dry seed is triphasic with a rapid initial uptake (Phase I) followed by a plateau (Phase II). A further increase in water uptake occurs only after germination is completed as the embryonic axes elongate (Phase III). The influx of water into the cells of dry seeds during phase I results in temporary structural disturbances particularly to membranes which lead to an immediate and rapid leakage of solutes into the surrounding immbibition solution. After a short time of rehydration the membranes return to its stable configuration which stops the solute leakage. The first changes after immbibition is the respiratory activity which is detected within minutes. This depends upon the availability of the oxygen. Germinating seeds of many sp. frequently produce ethanol. This is often the result of an internal deficiency of oxygen that is caused by restrictions to the diffusion of oxygen by the structure that surrounds the seed.

Radicle extention is the indication of the termination of the germination and marks the seedling growth. It also includes the DNA synthesis and cell division. Radicle extension is the turgor driven process that causes the radicle's cell wall to expand. The seed tissue surrounds the radicle tip should weaken thus allowing it to elongate and emerge out. After that the mobilization of the stored preserves start and continue until the seedling establishes itself by photosynthesizing the require assimilates.

Seeds are the delivery system in agriculture. High quality seed leads to excellent seedling performance in the field. It is the ultimate basis of successful companies that breed crop plants for seed production. **Seed quality** is a complex trait that is determined by interactions between multiple genetic factors and environmental conditions. Modern approches to improve seed quality therefore combine classical genetics, plant molecular biology and a variety of seed technologies. These "seed biotechnologies" enhance physiological quality, vigor and synchronity to establish a crop in the field under diverse environmental conditions.

One of the largest concerns of world agriculture as well as the home grower is the decrease in germination time and increase in germination percentage since both of these factors can bring great benefits. Some seeds especially some flowers and herbs are often quite difficult to germinate and using certain techniques to increase the rate and speed in which the sprout has been the focus of a large amount of scientific research.

On the basis of above scenario the present chapter has been prepared which deals the seed priming technology with special reference to disease resistance. However, the seed priming comes under seed enhancement technology by which one can improve/ enhance the quality of seed and its vigor.

Seed Treatment Technology

It is an important interface between seed producers and crop production industry. Its aim is to allow the seed treatment product to be used in such a form that represents highest quality in the market. It is also referred as seed enhancements and seed treatments include priming, pelleting, coating, artificial seeds, and other novel seed treatment methods of applied seed biology. Our basic and applied seed research projects focus on embryo growth and on the different seed covering layers (*e.g.* testa, endosperm, pericarp), which are determinants of seed quality and exhibit the biodiversity of seed structures. Seed germination is controlled by environmental factors (light, temperature, water) and on plant hormones as endogenous regulators (gibberellins, abscisic acid, ethylene, auxin, cytokinins and brassinosteroids). The utilization of plant hormones and inhibitors of their biosynthesis and action in seed treatment technologies affects seed germination and seedling emergence. The genes, enzymes, signaling components and down-stream targets of some plant hormones provide molecular marker for seed quality and seedling performance (The Seed Biology Place).

Important methods have been developed to enhance seed and seedling performance through the addition of chemicals to protect the seed from pathogens and/or to improve germination. Different techniques have been developed by various commercial seed industries for enhancing the seed quality and those are known as *Film Coating, Seed pelleting and Seed Coating.*

Film-Coating

Film-coating methods allow the chemicals to be applied in a synthetic polymer that is sprayed onto the seeds and provide a solid, thin coat covering them. The advantage of the polymers is that they adhere tightly to the seed and prevent loss of active materials like fungicides, nutrients, colorants or plant hormones. Some novel applications of film coating are used to modify imbibition and germination. They can confer temperature-sensitive water permeability to seeds or affect gaseous exchange. By this they control the timing of seed germination and seedling emergence. Certain temperature-dependent water-resistant polymers can delay imbibition until the climatic conditions become suitable for continued seedling growth. Film coatings form a very thin film around the seed which results in no change in shape, size and weight of that particular seed.

Seed Pelleting

Seed pelleting adds thicker artificial coverings to seeds, which can be used to cover irregular seed shapes and add chemicals to the pellet matrix, *e.g.* of sugar beet or vegetable seeds. The peliet matrix consists of filling materials and glue. Loam, starch, tyllose (cellulose derivative) or polyacrylate/polyacrylamide polymers are commercially used. Seed pelleting is also used to increase the size of very small horticultural seeds. This provides improved planting features, *e.g.* singulate planting, the use of planting machines, or precise placement and visibility in/on the soil.

Multilayer Coating

It is a highly sophisticated method allowing sequential application of multilayer materials, including the incorporation of fungicides and insecticides.

Artificial Seeds

Cell culture and regeneration techniques allow the mass production of somatic embryos. This can be used to generate genetically identical seedlings of poplar, orchids and other species. Somatic embryos can be packed in a suitable gel-type matrix (agar-agar, gums, dextrans) and covered with an artificial seed coat (Na alginate). These *artificial seeds* provide an important packaging system.

Seed Priming

It is the most important physiological seed enhancement method. Seed priming is an hydration treatment that allows controlled imbibition and induction of the pregerminative metabolism ("activation"), but radicle emergence is prevented. The hydration treatment is stopped before dessication tolerance is lost. An important problem is to stop the priming process in the right moment; this time depends on the species and the seed batch. Molecular marker can be used to control the priming process. Priming solutions can be supplemented with plant hormones or beneficial microorganisms. The seeds can be dried back for storage, distribution and planting. Germination speed and synchrony of primed seeds are enhanced (see figures below) and can be interpreted in the way that priming increases seed vigor (short or no "activation" time). A wider temperature range for germination, release of dormancy and faster emergence of uniform seedlings are achieved. This leads to better crop stands and higher yields. A practical drawback of primed seeds is often a decrease in storability and the need for cool storage temperatures.

Several types of seed priming are commonly used to improve germination and seedling vigor those are categorised as under *Osmopriming* (osmoconditioning): It is the standard priming technique. Seeds are incubated in well aerated solutions with a low water potential, and afterwards washes and dried. The low water potential of the solutions can be achieved by adding osmotica lie mannitol polyethyleneglycol (PEG) or salts lie KCl. *Hydropriming* (drum priming): It is achieved by continuous or successive addition of a limited amount of water to the seeds. A drum is used for this purpose and the water can also be applied by humid air. 'On-farm steeping' is the cheep and useful technique that is practized by incubating seeds (cereals, legumes) for a limited time in warm water.

Matrixpriming (Matriconditioning)

It is the incubation of seeds in a solid, insoluble matrix (vermiculite, diatomaceous earth, cross-linked highly water-aosorbent polymers) with a limited amount of water. This method confers a slow imbibition.

Halopriming

It is a presowing soaking of seeds in salt solution, which enhance germination and seedling emergence uniformly under adverse environmental condition (Bose and Mishra,1999; Ashraf *et al.*, 2003; Basra *et al.*, 2005a).

Biopriming

It involves coating of seeds with a bacterial biocontrol agent like *Pseudomonas aureofaciens* and hydrating for 20 h under warm conditions (23°C) in moist vermiculite or on moist germination blotters in a self-sealing plastic bag. The seeds are taken out from the solution before radical emergence (Callan *et al.*, 1990). It integrates the biological and physiological aspects of disease control was recently used as alternative method for controlling many seed and soil borne pathogens.

Pregerminated Seeds

It is only possible with a few species. In contrast to normal priming, seeds are allowed to perform radicle protrusion. This is followed by sorting for specific stages, a treatment that reinduces dessication tolerance, and drying. The use of pre germinated seeds causes rapid and uniform seedling development.

In general, priming offers the opportunity to germinate a lot of seed at much higher speeds without detrimental effects in germination percentages. For example, a two day treatment of parsley seeds with a PEG 6000 (Poly Ethylene Glycol) solution can reduce germination times substantially, from a few weeks to just a few days. Other seeds such as coriander might also benefit from similar treatments with PEG or treatments with NaCl solutions. In general one can use this technique by doing 3 small experiments best for particular seed variety and germination conditions. In the first experiment the seeds are simply soaked in water for 24 hours, another in which seeds are placed in a 200mg/L NaCl solution and in the third one the seeds are submerged in a PEG 6000 20 per cent solution, then the seeds are air-dried after the treatments. After comparing the results of these experiments with a control with no priming will be able to see which priming technique is better for that particular seed lot and most effectively increases the seed germination rates.

To sum it up priming of seeds is a very efficient technique to increase the speed of germination without sacrificing germination rates. These methods are not very useful for seeds such as lettuce or tomato–which germinate easily–but they are invaluable for plants such as parsley, coriander or carrots which are generally much harder to germinate. If crop grower have some seeds that take long time to germinate then they can set up some priming experiment that might be the best thing to do for them.

Priming in the traditional sense, soaking of seeds in water before sowing, has been the experience of farmers in India in an attempt to improve crop stand establishment but the practice was without the knowledge of the safe limit of soaking duration (Harris 1996). Moreover, Harris *et al.* (1999), promoted a low cost, low risk technology called *'on-farm seed priming'* that would be appropriate for all farmers, irrespective of their socioeconomic status. On-farm seed priming involves soaking the seed in water, surface drying and sowing the same day. The rationale is that sowing seed decrease the time needed for germination and allow the seedling to escape deteriorating soil physical conditions.

According to, Khan (1992), osmotic conditioning in its modern sense, aims to reduce the time of seedling emergence, as well as synchronize and improve the

Table 31.1: Influence of Seed Priming on Various Physico-Chemical Parameters of a Number of Plants/Crops

Sl.No.	*Types of Priming*	*Name of the Crop*	*Parameters*	*Name of the Scientist and Year*
1.	Hydro-Priming	Pinto bean (*Phaseolus vulgaris* L.); soybean; lucerne	Yield Attributes, improves drought tolerance; better crop stand	Ghassemi-Golezani *et al.* (2010); Xiaohuan *et al.* (2009); YanRong (2008)
2.	Hydro-Priming and osmohardening/ osmopriming	Rice; fine rice; chickpea; *Bromus tomentellus* L. and *B. inermis* L.; soybean	Nutrient homeostasis and reserve metabolism; physiological and biochemical aspects; increasing crop yield by modulating sucrose metabolism; seed germination and early growth characteristics; germination and seedling vigour; initial seedling development;	Farooq *et al.* (2010); Basra *et al.* (2005); Kaur *et al.* (2005); Tavili *et al.* (2010); Rouhi *et al.* (2011); Peske *et al.* (2010)
3.	Osmoconditioning/ osmopriming	Soybean; pepper; *Poa pratensis; Trifolium repens;* cucumber; rice; coffee; papaya; onion, carrot, tomatoes and lettuce seeds; sweetcorn	Chilling injury (Changes in plasmamembrane composition); germination and nuclear replication of seeds; germination seedling emergence and growth; germination studies in low temperature treatment; seed and seedling characters; moisture content, germination vigour and health test, seed viability and vigor; beneficial for seed germination and seedling development; higher values of germination	Zhuo *et al.* (2009); Lanteri *et al.* (1996); Lespinay *et al.* (2010); Peyvast *et al.* (2010); Golshani *et al.* (2010); Braz *et al.* (2008); Lopes and Souza (2008); Franco and Cattivello (2007); Oliveira *et al.* (2007); Krishnotar *et al.* (2009)
4.	Hydrohardening and halohardening	Rice; wheat	Germination, seedling vigour, nitrogen metabolism, stress metabolism in respect to low temperature; physiological and yield attributes	Mondal *et al.* (2011); Sharma and Bose (2009); Anaytullah and Bose (2007)
5.	Hydro, Hormonal and halopriming	Canola; *Phyllanthus niruri;* Pearl millet hybrid	Salt and osmotic stress; germination studies; yield attributes	Gao *et al.* (1999); Ponni (2008); Hooda and Singh (2006)

Contd...

Table 31.1–*Contd...*

Sl.No.	Types of Priming	Name of the Crop	Parameters	Name of the Scientist and Year
6.	Halopriming	Sesame; mustard; cotton; sweet pepper; tomato and chilli; vegetables and field crops; green gram;	Seed germination and seedling growth; crop growth and yield attributes; field emergence index and yield attributes; germination studies under low temperature and seed accelerated aging; growth and yield attributes; improved germination and seedling growth; growth and physiological parameters (LAI, CGR, RGR and NAR) and yield attributes; rapid and uniform field emergence maximum field emergence, minimum days to 50 per cent flowering and days to harvest maturity	Kumari *et al.* (2002); Pandey and Bose (2006); Sudarshan *et al.* (2006); Kikuti *et al.* (2005); Barlow and Haigh (1987); Sarkar and Pal (2006); Verma *et al.* (2006)
7.	Osmoconditioning and pesticide treatments	Beat; rice	Seedling vigour and seedling protection against microorganism; good seedling vigour in moisture stress	Dias *et al.* (2009); Mohanasarida and Mathew (2005)
8.	Pre-conditioning	Carrot	Increase the incidence and inoculum density of *Alternaria* sp.	Magalhaes *et al.* (2004)

germination percentage, by subjecting the seeds to a certain period of imbibitions using osmotic solutions. The seeds normally begin water uptake on contact with this solution and stop the process as soon as they become balanced with the water potential of the solution.

Table 2. Crops in which seed priming has increased yields, the countries involved and the references where the methods used and the results obtained can be found.

Crop	*Countries*	*References*
Wheat	India, Nepal, Pakistan	Harris *et al.* (2001b); Rashid *et al.* (2002); Sharma *et al.* (2009)
Upland rice	India, Nigeria, Sierra Leone, Gambia, Ghana, Cameroon	Harris *et al.* (1999); (2002); Harris (2003); Bose *et al.* (2010)
Maize	India, Nepal, Pakistan, Zimbabwe	Harris *et al.* (1999); (2001a); (2001c), Krishnotar *et al.* (2009); Bose *et al.* (1982)
Sorghum	Pakistan, Botswana, Zimbabwe	Harris (1996); Chivasa *et al.* (1998); (2001); Rashid *et al.* (2002)
Pearl millet	Pakistan, India	Harris and Mottram (2004)
Finger millet	India	Kumar *et al.* (2002)
Chickpea	Bangladesh, India, Nepal, Pakistan	Harris *et al.* (1999); Musa *et al.* (2001); Rashid *et al.* (2002)
Mungbean	Pakistan	Rashid *et al.* (2004b)
Cowpea	Senegal	Braconnier and Bouru (2004)
Mustard	India	Pandey and Bose (2006)

The beneficial effects of these priming treatments reflected in greater cellular membrane integrity, counter action of lipid peroxidation and free radical chain reaction often found to be directly correlated with the maintenance of viability and reduce moisture uptake by hydrated-dehydrated seed (Dollypan and Basu, 1985), antipathogenic effects (Powell and Mathews, 1986), repair of biochemical lesions by the cellular enzymatic repair system (Villers and Edgcumbe, 1975) and metabolic removal of toxic substances (Basu *et al.*, 1973) and counteraction of free radical and lipid peroxidation reactions (Rudrapal and Basu, 1982).

Kumari *et al.* (2002) reported that pre-soaked hardening seeds of sesame resulted in good germination and seedling growth. The benefits of seed priming in all crops included fast emergence, more and uniform crop stands, less to re-sow, more vigorous plants, drought tolerance, earlier tolerance, earlier flowering, earlier harvest maturity and higher yield.

Bose and her co-workers did a lot of work on seed priming technology since 1980s to till date. They have used salts of magnesium as well as nitrate for this purpose. They observed priming can reduce germination time in laboratory as well as in field conditions by improving the germination related hydrolyzing enzymes like proteases and α-amylase and increases the rate of formation of soluble sugars and soluble nitrogen in endosperm of germinating maize, wheat and rice seeds respectively (Bose *et al.*, 1982, Anayatullah and Bose, 2006 and Mondal *et al.*, 2011).

They also observed that seed priming effects can be carried over to vegetative and reproductive phases of plants life. They also introduced a new technique called nitrate seed hardening technology where the seeds primed with various salts of nitrate and then dehydrated and used for germination purpose after coming of proper season. The group claimed in series of their studies that this type of priming improved germination physiology, several growth factors of plant development, nitrogen and antioxidant metabolism, nitrogen and water harvesting capacity and finally the yield potential in mustard, wheat, maize and rice (Bose *et al.*, 2007; Anayatullah 2007, Krishnotar *et al.*, 2009; Mondal *et al.*, 2011)

What is Disease?

A disorder of structure or function in a human, animal, or plant, especially one that produces specific signs or symptoms or that affects a specific location and is not simply a direct result of physical injury. It may be caused by external factors, such as infectious disease, or it may be caused by internal dysfunctions, such as autoimmune diseases. This represents the definition of disease in general and the present chapter is a concern of plant diseases, caused mainly with aid of fungus, bacteria, viruses and nematodes. The common diseases of plants are wilt, blight, blast, rust, canker, decay, heart rot, root diseases, nursery diseases etc.

The life cycle of a plant passes through various biotic and abiotic stresses and plants are being attacked at different stages of growth by a number of disease causing organism. These organisms caused large crop loses which leads to human hunger and malnutrition. So for the control of plant diseases the major objectives have to be taken by the plant breeder, pathologist, physiologist and the agricultural chemical industry.

Plant can resist pathogen attack by different ways broadly called as plant disease resistance. **Plant disease resistance** derives both from **pre-formed defenses** and from **infection-induced responses** mediated by the plant immune system. Relative to a disease-susceptible plant, **disease resistance** is often defined as reduction of pathogen growth on or in the plant, while the term **disease tolerance** describes plants that exhibit less disease damage despite similar levels of pathogen growth. Disease outcome is determined by the three-way interaction of the pathogen, the plant, and the environmental conditions (an interaction known as the **disease triangle**). Defense-activating compounds can move cell-to-cell and systemically through the plant vascular system, but plants do not have circulating immune cells so most cell types in plants retain the capacity to express a broad suite of antimicrobial defenses. Although obvious *qualitative* differences in disease resistance can be observed when some plants are compared (allowing classification as "resistant" or "susceptible" after infection by the same pathogen strain at similar pathogen inoculum levels in similar environments), a gradation of *quantitative* differences in disease resistance is more typically observed between plant lines or genotypes. Plants are almost always resistant to certain pathogens but susceptible to other pathogens; resistance is usually pathogen species-specific or pathogen strain-specific (Bhattacharya and Vijaylaxmi, 2008).

Plant has the ability to resist itself by applying different mechanisms like **defense at the perimeter** which includes two strategies *i.e.* keeping the potential invaders out of attacking range and maintaining protective barriers that the invaders do not reach the plant surface (Campbell *et al.*, 1980). These structures are **trichomes** [Blight resistance variety of *Cicer arietinum* has more glandular hairs that secrete malic acid on their leaves than the varieties susceptible to *Mycospharella* blight (Goodman *et al.*, 1967)], **cuticle as electrically charged surface** [a negative charge usually develops on the leaf surface due to presence of fatty acids, many air borne spores also contain negative charge and repelled by leaf surfaces (Gregory, 1972)], **as a toxic barrier** [spore germination is strongly inhibited by the cuticular extraction of chrysanthemum in *Cladosporium fulvum* and *Botrytis cinerea* (Blakeman and Atkinson, 1976)], **as a mechanical barrier, root cap and mucilage** (it prevents wounding and reduces the potential for invasion by soil born pathogen), **seed coat** (effective barrier against penetration by many pathogens), **extruded chemicals** [phenolic compounds have been identified in the exudates from the seeds of several plants like sugar beets (Heydecker and Chetram, 1971), peas (Kraft, 1974) and peanuts (Reddy *et al.*, 1977)], **hypersensitive responses** against biotrophs like rust, powdery mildews and downy mildews (different enzymes were synthesized in hypersensitive tissues against the pathogen elicitors and prevent the spread of the pathogen, the enzymes are ribonuclease, peptidase, glucose-6-phosphate dehydrogenase, 6-phosphogluconate dehydrogenase, polyphenol oxidase, peroxidase, phenylalanine ammonia lyase, cytochrome oxidase etc). Some physical defenses are also induced by pathogens like interaction at the cell wall surface, calcium accumulation, cell wall thickening, and deposition of additional materials.

For controlling soil borne pathogens depends mainly on fungicidal applications, that causing hazards to the human health and environment. Soil amendment and bio-priming seed treatment are gaining importance in management of many plant pathogens as another alternative to chemical fungicides in recent times. Also, seed coating with bio-control agents was the most effective treatment for controlling root rot diseases as shown by Callan *et al.* (1991).

Indian farmers reported that primed **chickpea** suffered less damage from pod borers (Harris *et al.*, 1999) and damage in Bangladesh was much reduced but the apparent difference was not statistically significant (Musa *et al.*, 2001). However, damage in farmers' trials caused by collar rot (*Sclerotium rolfsii*) in Bangladesh was significantly reduced by priming seeds overnight, by 45 per cent in 1998-99 (30 trials) and by 30 per cent in 1999-00 (35 trials) (Musa *et al.*, 2001).

In an on-station trial in Peshawar, Pakistan in 2002, Rashid *et al.* (2004a) showed that primed seeds of **mungbean** cv. NM 92 for 8 h in water resulted in a significant five-fold increase in grain yield relative to a non-primed crop. This was associated with a large difference in the severity of symptoms of mungbean yellow mosaic virus (MYMV) assessed using a visual scoring index. More than 70 per cent of the non-primed plants had severe or lethal symptoms whereas only 14 per cent of the primed plants were similarly affected. Only 9 per cent of non-primed plants showed no disease symptoms in contrast to 32 per cent of primed plants. Rashid *et al.* (2004b) also observed similar differences in MYMV infection in other on-station mungbean priming trials.

Downy mildew disease, caused by the obligate biotroph *Sclerospora graminicola* (Sacc.) Schroet. is a major constraint to **pearl millet** yields. A standard greenhouse screening method (Jones *et al.*, 1995) was used to investigate the effect of seed priming on the disease resistance of pearl millet. Priming seeds in water for 8 h before sowing significantly reduced the incidence of downy mildew disease in seedlings of a highly susceptible cultivar from about 80 per cent to less than 60 per cent (Harris *et al.*, 2005). The effect was confirmed in additional glasshouse experiments (unpublished). Although the screen would not allow plants to be assessed at later stages of growth, there is a high degree of correlation between performance of cultivars in the screen and their resistance to downy mildew in the field (Jones *et al.*, 2002). However, a preliminary trial at ICRISAT, India in 2006 failed to confirm this effect in the field.

EI-Mohamedy *et al.* (2006) evaluated the efficacy of soil amendment with *Tricoderma harzianum* formulated on sugarcane bagasse and/ or bio-priming seed treatment in controlling cowpea root rot pathogens under greenhouse and field conditions. The percentage of root rot diseases caused by *Fusarium solani, Rhizoctonia solani and Macrophomina phaseolinae* were reduced significantly. The most effective treatments were bagasse + *T. harzianum* (10 per cent), bio-priming and bagasse + *T. harzianum* (5 per cent) they reduced *Fusarium* root rot by 73.9, 60.8 and 56.5 per cent, *Rhizoctonia* root rot by 78.6, 75, 71.4 per cent, *Macrophomina phaseolinae* charcoal rot by 70.8, 62.5 and 62.5 respectively.

Nayaka *et al.* (2008), attempted the use of *T. harzianum* as seed treatment for the controlling of maize ear rot and management of fumonisin (synthesized by *Fusarium verticillioides*) accumulation in maize seeds. Seed treatments with *T. harzianum* improves the seed germination and emergence, vigor index, plant height, yield and 1000seed weight of maize. Again it reduces the incidence of ear rot disease and the level of Fumonisins in maize samples.

Ratnam *et al.* (2004) observed that exogenous application of salicylic acid (SA at 1.5mM) and benzothiadiazole (Bion at 5mM) on sunflower (cv. Modern) leaves induced systemic resistance against *Alternaria helianthi.*

Salicylic acid (SA) was found in inducing resistance in groundnut plants against *Alternaria alternate.* Foliar application of SA at 1mM significantly reduced leaf blight disease intensity and increased pod yield under glasshouse conditions. The changes in the activities of phenylalanine ammonia lyase, chitinase, beta-1,3 glucanase and in phenolic content on groundnut after application of SA and inoculation with *A. alternate* were also studied and observed that in SA-treated leaves an increase in phenolic content five days after challenge inoculation with *A. alternate* in groundnut plants pretreated with SA (Chitra *et al.*, 2008).

Rao *et al.* (2009), did an experiment to test the efficacy of integrated seed treatment options for the management of *Alternaria* blight of sunflower. Seeds of sunflower hybrid were treated with carbendazim + iprodione (Quintal) at 0.3 per cent along with different organic solvents as priming agents such as polyethylene glycol (PEG 6000), acetone, dichloromethane (DCM) and glycerol and compared with a treatment with water. *Pseudomonas fluorescens* Migula was used for bio-priming of seeds with priming agents such as vermiculite, jelly, moist blotters, salicylic acid and compared with direct seed treatment. 100g of sunflower seeds were soaked for 24h at 25± 2°C in 500 ml solutions of acetone, glycerol, dichloromethane, PEG (30 per cent v/v) and water containing 0.3 per cent of Quintal separately and dried under shade. In bio-priming, 100g of seeds were treated with the bacterial biocontrol agent *Pseudomonas fluorescens* using priming agents such as vermiculite, jelly and moist blotter. Results showed that the seed treatment with carbendazim + iprodione at 0.3 per cent in PEG alone with foliar spray of hexaconazole recorded least percent disease indexes of 19.24, 28.86 and 37.74 per cent at 45, 60 and 75 DAS respectively and this treatment also recorded highest yield of 17.12q/ha with test weight and head diameter of 5.51g and 25cm respectively. Again Moeinzadeh *et al.* (2010) reported that bio-priming of sunflower seed with *Pseudomonas fluorescens* improve seed invigoration and seedling growth.

Kuril (2010) reported that the plants obtained from hardened seeds [$Mg(NO_3)_2$, Salicylic acid (SA), $Mg(NO_3)_2$ + SA] has less percent disease index [PDI (per cent)] as compared to non-hardened control set. $Mg(NO_3)_2$ + SA treatment was found to have lowest PDI (per cent) towards *Alternaria brassicae* as indentified by the method of Ellis (1971). $Mg(NO_3)_2$ alone can also lower down the PDI (per cent) in respect to SA treatment where later has less PDI (per cent) than control at both the sowing dates. He also reported that $Mg(NO_3)_2$ and SA either alone or in combination were able to improve the yield potential (seed as well as oil content) and disease (reduced PDI of *Alternaria brassicae*) resistance capability in hardened sets.

In the present chapter 'Seed priming in respect to disease resistance' elucidates that a voluminous work has been done in the past 2 decades on seed priming to improve various physiological aspects. Starting from seed germination, nitrogen, carbohydrate and antioxidant metabolisms including yield in a number of field crops and vegetables. But the reviewing of literature suggests that a very measure quantity of studies is done with amelioration of disease resistance capacity in plants by using primed seeds. Hence, the pathological studies in plants by using various kinds of seed priming may open a future channel for promising research area in the field of Agriculture.

References

Anayatullah (2007). Effect of seed hardening treatment with nitrate salts on physiological and yield attributes in wheat (*Triticum aestivum* L.). Ph.D. Thesis, I.Ag.Sc., B.H.U., Varanasi.

Anaytullah, and Bose, B. (2007). Nitrate-hardened seeds increase germination, amylase activity and proline content in wheat seedlings at low temperature. *Physiol. Mol. Biol. Plants.***13:** 199-207.

Ashraf, M., Kausar, A. and Ashraf, M.Y. (2003). Alleviation of salt stress in pearl millet [*Pennisetum glaucum* (L.) R. Br.] through seed treatments. *Agronomie, 23:* 227–234.

Barlow, E.W.R. and Haigh, A.M. (1987). Effect of seed priming on emergence, growth and yield of UCs 82B tomatoes in the field. *Acta Horticulturae.* **200:** 153-164.

Basra, S.M.A., Afzal, I., Anwar, S., Shafique, M., Haq, A. and Majeed, K. (2005a). Effect of different seed invigoration techniques on wheat (*Triticum aestivum* L.) seeds sown under saline and non-saline conditions. *J. Seed Technol.* **28:** 36–45.

Basra, S.M.A., Farooq, M., Tabassam, R. and Ahmad, N. (2005). Physiological and biochemical aspects of pre-sowing seed treatments in fine rice (*Oryza sativa* L.). *Seed Sci. & Technol.*, **33**: 623-628.

Basu, R.N., Chattopadhyay, K. and Pal, P. (1973). Maintenance of seed viability in rice (*Oryza sativa* L.) and Jute (*Corchorus capsularis* L. and *C. olitorius* L.). *Indian Agric.* **18:** 76-79.

Bewley, J.D. and Black, M. (1978). In *"Physiology and Biochemistry of Seeds in relation to Germination of Seed"*. Vol. I, Springer- Verlag Berlin Heidelberg, New York.

Bhattacharya, A. and Vijaylaxmi (2008). Physiology and biochemistry of disease resistance in plants. In *"Development in Physiology, Biochemistry and Molecular Biology of Plants"*, Eds. B. Bose and Hemantaranjan, A., New India Publishing Agency, New Delhi, pp.33-93.

Blakeman, J.P. and Atkinson, P. (1976). Evidence for a spore germination inhibitor co-extracted with wax from leaves. In *"Microbiology of Aerial Plant Surfaces"*, Eds. C.H. Dickinson and Preece, T.F., Academic Press, New York, pp. 441-449.

Bose, B., Kumar, R., Kuril, S.K. and Srivastava, A.K. (2007). Hardening of mustard seeds with Mg(NO3)2 increases seed germination, vegetative growth, nitrogen assimilation and yield. *Brassica,* 9 **(1-4):** 33-38.

Bose, B. and Mishra, T. (1999). Influence of pre-sowing soaking treatment in Brassica juncea seeds with Mg salt on growth, nitrate reductase activity, total protein content and yield responces. *Physiol. Mol. Biol. Plants.* **5:** 83-88.

Bose, B., Nigam, V.P. and Sharma, M.K. (2010). Seed soaking treatment with nitrate salts vis a *viz.*, physiology of germination and seedling vigor in rice (Oryza sativa L. var. Ratna). *International Journal of Agricultural Sciences.* **6(2):** 564-568.

Bose, B., Srivastava, H.S. and Mathur, S.N. (1982). Effect of some nitrogenous salts on nitrogen transfer and protease activity in germinating *Zea mays* L. seeds. *Biol. Plantarum.* **24:** 89-95.

Braconnier, S. and Bourou, S. (2004). Etude du pré-trempage des graines de niébé dans l'eau comme solution technique simple pour améliorer la tolérance à la sécheresse du niébé. Centre d'étude régional pour l'amélioration de l'adaptation à la sécheresse (CERAAS), Thies, Senegal.

Braz, M.R.S., Rossetto, C.A.V. (2008). Stored coffee seeds germination and vigour as affected by physiological conditioning. *Ciencia Rural.* **38(7):** 1849-1856.

Callan, N.W., Mathre, D.E. and Miller, T.B. (1990). Bio-priming seed treatment for biological control of Pythium ultinum pre-emergence damping-off in the sweet corn. *Plant Dis.* **74:** 368-371.

Callan, N.W., Mathre, D.T. and Miller, J.B. (1991). Yield performance of sweet corn seed bio-primed and coated with *Pseudomonas flurescence* AB 254. *Hort. Science.* **26:** 1163-1165.

Campbell, C.L., Huang, J.S. and Payne, G.A. (1980). Defence at the perimeter: The outer walls and the Gates. In " *Plant Diseases: An Advanced Treatise, Vol.V*" Eds. J.G. Horsfall and E.B. Cowling, Academic Press, New York, pp.103-120.

Chitra, K., Ragupathi, N., Dhanalakshmi, K., Mareeshwari, P., Indra, N., Kamalakannan, A., Sankaralingam, A. and Rabindran, R. (2008). Salicylic acid induced systemic resistant on peanut against *Alternaria alternate. Archives of Phytopathology and Plant Protection.* **41(1):** 50-56.

Chivasa, W., Harris, D. and Nyamudeza, P. (2001). On-farm seed priming: a key technology to improve crop establishment and yield in semi-arid tropics. *Sorghum and Millet Improvement Network News.* **3 (1):** 17-18.

Chivasa, W., Harris, D., Chiduza, C., Nyamudeza, P. and Mashingaidze, A.B. (1998). Agronomic practices, major crops and farmers' perceptions of the importance of good stand establishment in Musikavanhu Communal Area, Zimbabwe. *Journal of Applied Sciences in Southern Africa.* **4 (2):** 9-25.

Dias, M.A., Aquino, L.A., Dias, D.C.F.S. and Alvarenga, E.M. (2009). Physiological quality of suger beet (*Beta vulgaris* L.) seeds under osmotic conditioning and treatments with fungicide. *Revista Brasileira de Sementes.* **31(2):** 188-194.

Dollypan and Basu, R.N. (1985). Mid-storage and pre-sowing seed treatments for lettuce and carrot. *Scientia Hort.* **33(6):** 1026-1027.

El-Mohamedy, R.S.R., Abd Alla, M.A. and Badiaa, R.I. (2006). Soil amendment and seed bio-priming treatments as alternative fungicides for controlling root rot diseases on cowpea plants in Nobaria province. *Research Journal of Agriculture and Biological Sciences.* **2(6):** 391-398.

Ellis, B.E., Amichein, N. (1971). The "H-Shift" during aromatic orthohydroxylation in higher plants. *Phytochemistry.* **10:** 3069-3070.

Farooq, M., Basra, S.M.A, Wahid, A. and Ahmad, N. (2010). Changes in nutrient-homeostasis and reserves metabolism during rice seed priming: consequences for seedling emergence and growth. *Agricultural Sciences in China.* **9(2):** 191-198.

Franko, G. and Cattivello, C. (2007). Recent advances in seed conditioning of horticultural species with products of natural origin. *Notiziario ERSA.* **20(4):** 32-40.

Gao, Y.P., Young, L. Bonham-Smith, P. and Gustai, V. (1999). Characterization and expression of plasma and tonoplast membrane aquaporins in primed seed of *Brassica napus* during germination under stress conditions. *Plant Mol. Biol.* **40:** 635-644.

Ghassemi-Golezani, K., Chadordooz-Jeddi, A., Nasrullahzadeh, M. and Moghaddam, M. (2010). Influence of hydro-priming duration on field performance of pinto bean (*Phaseolus vulgaris* L.) cultivars. *African Journal of Agricultural Research.* **5(9):** 893-897.

Golshani, M., Pirdashti, H., Saeb, K., Babakhani, B. and Heidarzade, A. (2010). Response of seed germination and seedling emergence of rice (*Oryza sativa* L.) genotypes to different osmopriming levels. *World Applied Sciences Journal.* **9(2):** 221-225.

Goodman, R.N., Kiraly, Z. and Zaitlin, M. (1967). In *"Biochemistry and Physiology of Infectious Plant Diseases"* Ed.Van Nostrand-Reihold, Princeton, New Jersey.

Gregory, P.H. (1972). In *"The Microbiology of the atmosphere"*, Leonard Hill, London.

Harris, D. (1996). The effects of manure, genotype, seed priming, depth and date of sowing on the emergence and early growth of Sorghum bicolor (L.) Moench in semi-arid Botswana. *Soil & Tillage Research.* **40:** 73-88.

Harris, D. (2002). Getting started quickly: crop establishment and 'on-farm' seed priming. Invited lecture, Cornell University, 29th January 2002. Ithaca, New York, USA.

Harris, D. (2003). Reducing risk and increasing yields from rainfed crops in Africa using 'on-farm' seed priming. Pages 87-88 in Abstracts: 'Harnessing Crop Technologies to Alleviate Hunger and Poverty in Africa', 6th Biennial Conference of the African Crop Science Society, Hilton Nairobi Hotel, Kenya, 12th–16th October, 2003.

Harris, D. and Mottram, A. (2004). Practical hydration of seeds of tropical crops: 'on-farm' seed priming. In 'Seed Science and Technology: Trends and Advances', ed. A.S. Basra. The Howarth Press (in press).

Harris, D., Breese, W.A. and Kumar Rao, J.V.D.K. (2005). The improvement of crop yield in marginal environments using 'on-farm' seed priming: nodulation, nitrogen fixation and disease resistance. *Australian Journal of Agricultural Research.* **56 (11):** 1211-1218.

Harris, D., Joshi, A., Khan, P.A., Gothkar, P. and Sodhi, P.S. (1999). On-farm seed priming in semi-arid agriculture: development and evaluation in maize, rice and chickpea in India using participatory methods. *Experimental Agriculture.* **35:** 15-29.

Harris, D. Kumar, Rao, J.V.D.K. and Kumar, J. (2001a). 'On-farm' seed priming. *Agricultural Research and Extension Network Newsletter.* **44:** 3.

Harris, D., Pathan, A. K., Gothkar, P., Joshi, A., Chivasa, W. and Nyamudeza, P. (2001b). On-farm seed priming: using participatory methods to revive and refine a key technology. *Agricultural Systems.* **69 (1-2):** 151-164.

Harris, D., Raghuwanshi, B.S., Gangwar, J.S., Singh, S.C., Joshi, K.D., Rashid, A. and Hollington, P.A. (2001c). Participatory evaluation by farmers of 'on-farm' seed priming in wheat in India, Nepal and Pakistan. *Experimental Agriculture.* **37 (3):** 403-415.

Heydecker, W. and Chetram, R.S. (1971). Water relations of beet root seed germination. I. Microbial factors with special reference to laboratory germination. *Ann. Bot* (London) [NS], **35:** 17-29.

Hooda, R.S. and Singh, H. (2006). Evaluation of some crop establishment techniques for pearl millet (*Pennisetum glaucum* L. Emend. Stuntz) under sub-optimal conditions. *Haryana J. Agronomy*. **22(1):** 21-23.

Jones, E.S. Breese, W.A. Liu, C.J. Singh, S.D. Shaw, D.S. and Witcombe, J.R. (2002). Mapping quantitative trait loci for resistance to downy mildew in pearl millet: field and glasshouse screens detect the same QTL. *Crop Science*. **42:**1316–1323.

Jones, E.S., Liu, C.J., Gale, M.D., Hash, C.T. and Witcombe, J.R. (1995). Mapping quantitative trait loci for downy mildew resistance in pearl millet. *Theoretical and Applied Genetics*. **91:**448–456.

Kaur, S., Gupta, A.K. and Kaur, N. (2005). Seed priming increases crop yield possibly by modulating enzymes of sucrose metabolism in chickpea. *J. Agronomy & Crop Science*. **191:** 81–87.

Khan, A.A. (1992). Pre-plant physiological conditioning. *Hort. Rew.* **13:** 131-181.

Kikuti, A.L.P., Kikuti, H. and Minami, K. (2005). Physiological conditioning in sweet pepper seeds. *Revista Ciencia Agronomica*. **36(2):** 243-248.

Kraft, J.M. (1974). Influence of seedling exudates on the resistance of pea to *Fusarium* and *Pythium* root rot. *Phytopathology*. **64(2):** 190-193.

Krishnotar, Bose, B., Srivastava, A.K. and Shahi, J.P. (2009). Response of rabi maize crop to seed invigoration with magnesium nitrate and distilled water. *Indian J. Plant Physiol.*, **14:** 71-77.

Kumar, A., Gangwar, J.S., Prasad, S.C. and Harris, D. (2002). 'On-farm' seed priming increases yield of direct-sown finger millet (Eleusine coracana) in India. *International Sorghum and Millets Newsletter*. **43:** 90-92.

Kumari, V.K., Sivaprasad, A., Hari, N. and Gopalakrishnan, P.K. (2002). Effect of presowing hardening treatment with various chemicals on germination and seedling growth of important oil yielding crop *Sesamum indicum* var. Kylm-1. *Advances in Plant Sciences*. **15(1):** 345-348.

Kuril, S.K. (2010). Influence of seed hardening with $Mg(NO_3)_2$ and salicylic acid on timely and late sown mustard (*Brassica juncea* L. Czern and Coss.) varieties. Ph.D. Thesis, I. Ag. Sc., B.H.U., Varanasi.

Lanteri, S., Nada, E., Belletti, P., Quagliotti, L. and Binof, R.J. (1996). Effects of controlled deterioration and osmoconditioning on germination and nuclear replication in seeds of pepper *(capsicum annuum* l.). *Annals of Botany*. **77:** 591-597.

Lespinay, A.D., Lequeux, H., Lambillotte, B. and Lutts, S. (2010). Protein synthesis is differentially required for germination in Poa pratensis and *Trifolium repens* in the absence or in the presence of cadmium. *Plant Growth Regulator*. **61:** 205-214.

Lopes, H.M. and Souza, C.M. (2008). Effects of gibberellic acid and drying in osmotic conditioning on viability and vigor of papaya (Carica papaya L.) seeds. *Revista Brasileira de Sementes*. **30(1):** 181-189.

Magalhaes, F.H.L., Machado, J., Da, C., Vieira, M., Das, g.G.C., Guimaraes, R.M., Oliveira, J.A., Ledo, C.A. and Da, S. (2004). Performance of carrot seeds infected by Alternaria species after physiological conditioning with the thiram. *Ciencia e Agrotechnologia*. **28(5):** 1007-1014.

Moeinzadeh, A., Sharif-Zadeh, F., Ahmadzadeh, M. and Tajabadi, F.H. (2010). Bio-priming of sunflower (*Helianthus annuus* L.) seed with *Pseudomonas fluorescens* for improvement of seed invigoration and seedling growth. *Australian Journal of Crop Science*. **4(7):** 564-570.

Mohanasarida, K. and Mathew, J. (2005). Seed priming on crop establishment and seedling vigour in semi-dry rice (*Oryza sativa*). *Research on Crops*. **6(1):** 23-25.

Mondal, S., Vijai, P. and Bose, B. (2011). Role of seed hardening in rice variety Swarna (MTU 7029). *Reseach J. Seed Sci*. **4(3):** 157-165.

Musa, A.M., Harris, D., Johansen, C. and Kumar J. (2001). Short duration chickpea to replace fallow after aman rice: the role of on-farm seed priming in the High Barind Tract of Bangladesh. *Experimental Agriculture*. **37 (4):** 509-521.

Nayaka, S.C., Niranjana, S.R., Shankar, A.C.U., Raj, S.N., Reddy, M.S., Prakash, H.S. and Mortensen, C.N. (2008). Seed biopriming with novel strain of Trichoderma harzianum for the control of toxigenic Fusarium verticillioides and fumonisins in maize. *Archives of phytopathology and plant protection*. 1-19.

Oliveira, A., Dos, S., Mann, S.R., Santos, M., Da, F., Gois, I.B., Barretto, M.C. and De, V. (2007). Osmotic conditioning in seeds of sweet maize under storing. *Revista Ciencia Agronomica*. **38(4):** 444-448.

Pandey, D. and Bose, B. (2006). Influence of pre-sowing seed treatment with nitrate salts and different sowing dates on performance of mustard. *Indian J. Plant physiol*. **11(3):** 261-265.

Peske, F.B., Novembre, A.D. and Da, L.C. (2010). Physiological conditioning of pearl millet seeds. *Revista Brasileira de Sementes*. **32(4):** 132-142.

Peyvast, G.H., Olfati, J.A., Piri, M. and Mahdieh, M.B. (2010). Priming effect on cucumber germination at low temperatures. *Acta Horticulturae*. **871:** 615-616.

Ponni, A.S. (2008). Influence of seed treatment on germination and seedling vigour of *Phyllanthus niruri*. *Acta Horticulturae*. *782:* 155-160.

Powell, A.A. and Mathews, S. (1986). Cell membranes and seed leachate conductivity in relation to the quality of seeds for sowing. *J. Seed Tech*. **10:** 81-100.

Rao, M.S.L., Kulkarni, S., Lingaraju, S. and Nadaf, H.L. (2009). Bio-priming of seeds: A potential tool in the integrated management of *Alternaria blight* of sunflower. *Helia*. **32 (50):**107-114.

Rashid, A., Harris, D., Hollington, P.A. and Ali, S. (2004a). On-farm seed priming reduces yield losses of mungbean (Vigna radiata) associated with mungbean yellow mosaic virus in the North West Frontier Province of Pakistan. *Crop Protection*. **23:** 1119-1124.

Rashid, A., Harris, D., Hollington, P.A. and Rafiq, M. (2004b). Improving the yield of mungbean (Vigna radiata) in the North West Frontier Province of Pakistan using on-farm seed priming. *Experimental Agriculture.* **40 (2):** 233-244.

Rashid, A., Harris, D., Hollington, P.A. and Khattak, R.A. (2002). On-farm seed priming: a key technology for improving the livelihoods of resource-poor farmers on saline lands. Pp 423-431 in: R. Ahmad and K.A. Malik (eds.) *'Prospects for Saline Agriculture'*. Kluwer Academic Publishers. The Netherlands.

Ratnam, M.V.S., Reddy, P.N., Rao, S.C. and Raju, R.B. (2004). Systemic induced resistance in sunflower to *Alternaria* leaf blight by foliar application of SA and Bion. *Journal of Oilseeds Research.* **21(1):** 104-107.

Reddy, M.N., Ramagopal, G. and Rao, A.S. (1977). Phenolic acid in groundnut root exudates. *Plant Soil.* **46(3):** 655-658.

Rouhi, H.R., Surki, A.A., Sharif-Zadeh, F., Afshari, R.T., Aboutalebian, M.A. and Ahmadvand, G. (2011). Study of different priming treatments on germination traits of soybean seed lots. *Not. Sci. Biol.* **3(1):** 101-108.

Rudrapal, A.B. and Basu, R.N. (1982). Use of chlorine and bromine in controlling mustard seed deterioration. *Seed Res.* **9:** 188-191.

Sarkar, R.K. and Kaur, N. (2006). Effect of hydro and osmopriming of chickpea (*Cicer arietinum* L.) seeds on enzymes of sucrose and nitrogen metabolism in nodules. *Plant Growth Regulator.* **49(2/3):** 177-182.

Sharma, M.K. and Bose, B. (2006). Effect of seed hardening with nitrate salts on seedling emergence, plant growth and nitrate assimilation of wheat (Triticum aestivum L.). *Physiol. Mol. Biol. Plants.* **12:** 173-176.

Sharma, M.K., Bose, B. and Srivastava, A.K. (2009). Effect of seed hardening with nitrate salts on physiological attributes at ear head emergence stage and yield of wheat (*Triticum aestivum* L.). *International Journal of Agricultural Sciences*, **5(2):** 439-442.

Sudarshan, G., Ankaiah, R., Meenakumari, K.V.S., Reddy, B.G. and Reddy, N.M. (2006). Effect of invigoration treatments of aged seeds on plant growth, development and yield in cotton hybrids. *National Seed Seminar*, Abstract ANGRAU, Hyderabad, P.49.

Tavili, A., Zare, S., Moosavi, A.S. and Enayati, A. (2010). Effects of Priming Techniques on Seed Germination and Early Growth Characteristics of *Bromus tomentellus* L. and *Bromus inermis* L. *Not. Sci. Biol.* **2 (1):** 2010, 104-108. The Seed Biology Place. Website Gerhard Leubner Lab University Freiburg, Germany.

Verma, S.S., Punis, R.C. and Dahiya, O.S. (2006). Pre-sowing seed treatment for better crop establishment in mungbean. *National Seed Seminar.* Abstract XII ANGRAU, Hyderabad, p.145.

Villiers, T. A. and Edgcumbe, D.J. (1975). On the cause of seed determination in dry storage. *Seed Science and Technology.* 3: 761-774.

XiaoHuan, Y., YuGuo, W., WenXiu, Y., HongFu, W. and JinHu, M. (2009). Effect of seed priming on physiological characteristics of soybean seedling under water stress. *Zhongguo Shengtai Nongye Xuebao/ Chinese J. Eco-Agriculture.* **17(6)**:1191-1195.

YangRong, L.H.X.G.Z.G.W. (2008). Optimal conditions for hydropriming in lucerne seeds. *NewZealand J. Agricultural Research.* **51(1)**:69-75.

Zhuo, J., Wang, W., Lu, Y., Sen, W. and Wang, X. (2009). Osmopriming-regulated changes of plasma membrane composition and function were inhibited by phenylarsine oxide in soybean seeds. *Journal of Integrative Plant Biology.* **51 (9)**: 858–867.

Chapter 32

Major Disease of Tomato and their Management

Kartikeya Srivastava, Ravindra Kumar and Jyoti Pandey

Department of Genetics & Plant Breeding, Institute of Agricultural Sciences, Banaras Hindu University, Varanasi – 221 005, U.P.

Tomato (*Solanum lycopersicon*) is one of the most widely grown vegetables crop in the world grown for its edible fruit. The cultivated tomato originates from wild plants found in the Andean regions of Peru. The tomato was first domesticated in Mexico and it is believed that the Spanish explorer Cortez may have been the first to transfer it to Europe in the mid 16th century. It was grown for the beauty of its fruit, which was not often eaten. It was only in the 20th century that its importance as an edible fruit emerged. Field grown tomatoes are widely grown in warm temperate and tropical climates, and glass house tomatoes are produce in many additional regions. The tomato plant is very versatile and the crop can be divided into two categories: fresh market tomatoes and processing tomatoes.

There are several different types of tomato disease problems in India. The most common ones are spots or blights of leaves or fruit, rots, wilts, virus diseases and non-parasitic disorders. Tomatoes suffer from many fungal, bacterial, and viral diseases. This tutorial will discuss only a few of the most significant diseases.

Fungal Leaf and Fruit Spots or Blights

Blights cause serious spotting of tomato leaves in India. Blight is the most common disease and is a serious problem in many years. Early blight is also fairly common. These diseases usually begin after the first fruits have set. Blights, which may occur only in unusually cool, wet weather. All are fungal diseases spread by airborne spores which require dew or rain to infect the plants. These diseases build up rapidly in wet weather and cause dark leaf spots followed by yellowing and defoliation. Some of these diseases also produce spots on the fruits.

Early Blight

Early blight, caused by the fungus *Alternaria solani*, appears first on the lower leaves, usually after a heavy fruit set. Early Blight can affect the foliage, stems and fruit of tomatoes. Brown to black circular lesion on mature leaves. The leaf area around each target spot turns yellow, and soon the entire leaf turns yellow and drops. The disease spreads rapidly to the upper leaves in rainy weather at temperatures of 23 to 29°C, causing severe defoliation. Such lesion may be surrounded by chlorotic tissue. Affected leaves may die prematurely, exposing the fruit to sun scald. Dark spots with concentric rings develop on older leaves first. The surrounding leaf area may turn yellow.

Management

Early Blight fungus overwinters in plant residue and is soil-borne. It can also come in on transplants. Remove affected plants and thoroughly clean fall garden debris. Use the disease free and healthy seeds for raising seedling. Copper and/or sulphur sprays can prevent further development of the fungus. Seeds treat with Thiram and Bavistin amount of 2.5g/kg seeds.

Septoria Leaf Spot (Septoria Blight)

Septoria leaf spot, caused by the fungus *Septoria lycopersici*, usually appears on the lower leaves after the first fruits set. Problems growing tomatoes are often the result of weather conditions. The first symptoms appear later in the season on the older leaves near the ground where small, water-soaked spots can be seen. Rapid leaf loss reduces fruit yield and quality. Exposed fruits may be more susceptible to sunscald. Rainy weather with temperatures of 20° to 25°C favours disease development and a rapid spread of the disease to the upper leaves.

The disease overwinters on residues from diseased plants. The spots soon become nearly circular and have gray centers surrounded by darker margins. Later the centers show tiny dark specks in which the spores of the fungus are produced. The spots are smaller and more numerous than those found in early blight. If a leaf develops many spots, it usually dies and drops from the plant. As the disease spreads, there is a progressive loss of leaves until only a few are left at the top of the plant and the fruit become exposed to sunscald. Wet weather favours fungus growth, spread, and subsequent disease development

Management

Products providing good to excellent control are Iprodione 0.2 per cent, and Flint. A good control product has been Ridomil/Bravo0.2 per cent. Fair control has been achieved with Kocide and Copper Sulphate. There has been poor control with Pentathlon and Mancozeb.

Tomato diseases are rarely fatal, if the proper management is employed. It is important to catch any tomato disease early, before it spreads to all of your tomato plants and possibly other plants in the same family, such as potatoes, eggplants and peppers. Here are some common tomato diseases, their symptoms and what to do if tomato diseases threaten your home vegetable garden.

Management of Leaf Spots

The danger from leaf diseases is reduced by rotating the planting areas in your garden; plant tomatoes in the same place only once in three or four years. Remove and destroy tomato vines in the fall. Use cultivation (plowing or rototilling) to bury the remaining crop refuse. Use healthy transplants. Remove badly diseased lower leaves, as these are a source of leaf spot fungus spores that help spread the disease. Water the plants at the base. Use a garden hose to trickle water into a shallow trench or depression in the soil. Avoid watering with overhead sprinklers in late afternoon or evening; if the plants stay wet all night, leaf spot infections are likely to occur. Use fungicides when needed. These diseases spread rapidly, and are difficult to control once established. 0.2 per cent Chlorothalonil fungicide (Ortho Multi-Purpose Fungicide) must be applied to the plant before the disease first appears or at the first sign of disease. Since timing is critical, it may be preferable to start a preventive spray program when the first fruits are marble-sized. Usually good control is achieved if you begin spraying about mid-July. Mancozeb and maneb fungicides can be applied to commercial tomatoes but are no longer available for home gardeners. Chlorothalonil can be applied up to the day of harvest. Captan fungicide is available, but should not be used because it does not control *Septoria* leaf spot. Check the list of ingredients to determine the active ingredients in fungicides or home garden fungicide-insecticide mixtures. Copper sprays are somewhat affective at halting the spread of symptoms.

Late Blight

Late blight, caused by the fungus *Phytophthora infestans*, is a most problem in India, occurring only in moist weather with cool nights and moderately warm days. Night temperatures of 10-13°C and heavy dew favour late blight. Daytime temperatures above 30°C are unfavourable to late blight development. Stem and petioles can also develop the water- soaked and later brown to gray lesions. This affects both the leaves and fruit of tomatoes. It first appears on the leaves and stems, and later on the fruits. Symptoms appear on leaves as pale green, water-soaked spots, often beginning at leaf tips or edges. Late Blight is the disease responsible for the Irish Potato Famine. irregularly shaped gray spots appear on leaves.

Management

The disease is soil borne and can be managed by using integrated approaches to crop production and protection. Use the disease free and certified seeds only. Late blight, caused by a fungus Phytophthora infestans, is one of the most devastating diseases of tomato worldwide. For the chemical control spray of mancozeb (2.5 g/ liter) and Copper hydroxoide (2g/lit), chlorothalonil (2g/liter) and mancozeb (2.5 g/ liter) fungicides are the standard protectants used for control. They are usually applied every seven to ten days for best protection. Begin chemical control programs before symptoms appear.

Powdery Mildew (*Oidium lycopersicum*)

Powdery mildew of tomato is caused by the fungus *Oidium lycopersicum*. Symptoms first appear as light green to bright yellow spots on the upper surface of the leaf. These spots usually don't have very distinct margins and gradually become more noticeable as they develop the white, powdery appearance typical of powdery mildews. However, this is where this disease differs from most other powdery mildews that we encounter. The powdery mildew of tomato is apparently much more aggressive than other mildews. Once leaves are infected, they quickly brown and shrivel on the plant. This rapid death of infected leaves and defoliation of plants is not typical of most mildews. The fungus is readily spread to nearby leaves or plants since abundant, powdery spores are produced and are easily carried by air currents or production activities. The conditions that favour the development of disease include relative humidity levels greater than 50 per cent (optimum RH > 90 per cent) and temperatures ranging from 10-35 °C (best below 30°C). Unlike the situation with most other fungal diseases, free water on leaf surfaces not necessary for infection. The tomato powdery mildew fungus has a very broad host range which includes rosemary, pepper, eggplant, and many bedding plants. The organism can survive in weed hosts as mycelium and in living plants between crops.

Management

Wider spacing and staking seems to be most important cultural control for powdery mildew. Integrated pest management practices used to reduce disease incidence include crop rotation, plastic mulch, drip irrigation, keeping plants dry, sanitation, variety selection, ventilation and staking.Sulfur and other protectant fungicides are available for control of this disease.

Foliar Diseases

Gray Leaf Spot

Gray Leaf Spot affects only the leaves of tomatoes, starting with the oldest leaves. Symptoms: Small, dark spots that can be seen on both the top and bottom surfaces of the leaves. The spots enlarge and turn a greyish brown. Eventually, the centers of the spots crack and fall out. Surrounding leaf areas will turn yellow and the leaves will dry and drop. Fruit production is inhibited. Greasy looking, A ring of white mould can develop around the spots, especially in wet weather. The spots eventually turn dry and papery. Blackened areas may appear on the stems. The fruit also develop large, irregularly shaped, greasy gray spots.

Management

Warm, moist conditions worsen gray leaf spot problems. Remove all affected plants and fall garden debris. Select the resistant varieties. Cool, wet weather encourages the development of the fungus. Copper sprays offer some control. Serenade works best as a deterrent, rather than a cure.

Southern Blight

Southern Blight manifests as a white mould growing on the stem near the soil line. Dark, round spots will appear on the lower stem and both the outer and inner stem will become discoloured. Southern Blight fungus girdles the tomato stem and prevents the plant from taking up water and nutrients. Young plants may collapse at the soil line.

Management

Crop rotation seems to help. There has also been some evidence that extra calcium and the use of fertilizers containing ammonium offer some protection.

Wilts

Wilts are caused by fungi that infest the soil, remaining there for years.

Verticillium Wilt

Verticillium Wilt This name can be misleading, as sometimes the leaves will turn yellow, dry up and never appear to wilt. Verticillium wilt is caused by a soil-borne fungus *Verticillium albo-atrum* and *V. dahliae*. and it can affect many different vegetables. The fungus can persist in the soil for many years, so crop rotation and selection of resistant varieties is crucial. Wilting during the hottest part of the day and recovering at night, yellowing and eventually browning between the leaf veins starting with the older, lower leaves and discoloration inside the stems. Verticillium Wilt inhibits the plants ability to take in water and nutrients and will eventually kill the plant. Verticillium wilt is more pronounced in cool weather.

Fusarium Wilt

Fusarium wilt is caused by *Fusarium oxysporum* f. sp. *lycopersici*. The first symptoms are a yellowing and drooping of lower leaves on a single stem. Typically the wilting will begin on one side of the tomato plant's leaves or shoots and spread to the other side as infection worsens. Full grown tomato plants have the biggest problem with severe wilting; it is not as severe in younger shoots. A progressive yellowing and wilting of the leaves occurs, and the plants may die. When the stem is cut open, the vascular (water conducting) tissues under the surface are frequently discoloured brown. This vascular browning extends from the roots to the upper portions of the plant and into the leaf petioles. This fungus infects only tomato.

Wilt Management

The first line of defence against wilt is to use disease-free seedlings. This avoids introducing wilt fungi into the garden. Remove and destroy wilted plants and all debris of tomato and other susceptible crops at the end of the growing season. Rotate tomato areas; grow tomatoes in the same part of the field only once in four years. The benefit of rotation is less with wilts than with other diseases since both wilt fungi survive for years in the soil and the Verticillium fungus attacks many different crops. Fungicides for control of leaf blights have no effect on the wilt diseases, which are internal infections. Once soil is infested and further rotation impossible, the only possible management practice is use of resistant varieties.

Bacterial Leaf and Fruit Spots

Bacterial Speck and Bacterial Spot

Bacterial speck, caused by the bacterium *Pseudomonas syringae* pv. *tomato*. These specks are slightly raised and may occur on fruit or leaves. The bacterium is seed-borne. Foliar symptoms are dark brown to black spot on leaves and stem. Leaf spots can be circular to angular in shape, and individual smaller spots. These symptoms are appearing top and bottom side of the leaf. Bacterial speck can result in stunted growth, delayed crop maturity and reduced yield. Bacterial spot, caused by the bacterium *Xanthomonas campestris* pv. *vesicatoria*, produces circular scabby spots on immature fruits and on leaves. The spots are 1/8 inch across. Bell peppers also may be attacked. The bacterium is seed-borne and is often carried on diseased transplants. It can also occur in weeds of the nightshade family and on volunteer tomatoes. It overwinters in soil and on old tomato vines and pepper plants. Bacterial spot is favoured by warm temperatures (20-35° C), high humidity, long dew periods, and driving rain. Night temperatures of 24 to 30° C are especially favourable for infection by the bacterial spot pathogen.

Management

Manage bacterial speck and spot by using disease-free tomato transplants from a reliable source or by starting your own plants from disease-free seed. Grow them in sterilized potting medium. Rotate tomato and peppers in the garden; do not plant tomatoes or peppers in the same part of the garden more than once in four years. Dispose of old tomato and pepper plants. Control all weeds, especially those in the nightshade (potato/tomato) family. If either disease appears, copper fungicides will slow the spread of disease, but they will not give complete control if the disease is well established and wet weather persists.

Bacterial Canker

Bacterial canker is caused by *Clavibacter michiganensis* subsp. *michiganensis*. The canker bacteria enter the plant through natural openings and wounds, including root wounds. Pruning or transplant clipping operations can introduce the bacteria directly into the vascular system, resulting in the more serious systemic infections. Infections spread through splashing water, wind-driven rain and the fine water droplets or aerosols produced during storms. In the field, bacteria transfer by machinery or workers is probably not as significant as in the transplant greenhouse where plant density is high and growth conditions for the bacterium are optimal. Symptoms usually affect the lover leaves first and consist of curling of leaves and branches, chlorosis of leaves and brown necrosis. Sometimes the leaflets on one side of a leaf wilt. The disease progresses up the plant. Diseased seedlings are the primary source of infection. The canker bacteria also survive in the soil, on old tomato stakes, on dead stems, and on members of the nightshade family. They are spread by rain, splashing water, pruning, and by gardeners and their tools. Temperatures of 24 to 32°C and low light intensity favour the disease.

Management

Management practices include using disease-free seed, pasteurized potting mix, and sterilized pots or buying disease-free seedlings. Control weeds in the nightshade family. Avoid working among the tomato plants when they are wet. Use new or pasteurized stakes and trellises. Disinfect pruning tools using a mixture of 10 per cent household bleach and 90 per cent water — be sure to wash the tools with water, dry and oil them before putting them away.

Rots

Soil rot, a fruit rot caused by the soil-borne fungus *Rhizoctonia solani*, occurs during rainy periods. A soft brown rot develops, usually on the sides of fruits which are touching the ground, or where soil is splashed up. The spots are sunken, large, and may have a closely-spaced target pattern. Often the surface of the fruit cracks open. Staking may be used to reduce soil rot by keeping fruit off the ground. In larger plantings, reduce soil rot by using mulch as described in the section on blossom end rot.

Timber Rot

Timber rot, caused by the white mould fungus, *Sclerotinia sclerotiorum*, has not been seen on tomatoes. Since white mould has increased dramatically on susceptible crops (sunflower, dry beans, and others) as well as on ornamental flowers, it may occur on tomatoes in years with abundant rainfall. Infected tomato plants may be infected on the main stem near the soil line or the stem may be infected at a node (the point where a branch is attached to the stem). A mushy brown rot develops and eventually the plant above this area wilts and dies. If the rotted stem is split open, hard, black pebble-like fungus bodies are found imbedded in a cottony mass of fungus growth. Timber rot is favoured by rainy weather which keeps the soil wet for several weeks, as well as by rain, fog or overhead sprinkling. Infection may occur at temperatures of 15 to 2° C. There is no satisfactory management for timber rot. To help reduce damage, do not plant tomatoes in areas where susceptible crops (beans, ornamental flowers, melons, etc.), have been grown and do not provide excessive overhead irrigation.

Nematode Disease

Root Knot Nematode

Affected plants become stunted and yellow and have a tendency to wilt in hot weather. The roots are severely distorted, swollen and bear knots or galls caused by *Meloidogyne incognita, M. javanica,* and *M. hapla*. The female deposits eggs in or on the roots, or in decaying root debris. The juveniles hatch from the eggs and move toward root tips or minor wounds. They feed on root tissue. All stages of plant growth are attacked.

Management

Use resistant varieties of tomatoes to root knot. Rotate the tomato crop (susceptible) with other crops such as grasses or brassicas (tolerant), followed by onion (resistant) and then dry fallow during hot, dry weather if possible. In small vegetable plantings interplanting with French marigold (*Tagetes patula*) or African marigold (*T. erecta*) is very effective in lowering the nematode density in soil. Neem karnel-cakes and animal manures have high nitrogen contents of 2–7 per cent and are the most nematicidal amendments must be applied at 4–10 t/ha to be effective. Chemicals such as carbofuran, oxamyl,and others may be applied as granular or liquid formulations, and incorporated into the top few centimeters of soil.

Viral Diseases

Virus diseases cause a mottling and distortion of foliage (leaves) and sometimes cause a mottling of fruits. Tobacco mosaic, cucumber mosaic and spotted wilt have occurred sporadically in recent years.

Tobacco Mosaic

Tobacco mosaic virus (TMV) is distributed worldwide and may cause significant losses in the field and greenhouse. TMV is one of the most stable viruses known, able to survive in dried plant debris as long as 100 years. Many strains of TMV have been reported and characterized. TMV can be seed borne in tomato. It is readily transmitted mechanically from plant to plant by gardeners' hands, their clothing, and their tools and may be present in tobacco products. Infection by the tobacco mosaic virus causes leaves to be mottled light and dark green. The plants are stunted and the leaves rough, occasionally fern-like, and the edges turned down. The symptoms in tomato vary greatly in intensity depending upon the variety, virus strain, and time of infection, light intensity, and temperature. High temperatures, for example, may mask foliar symptoms. The most characteristic symptom of the disease on leaves is a light- and dark-green mosaic pattern.

Management

With the use of TMV resistant or -tolerant varieties, plants may be infected by some strains whose symptoms are latent. Ordinarily the fruit from infected plants do not show mosaic symptoms, but may be reduced in size and number.

Cucumber Mosaic

Cucumber mosaic virus (CMV) is the second most important virus disease of tomato in the state. CMV has an extensive host range and is transmitted by aphids in a no persistent manner. Early symptoms of cucumber mosaic are yellow, bushy and stunted plants. Later symptoms are similar to tobacco mosaic except that leaves are frequently elongate, narrow, and distorted to form a "shoestring" effect. These symptoms are distinguished from 2,4-D injury, because the leaf is not thick and leathery. Cucumber mosaic virus also causes a fern-leaf effect on some tomato leaves. Cucumber mosaic virus is spread primarily by aphids, but can also be spread mechanically. The virus also infects cucumber, melon, pepper, various flowers, and many weeds.

Management

Because of the wide host range for CMV, sources of inoculum for field plantings are numerous. Avoid planting near weedy border areas or isolate tomatoes from such areas by growing taller, nonsusceptible barrier crops such as corn. No CMV-resistant or -tolerant varieties are currently available.

Mosaic Management

Avoid all tobacco products while working with tomato plants. Use only vigorous, healthy-looking plants. Remove virus-diseased plants as soon as symptoms appear. Wash hands and tools with detergent immediately after contact with virus diseased plants. Detergent is more effective than soap because detergent inactivates the virus. Control all weeds within 150 feet of garden areas. Control insects. Avoid handling other susceptible plants (eg. flowers, peppers, cucumbers and melons) while working with tomatoes. Do not grow tomatoes next to susceptible crops such as potatoes, peppers, cucumbers and melons.

Spotted Wilt

Tomato spotted wilt virus (TSWV) causes leaves to develop irregularly shaped to circular, black, small spots. Older leaves may turn brown, droop and die. Young shoots may develop dark streaks, progressing to a top dieback and eventually leading to plant death. Raised yellowish spots up to 1/2 inch across develop on green fruits. As the fruits ripen, these spots become striking with concentric rings of yellow or brown alternating with green and later pink or red. Early infection may lead to premature death of the plant, while later infections may lead to symptoms on the developing fruits. No satisfactory management plan is available, but control of weed hosts and separation of vegetable hosts from flower beds may reduce, but not eliminate, the problem. When purchasing tomato plants, home gardeners should select vigorous and apparently healthy plants and avoid plants that appear stunted or have an abnormal colour.

Nursery Disease: Damping Off

Several various types of fungus which attack and cause seeds or seedlings to die. The disease is found in soil and water but once spores develop they spread by air. It's a fungal disease so all the yucky conditions that spores thrive in with an emphasis on dark, wet and humid! Lack of ventilation/air circulation is also a cause. So in the context of tomato growing triggers would be:

Causal Organism

Pythium spp., *Phytophthora* spp., *Fusarium* spp., *Rhizoctonia solani*, and some other fungi are responsible for causing damping off in most of nursery crops.

Over watering: If the nurseries are waterlogged, air/ oxygen unable to circulate. Excessive humidity in a propagator or greenhouse, too many seedlings in one container or overcrowding means air could not circulate round the seedlings and that's it's easy for the disease to spread. Low or poor lighting levels also lead to development of lanky seedlings.

Affected plants usually occur in patches in nursery beds or in low parts of sloped fields. In level fields, affected plants are generally found in scattered areas. In *pre-emergence* damping off, the seeds fail to emerge after sowing. They become soft, mushy, turn brown, and decompose as a consequence of seed infection. In *post-emergence* damping-off, the seedling emerges from the soil but dies shortly afterwards. The affected portions are pale brown, soft, water soaked, and thinner than non-affected tissue. Infected stems collapse. Stunting of plants due to root rot or collar rot may also occur. Symptoms may vary with age and stage of development of the tomato plant. Infection by the pathogen may occur much later after emergence; in this case, the infection is usually not lethal but plant growth and yield may be reduced. Plants severely affected in the root region may wilt in warm or windy weather. Extensive root rotting causes symptoms of nutrient deficiencies in the plant since nutrients cannot move sufficiently from the soil up through the plant. Since several pathogens can cause similar symptoms, pathogen isolation and identification is needed to confirm diagnosis.

Management

If possible, use plug transplants and a soilless pathogen-free growth medium to avoid damping off. Mixing of local field soil or manure with soilless growth medium may result in severe damping off. Water seedlings only when the soil or growth medium is dry, preferably in the morning to allow drying to occur by the late afternoon. Avoid contact with ground soil or other sources of contamination. Pots or transplant containers should be new or treated recently with a disinfectant (10 per cent household bleach) or fungicide. Keep seed flats raised, away from splashing water and away from dirty benches or floors. Treat surfaces with a disinfectant before placing flats there.

For seedbeds, choose well-drained locations. Keep the seedbed well ventilated and dry. Sow on raised beds. Avoid overcrowding of plants and the movement of infested soil or contaminated plant material into the nursery bed. Solarization of nursery beds kills the soil borne microbes, therefore, solarization of nursery beds should be followed during summer season. Workers should clean their hands and tools before handling healthy plants. Water plants in the late morning. Surface irrigation ponds may be a source of fungal contamination.

Physiological Leaf Roll

Leaf roll develops in rainy periods. The lower leaves roll upward until the edges touch and become thick and leathery. Leaf roll is favoured by cool wet weather, excessive fertilizer, and pruning. It has no serious effect on plant growth or yield, and usually no management practices are required.

Catface

This abnormality generally develops on the blossom end of the fruit, resulting in puckered, swollen protuberances and deep cavities. Cool weather at blossom time may cause this abnormality. Some varieties are more susceptible to catface than others.

Growth Cracks

Cracks frequently develop on the stem end of fruits ripened during hot, rainy weather and also due to calcium deficiency. These conditions promote rapid growth. Cracks may radiate away from the stem end or form in concentric bands around the stem end. Growth cracks are most severe when wet weather follows a dry period. Varieties differ in their susceptibility to growth cracks, and this information is frequently part of the variety description

Blossom-End Rot

Blossom-end rot is caused by calcium deficiency. It starts as a water-soaked spot, which enlarges becoming brown; the surface of the spot becomes dark, sunken, leathery and dry, when wet secondary infection may occur.

Management

Use Calcium ammonium nitrate as source of nitrogen in fertilizer. Test the soil of field to check the calcium deficiency. Add calcium sulphate at the dose of 10 kg /ha. Spray any nutrient mixture *i.e.* Tracil, 0.2 per cent solution or calcium chloride 0.2 per cent solution.

Herbicide Injury

Tomatoes are very sensitive to injury by 2,4-D and related growth regulator herbicides. Leaves may become cupped, or veins may become parallel with leaves thick and leathery. Fruits may be catfaced or may only partially ripen. Growth may be stunted and abnormal or twisted. Symptoms of 2,4-D injury on tomatoes may resemble symptoms of virus diseases, except that 2,4-D usually causes the leaves to be thickened. Herbicides may drift when being applied to lawns; if a volatile ester is used, the fumes may cause injury several days after treatment. Never mulch tomato plants with grass clippings if the lawn was treated with 2,4-D during the previous two or three weeks. Do not use lawn-type weed killers close to the vegetable garden.

General Methods of Disease Management

There are some general methods used for Reducing disease problems in tomato crop.

- ☆ Remove all plant material or residues from the greenhouse/ cultivation area before starting a new crop.

- ☆ Treat the seeds with fungicides (Thirum or Captan) and insecticides (imidacloprid).
- ☆ Select vigorous, healthy plants from nursery.
- ☆ Crop rotation must for tomato crop.
- ☆ Tomatoes crop in the same place only once in three to four years.
- ☆ Disease-resistant tomato varieties grown.
- ☆ Remove and destroy all plant refuse in the fall and use deep cultivation to bury any remaining refuse.
- ☆ Drainage facility should be available in the field.
- ☆ Control weeds in and around the garden plots. Weeds not only compete with vegetables for soil moisture and nutrients but also serve as hosts for insects and disease organisms.
- ☆ Appearing first sign of leaf spot diseases apply a recommended fungicide according to label directions.
- ☆ Many insect pest transmit Viral and mycoplasma disease from plant to plant, therefore, control insect pest timely.
- ☆ Remove abnormal appearing plants as soon as they are observed. Virus diseases may be involved. To reduce the spread of viruses, wash hands and tools with a detergent after handling diseased or unusual looking plants.
- ☆ Do not use tobacco products while handling tomato plants. These products may carry viruses, especially tobacco mosaic virus.
- ☆ Use plastic or organic mulches to reduce disease and blossom-end rot problems.
- ☆ Choose a sunny location for tomato nursery as well as cultivation. It is less likely to have leaf disease problems in a sunny location than in a semi-shady one.

References

Barksdale, T.H. and Stoner, A.K. (1973). Segregation for horizontal resistance to tomato early blight. *Plant Disease Reporter*, **57**: 964–965.

Bose, T.K., Some, M.S. and Kabir, J. (1993). "Vegetable Crops" Naya Prokash, Kolkata, p. 113-280.

Efrat, Glick, Yael, Levy and Yedidya, Gafni. (2009). *Plant Protect. Sci.*, **45, (3):** 81–97.

Hand Book of Horticulture (2007). ICAR, Pusa Anusandhan Bhavan, New Delhi.

Hari Har Ram (1997). "Vegtable Breeding Principles and Practices" Kalyani publication, Ludhiana 183-206.

http://en.wikipedia.org/wiki/Tomato

http://gardening.about.com/od/vegetablepatch/a/TomatoProblems.htm

http://www.avrdc.org/pdf/tomato/nematode.pdf

http://www.biovision.ch/fileadmin/pdf/e/projects/tomato_6-02-08.pdf

Kamble, S.B., Pawar, D.R., Sankeshwari, S.B., Parab. M.M. and Arekar, J.S. (2007). "Field screening of different tomato varieties and advanced lines against early blight of tomato." *Agricultural Science Digest*, **27**: 267-269.

Koike, Gladders, Paulus. Vegetable Diseases. Manson Publishing Ltd., London.

Lanna Filho, R. Romeiro, R. da S. Alves, E. (2010). "Complete Reference" Bacterial spot and early blight biocontrol by epiphytic bacteria in tomato plants. Pesquisa Agropecuaria Brasileira. 45: **12**, 1381-1387.

Marisa Maierol and Timothy J. Ng (1990). Genetic Resistance to Early Blight in Tomato Breeding Line. HortScience **25(3)**: 344-346.

Pandey, K.K., Pandey, P.K., Kalloo G . and Banerjee, M.K. (2003). Resistance to early blight of tomato with respect to various parameters of disease epidemics. J Gen Plant Pathol, **69**: 364–371.

Patel, S.J., Subramanian, R.B. and Jha, Y.S. (2011). "Complete Reference" Biochemical and molecular studies of early blight disease in tomato. Phytoparasitica, **39**: 3, 269-283.

Reddy, P.P. (2010). Fungal Diseases and their Management in Horticultural Crops. Scientific Publishers (India).

Sharon, M. Douglas (2003). New Haven, CT 06504 web: www.avrdc.org

Written by Ray Cerkauskas, Visiting Scientist from Agriculture and Agri-Food Canada.

Edited by Tom Kalb.Published by AVRDC–The World Vegetable Center; P.O. Box 42, Shanhua; Taiwan 741; ROC.

deebakamil@gmail.com

Chapter 33

Biochemical Investigation of Phenolic Compounds with Reference to some Cyanobacteria

Neerja Asthna and Ranjana Kumari

R.D.S.College, B.R.A.Bihar University, Muzaffarpur, Bihar

Phenolic compounds are well known secondary metabolite of lower as well as higher plants ranging from algae to angiosperms (*cf.* Walker, 1975). Plants need phenolic compound for pigmentation, growth, reproduction, résistance to pathogens and for many other factors. Therefore they represent adaptive characters that have been subjected to natural selection during evolution. This compound needs comprehensive and concentrated studies because of their various significances. Recently much attention has been focused on microalgae as sources of novel and biologically active compounds. However the occurrence of phenolic compounds in blue green algae is less documented than that of higher plants. Algal phenolic compounds were reported to be potential candidate to combat free radicals which are harmful to our body and food system (Estrada *et al.*, 2001). In vitro studies demonstrated that blue green algal phenolics have several physiological and therapeutic properties, and these compounds also protect the blue green algae from bacteria and other herbivorous organisms (*cf*;Salisbury and Ross, 1986). Recently it has been reported that *Arthrospira pletensis* shows free radical scavenging activity due to presence of hydroxyl groups in the chemical structure of phenolic compounds that can provide the necessary components as radical scavenger (Halimoon, Normala

* Corresponding Author E-mail: deebakamil@gmail.com

and Ali, Roslina, 2009).Considering the importance of phenolic compounds which can be used as a diet supplements in functional food or in medicine, and as a natural free radical scavenger can reduce the problem of climate change by absorption of Co? from the atmosphere. Studies show that blue green algae are rich in phenolic compounds. Farming of selected microalgae as an important alternative material to extract natural antioxidant compounds to delay or prevent the oxidative damage, to reduce stress and anxiety, reduce the nephrotoxicity of pharmaceuticals and also effect the climate change.

This research work is focused on quantitative estimation of intra cellular phenolic contents during different growth phases of the following four taxa of blue green algae. The age dependent quantitative estimation also enables us to determine the concentration of phenolic compounds at different growth phases of these four algae, so, that optimal utilization of this bioactive product can be done. The following four blue green algal taxa are quite common in Muzaffarpur. Due to their faster rate of growth and easy availability in this area, they were selected for the present study and collected from different places of Muzaffarpur and proper identification was done before study.

Method of Isolation

The four taxa under investigation were followings:

- ✰ *Eucapsis minuta*, Frich.
- ✰ *Anabaena laxa*.
- ✰ *Arthrospira pletensis*, var. *tenuis* (Rao, C.B.) Comb. Nov.
- ✰ *Scytonema coactile*, Motagne ex Born. Fla

Isolation and Purification of Algae

The methods suggested by Stein (1973) were adopted for the isolation and purification of blue green algae under investigation. These methods are called capillary pipette and streak plate methods. Only unicellular form, *Eucapsis minuta* was isolated by the method called streak plate method otherwise all the other three forms *Anabaena laxa, Arthrospira pletensis and Scytonema coactile* were isolated by capillary pipette method. After obtaining the pure and unialgal culture two of them *Eucapsis minuta* and *Arthrospira pletensis* maintained in Hughes(+) medium and other two *Anabaena laxa* and *Scytonema coactile* were grew properly in Hughes (?) medium. Culture of each of the four organism under taken for investigation was regularly exposed to 3300lux light in a daily cycle of hours light and 6 hours darkness by fluorescent tubes fitted at the distance of 20" from culture vessels. Proper temperature and pH were also ascertained.

Source of Inocula

The inoculation of an unicellular form *Eucapsis minuta* and three filamentous form : *Arthrospira pletensis, Anabaena laxa* and *Scytonema coactile* were done after centrifugation at 3000rpm, for the three filamentous forms the centrifuged, and washed algal mass of each was broken into pieces by the sterile glass beads to make homogeneous suspension.

The amount of inoculate used for every fresh inoculation was so adjusted that after inoculation culture read 0.05O.D at the wave length of 66onm. The O.D was measured by U.V visible recording spectrophotometer (U.V~160 Shima dzu, Japan

Growth Phase of Organisms

The determination of growth phase of organisms is the pre requisite for present study, as it aims at to determine the quantitative variation of phenolic compounds at different growth phases of organism. The growth phase of Arthrospira platensis was determined by dry weight method.

It was found that the growth phase of uni cellular form *Eucapsis minuta* was determined by measuring optical density of its homogenous suspension in culture medium at a wave length of 660nm by UV Visible recording spectrophotometer against the blank of growth medium. Determination of growth phases of fillamentous forms *Arthrospira pletensis, Anabaena laxa* and *Scytonema coactile* was done by dry weight method.

Details of Growth Phases of Organisms

Eucapsis Minuta

Lag phase of about 5~6 hours followed by a long exponential phase of 28 day from day 1 to days 28. The static phase remained for 7 days from day 28 to day 35. Thereafter decline phase set in.

Anabaena Laxa

Lag phase of 3 days from day 1 to day 3 followed by exponential phase of 19 days from day 4 to day 22. The static phase remained for 6 days from day 23 to day 28. And then decline

Arthrospira Pletensis

Lag phase of 4 days from day 1 to day 4 followed by exponential phase of 17 days from day 5 to day 21. The static phase remained for 5 days from day 22 to day 26. Thereafter decline phase started.

Scytonema Coactile

Lag phase of 5 days from day 1 to day 5 followed by exponential phase of 16 days from day 6 to day 21. The static phase remained for 7 days from day 22 to day 28. And then decline phase established.

Data Analysis

All analysis were performed in triplicate and data reported as mean ± standard deviation, analysis of variance was also done.

Estimation of Intra Cellur Phenolic Contents

The method suggested by Singh *et al.*(1979) was applied to calculate the quantitative variation of extra cellular contents of phenolic compounds of the blue green algae undertaken for study.

Reagents

- ✰ 0.3 per cent Ferric chloride in 0.4N HCL.
- ✰ 0.3 per cent Potassium Ferricyanide.
- ✰ Colour reagent prepared by mixing equal volume of 0.3 per cent ferric chloride solution in 0.4 per cent HCL and 0.3 per cent potassium ferricyanide solution.

To estimate the intra cellular phenolic contents 100mg dried sample of organism was taken starting from the 10th day to the 45th day of culture age at the interval of every 5 days. And these samples were crushed in 10ml of 80 per cent hot ethyl alcohol to extract the phenolic compounds. The extract was then centrifuged for 15 minutes at 3000rpm. From each of these extract 1ml supernatant was taken and 4ml distilled water was added to it. This preparation treated with 5ml of colour reagent.

The phenolic contents were then estimated by UV~Visible Recording Spectrophotometer at the wave length of 675nm. Tannic acid was used as standard.

Observation

The quantitative estimation of each of the four blue green algae under investigation was done in cell extract at the interval of every five days starting from 10th day of culture age to 45th day (which corresponds to early exponential to late decline phase with the intermediate static phase.)

The data obtained with regard to the study of intra cellular phenolics in each sample of every organism has been displayed in Table. Indicated that on early exponential phase (i;e on 10th day) the intra cellular contents of total phenolics was maximum in *Arthorospira platensis* 3.15 per cent (mg/ 100mg dry weight) and minimum in *Anabaena laxa*1.89 per cent. The Table also shows that there was a gradual decreasing tendency in amount of phenolic contents with the increase of the age of culture. On the 45th day (late decline phase) the amount of intra cellular phenolics was found to be considerably low.

The literary evidences of age dependent variability of total intra cellular phenolic contents are very insufficient, particularly with reference to algae. Though some reports have been made by Towers (1964), Moor and Pecket (1972),Sinha (1980), Dogra and Sinha (1983), but all with regards to different parts of higher plants. According to Bernett ~Barnell(1945) Ramstad(1959) and Swain(1965), age is one of important factor which determines the accumulation of phenolic contents in different parts of the higher plants. The gradual declining tendency of total phenolic contents beyond the 10th day as found in the present result has been reported earlier with regard to blue green algae by Shukla (1989) and Shukla and Verma (1989). This gradual declining tendency in the amount of phenolics may be due to the oxidation of phenolic contents by the increasing activity of enzyme called esterases and peroxydases because these enzymes are believed to be responsible for the oxidation of phenolic derivatives in higher plants (cf. Swain 1965, Hosel~Barz, 1972).

At early exponential phase the amount of intra cellular phenolic content in the four blue green algae which are under investigation, are as follows *Arthrospira platensis* (3.15mg), *Eucapsis minuta* (2.72mg), *Scytonema coactile* (2.62mg) and *Anabaena laxa*(1.8mg) respectively in 100mg dry weight of algal mass. These amount of intra cellular phenolics are considerable compared to the reported maximum amount (1.33 per cent) in many of the higher plants used in the folk medicines by Santhales (cf. Dogra and Sinha, 1973). In young leaves of *Parthenium hysterophorus* it was 3.23 per cent (cf. Ambastha and Hamidi, 1980)

Survey of literature on phenolics, reveals that a good beginning has also been made with regard to blue green algae. Several encouraging results have been reported in connection with total intracellular phenolics of blue green algae(cf. Barclay.

Table 33.1: Quantitative Estimation of Intracellular Phenolics in Taxa Under Investigation (mg /100mg dry weight)

Age of Culture (in days)	*A.laxa*	*E.minuta*	*A.platensis*	*S.coactile*
10	1.89±0.020	2.72±0.011	3.15±0.011	2.62±0.030
15	1.72±0.008	2.60±0.015	2.90±0.020	2.51±0.019
20	1.53±0.023	2.53±0.025	2.72±0.030	2.21±0.012
25	1.42±0.160	2.49±0.020	2.41±0.021	2.03±0.009
30	1.35±0.012	2.10±0.008	2.15±0.018	1.91±0.015
35	1.31±0.032	1.79±0.019	1.89±0.022	1.69±0.006
40	1.29±0.025	1.45±0.030	1.73±0.021	1.52±0.012
45	1.25±0.020	1.21±0.014	1.42±0.014	1.32±0.019

Purpose of the Research Work

- ☆ Biological evaluation of the nature of phenolics.
- ☆ Chemical analysis of biologically derived phenolics.
- ☆ Finding the alternative sources of phenolics
- ☆ Determining the growth phase of alga at which maximum concentration of phenolics is present for optimal utilization.

References

Ambasta, K.K. and Hamidi, M.K. (1988). Variability of total phenolics in different parts of *Parthenium hysterophorous* L. *Environment and Ecology* **6:** 484-885.

Angadi, S.B. and Bharti, S.G. (1983). Nitrogen fixation and production of extracellular nitrogen by Hapalosiphon confervaceous and Nostoc microscopicum, *Beitr. Pflanz.* 58 **(2)** : 205-210.

Anusuya, D.M. and Venketaraman, L.V. (1983). Supplementary value of the proteins of blue green algae *Spirulina platensis* to rice and wheat proteins. *Nutr. Rep. Int.* 28 **(5):** 1029-1036.

Abd, El., Baky, H.H. (2003). Over production of phycocaynin pigment in blue green alga *Spirulina sp*. And its inhibitory effect on growth of Ehrlich ascites carcinoma cells. *J. Med. Sci.*, **3:** 314-324

Abd, El., Baky, H. H., El, Baz.F.K. and El- Baroty, G.S. (2004). Production of antioxidant by the green alga *Dunaliella Salina. Int. J. Agric Biol.* **6:** 49-57.

Barclay, W.R., Kennish, J.M., Goodrich, V.M. and Fall, R. (1987). High levels of phenolic compounds in *Prochloron species* (Lond) **168 :** 426.

Screening of cyanobacteria from Biochemical studies on the Response of organic phosphorous Insecticide and Release of Extra Cellular Products (2007). Research Journal of Fisheries and Hydrobiology. 2 **(1):** 13-17.

Dogra, J.V.V. and Sinha, S.K.P. (1983). Variability of total phenolics in maturing leaves and fruits of three medicinal plants. *Herbs Hungarica* **22 :** 25-31.

Fogg, G.E. (1966). The extracellular product of algae. *Oceanogr. Mar. Biol.* **4:** 195-212.

Jain, J.L. (1983). In "Fundamentals of Biochemistry (eds) Published by S. Chand.& Company (Pvt.) Ltd. New Delhi. Pp. 390-394.

Jayraman, J. (1985). In Laboratory Manual in Biochemistry (eds) Published by Wiley Eastern Limited. New Delhi.

Pillinger., J.M., Cooper, J.A. and Ridge, I. (1994). V Journal of chemical ecology volume 20, Number 7, 1557-1569,Doi; 10, 1007/ BF02059880.

Kumar, A. (1986). Studies on extra and intra cellular products of some common algae of Muzaffarpur (Bihar).

Paul, R. and Ahluwalia, A.S. Biochemical studies during differentiation in 2 *Anabaena sp.Better biol pflanz* 61 **(3):** 327-336.

Rajangam, M. (1988). Phenolic acid and Tennin levels in mangroves of pichavaram. IBC **5:** 377-39

Rao, K.M., Reddy, Rako, A.V.N. and Krishna, H.R.B. (1978). Phenolic Compounds in chick pea (Cicer arietinuim L) *Indian J. Expt. Biol.* **16:** 1213-1214.

Sahay, A.P. (1990). Ph. D. Thesis, B. R. A.Bhiar University, Muzaffarpur.

Salisbury, F.B. and Ross, C.W. (1986). Plant physiology (eds). CBS Publishers and Distributers. Delhi. Pp 278-283.

Shekhar, K.M., Venkataraman, L.V. and Salimath, P.V. (1987). Carbohydrate composition and characterization of two kunusual sugars from the blue green alga. Spirulina platensis. *Phytochemistry* (OXF) 26 **(B):** 2267-2269.

Shukla, C.P. and Verma, B.N. (1989). Estimation of Total intracellular Phenolics and their leachates in culture filtrate during different growth phases of *Oscillatoria subrevis. Environment and Ecology* 7 **(3):** 730-732.

Shukla, C.P. and Verma, B.N. (1989). Estimation of Total intracellular phenolics during different growth Phases 0f Chroococus minor. *Biojournal* 1 **(2):** 111-112.

Singh, M., Singh, S.S. and Sanwal (1978). Anew colorimetric method for the determination of phenolics. *Indian J. Exp. Biol.* **16;** 712-714.

Stein, J.R. (1973). Culture methods and growth measurements. In Handbook of phycological methods, Cambridge Univ. Press.

Vincent, W.A. (1969). Algae for food and feed. Process *Biochemistry* **4:** 45.

Zheng, W. and Wang,Y. (2001). Antioxident activity and phenol compounds in selected herbs. *J. Agric Food Chem.* **49:** 5165-5170.

Chapter 34

Multi-omics Strategies to Identify the Unidentified: Current Approaches in Molecular Microbial Diversity Analysis

Dhananjaya P. Singh**, Kamlesh K. Meena, Udai B. Singh, Lalan Sharma and Dilip K. Arora

National Bureau of Agriculturally Important Microorganisms, Indian Council of Agricultural Research, Kushmaur, Maunath Bhanjan – 275 101 Uttar Pradesh

Microorganisms are silently playing valuable role in the biogeochemical cycles of the atmosphere since the early days of their evolution and it is because of their activities, this earth has been transformed into today's situation. These tiny and often unseen organisms are responsible for recycling of nutrients and organic compounds and contribute to plant, animal and human nutrition, soil structure and fertility, soil health and ecosystem functioning. Prokaryotes, the first life forms of the earth are considered as the ancestors of all kinds of organisms. These life forms with cells having hereditary information not bound by the nucleus (*karyon*) have existed twice longer (4 Gyr) than higher eukaryotic organisms with DNA bound in nucleus such as fungi, plants and animals (2 Gyr). Although the type, composition and the microbial

* Corresponding Author E-mail: dpsfarm@rediffmail.com

processes may vary from habitat to habitat like sea, fresh water and the soil, the molecular and biochemical characteristics, functional activities and the quality of life-processes are almost similar. The prokaryotic abundance within the ocean and soil are estimated to be 1.2×10^{29} and 2.6×10^{29} while that on the ocean and terrestrial sub-surfaces are 3.5×10^{30} and $0.25\text{-}2.5 \times 10^{30}$, respectively (Whitman *et al*, 1998). This estimation of the unseen majority of prokaryotes represents the likely largest living surface on the earth and the enormous interface between the living (biotic) and abiotic world is one of the reasons for their significance in regulating several biogeochemical cycles and transforming organic matter. Besides their importance, however, our understanding to the biology, diversity, identification, properties, activities and functions of these most powerful organisms are still at a very beginning stage and this may be because of our inability to culture, identify, characterize and name them for a particular biological purpose.

A large number of microbes are associated with plants and animals, not only as pathogens but as associative organisms to make benefit of each other. Such associations are under strict scientific investigation to get answers about significance of the multitrophic interactions with plants and animals and survival and performance under a given ecological niche. The kind of associated microbial species and their pattern of interactions with other microbes in ecological niche, their hosts and non-hosts, within biotic and abiotic component represent a classical ecological relationship that constitutes the basis of cooperative and constitutive livelihood in the nature leading to mutual benefits and losses in many ways. Associations of microbes with the hosts and other habitats are critical determinants for many issues related to the quality of the ecological success, impact of environment, global climate change, production of greenhouse gases, quality of human, plant and animal health, and finally loss or gain in agricultural productivity and food. This is why the identification of different microbial communities in various habitats is of prime importance for taking over maximum benefit of these communities for man kind and for their conservation for longer duration.

Habitat-wise Microbial Diversity and Analysis Strategies

Microorganisms find their existence in diverse environmental conditions, they can thrive well in most of the normal, sub-normal and extreme habitats such as extremophilic physiological environments like colds, high temperature, frost, hot springs, droughts, acidic and salinity conditions. These organisms have many unexplored and hidden physiologies to tell since they have faced a lot of diverse conditions during their evolutionary track and this make their communities so diverse that even in the era of most modern instrumentation backup, we could only predict their presence up to only 1 per cent and rest 99 per cent are predicted to be unexplored. One gram of the soil may harbor up to 10 billion microorganisms of possible of thousands of different species and with this huge microbial dynamics, soil ecosystems are to a large extent uncharted. This is also true for the aquatic systems, especially marine ecosystems which cover more than 70 per cent of the Earths' surface. Microbial diversity poses great complexity, divergence and variability at different levels of biological parameters, especially in terms of genetic variability within taxonomic groups (genera and species), number (species richness in confined region), relative abundance (evenness) of taxons and functional groups (guilds) in communities. Spatial and temporal patterns of microbial diversity are also obscure and therefore, estimating prokaryote diversity in natural ecosystems is a priority in current ecological research. Other important aspects to be addressed are the range of the processes, complexities of the multitrophic interactions and final benefits (functional aspects) of the whole community level characterization to the plant, soil, and other organisms living together.

The hidden base of the plant system i.e. the roots thrive below-ground in a diverse, ever changing environment of microbial communities including bacteria, fungi and other microbes living together and approaching each other for their own benefits. Thee area of soil surrounding the root zone is an unique place of physical, biochemical, molecular and ecological interface through which the roots and microbes in the surroundings are gaining their nutrition and health supplements. This zone, the so-called rhizosphere is supposed to be self-regulatory by secreting/excreting numerous chemicals in the surrounding soils, which, then attracts many microbial communities to proliferate there, and in return, keep plants healthy by many known mechanisms. It is estimated that nearly 5-21 per cent of the photosynthetically fixed carbon materials is eventually excreted/secreted in the rhizosphere by the roots in the form of exudates. Because of a rich source of nutrition, rhizosphere constitutes an important region for different microbes to thrive and establish their communities there. Since this region is responsible for many of the ecosystem services including acquisition, mobilization and distribution of nutrients, decomposition of organic matters, cycling of soil minerals, aeration and aggregation of soils, filtration and bioremediation of pollutants, plant growth promotion, suppression of diseases and causal agents and production of release of greenhouse gases, identification of associated microbial communities can uncover many unknown mysteries about the function and association of below ground life forms.

In almost all ecosystems, whether it be soil, aquatic environment or above and below-ground plant-surroundings, so many questions remain unanswered. What is the total number of microbial species on the Earth? How many bacterial species can live in a micro-niche or in a liter of marine or fresh-water or in a kg of fertile soil or sediment? Is there a vertical and longitudinal divergence of prokaryotic diversity? Is the microbial diversity, coupled with other organisms living together in the ecosystem, display idiosyncratic relationship? What is the root cause of so many diverse genera and species? All these and other related unanswered questions provide evidence of this vast but unchecked field of microbiological-ecological research. The development of accurate, rapid and universally adopted methods for the determination of microbial diversity is therefore, essential and highly desirable to construct solid, reproducible and consistent data sets enabling large spatial and temporal studies possible. The parameters for studying microbial diversity need to include multiple methods taking into account a great integration of the available taxonomic background and datasets and holistic approaches at the community level and partial approaches targeting structural or functional aspects.

The assessment of microbial diversity in nature usually involves two aspects, species richness (number of species present in a sample) and species evenness (distribution of relative abundance of species). To estimate species richness, we widely rely on the assignment of 16S sequences into Operational Taxonomic Unit (OTU or phylotype) clusters. The criterion used to define an OTU at the species level is the percentage of nucleotide sequence divergence; the cut-off values vary between 1 per cent, 3 per cent, or 5 per cent, largely depending upon the study and due to these inconsistencies, a reliable statistical comparisons or descriptions of species richness across studies are preferred. The total community diversity of a single environment (the α-diversity) is represented by rarefaction curves. These curves plot the cumulative number of OTUs or phylotypes captured as a function of sampling effort and, therefore, indicate only the OTU richness observed in a given set of samples. Nonparametric methods including Chao1 or ACE, are richness determinations of overall <u>*α*</u>-diversity. Shannon or Simpson indices i.e. the quantitative methods measure the evenness of the α-diversity. These estimators usually describe the diversity of the microbiota associated with a healthy or diseased state they are not informative of the phylogenetic diversity of an environmental sample.

In contrast to the α-diversity, the β-diversity is a measurement of a community structure comparison (taxon composition and relative abundance) between two or more environmental samples and can ne helpful in comparing similarities and differences in microbial communities in healthy and diseased conditions. A broad range of qualitative (presence or absence of taxa) and quantitative (taxon abundance) measures of community distance are available using several tools, including LIBHUFF, P-test, TreeClimber, SONS, DPCoA, or UniFrac. The robust unweighted UniFrac tool measures the phylogenetic distance between two communities as the fraction of phylogenetic tree branch lengths leading to a descendant from one o other community. UniFrac can detects differences in the presence or absence of bacterial lineages and more developed weighted UniFrac is the qualitative version of original UniFrac that can provide detection of differences in the relative abundance of bacterial lineages.

Omics Strategies for Microbial Diversity Analysis

Revolutionary advances in high-throughput DNA sequencing technologies in the present era have resulted in tremendous applications in the analysis of diverse microbial lineages across different environmental conditions. Out of 11,364 whole genome projects running on eukaryota, prokaryota (archaea and bacteria) and viruses, 7473 projects belong only to microbes with 1695 complete microbial genomes sequences available while 2247 being assembled and 3531 still remaining unfinished (NCBI database, 2011). The computational-driven annotation and comparative genome analysis of DNA sequences from microbial entities have uncovered many mysteries related to whole genome structures, gene function, biological pathways, metabolic and regulatory networks and evolutionary diversification among microbial lineages that has greatly enhanced our understanding of structural and functional microbial diversity and its implications. The elucidation of microbial diversity, its responses to normal, sub-normal and extreme environments, their interactions in the habitats and the factors responsible for the emergence of microbial communities during the course of evolution factors is a challenging task that necessitates functional characterization and accurate quantification of all levels of gene and gene products, mRNAs, proteins and metabolites as well as their multi-phasic interactions. In the past decades, significant efforts in improving high-end technologies pertaining to DNA sequencing, measurement of gene expression, protein profiling and modeling, and unbiased assessment of microbial metabolites (small molecules) responsible for multi-pronged interactions in the environment have been made. These efforts have led to the emergence of several new 'omics' research fields : genomics, transcriptomics, proteomics, metabolomics, interactomics etc. Unlike traditional methods, 'omics' approaches are high-throughput, data-driven, holistic and top-down methodologies that attempt to understand microbial biology as an 'integrated system' rather than a collection of different molecular species. These high-throughput 'omics' approaches generate large amounts of data and the analysis of these data often requires significant statistical and computational efforts. This article describes the developments that have been made in defining and understanding microbial diversity using multi-omics approaches.

Multi-omics strategies based on new molecular technologies in genomics, proteomics and metabolomics has shifted traditional techniques for microbial taxonomy, systematics, classification, identification, and characterization toward the methods based on the DNA finger printing, elucidation of specific gene sequences, characterization of proteomic or metabolomic components of the cell or cellular fluids to identify specific protein, fatty acid or biochemical components directly related to the microbial communities. Current genotypic and proteomics techniques for bacterial identification and characterization now complement conventional approaches. The new methods are rapid, offer high throughput and produce unprecedented levels of discrimination among strains microbes. Supported with the huge database of whole genomes sequences, nucleotide and protein sequences and large libraries of metabolites (small molecules) for referencing and comparative analysis, multi-omics methods are finding their applications in microbial molecular identification and phylogenetics. Existing challenges with these fields include developing appropriate standards and methods for routine application of techniques and enriching and integrating existing bioinformatics based-databases for comparative analysis and to store large amount of data being generated in the laboratories.

The beginning of the new microbial systematics era started with the introduction and application of new taxonomic concepts and techniques, by and large based on the genome, proteome and metabolome levels that are principally considered to possess the differential fingerprint of each and every organism. Progress has been made in numerical taxonomy and molecular taxonomy that is brought into effect after the emergence of the molecular biology resources. This has provided knowledge that comprises systematics of bacteria comprising great evolutionary interest. The composition and disposition of nucleotides in within the genomes are least susceptible to the environmental alterations than proteins and metabolites and this largely governs the divergence of microbial species and their functional characteristics in the environments. Molecular taxonomy is largely dependent on the studying variability among the DNA and RNA and the main techniques used in the molecular systematics comprise the building of restriction maps, DNA-DNA hybridization, DNA-RNA hybridization, sequencing of DNA sequencing of sub-units 16S and 23S of rRNA, RAPD, RFLP, PFGE etc. Recent developed molecular techniques in past years included HPLC-determination of deoxyribonucleic acid base composition and assessment of base sequence homology by means of DNA-RNA hybridization and DNA microarrays for genomic analysis, 2D gel electrophoresis followed by MALDI-TOF for proteomic analysis and isoenzymes, mass- and nuclear magnetic resonance (NMR) based metabolomic platforms for the comparative analysis. Understanding of the evolution, diversity, systematics and proper nomenclature of microbes is of paramount importance in biological sciences and molecular approaches to phylogenetic analysis led to the three-domain model system based on archaea, bacteria and eukarya has strengthened our understanding about the evolutionary diversification of microbes on the earth.

Methods for Microbial Diversity Analysis and Identification

The demand for precise identification coupled with the diversity of microbial communities have steadily increased with the realization that complex microbial diversity structure confers potential functionality that can be exploited in agricultural, environmental and medical perspectives. To overcome the problems associated with the characterization of both culturable and nonculturable microbial communities, several methods with high resolving power have been developed or are under development. Most of them that are based on the analysis of ribosomal RNA genes (rDNA) have opened the gateway to uncover microbial diversity to a certain extent and have yielded many sequences from many novel phylogenetic lineages. Likewise techniques based on molecular fingerprinting when coupled with the analysis of the functional activities and genes, also revealed and taxonomic variability. Such investigations have led to the understanding of the relationship between structural and functional diversity in a particular ecosystem.

Diversity and function of the major biological diversity sites on the earth poses great challenges. Methods to measure soil microbial diversity consisting of species richness and evenness, total number of species present and spatial and temporal distribution can be categorized in two broad categories: the biochemical based techniques and molecular approaches. However, all these approaches usually result in relative diversity of communities across the habitats and determination of what we call "true diversity" with so many unknowns is still a difficult task to be completed. Most widely used DNA-based techniques such as polymerase chain reaction are increasingly popular due to their specificity and speed compared to biochemical and culture-based methods. These methods allow the detection and identification of "viable but nonculturable" cells that are metabolically-active but non-dividing. However, even using these improved methods, the total number of bacterial species is not known and cannot even be estimated with any certainty.

Microbial identification methods based on genotypic characteristics (considering diversity within microbial genomes) can be divided in to two broad categories, (1) pattern- or fingerprint-based techniques and (2) sequence-based techniques. Pattern-based techniques produce a series of chromosomal DNA fragments from an organism and that fragments are separated by size to generate a profile or fingerprint unique to that organism and/or its close relatives. This information can help researchers in creating a library or set of databases of fingerprints from known organisms to which test organisms can be compared. When the profiles of two organisms match or nearly match, they can be considered related very closely at the strain or species level.

Genotyping methods have been particularly useful for microbial ecology-driven studies but are not very suitable for diagnostic purposes leading to the detection of specific species, strains or clones. A large volume of work in recent years focus on DNA-based methods for bacterial detection and identification, highlighting strategies for selecting taxa-specific loci and emphasizing the molecular techniques and emerging technological solutions for increasing the detection specificity and sensitivity. The massive and increasing number of available bacterial sequences in databases, together with already employed bioinformatics tools, hold promise of more reliable, fast and cost-effective methods for bacterial identification in a wide range of samples in the years to come. Also the validation and certification of these methods and their routine implementation is also becoming a prime concern for researchers today.

Efforts were made to overcome the problems of the application of one of many approaches in the analysis of microbial community structure analysis. Separation of polymerase chain reaction (PCR) amplicons using denaturing gradient gel electrophoresis (DGGE) was first used by the medical researchers to identify gene mutations but its application for the identification of microbial communities was started by Muyzer *et al.*, 1993. Microbial communities in different habitats like deep-sea hypothermal vents and hotsprings were initially explored. After that the technique became popular to be used for the analysis of the soil community structure. Researchers usually target specific population known to be as major inhabitants. Extraction of whole genome directly from the soils leads to the examination of a community without the limitation posed by culturing them. Polymerase chain reaction is a strong mean to increase the numbers of specific genetic target for its detection on gels. Using the rRNA genes, PCR provides phylogenetic information on populations comprising communities. Fingerprints produced by this method allowed spatial and temporal comparisons of soil communities within and between locations or among treatments. PCR-DGGE technique has been proven as a relatively rapid and high throughput tool to study and compare the community structure of different systems such as agricultural soils, rhizospheres, forests, grasslands, paddy soils, wetland and upland soils and Orchards. DGGE was successfully applied not only for the elucidation of microbial communities but for comparing them in different ecological conditions. Scientists are now finding a reliable method to address complex ecological questions about the spatial, temporal, and nutritional interactions faced by microbes in the soil and to obtain a "snapshot" of whole community structure, an approximation of the numbers of populations and their proportional representation within the total community (Okubo and Sugiyama 2009).

Effective and accurate assessment of total microbial community diversity is one of the primary challenges in modern microbial ecology is particularly true with regard to the detection and characterization of low-abundant unculturable populations. A novel strategy, GC fractionation coupled with denaturing gradient gel electrophoresis (GC-DGGE) combines mechanistically different community analysis approaches to enhance assessment of microbial community diversity and assessment of minority populations of microbes. This approach employs GC fractionation as an initial step to reduce the complexity of the community in each fraction. GC fractionation of total community DNA is independent of PCR amplification process and provides a relative abundance of bacterial populations although only with a low resolution. The output from this approach is a fractionated profile of the entire community that indicates relative abundance of DNA as a function of G+C content and inferential information regarding the taxa comprising the community. This technique has been successfully employed to study and compare microbial community structures in a variety of environments, including soils and sediments, bioreactors, and GI tracts of insects and animals. In addition, this technique physically fractionates total community DNA into aliquots that represent different G+C contents. These highly purified fractions are of high molecular weight and thus are suitable for additional molecular manipulations, including PCR amplification, DGGE analysis, and cloning. This reduced complexity facilitates subsequent detection of diversity in individual fractions. DGGE analysis of individual fractions revealed bands that were undetected or only poorly represented when total bacterial community DNA was analyzed. Also, directed cloning and sequencing of individual bands from DGGE lanes corresponding to individual G+C fractions allowed detection of numerous phylotypes that were not recovered using a traditional random cloning and sequencing approach.

Terminal Restriction Fragment Length Polymorphisms (t-RFLP) (also known as Terminal Restriction Fragment patterns) is a PCR-based fingerprinting method commonly used for comparative microbial community analysis and a potential tool for studying microbial community structure and dynamics. The method can be used to analyze communities of bacteria, archaea, fungi, other phylogenetic groups or subgroups, as well as functional genes. The method is highly rapid, reproducible, and yields a higher number of operational taxonomic units than other, commonly used PCR-fingerprinting methods. Size of terminal restriction fragments (T-RFs) can now be done using capillary sequencing technology allowing samples contained in 96- or 384-well plates to be sized in an overnight run. Many multivariate statistical approaches have been used to interpret and compare T-RFLP fingerprints derived from different communities. Detrended correspondence analysis and the additive main effects with multiplicative interaction model are particularly useful for revealing trends in T-RFLP data. Due to biases inherent in the method, linking the size of T-RFs derived from complex communities to existing sequence databases to infer their taxonomic position is not very robust. This approach has been used successfully, however, to identify and follow the dynamics of members within very simple or model communities. The T-RFLP approach has been used successfully to analyze the composition of microbial communities in soil, water, marine, and lacustrine sediments, biofilms, feces, in and on plant tissues, and in the digestive tracts of insects and mammals. The identification of specific elements in a TRF pattern is possible by comparison to entries in a good sequence database or by comparison to a clone library. As an added advantage when investigating complex microbial communities such as those in soils and intestines, TRF patterns are recognized as having better resolution than other DNA-based methods for evaluating community structure. The method is user-friendly and potential approach to evaluate microbial community analysis that is adding significant information to studies of microbial populations in many environments. The partial community level fingerprints derived from DGGE/TGGE banding patterns based on the number and the intensity of the DNA bands as well as the similarity between the treatments have been analyzed for the diversity studies. However, special care is required for the interpretation of results because of the limitations of the PCR and that of banding pattern separations. Excising the specific DGGE/TGGE bands from the gels, their re-amplification and sequencing or hybridization with specific primers on the membranes would provide more information on structural or functional diversity.

tRFLP is usually used for profiling of microbial communities based on the position of a restriction site closest to a labeled end of an amplified gene. Like other community analysis methods, tRFLP is also based on PCR amplification of a target gene where the amplification is performed with one or both the primers having their 5′ end labeled with a fluorescent molecule. In case both primers are labeled different fluorescent dyes are required. The method successfully excludes some of the limitations of RFLP and The method is based on digesting a mixture of PCR amplified variants of a single gene using one or more restriction enzymes and detecting the size of each of the individual resulting terminal fragments using a DNA sequencer. Several common fluorescent dyes such as TET, 6-FAM, ROX, TAMARA, and HEX, can be used for the purpose of tagging but the most widely applied dye is 6-FAM (phosphoramidite fluorochrome 5-carboxyfluorescein). The mixture of amplicons is then subjected to a restriction reaction, normally using a tetra-cutter restriction enzyme following which, the mixture of fragments is separated using either capillary or polyacrylamide electrophoresis in a DNA sequencer and the size of different terminal fragments are determined by the fluorescence detector. Because the excised mixture of amplicons is analyzed in a sequencer, only the terminal fragments (i.e the labeled end or ends of the amplicon) are read and by this way, t-RFLP is different from ARDRA and RFLP where all restriction fragments are visualized. In addition to these steps, the tRFLP protocol often includes a cleanup of the PCR products prior to the restriction and in case a capillary electrophoresis is used, a desalting step is also performed prior to running the samples. Since the method allows the detection of only the labelled restriction fragments, this leads to simplified banding pattern. The analysis of complex communities and diversity analysis is possible in terms of visible bands which represent a single operational taxonomic unit or ribotype. The whole banding pattern can also represent species richness and evenness along with the similarities between samples. However, like all PCR-based methods, t-RFLP may underestimate true diversity because only dominant species can be detected due to large quantity of available template DNA. In addition, different species could have different gene copy numbers and could bias the results (Liu *et al*, 1997). However, despite several limitations, t-RFLP is largely being used to measure spatial and temporal changes in bacterial communities, to study complex bacterial communities and to detect and monitor populations (Kirk *et al.*, 2004).

Single strand conformation polymorphism (SSCP), alternatively called single strand *chain* polymorphism, is conformational difference of single stranded nucleotide sequences of identical length as induced by differences in the sequences under certain conditions. This property allows distinguishing the sequences by means of gel electrophoresis, which separates the different conformations. Like DGGE/TGGE this technique was developed to detect known or novel polymorphisms or point mutations in DNA. Single stranded DNA, after denaturation, undergoes a 3-dimensional folding and may assume a unique conformational state based on its DNA sequence. The difference in shape between two single-stranded DNA strands with different sequences can cause them to migrate differently on polyacrylamide gel, even though the number of nucleotides is the same. SSCP holds all the limitations of the DGGE.

Restriction Fragment Length Polymorphism (RFLP) {also called as *a*mplified *r*ibosomal DNA *r*estriction *a*nalysis (ARDRA)} is an important and widely used method to study microbial diversity analysis. RFLP measures difference in homologous DNA sequences that can be detected by the presence of fragments of different lengths after digestion of the DNA samples in question with specific restriction endonucleases. Restriction fragment length analyses uses restriction enzymes (RE) to cut DNA at specific 4-6 bp recognition sites. Sample DNA is cut (digested) with one or more RE's and resulting fragments are separated according to molecular size using gel electrophoresis. Molecular size standards are used to estimate fragment size on the basis of ethidium bromide staining used to reveal the fragments under UV (260 nm) light. Differences result from base substitutions, additions, deletions or sequence rearrangements within RE recognition sequences. Restriction fragment length polymorphism (RFLP) is most suited to studies at the intraspecific level or among closely related taxa. Presence and absence of fragments resulting from changes in recognition sites are used identifying species or populations. The method relies on DNA polymorphism and as a molecular marker, is specific to a single clone/ restriction enzyme combination and therefore can be used for the bacterial community structure analysis. This technique is helpful in detecting structural changes in microbial communities but, is not a measure of the diversity or can detect specific phylogenetic groups.

Alike the principles of RFLP and t-RFLP, ribosomal intergenic spacer analysis (RISA) and automated ribosomal intergenic spacer analysis (ARISA) provide ribosomal based fingerprinting of the microbial communities. RISA provides estimates of microbial diversity and community composition without the bias imposed by culture-based approaches or the labor involved with small-subunit rRNA gene clone library construction. RISA was used originally to construct diversity in soils and more recently to examine microbial diversity in the rhizosphere and marine environments (Fisher and Triplett, 1999). The method exploits the variability in the length of the intergenic spacer (IGS) between the small (16S) and large (23S) subunit rRNA genes in the *rrn* operon. The 16S-23S intergenic region, which may encode tRNAs depending on the bacterial species, displays significant heterogeneity in both length and nucleotide sequence. Both types of variation have been extensively used to distinguish bacterial strains and closely related species. In RISA, the length heterogeneity of the intergenic spacer is exploited. The PCR product (a mixture of fragments contributed by community members) is electrophoresed in a polyacrylamide gel, and the DNA is visualized by silver staining. The result is a complex banding pattern that provides a community-specific profile, with each DNA band corresponding to at least one organism in the original assemblage. This approach has been used successfully to assess the structure of soil bacterial communities and to evaluate the changes that follow antibiotic treatment, mercury stress, and deforestation. An automated RISA (ARISA) method has been developed to improve both resolution and analysis and efficiently compared freshwater communities and differentiated these communities on a fine spatial scale. Like terminal restriction fragment length polymorphism (RFLP) and length heterogeneity RFLP, ARISA involves the use of a fluorescence-tagged oligonucleotide primer for PCR amplification and for subsequent electrophoresis in an automated system. This allows rapid investigation of bacterial community structure even when there are a large number of samples. Due to the high resolution of the gels and the high sensitivity of fluorescence detection, the number of peaks detected is much higher on ARISA profiles than on RISA profiles. Similarly, differences in the intensity of the bands can be estimated precisely, which allows a finer comparison of the profiles. However, the level of sensitivity might have some drawbacks because it may introduce a variability within profiles that has no biological origin (Ranjard *et al.*, 2001. Appl. Environ. Microbiol. 67:4479-487).

Many prokaryotic (and eukaryotic organisms) contain highly repetitive short DNA sequences of almost 1-10 base pairs repeated throughout their genomes. During the course of evolution, the difference in these sequences can act as diagnostics and can allow differentiation down to the species of strain level. This approach often termed as *rep*-PCR, is helpful in the identification of bacterial communities because it provides a genomic fingerprint of chromosome structure which is considered to be variable between strains. Highly repetitive sequences, also called microsatellites, are very informative and are useful to assess genetic variability, analyze mating systems and in genetic mapping. Therefore, fingerprinting of PCR amplified microsatellites can be compared using similarity indices to analyse difference at inter- and intraspecific level. However, the use of this method of the microbial diversity analysis is limited as per the complexity of the community but, with the help of this method, probe(s) can be designed and developed that can be able to detect changes in the microbial communities cause due to environmental alterations. In recent years, real-time PCR methods have been developed and described for the rapid detection and identification of several bacterial strains. Real-time PCR is a promising tool for distinguishing specific sequences from a complex mixture of DNA and therefore is useful for determining the presence and quantity of pathogen-specific or other unique sequences within a sample. Real-time PCR facilitates a rapid detection of low amounts of bacterial DNA accelerating therapeutic decisions and enabling an earlier adequate antibiotic treatment.

Sequence-based techniques mostly rely on determining the sequence of a specific stretch of DNA, usually but not always, associated with a specific gene, most often that of the most conserved gene. The approach is similar to genotyping in that a database of specific DNA sequences is generated. When compared with the test sequence, the degree of similarity or match between the two or more sequences determines how closely related the two organisms are? A number of computer algorithms have been created to compare multiple sequences to one another and generate a phylogenetic tree showing close lineages of the organism. Sequence comparisons of the ribosomal RNA (rRNA) gene to distinguish bacteria and archaea demonstrate is now being most widely applied to identify relationships among microorganisms. Both fingerprinting techniques and sequence-based methods have their strengths and weaknesses. Sequence based approaches e.g. analysis of the 16S rRNA gene are effective in establishing broader phylogenetic relationships among bacteria at the genus, family, order and phylum levels whereas fingerprinting-based methods are better to determine and distinguish strain- or species-level relationships although less reliable for establishing relatedness above the species or genus level. Coupled with each other, and with other phenotypic tests, these methods can create an omics-based polyphasic approach standard for describing new microbial species.

Usual molecular methods for the identification of microbial community structure and function include PCR-amplification based techniques along with the DNA reassociation, DNA-DNA and mRNA-DNA hybridization, DNA cloning and sequencing. DNA reassociation is a measure of genetic complexity in which total DNA from the environmental sample is extracted, purified, denatured and allowed to reanneal. The rate of hybridization or reassociation depends on the similarity of sequences and increasing complexity of DNA sequences that represents diversity, the rate of DNA reassociation decreases. The time required for the half of the DNA to reassociate (the half association value) may be used as a diversity index. The similarity between the environmental samples can be measured by the hybridization kinetics of the degree of similarity of the DNA. Nucleic acid hybridization using specific probes has remained one of the important molecular tools for measuring qualitative or quantitative bacterial ecology. Specific probes for species or domain of known nucleotide sequences can be tagged with fluorescent markers (derivatives of fluorescein or rhodamine) at 5′- end and qualitative dot-blot hybridization, which can be conducted at cellular level or *in situ* is used to analyze relative abundance of certain group of organisms which may represent changes in the population abundance. The method called fluorescent *in situ* hybridization (FISH) has successfully been used to analyze spatial distribution of the bacterial population in biofilms (Schramm *et al.*, 1996).

DNA microarray, a multiplex technology used in molecular biology consists of an arrayed series of thousands of microscopic spots of DNA oligonucleotides, called features, each containing picomoles ($10^{?12}$ moles) of a specific DNA sequence, known as *probes* (or *reporters*). This can be a small portion of a gene or other DNA element that are used to hybridize a cDNA or cRNA sample (called *target*) under high-stringency conditions. Probe-target hybridization is usually detected and quantified by detection of fluorophore-, silver-, or chemiluminescence-labeled targets to determine relative abundance of nucleic acid sequences in the target. Since an array can contain tens of thousands of probes, a microarray can accomplish many genetic tests in parallel and have dramatically accelerated many types of investigation. The technique is very useful in bacterial diversity studies with high specificity.

Reverse sample genome probing (RSGP), a method that used genome microarray can be used to analyze microbial community composition of the most dominant culturable species. Like DNA-DNA hybridization, RSGP and DNA microarrays has the advantage of not being confounded of PCR biases and of high specificity with the target gene sequences. However, it has the limitation to detect only the most abundant species. The method is equally applied to culturables and unculturables both and culturable species or cloned DNA fragments of unculturable microbes can also be used.

During late eighties, environmental microbiologists greatly realized the potentialities of the polymerase chain reaction (PCR) targeting the 16S rDNA for the design of innovative approaches to detect and identify microorganisms individually and to study microbial communities in diverse and complex environmental conditions. PCR-based techniques allow for the identification of microorganisms as well as the prediction of phylogenetic relationships regardless of their culturability. Likewise, 18S rDNA and internal transcribed spacer (ITS) regions are increasingly being used for the identification of fungal communities but the available databases are not as extensive as for the prokaryotes. Earlier, molecular based methods largely relied on the cloning of the target genes isolated from environment samples and their sequencing but, sequencing of the thousands of clones always remained a cumbersome job. Now, a large number of reports are coming out describing PCR-based methods, frequently complemented by sequencing or hybridization profiling to detect and identify microorganisms using taxa-specific genomic markers and to infer taxonomic and clonal microbial diversity (Albuquerque *et al.*, 2009). In such studies, DNA is extracted from the environmental samples, purified and the target DNA (16S, 18S or ITS) is amplified using universal or specific primers and finally resulting products are separated using different ways. Small-subunit 16S rRNA gene sequencing is a widely accepted tool for identifying bacterial isolates in diverse communities. rRNA molecules comprise several functionally different regions and some of these are characterized by highly conserved sequences, i.e., sequences that can be found among a wide range of bacteria. Other regions show highly variable sequences, i.e., nucleic acid sequences that are specific for a species or a genus. Thus, the 16S rRNA sequence of a species is a genotypic feature which allows the identification of microbes at the genus or the species level. In addition, molecular identification offers the possibility of recognizing yet undescribed taxa, because the similarity in ribosomal DNA (rDNA) sequences reflects phylogenetic relationships.

Techniques based on the identification and characterization of specific (marker) microbial metabolites can lead to the comparative and relative microbial diversity and if developed parallel to the molecular methods, can become very useful. Techniques based on plate counts, sole carbon source utilization pattern or community level physiology profiling (CLPP) and fatty acid methyl ester analysis (FAME) have been used traditionally to determine microbial diversity and identification. Selective plating of environmental samples and viable counts is a fast and inexpensive method and can provide information of active and heterotrophic component of a population. Although, the technique is being routinely used for assessing the diversity and isolating microbes by and large, the obvious limitations include its inability to detect the unculturable population. Also dominant and fast growing microbes appear immediately on the plates thereby excluding the possibilities of the minor organisms to appear. Variations among the microbes in relation to growth characteristics, temperature, nutrient conditions, colony-colony inhibition and inability to culture large number of bacterial or fungal isolates are certain other limitations which apparently influence the microbial diversity. The BiOLOG assay, originally developed as a way to characterize the functional ability of the bacterial isolates, has been widely used for soil and water samples. However, the method presents several problems to overcome before a reproducible result can be obtained. The high population of bacteria in the soil needs dilution but reducing the bacterial cells per ml of water again results in uneven distribution of rare bacteria from well to well. In addition, bacteria are not evenly distributed in the soil. So taking the samples from within the soil solutions only allows the characterization of those cells that are easily removed from the soil surfaces and therefore, the method need improved methods to extracts as much as possible number of cells from the community habitats and further validation.

A technique based on the utilization pattern of sole carbon source by Gram-negative (GN) or Gram-Positive (GP) bacteria was developed using a 96 well microtiter plates, 95 of them containing different carbon sources and one blank (control) to assess the potential functional diversity. GN or GP plates are available from Biolog (Hayward, USA) and can be used with growth medium, a tetrazolium salt (metabolic marker changing colour as the substrate is metabolized) and site-specific carbon sources to analyze samples. Many fungal species are unable to reduce tetrazolium salt and therefore, fungal specific plates SFN2 and SFP2 were developed. Bacterial or fungal population inoculated in the plates was monitored for their ability to utilize carbon sources and the rate at which these substrates were utilized. Multivariate analysis of the data can lead to the relative difference between soil functional diversity in contaminated sites, rhizosphere, arctic and stress soils and soils inoculated with the microbial inoculants. Anaerobic community level physiological profiling and denaturing gradient gel electrophoresis (DGGE) was used to separate microbial communities from the polluted sites (aquifers). Similarly, in API systems the API strips are available with various carbon sources that can be used to measure functional and metabolic diversity.

Fatty acid methyl ester analysis (FAME) does not rely on culturing of microbes. Fatty acid signatures exist through out the microbial communities that can differentiate major taxonomic groups very consistently. Therefore, changes in fatty acid profiling can represent changes in microbial communities. FAME analysis can be conducted by extracting fatty acids directly from the soils and after methylation, can be analysed using gas chromatography. Multivariate analysis in the differences of fatty acid composition of different soils can reflect changes in the composition of bacterial or fungal population and identify signature fatty acids of different group of microorganisms. FAME analysis coupled with CLPP is widely being used to define the microbial communities in the soil rhizosphere and other habitats. While there are some critics over the CLPP method as far as its concern over the culturability of microbes, it nevertheless used by the scientific communities all over the world for

Table 34.1: Molecular Methods for the Analysis of Microbial Communities

Sl.No.	*Methods*	*Tasks Performed*
1.	**DNA based methods**	
	(*a*) DNA-DNA reassociation kinetics	From environmental samples, total DNA is extracted, purified, denatured and allowed to reaneal. The time required for half of the DNA to reassociate ($C_o t_{1/2}$) is used as a measure of diversity index.
	(*b*) Nucleic acid hybridization	Nucleic acid hybridization using specific probes is a qualitative and quantitative tool that can be performed with the extracted DNA or RNA or *in-situ*. From the total nucleic acid of an environmental sample, rRNA sequences of interest are quantified relative to the total rRNA by dot-blot hybridization which is used to determine relative abundance of a group of microbes using specific or universal nucleotide primers.
	(*c*) Fluorescent *in-situ* hybridization	Fluorescent-labeled primers are added and allowed to hybridize. Excess dye is washed away and the hybridized cells are detected
	(*d*) DNA Microarrays	An array containing thousands of DNA sequences with high specificity or even specific target genes is cross-hybridized with DNA from pure cultures. DNA samples with less than 70 per cent hybridization represent different species in the environmental sample.
	(*e*) Metagenome sequence analysis	The technique is used to identify bacteria species or to assess structural as well as functional bacterial diversity. Method is not confined to PCR biases and can contain thousands of target genes.
2.	**PCR dependent methods**	
	(*a*) Denaturing Gradient Gel Electrophoresis	Denaturing gradient gels are used to detect non-RFLP polymorphisms. The small (200-700 bp) genomic restriction fragments are run on a low to high denaturant gradient acrylamide gel. Initially the fragments move according to molecular weight but finally partial melting due o denaturation results in a mobility shift that can differentiate slightly different sequences and alleles are detected by such differences.
	(*b*) Terminal-fragment length polymorphism (T-RFLP)	T-RFLP is based on digesting a mixture of PCR amplified variants of a single gene using one or more restriction enzymes and detecting the size of each of the individual terminal fragments using DNA sequencer.
	(*c*) Single strand conformational polymorphism (SSCP)	**SSCP** (or single-strand *chain* polymorphism) is conformational difference of single-stranded nucleotide sequences of identical length as induced by differences in the sequences. This ability can be used to differentiate nucleotide sequences on gel electrophoresis that separates different conformations.

Contd...

Table 34.1–*Contd...*

Sl.No.	*Methods*	*Tasks Performed*
(*d*)	Ribosomal intergenic spacer analysis (RISA)	RISA is a method of microbial community analysis and provides a means of comparing differing environments or treatment impacts without the bias imposed by culture- dependent approaches. It involves PCR amplification of a region of the rRNA gene operon between the small (16S) and large (23S) subunits called the intergenic spacer region ISR.
(*e*)	Automated ribosomal intergenic spacer analysis RISA (ARISA)	ARISA is an automated method developed for the rapid estimation of microbial diversity and community composition. Afer DNA isolation, PCR amplification of the 16S-23S inergenic spacer regions in the rRNA operon was performed with a fluorescence-labelled forward primer. In order o understand he potential biases in diversity estimates provided by ARISA. His is a sensitive, highly reliable and reproducible method for freshwater microbial community analysis.
(*f*)	Sequence analysis of 16S rRNA gene clone analysis	**16S ribosomal RNA** (or 16S rRNA; (1500 nucleotides) is a component of the 30S subunit of prokaryotic ribosomes. The genes coding for it are referred to as **16S rDNA** and are used in reconstructing phylogenies. 16S rRNA gene is used for phylogenetic studies as it is highly conserved between different species of bacteria and archaea. Sequence analysis of the 16S rRNA sequences is done with the help of several primers, called "universal primers." These primers target the conserved region of 16S rRNA gene and amplify the target in parts.

their interest in functionally based microbial community assay. The method is practically inexpensive, rapid, reliable and easy to use and when compared with the other community assays, has the ability to discriminate among a wide range of soil microbial communities.

Differences in the G+C (guanine + cytosine) content of DNA can also be used as a basis of bacterial diversity. Microorganisms widely differ in their G+C composition and the taxonomically related groups generally differ from one another by 3-5 per cent . The method is reliable, includes whole genomic DNA, not influenced by PCR biases and quantitative and can uncover rare members in the microbial population although, it is a coarse method.

Metagenomics: Identifying the Unculturables

Efforts have been made since the first report that DNA can be extracted from the soil and subsequently characterized, led to the information that as many as 6000 to 10000 different genomes can be found in 1g of soil. The genetic complexity of the microbial community can be assessed by reassociation of community DNA. Such broad scale analysis has revealed that the community genome size equals the size of 6000 to 10000 *Escherichia coli* genomes in unperturbed organic soils and 350 to 1500 genomes in aerable or heavy-metal polluted soils (Torsvik and Ovreas 2002). The total genomic complexity is a function of the diversity based on genetic information and finally leads to overall taxonomic (structural) and functional variability at the community level. Therefore, a long-standing challenge for the study of the microbial diversity in the aquatic, soils and rhizosphere ecosystems remains in developing effective methods that can be used to describe diversity, function and abundance of soil and plant associated-microbial population.

It is now a matter of wide consideration that the application of cultural methods for the unbiased recovery of microbes from the environmental samples has limited applicability and much of the true extant natural microbial genetic diversity remains unexplored (Cowan *et al.*, 2005). A wider understanding of the microbial communities therefore, can only be made by exploiting the wealth of genetic information through environmental nucleic acid analysis, collectively called the 'metagenome' (Hugenholtz, *et al.*, 1998). The phylotypic analysis of community DNA preparations through metagenomic technologies over the past 10 years has provided access to most of the prokaryotic genetic diversity lying with the environmental samples (Rappe and Giovannoni, 2003). The uncultured microbial majority. Annu. Rev. Microbiol. 57: 369-394). Metagenomics is a relatively new approach to address the genomics of unculturable microorganisms (Stevenson *et al.*, 2004) and involves extraction of total DNA from environmental samples, cloning of the DNA into a viable and suitable vector, transforming the clones into a host, usually *Escherickia coli* and screening the resulting transformants (Schloss and Handelsman 2003; Bohannan and Hughes 2003). The technology is potentially applicable for community-wide analysis of metabolic and biogeochemical functions because, theoretically, a metagenomics database is supposed to contain almost all kinds of nucleotide sequences for all the genes that a particular microbial community holds.

The existence of complex communities of unexplored species in the environment was considerably ensured during 1980s from the early metagenomic studies that revealed presence of large group of often uncultured and thus unsequenced organisms in many environments. Despite their ubiquity and importance in nature, complex microbial communities are poorly identified, understood and characterized because of the limitations of the traditional microbiological methods or molecular methods in both scale and precision and also our inability to detect patterns of the enormous data of sequences generated due to high throughput genomics technologies (Janet *et al.*, 2002). High-throughput genomics in recent years have led to the characterization of microbial communities in diverse environments (Justin *et al.*, 2010).

Conclusion

Major problem associated with the soil microbial diversity really lies in spatial heterogeneity of organisms, inabilities to culture them, limitations of molecular and biochemical methods of identification, taxonomic ambiguity of microbes and lack of understanding of genetic polymorphisms in many microbial groups. Measuring simply 1 to 5 g of soil samples in several replications taken through random sampling from a specific habitat and concluding the results in terms of the total community seems unwise. The most obvious problem with this approach lies in the innate heterogeneity of the soil along with the spatial distribution of the microbes Therefore, to address spatial heterogeneity correctly, multiple spatial parameters need to be scaled up in terms of periodical sampling intervals and variation in sampling depth range. It has been widely reported that the microbial communities may have several nested levels of organizations that are directly dependent on the physical and chemical properties of the soil, the climate conditions and the type of agricultural practices employed (Franklin and Mills 2003). Small scale (1 to 5 g) and narrow samplings may bias the results and thereby, favouring the detection of only dominant genera or species. Therefore, taking as much small samples as possible from a definite habitat considering that the same is containing many microhabitats could be a rather better approach leading to a real picture of microbial population. Not only the soil, but plants also influence the microbial communities and their spatial density for example, root rhizosphere is a real niche for the maximum spatial distribution of soil bacteria and fungi and can increase microbial population up to two-fold over the bulk soil. The sampling methods therefore, should be designed in such a way that could reduce the variability in samples, increase the statistical power and provide a more representative sampling regime.

The immense phenotypic and genetic diversity lying with the soil microbes poses another problem of characterization of microbial communities and subsequently, identification of individual microbes. It has been realized that at least 99 per cent of the bacteria observed under the microscope can not be cultured by common laboratory approaches. The 1 per cent, that can be cultured, may become the representative of the entire population while rest 99 per cent are simply lying in a physiological state that restrict them to culture. Alternatively, it is also likely that the 99 per cent are phenotypically and genotypically so different from the culturable 1 per cent and only the minor population is being represented. Both the circumstances are ground realities to be addressed in a more concrete, precise and integrated manner. Characterizing both culturable and non-culturable microbial diversity is an intricate and difficult task in which all current and modern methods fall short.

Molecular techniques employed for the characterization of microbial communities and taxonomic identification of individual microbes owes their own limitations. Lysis efficiency of microbes varies from cell to cell and species to species. For example, bacterial population existing on the surface or beneath the subsurface of the soil or plants needs special care for extraction (cell lysis) because gentle extraction may extract gram-negative bacterial population but not gram-positive ones. At the same time, harsh treatment can extract both gram-negative as well as gram-positive bacterial cells but, could lead to the shearing of their DNA. Likewise for fungus, spores can be lysed differently than the mycelia. Therefore, optimal cell lysis and extraction of genetic material is the very basic criteria for the unbiased molecular based diversity studies. Similarly the methods employed for the isolation of DNA or RNA can also bias diversity studies. Harsh treatments could lead DNA to shear while differential methods of extraction may cause qualitative or quantitative problems. Contaminants like humic acid, phenols etc. existing in the environmental samples can interfere with the result while purification steps could cause potential losses to the quantity of DNA or RNA leading to biasing in molecular diversity studies. PCR amplification of target genes can also bias molecular approaches for diversity analysis. In general, 16S rRNA, 18S rRNA, and/or ITS regions are targeted using selective primers. In such conditions, issues related to differential PCR amplifications including different affinities of primers to the templates, copy numbers of the target genes, sequences with lower G+C content, hybridization efficiency and primer specificity should be considered.

The problem of defining microbial species in the communities for taxonomic perspectives is also a crucial one. There exists no marked official definition of a bacterial or fungal species (Kirk *et al.*, 2005, Journal of Microbiological Methods, 58: 169-188). The genetic plasticity of bacteria frequently allows gene transfers through plasmids, bacteriophages and transposons, further complicates the process of identification. Similar problems also exist with fungal taxonomy. Most of the current evidences of fungus identification lies with their sexual stages and spore size, shape, colour and texture, Molecular approaches using the restriction analysis of the internal transcribed spacer (ITS) region, 18S rDNA and restriction fragment length polymorphism (RFLP) were commonly used to identify the fungi but the databases still not sufficiently developed.

References

Albuquerque, Pedro, Mendes, Marta, V., Santos, Catarina L., Tavares, Fernando, Mordas – Ferreira, Pedro (2009). DNA signature-based approaches for bacterial detection and identification. *Science of Total Environment* **407:** 3641-3651.

Bohannan, B. J. M. and Hughes, J. (2003) New approaches to analyzing microbial diversity data. *Curr. Opin. Microbiol.* **6:** 282-287.

Christian, S., Riesenfeld, Patrick D., Schloss, and Handelsman, Jo (2004). Metagenomics: Genomic analysis of microbial communities. *Annu. Rev. Gen.* **38:** 525-552.

Cowan, D., Meyer Q, Stafford W, Muyanga S, Cameron R, Wittwer P. (2005). Metagenomic gene discovery: past, present and future. *Trends in Biotechnology,* **23:** 321-329.

Fisher, M. M. and Triplett, E. W. (1999). Automated Approach for Ribosomal Intergenic Spacer Analysis of Microbial Diversity and Its Application to Freshwater Bacterial Communities. *Appl. Environ. Microbiol.* **65:** 4630-4636.

Franklin, R. B. and Mills, A. L. (2003). Multi –scale variation in spatial heterogeneity for microbial community structure in an eastern Virginia agricultural field. *Abst., FEMS Microbiol. Ecol.* **44:** 335-346.

Grayston, S. J., Campbell, C. D., Bardgett, R. D., Mawdsley, J. L., Clegg, C. D., Ritz, K., Griffiths, B. S., Rodwell, J. S., Edwards, S. J., Davies, W. J., Elston, D. J. and Millard, P. (2004). Assessing shifts in microbial community structure across a range of grasslands of differing management intensity using CLPP, PLFA and community DNA techniques. *Appl. Soil Ecol.* **25:** 63–84.

Handelsman, J. (2004). Metagenomics: Application of Genomics to Uncultured Microorganisms. *Microbiol. Molecular Biol.* **68:** 669-685.

Hugenholtz, P., Goebel, B. M. and Pace, N. R. (1998). Impact of Culture-Independent Studies on the Emerging Phylogenetic View of Bacterial Diversity. *J. Bacteriology.* **180:** 4765-4774.

Janet, E. H., Seipp, R. P., Betts, M., Hawkins, L., Kessel, A. G. V., Crosby, W. L. and Hemmingsen, S. M. (2002). Extensive Profiling of a Complex Microbial Community by High-Throughput Sequencing. *Applied Environ. Microbiol.* **68:** 3055-3066.

Justin, K., Liu, Z., Lozupone, C., McDonald, D., Fierer, N. and Knight, R. (2010). Microbial community resemblance methods differ in their ability to detect biologically relevant patterns. *Nature Methods,* **7:** 813-819.

Kirk, J. L., Beaudette, L. A., Hart, M., Moutoglis, P., Klironomos, J. N., Lee, H. and Trevors, J. T. (2004). Methods of studying soil microbial diversity. *J. Microbiol Methods,* **58:** 169-188.

Liu, W. T., T. L. Marsh, H. Cheng, and L. J. Forney. (1997). Characterization of microbial diversity by determining terminal restriction fragment length polymorphisms of genes encoding 16S rRNA. *Appl. Environ. Microbiol.* **63:**4516-4522.

Okubo, A. and Sugiyama, S-ichi. (2009). Comparison of molecular fingerprinting methods for analysis of soil microbial community structure. *Ecological Research,* **24:** 1399-1405.

Ranjard, L., F. Poly, J.-C. Lata, C. Mougel, J. Thioulouse, and S. Nazaret. (2001). Characterization of bacterial and fungal soil communities by automated ribosomal intergenic spacer analysis fingerprints: biological and methodological variability. *Appl. Environ. Microbiol.* **67:** 4479-4487.

Rappe, S. M., and Giovannoni, S. J. (2003). The uncultured microbial majority. *Annu. Rev. Microbiol.* **57:** 369-394.

Schramm, A., Larsen, L. H., Revsbech, N. P., Ramsing, N. B., Amann, R. and Schleifer, K. H. (1996). Structure and function of a nitrifying biofilm as determined by *in situ* hybridization and the use of microelectrodes. *Appl. Envir. Microbiol.* **62:** 4641–4647.

Schloss, P.D. and Handelsman, J. (2003). Biotechnological prospects from metagenomics. *Current Op. Biotech.* **14:** 303-310.

Sharma, S., Kaul, H. and Schloter, M. (2006). *In* Molecular approaches to soil, rhizosphere and plant microorganism analysis. CABI, UK. 1-17.

Steele and Streit (2006). *In* Molecular approaches to soil, rhizosphere and plant microorganism analysis. CABI, UK. 1-17.

Stevenson, B., Eichorst, S., Wertz, J., Schmidt, T. and Breznak, J. (2004). New strategies for cultivation and detection of previously uncultured microbes. *Appl. Environ. Microbiol.* **70:** 4748–4755.

Torsvik, V. and Ovreas, L. (2002). Microbial diversity and function in soil: from genes to ecosystems. *Current Opinion in Microbiology,* **5:** 240-245.

Whitman, W. B. (1998). Prokaryotes: The unseen majority. *Proceedings of the National Academy of Sciences,* **95:** 6578–6583.

Chapter 35

Stable Isotope Probing: A Technique for the Microbial Diversity Analysis of Uncultivable C-1 Compound Consuming Microbes

Kamlesh K. Meena, Dhananjaya P. Singh, Manish Kumar, Alka Singh and Dilip K. Arora

National Bureau of Agriculturally Important Microorganisms, Indian Council of Agricultural Research, Maunath Bhanjan – 275 101, Uttar Pradesh

Introduction

One of the biggest challenges that microbial ecologists face is the identification of microorganisms carrying a specific set of metabolic processes in the natural environment. The best way to address this question is to cultivate microbial strains in the laboratory using growth media containing specific substrate and to identify the cultivated bacteria at the physiological, biochemical and more recently molecular level. The metabolic properties of these bacterial isolates could be used to infer the potential role of microorganisms *in situ* in the environment. An important limitation to this approach has been that most microorganisms in the natural environment are not easily cultivable in the laboratory. However, it is the duty of microbiologists to use new molecular biological data to assist in enriching and cultivating so-called 'unculturable' microorganisms in the laboratory perhaps not in pure culture as has been the tradition, but in defined mixed cultures or consortia too. In the absence of methods to cultivate many environmental organisms, several techniques have been developed to enable microbial ecologists to identify the activities of microorganisms. Cultivation-independent methods to study microbial community composition were pioneered in the 1980s and have provided insight into the enormous bacterial diversity. Although these methods, such as the polymerase chain reaction (PCR), dot-blot and fluorescence *in situ* hybridization (FISH) offer microbiologists the means to detect and identify novel prokaryotes in the environment, they do not necessarily enable us to infer their functions. In the last decade, techniques have been developed that combine cultivation independent identification of microorganisms with metabolic analyses. One such method stable isotope probing (SIP), has been developed and standardized with divers environmental samples.

Stable isotope probing (SIP) is a powerful technique which is used for detection and identification of active microorganisms in environmental samples that are biologically active in taking up a specific radio-labeled carbon substrate. The isotope probing is an approach that makes it possible to study the function and activity of microorganisms in their natural environment. SIP has only been carried out using ^{13}C, but there is potential for some applications to use other stable isotopes such as ^{15}N. SIP was first applied in the analysis of phospholipid fatty acids (PLFA) that can be extracted from an environmental sample and analysed by isotope-ratio mass spectrometry (IRMS). Groups of microorganisms often have signature PLFA molecules and therefore, in some cases, it is possible to identify the microorganisms that have incorporated the ^{13}C-substrate. A good example of how this technique can be used to link microbial populations to specific biogeochemical processes is the use of ^{13}C-acetate and ^{13}C-methane to study active sulphate reducers and methanotrophs, respectively. The disadvantage of this technique is that nothing is known about the PLFA patterns of microorganisms for which there are no cultivated representatives. Although PLFA analysis offers great sensitivity, the use of labeled nucleic acids as biomarkers has the potential to identify a wider range of bacteria with a greater degree of confidence.

DNA-SIP and Methylotrophs

DNA-SIP is dependent on the commercial availability of compounds that are highly enriched in ^{13}C. The earliest substrates that were used in DNA-SIP were $^{13}CH_3OH$ and $^{13}CH_4$ which contained >99 per cent ^{13}C atom. These one-carbon compounds are substrates for a group of bacteria known as methylotrophs. In DNA-SIP studies with $^{13}CH_3OH$ and $^{13}CH_4$, it was possible to target 'functional' genes in addition to the 16S rRNA gene, as methylotrophic bacteria have been relatively well-studied and the genes that encode enzymes involved in methanol and methane oxidation are known. The genes amplified by PCR can also be analysed using microarrays as described for the *pmoA* gene which encodes a subunit of the particulate methane monooxygenase enzyme and *mxaF* encoding a subunit of the methanol dehydrogenase enzyme. Although PCR is the easiest method to obtain information from a DNA-SIP experiment, it is also possible to directly clone ^{13}C-DNA and then sequence the cloned DNA, which might reveal genes of known function that have been found in cultivated microorganisms.

It seems that some *Beijerinckia* species can grow on methanol in the laboratory, therefore illustrating the utility of the DNA-SIP technique for identifying new metabolic capabilities in bacteria present in the environment that had previously been considered to be non-methylotrophic. DNA-SIP techniques have been used to study the functionally active populations of methane-oxidizing bacteria (methanotrophs) in peat soil and acidic forest soil. Initial studies involved relatively long incubation times (~40 days), which had the drawback that continued exposure of soil organisms to a specific ^{13}C-labelled substrate might result in detection of organisms that do not directly assimilate this substrate but instead use ^{13}C-labelled intermediates or by-products generated by the primary consumer organisms. However, more recent studies using $^{13}CH_4$ with samples from unusual environments, such as the enclosed ecosystem at the Movile Cave in Romania and Russian soda lake sediments, have used shorter incubation times and have yielded heavy DNA. When analysed by PCR with rRNA and functional gene probes for methanotrophs, this heavy DNA resulted in a high proportion of DNA sequences recognizable as those from extant methanotrophs. The effectiveness of working with ^{13}C-labelled one-carbon compounds has also been demonstrated in studies using $^{13}CO_2$ to identify nitrifiers in freshwater sediments, and $^{13}CH_3Br$ and $^{13}CH_3Cl$ to investigate the active populations of methyl-halide degrading bacteria in soils. In the latter studies, PCR primers were used that target the *cmuA* gene, which encodes a methyltransferase that is essential for growth on methyl halides, thereby linking global cycling of methyl halides to a poorly characterized group of bacteria. Wagner and colleagues demonstrated the usefulness and power of combining SIP with other techniques in their study of denitrification by microbial communities in activated sludge. Again, a methylotrophic substrate, $^{13}CH_3OH$, was used in DNA-SIP, and DNA that was enriched with ^{13}C- was used as a template to amplify 16S rRNA genes by PCR. 16S rRNA gene sequences that were closely related to those of the methylotroph genera *Methylophilus* and *Methylobacillus* were abundant in the heavy DNA. These sequence data were then used to design FISH probes, and after uptake of $^{14}CH_3OH$ under denitrifying conditions, FISH–micro autoradiography was used to prove that the bacteria identified in SIP experiments were important denitrifiers in the activated sludge under study.

This protocol provides visual step-by-step explanations of the protocol for density gradient ultracentrifugation, gradient fractionation and recovery of labeled DNA. The protocol also includes sample SIP data and highlights important tips and cautions that must be considered to ensure a successful DNA-SIP analysis.

Principle

The ^{13}C- labeled specific substrates were added to soil samples or enrichment cultures and, after an incubation period, the DNA was isolated and subjected to caesium chloride (CsCl) buoyant density-gradient centrifugation with ethidium bromide. Although the buoyant density of DNA varies with its GC (guanine and cytosine) content, 13C-DNA ('heavy') separates with remarkable efficiency from the 12C-DNA ('light'), forming a second band in the gradient that is visible under ultraviolet illumination. The two DNA fractions can be retrieved separately from the gradient with a needle and syringe or by fractionation. The heavy ^{13}C-DNA can be purified away from the light ^{12}C-DNA by a further round of centrifugation if necessary. The fraction that contains the 13C-DNA harbors the combined genomes of the microorganisms in the environmental sample that are active and have incorporated the labeled substrate into their nucleic acids. Once ^{13}C-DNA has been isolated, it can be used as a template in PCR, with general primer sets that amplify rRNA or target 'functional' genes of most known Bacteria, Archaea or Eukarya. The analysis of the rRNA or target 'functional' gene PCR products enables the identification of microorganisms that have assimilated the ^{13}C-substrate.

Material and Methods

Preparation of Reagents

DNA-SIP requires the use of reagents that should be prepared in advance of the actual procedure. The directions for preparing each reagent are listed in this section

Cesium Chloride (CsCl) Solution for Preparing SIP Gradients

1. Prepare a 7.163 M CsCl solution by gradually dissolving 603.0 g of CsCl in double distilled water (ddH$_2$O) to a final volume of 500 mL. Aliquot the final solution in sealed aliquots. The sealed aliquots can be stored indefinitely at room temperature. The seals help prevent evaporation and CsCl "crust" formation. Determine the density of the solution by weighing triplicate 100-iL aliquots, or by using a digital refractometer that has been carefully calibrated for CsCl solutions. At room temperature, the final density of this solution typically ranges from 1.88-1.89 g/ml. The density varies slightly each time a new stock is prepared.
2. Cesium chloride solution for preparing gradients with ethidium bromide (EtBr) - Combine 250 g of CsCl with 250 mL of sterile ddH2O water. Aliquot this solution into separate serum vials that have been crimp-sealed with butyl rubber seals.
3. Gradient Buffer - Combine 50 ml of 1 M Tris-HCl, 3.75 g KCl and 1 ml of 0.5 M EDTA to 400 ml of water. Dissolve the KCl, then add ddH2O to 500 ml. Filter-sterilize and autoclave. The final solution is 0.1 M Tris, 0.1 M KCl and 1 mM EDTA.
4. TE Buffer - Prepare a solution of 10 mM Tris-HCl (pH 8.0) and 1 mM EDTA (pH 8.0) in sterile ddH2O water, using autoclaved stock solutions of 1 M Tris-HCl (pH 8.0) and 0.5 M EDTA (pH 8.0). Filter sterilize and autoclave.
5. 70 per cent Ethanol - Combine 350 ml of high purity ethanol with 150 ml of sterile ddH$_2$O water.

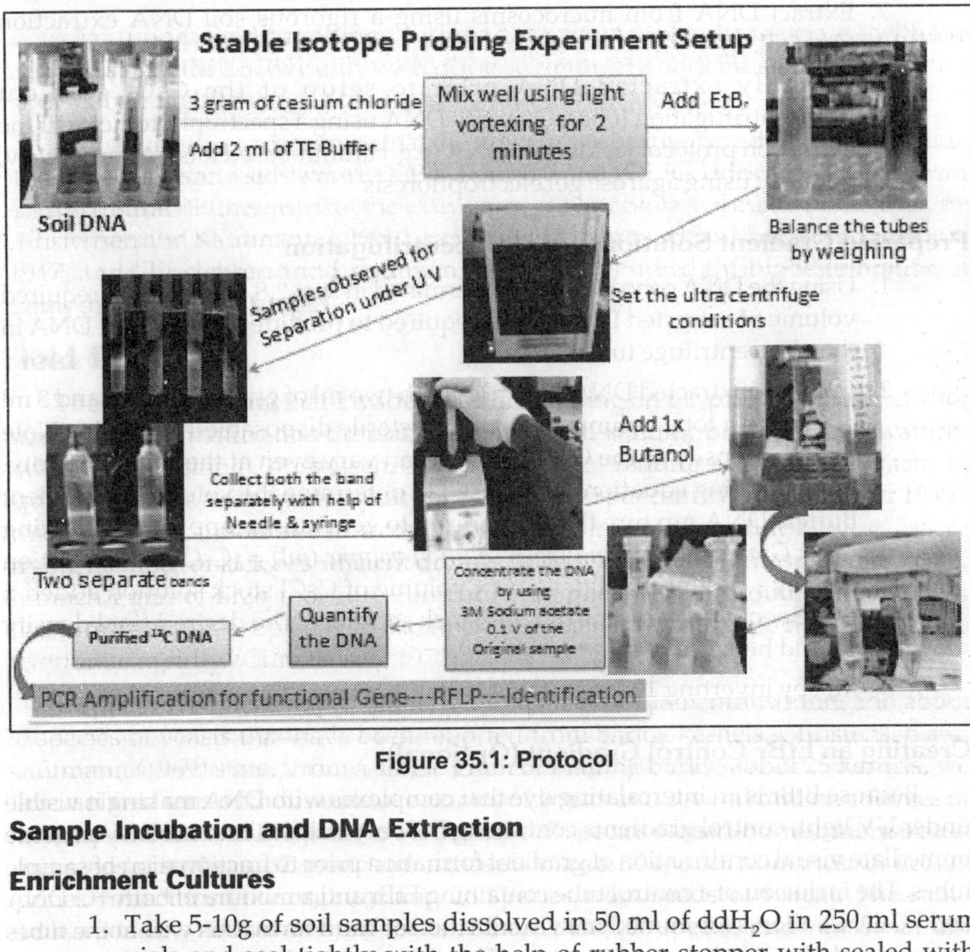

Figure 35.1: Protocol

Sample Incubation and DNA Extraction

Enrichment Cultures

1. Take 5-10g of soil samples dissolved in 50 ml of ddH_2O in 250 ml serum vials and seal tightly with the help of rubber stopper with sealed with aluminum stopper and simultaneously take a control by autoclaving at 121 lbs per 15 min.
2. The enrichment cultures were set up on same day with two different medium *i.e* AMS (Ammonium Mineral Salt) and NMS (Nitrate mineral salt) with a concentration of 0.1X and a total of 25 ml ^{13}C labeled methane was also used for the enrichment.
3. First observe for the methane concentration just after the enrichment with help of gas chromatograph.
4. Observe for the utilization of methane at an interval of 7 days till it is showing utilization.
5. Harvest the soil samples from the incubation bottles using centrifugation at 5000 rpm for 10 minutes.
6. Discard the supernatant and use the pellet (soil)

7. Extract DNA from microcosms using a rigorous soil DNA extraction protocol.
8. Quantify extracted DNA prior to setup of the CsCl gradient ultracentrifugation tubes. Quantify DNA using a spectrophotometer, if the extraction protocol yields only DNA (e.g. column-based kits). Alternatively, quantify using agarose gel electrophoresis

Preparing Gradient Solutions for Ultracentrifugation

1. Using the DNA concentrations determined in step 2.8, calculate the required volume of extracted DNA that is required to provide 0.5 ìg - 5 ìg of DNA in the ultracentrifuge tubes.
2. Combine extracted DNA (0.5 - 5 ìg) with two ml of gradient buffer and 3 ml of CsCl to a total volume of ~3 ml in a sterile disposable 15-ml tube. Note that the density of the CsCl solution can vary even at the same molarity. The following equation can be used to determine the volume of Gradient Buffer/DNA mixture that is required to generate an appropriate mixing ratio: *Gradient buffer and DNA solution volume (ml)* = (CsCl stock solution density - desired final density) x volume of CsCl stock solution added x 1.52 Specify the volume of CsCl stock solution . The desired final density should be 1.725 g/ml.
3. Mix by inverting 10 times. DNA is stable at room temperature in CsCl.

Creating an EtBr Control Gradient (Optional)

Because EtBr is an intercalating dye that complexes with DNA making it visible under UV light, control gradients containing EtBr are helpful because they provide immediate visual confirmation of gradient formation prior to fractionation of sample tubes. The inclusion of a control tube containing EtBr and a mixture of both ^{12}C-DNA and ^{13}C-DNA allows for immediate visualization of band formation within the tubes upon completion of ultracentrifugation. This is important because a ruptured tube during ultracentrifugation or improperly programmed run conditions can result in failed gradient formation. Bound to DNA, EtBr lowers the density of the DNA and as a result, a different protocol is followed to prepare gradients. Note that other nucleic acid stains can be used instead of EtBr but the protocol will require optimization with other fluorophores.

1. The control gradient requires two volumes of genomic DNA: one fully labelled with stable-isotope and one without label.
2. Combine a 5 -10 ìg quantity of both the ^{12}C-DNA and ^{13}C-DNA with Gradient Buffer to a final volume of 1.00 ml in a disposable 15-ml screw-cap tube.
3. Add 3.00 g of solid CsCl to the same tube. Mix by inversion.
4. Add 50 ìl of a EtBr solution and 3 ml of a CsCl stock solution to the same screw-cap tube.
5. An additional "blank" control solution containing EtBr will also be required to counterbalance the solution.

Ultracentrifugation

1. Using a bulb and Pasteur pipette, carefully fill ultracentrifuge tubes with gradient solutions prepared in step. Label the tubes on the tube shoulder with a fine permanent marker.
2. When all of the required tubes are filled with sample solutions, record the precise mass of each tube. Pair tubes and balance them. For balancing, find nearly matched pairs and add or remove minute quantities of solution until they are balanced, keeping the solution level as close to the base of the tube necks as possible. Note that for weighing tubes, we use an inverted 15-ml screw-cap tube that has been cut in half as a tube holder for the balance.
3. Seal the tubes using a 'tube topper' according to the manufacturer's instructions.
4. Check that the tubes are sealed properly by inverting them and applying moderate pressure. Weigh the tubes again to check that they are still balanced after sealing.
5. Check each rotor well carefully to ensure that the wells are clean and free of debris or dust that might puncture the tubes during ultracentrifugation.
6. Insert the tubes into the rotor with the balanced pairs opposite one another. Record the rotor location of each sample because the ultracentrifugation process can cause marker labels to be damaged or erased. Carefully seal the rotor wells as indicated by the manufacturer.
7. Load the rotor into the ultracentrifuge. Close the ultracentrifuge door and apply a vacuum. If using a TLA 110 rotor, set the rotation speed to 80000 rpm, the temperature at 20°C, and ultracentrifugation time for 72 hours. Select vacuum, maximum acceleration, and turn off the brake (ensures gradient not disrupted by deceleration). Note that turning off the brake will add an additional 1-2 hours to the run time. Also note that shorter run times may not achieve sufficient band resolution. Long ultracentrifugation runs with low speed are recommended, as they lead to greater resolution of distinct nucleic acid bands.
8. Immediately upon completion of the ultracentrifugation procedure, remove the rotor carefully. Avoiding any tilting or bumping of the rotor, gently remove tubes from the rotor to avoid disturbing the gradients within the tubes. In rare circumstances, a tube will burst during the run. If so, there is a chance that the gradients in the other tubes did not form properly. If a control gradient was included, check this tube carefully under UV light to confirm gradient formation. If the gradient has not formed properly in the control tube, it is best to repeat the experiment. Note that the EtBr control tube and its blank control may be stored in the dark and reused for up to six months. Take care to clean the rotor carefully according to the manufacturer's instructions once the burst tube has been removed. Do not use metal brushes or abrasive cleaners to clean rotor wells in order to avoid scratches to the rotor wells. Rotor-specific brushes and cleaning solution can be purchased from Beckman.

Gradient Fractionation

The simplest method that is currently used to recover DNA from the ultracentrifuge tubes is needle extraction. In this method we used the needle and syringe to collect the two bands separately.

1. Fix one of the ultracentrifugation tubes to a clamp stand and take out the light weight band of ^{12}CDNA first and collect it in a 15 ml sterile plastic tube.
2. Take out the heavy isotope labeled ^{13}C-DNA band by using another new needle and syringe by same way as collected ^{12}C-DNA band from the ultracentrifuge tube and collect in fresh sterilized 15 ml plastic tube.

DNA Purification and Precipitation

The collected DNA is having a mixture of heavy isotope ^{13}C- and EtBr. for purification of DNA we generally uses the following procedure

1. Add 2ml of the 1xbutanol and mix it, the butanol will come up along with EtBr and two separate layer can visualized one pink on top and another transparent colorless.
2. Picked out the pink layer with help of micro-pipette carefully.
3. Repeat the step one and two till disappearing of the pink layer
4. Add 1/10 volume of 3m sodium acetate for recitations of the 13
5. Centrifuge at 13,000 g for 30 minutes with the back of the tubes facing outwards for a consistent tube orientation in the rotor. Carefully aspirate and discard the supernatant. A pellet should be visible but can be very difficult to see at this stage. Work under a bright light source (e.g. desk lamp) to assist in visualizing the pellet.
6. Wash the pellet with 500 ìl of 70 per cent ethanol. Centrifuge at 13,000 g for 10 minutes. Carefully aspirate and discard the supernatant. The pellet will usually be more visible for this step, but will dissociate from the tube wall more easily.
7. Allow the pellet to dry at room temperature for 15 minutes.
8. Suspend each pellet in 50 ìl of TE buffer. Run 5 ìl of each fraction on an agarose gel according to standard lab protocols.

Fraction Characterization

The method used to characterize gradient fractions to assess the success of a SIP incubation will vary depending on the lab and availability of equipment. Using a fingerprinting method for targeting the 16S rRNA gene is a common approach and methods such as terminal restriction fragment length polymorphism (T-RFLP) or denaturing gradient gel electrophoresis (DGGE) are appropriate. Following the protocol described above, expect the light DNA to be associated with fractions 9-11 (~1.705-1.720 g/ml) and the heavy DNA fingerprints to be associated within fractions 5-8 (~1.720-1.735 g/ml). Unique fingerprints associated with fractions 5-8 of stable-isotope incubated samples, but not with native-substrate incubated controls provides strong evidence linking specific organisms with the metabolism of particular labeled substrate. If insufficient labeled DNA remains for some applications (hybridization, metagenomics), multiple displacement amplification may be used to produce greater quantities 13-15 but this can introduce chimeras into the amplified DNA.

References

Bodrossy, L., Stralis-Pavese, N., Konrad-Köszler, M., Weilharter, A., Reichenauer, T. G., Schofer, D. and Sessitsch, A. (2003). Development and validation of a diagnostic microbial microarray for methanotrophs. *Environ. Microbiol.* **5**: 566–582.

Hutchens, E., Radajewski, S., Dumont, M. G., McDonald, I. R. & Murrell, J. C. (2004). Analysis of methanotrophic bacteria in Movile Cave by stable isotope probing. *Environ. Microbiol.* **6**: 111–120.

Lin, J.-L., Radajewski, S., Eshinimaev, B. T., Trotsenko, Y. A., McDonald, I. R. and Murrell, J. C. (2004). Molecular diversity of methanotrophs in Transbaikal soda lake sediments and identification of potentially active populations by stable isotope probing. *Environ. Microbiol.* **6**: 1049–1060.

McDonald, I. R.. Warner, K. L., McAnulla, C., Woodall, C. A., Oremland, R. S. and Murrell, J. C. (2002). A review of bacterial methyl halide degradation: biochemistry, genetics and molecular ecology. *Environ. Microbiol.* **4**: 193–203.

Miller, L.G., Warner, K.L., Baesman, S.M., Oremland, R.S., McDonald, I.R., Radajewski, S., and Murrell, J.C. (2004). Degradation of methyl bromide and methyl chloride in soil microcosms: Use of stable C isotope fractionation and stable isotope probing to identify reactions and the responsible microorganisms. *Geochim. Cosmochim. Acta*, **68**: 3271-3283.

Morris, S. A., Radajewski, S., Willison, T. W. and Murrell, J. C. (2002). Identification of the functionally active methanotroph population in a peat soil microcosm by stable isotopeprobing. *Appl. Environ. Microbiol.* **68** : 1446–1453.

Radajewski, S., Webster, G., Reay, D. S., Morris, S. A., Ineson, P., Nedwell, D. B., Prosser, J. I. and Murrell, I. C. (2002). Identification of active methylotroph populations in an acidic forest soil by stable-isotope probing. *Microbiology,* **148**: 2331–2342.

Rappé, M. S. and Giovannoni, S. J. (2003). The uncultured microbial majority. *Annu. Rev. Microbiol.* **57**, 369–394.

Schloss, P. D. & Handelsman, J. (2004). Status of the microbial census. *Microbiol. Mol. Biol. Rev.* **68**: 686–691 (2004).

Whitby, C. B., Hall, G., Pickup, R., Saunders, J. R., Ineson, P., Parekh, N. R. and McCarthy, A. (2001). 13C incorporation into DNA as a means of identifying the active components of ammoniaoxidizer populations. *Lett. Appl. Microbiol.* **32**: 398–401.

Zengler, K., Toledo, G., rappe, M., Elkins, J., Mathur, E. J., Short, J. M. and Keller, M. (2002). Cultivating the uncultured. *Proc. Natl. Acad. Sci. USA* **99**: 15681–15686.